T0213224

COMPACT HANDBOOK OF
Computational
Biology

Edited by

Andrzej K. Konopka
M. James C. Crabbe

CRC Press
Taylor & Francis Group
Boca Raton London New York

CRC Press is an imprint of the
Taylor & Francis Group, an **informa** business

CRC Press
Taylor & Francis Group
6000 Broken Sound Parkway NW, Suite 300
Boca Raton, FL 33487-2742

First issued in paperback 2019

© 2004 by Taylor & Francis Group, LLC
CRC Press is an imprint of Taylor & Francis Group, an Informa business

ISBN-13: 978-0-8247-0982-2 (hbk)
ISBN-13: 978-0-367-39391-5 (pbk)

Library of Congress Cataloging-in-Publication Data
A catalog record for this book is available from the Library of Congress.

Visit the Taylor & Francis Web site at
http://www.taylorandfrancis.com

and the CRC Press Web site at
http://www.crcpress.com

Foreword

Computational biology is a relatively new but already mature field of academic research. In this handbook we focused on the following three goals:

1. Outlining pivotal general methodologies that will guide research for the years to come.
2. Providing a survey of specific algorithms, which have been successfully applied in molecular biology, genomics, structural biology and bioinformatics.
3. Describing and explaining away multiple misconceptions that could jeopardize science and software development of the future.

The handbook is written for mature readers with a great deal of rigorous experience with either doing scientific research or writing scientific software. However no specific background in computer science, statistics, or biology is required to understand most of the chapters. In order to accommodate the readers who wish to become professional computational biologists in the future we also provide appendices that contain educationally sound glossaries of terms and descriptions of major sequence analysis algorithms. This can also be of help to executives in charge of industrial bioinformatics as well as to academic teachers who plan their courses in bioinformatics, computational biology, or one of the "*omics" (genomics, proteomics, and their variants). As a matter of fact we believe that this volume should be suitable as a senior undergraduate or graduate-level textbook of computational biology within departments that pertain to any of the life sciences. Selected chapters could also be used as teaching materials for students of computer science and bioengineering.

This handbook is a true celebration of computational biology. It has been in preparation for a very long time with multiple updates and exchanges of chapters. However the quality and potential usefulness of the contributions make us believe that preparing the ultimate final version was the time very well spent.

Our thanks go to all the authors for their masterful contributions and devotion to the mission of science. We would also like to extend our most sincere thanks to our publishing editor Anita Lekhwani and the outstanding production editor Barbara Methieu. Their almost infinite patience and encouragement during inordinately long periods of collecting updates and final versions of the chapters greatly contributed to timely completion of this project.

Andrzej K. Konopka and M. James C. Crabbe

Contents

Contributors

Philipp Bucher Bioinformatics, Institut Suisse de Recherches Experimentales sur le Cancer (ISREC), Epalinges s/Lausanne, Switzerland

M. James C. Crabbe University of Reading, Whiteknights, Reading, England

Maxime Crochemore Institut Gaspard-Monge, University of Marne-la-Vallée, Marne-la-Vallée, France

Jaap Heringa Centre for Integrative Bioinformatics VU (IBIVU), Faculty of Sciences and Faculty of Earth and Life Sciences, Free University, Amsterdam, The Netherlands, and MRC National Institute for Medical Research, London, England

Hidetoshi Kono University of Pennsylvania, Philadelphia, Pennsylvania, U.S.A.

Andrzej K. Konopka BioLingua Research Inc., Gaithersburg, Maryland, U.S.A.

Frederique Lisacek Génome and Informatique, Evry, France and Geneva Bionformatics (GeneBio), Geneva, Switzerland

Izabela Makalowska   Pennsylvania State University, University Park, PA, U.S.A.

Wojciech Makalowski   Pennsylvania State University, University Park, PA, U.S.A.

Graziano Pesole   University of Milan, Milan, Italy

Cecilia Saccone   University of Bari, Bari, Italy

Marie-France Sagot   INRIA, Laboratoire de Biométrie et Biologie Évolutive, University Claude Bernard, Lyon, France

Akinori Sarai   RIKEN Institute, Tsukuba Life Science Center, Tsukuba, Japan

Peter Schuster   University of Vienna, Vienna, Austria

Peter F. Stadler   University of Leipzig, Leipzig, Germany

William R. Taylor   National Institute for Medical Research, London, England

1

Introduction: Computational Biology in 14 Brief Paragraphs

Andrzej K. Konopka
BioLingua Research Inc., Gaithersburg, Maryland, U.S.A.

M. James C. Crabbe
University of Reading, Whiteknights, Reading, England

1. Computational biology has been a marvelous experience for at least three generations of the brightest scientists of the second half of the twentieth century and continues to be so in the twenty-first century. Simply speaking, computational biology is the science of biology done with the use of computers. Because computers require some specialized knowledge of the cultural and technical infrastructure in which they can be used, computational biology is significantly motivated (or even inspired) by computer science and its engineering variant known as information technology. On the other hand, the kind of data and data structures that can be processed by computers provide practical constraints on the selection of computational biology research topics. For instance, it is easier to analyze sequences of biopolymers with string processing techniques and statistics than to infer unknown biological functions from the unknown three-dimensional structures of the same biopolymers by using image processing tools. Similarly, it is advisable to study systems of chemical reactions (including metabolic pathways) in terms

of rates and fluxes rather than in terms of colors and smells of substrates and products.

2.   Computational biology is genuinely interdisciplinary. It draws on facts and methods from fields of science as diverse as logic, algebra, chemical kinetics, thermodynamics, statistical mechanics, statistics, linguistics, cryptology, molecular biology, evolutionary biology, genetics, embryology, structural biology, chemistry, and even telecommunications engineering. However, most of its methods are a result of a new scientific culture in which computers serve as an extension of human memory and the ability to manipulate symbols. The methodological distinctiveness of computational biology from other modes of computer-assisted biology (such as bioinformatics or theoretical biology in our simplified classification) is indicated in Figure 1.

3.   All fields of science (particularly biology) employ a clear distinction between their material and symbolic aspects. The material aspects of life sciences are related to energy- and rates-dependent (i.e., physical) properties of substances of which biological objects of interest are made. That includes media via which substances and objects can interact with each other in an energy- and rates(forces)-dependent manner. The symbolic aspects are related, on the one hand, to mechanisms by which biological objects (such as DNA) are involved in the functioning of larger systems (such as cells) and, on the other hand, to the nature of information (meaning) of these biological objects for the mechanisms of which they appear to be a part.

4.   Most scientists accept the duality between material and symbolic properties of biological systems. We do not really know why it is convenient to distinguish energy- and rates-based descriptions of matter (i.e., material aspects) from energy- and rate-independent models of organization of matter (i.e., its symbolic properties). The fact is, though, that both modes of describing Nature are highly effective in terms of understanding observable objects (or systems), processes, and phenomena.

5.   Computational biology naturally draws on symbolic aspects of systems representation because, as mentioned earlier, data structures need to be accessible for computers. Another reason for exposure of symbolic aspects of descriptions of biological systems is the fact that living things are a result of (often long) evolutionary history that remains in the memory of their present form. Yet another reason is the fact that both genetics (studies of inheritance between generations of individuals) and developmental biology (studies of implementation of a "master plan" to build an individual organism) have explored symbolic aspects of living matter for centuries.*

---

* The symbolic essence of living things has been generally postulated since ancient times (Aristotle, 3rd century BC) as a set of characteristics that needs to be supplied in order to distinguish a functional system from a (dysfunctional) collection of its components.

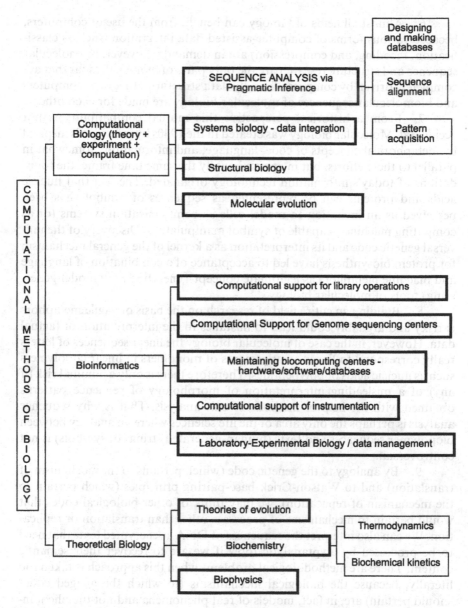

FIGURE 1 Simplified classification of computer-assisted methods of life sciences. Bioinformatics and theoretical biology appear to be methodologically different from computational biology as well as different from each other.

6.   Almost all fields of biology can benefit from the use of computers, because various forms of computer-assisted data integration (such as classification, recoding, and compression) are in demand. However, biomolecular sequence and structure research are the only parts of biology thus far that are completely driven by computer-friendly data structures. In a way, computers and biomolecular sequences of molecular biology are made for each other.

7.   From a historical perspective this mutual compatibility is not accidental. Molecular biology was created in the 1940s by individuals inspired by metaphorical concepts of code, language, and information. Somewhat in parallel to these efforts, but in approximately the same time frame, the foundations of today's information technology originated. The fact that nucleic acids and proteins can be represented as sequences of symbols has been perceived as an invitation to model cells as communication systems (or as computing machines) capable of symbol manipulation. Discovery of the universal genetic code and its interpretation as a kernel of the general mechanism for protein biosynthesis have led to acceptance of a combination of language and machine metaphors as legitimate concept-generating (and model-generating) tools of molecular biology.

8.   Building an entire field of research on the basis of vague metaphors is risky and arguably could lead to mistakes in the interpretation of factual data. However, in the case of molecular biology the linear sequences of letters really correspond to linear arrangements of monomers in linear copolymers such as nucleic acids or polypeptides. Therefore there is only a minimal risk (if any) of a misleading interpretation of morphology of sequence patterns obtained with the help of methods of text analysis. That is why sequence analysis is perhaps the only area of the life sciences where an analogy between biopolymer sequences and texts (human-generated strings of symbols) is not controversial.

9.   By analogy to the genetic code (which pertains to the mechanism of translation) and to Watson-Crick base-pairing principles (which pertain to the mechanism of replication), we may think of other biological codes that would pertain to mechanisms of processes other than translation or replication. We can also interpret the sequences of DNA as encrypted texts that need to be processed by cryptanalytic tools if we are to recover their semantic content. There are methodological problems when this approach is taken too literally, because the biological mechanisms (to which the alleged codes should pertain) are, in fact, models of real phenomena and not the phenomena themselves. That is why a huge part of research in computational biology (particularly in sequence analysis) is devoted to the evaluation of the adequacy of models and validation of principles of modeling.

10.   Attempts to validate models led to the concept of sequence or structure motifs replacing the idea of functional code words. For a (sequence)

pattern to be a motif (classification code word) requires only a good correlation between its presence in sequences and postulated cellular mechanisms (models of cell functioning). In contrast, a pattern is a functional code word regarding an experimentally validated cellular mechanism only when it (the pattern) has the same effect on the mechanism each time it is present (i.e., if there is deterministic dependence between pattern and mechanism, not just a correlation of observations).

11. Laboratory-experimental verification of postulated cellular mechanisms is usually out of immediate reach. That contributes to extreme vagueness of the concept of biological function at the molecular level. Because of this imprecision in defining function, the idea of functional codes does not have realistic reference in factual reality most of the time.

12. Searching for motifs in large collections of functionally equivalent sequences (FESs) is a way of avoiding a mechanistic definition of function. Instead of waiting for laboratory biologists to elucidate and confirm a given mechanism we just assume the functional equivalence as a null hypothesis to be tested. Within this paradigm the function is simply a name of the collection of sequences or a name-equivalent descriptor such as a pointer that singles out a given FES from other FESs. To decide if a sequence pattern is a motif with respect to a given FES, we need to apply a measure of significance. That is, the pattern must be significant in a measurable, quantitative way.

13. Beyond sequence research, computational biology is concerned with modeling structure in a way that could be predictable from sequences of biopolymers. The task is truly challenging because sequences alone do not seem to contain enough information to prompt the existing general models of folding to predict structure with any degree of reliability. That is why most successes of structural computational biology have been in the areas of structure classification (such as databases of protein folds). Predictions of specific cases of biopolymer folding (such as 5sRNA) have thus far been based on methods that explore a very large case-oriented knowledge base. However, progress in devising universal methods can be noted in the area of redefining alphabets of meaningful structural units for newer generations of models of folding. In this respect, structural computational biology comes close to sequence analysis because both fields rely on a general paradigm of pragmatic inference. However, in other respects the two fields differ. For instance, the general, long-term motivation of structural computational biology is the desire to understand origins of living matter in terms of prebiotic evolution of ensembles of chemical entities. In contrast, biomolecular sequence analysis is generally motivated by questions concerning existing biological functions and their evolution via modification of sequences.

14. The field of molecular evolution is reportedly the most mature area of computational biology. It originated soon after the first protein sequences

were known in the 1950s and has continued its prominent presence up to this day. Data on several fully sequenced genomes (including a draft of the human genome) will almost certainly lead to new discoveries concerning evolutionary relationships between DNA fragments that do not obviously code for proteins (such as various "regulatory" regions) as well as further elucidation of the evolutionary history of today's phenotypes. As far as devising methods for sequence comparison is concerned, the field of molecular evolution is close to sequence analysis because both disciplines employ protocols of pragmatic inference. However, the methods employed to draw historic interpretation from sequence-related facts still remain specific to evolutionary biology and much less relevant to sequence research.

2

Introduction to Pragmatic Analysis of Nucleic Acid Sequences*

Andrzej K. Konopka
BioLingua Research Inc., Gaithersburg, Maryland, U.S.A.

1. INTRODUCTION

The chapter is designed as a general introduction to sequence analysis for scientists and software developers who desire to write their own computer programs to study nucleic acid sequences. It is also intended for practically minded readers who would like to understand fundamental principles of algorithms used in sequence research and the reasons specific methods were preferred over the alternatives. The unifying paradigm in this respect is the idea of *pragmatic inference*: an organized ensemble of protocols [9–12] that on the one hand allows one to construct materially adequate generalizations based upon instances of observable facts (validated induction) and on the other hand generates predictions of novel potentially observable facts that cannot be attained by in-

---
*Most of this chapter is based on the author's past reference work [1–11] as well as on instructional sessions given to software users and customers of the author's organization (BioLingua Research Inc.) during recent several years. Large parts of the material covered here were also taught to graduate and senior undergraduate students of biology, medicine, and computer science in the period between 1999 and 2004.

duction alone (unverified discoveries).* More specific methods (primarily frequency count analysis) for making biological predictions are outlined in considerable detail. However, the coverage of sequence alignment and database searches (including pattern-matching algorithms) is reduced to the minimum because these topics are fully addressed in other chapters of this volume [see Chapters by Crochemore and Sagot (3), Heringa (4), Taylor (6), and Lisacek (8)].

Readers interested only in running existing computer programs may elect to go straight to the appendix devoted to an annotated glossary of software (Konopka and Heringa; this volume) that is placed at the end of this volume. Appendices to some of the chapters in this volume also contain remarks about the software that pertains to their content. However, despite the availability of software-oriented appendices I strongly advise reading the full text of chapters in this handbook before trying to either use sequence analysis programs or write such codes on one's own. For one thing, the regular text explains the limitations and constraints under which the utility software operates. Knowledge of these constraints is in turn essential for interpretation of output of any sequence analysis program. The second reason is the need to understand the nature of questions about data structure that software developers have asked before they wrote programs. Without such understanding one cannot decide which software tool can be used for a given purpose. The third reason is the fact that sequence analysis software is often available with only very rudimentary, if any, instruction on its use. Some understanding of the principles residing behind algorithms is necessary for anyone who would like to either use or write such software.

Like linguistics and cryptology, sequence analysis is about rewriting strings of symbols from one alphabet to another. It is hoped that such rewriting can reveal scripting systems that will be well correlated with plausible models of biologically relevant processes. In this general sense, nucleic acid and protein sequence analysis is devoted to sequence segmenting and annotation. In a less general (but also more imprecise) sense, we can think of sequence analysis as redefining regions in sequences (substrings, subsequences) in a way that is compatible with possible laboratory observations of the biological roles of these subsequences. From this perspective, sequence analysis has a predictive power as well as the quality of generating falsifiable predictions. By today's standards, both attributes make biomolecular sequence analysis a genuine field of science.

There is no doubt that an enormous potential of frequency count–based methods of alphabet (and mechanism) acquisition has not even begun to be seriously explored, but early preliminary research results are truly encourag-

---

*This paradigm has been practiced for at least 300 years of modern science but clearly described, without much of the earlier hand waving, only during the last decade in the context of computer-assisted sequence analysis [9,10,12].

ing (see, for instance, Refs. 9, 11, 13, and 14 for more details that illustrate this point). The majority of these methods require the design of pattern acquisition protocol and then, via implementation of such protocol, discovery of the representation of sequence data that appears most appropriate for answering a biological question at hand. Sequence alignment and database searches are very important too. However, they can be performed only after the problem-oriented alphabet in which searches will be made and an appropriate scoring system have been determined.

This chapter covers the methodological foundations of sequence analysis that, to my knowledge, would be difficult to find anywhere else. In contrast to other sources known to me, it does not focus on the details of using computer technology (such as manuals of use of bioinformatics tools). Instead it exposes leading metaphors and styles of thinking that are needed in today's practice of biomolecular sequence data handling and interpretation. The approach is strictly practical, and therefore there is no emphasis on details of combinatorics of strings that are mathematically interesting but methodologically ancillary. In other words, we assume that for all (our) practical purposes all necessary mathematics can be invented "on the spot," perhaps with intelligent inclusion of already existing mathematical methods. More theoretically inclined readers are welcome to consult references 11 and 15–19, which cover combinatorics and statistics in more detail.

2. LEADING METAPHORS AND CONCEPTS BEHIND SEQUENCE ANALYSIS

Science of the last 60 years has been inspired by metaphors of language and machines which in turn have led to metaphorical concepts of information, code, and symbol manipulation. The initial opposition of physicists to the alleged misrepresentation of their field by various "philosophers"* has

---

* "Philosopher" was used as a pejorative term by some twentieth century physicists to denote nonphysicists as well as renegade physicists rejected by political structures of their field. The idea of contrasting "philosophers" with scientists very likely begun in the 1930s when the pro-Nazi German scholars advocated practicality of what they called "Aryan physics" as opposed to the non-Aryan "philosophizing" or "theorizing" about physics. The attitude of using the term "philosopher" as an implicit death sentence for the enemies of tyrants of the time was reportedly present during the Holy Inquisition in Europe as well as during the Cultural Revolution in China. Despite the fact that isolationistic, discriminatory attitudes of this kind have long been discredited, instances of using the term "philosopher" for the purpose of pejorative labeling politically inconvenient colleagues continue to recur. Occasional substitution of the word "philosopher" for "theoretician"—frequently utilized by a fraction of biomedical savants within the academic/ industrial complex of our time—does not make today's labeling techniques less reprehensible or less damaging for our culture than the past prototypes were for the intellectual ambience of their time. See also Refs. 20–23 for comments on the subject of contrasting "theory" with "practice."

effectively been abandoned and replaced by a desire to model complex material systems such as living organisms. In order to follow this wish one needed to take into account not only the material aspects but also the symbolic aspects of complex systems. In this superficial sense the paradigms of life sciences and physics came close to each other in terms of attempting to model organisms.

Formal logic, electronic computers, theories of codes and communication, and formal linguistics had all been actively pursued [24–29] by the time of the birth of molecular biology in the mid-1940s [30–32] (see also Refs. 2, 8, 9, 11, 33, and 34 for sequence analysis–centered materials). In addition, the Second World War (assumed to have ended in 1945 for non-central and non-eastern Europe [35]) led to the involvement of many educated individuals in the cryptological services of their countries [36–40]. The spectacular successes of WWII (and WW1, for that matter) cryptanalysts with the use of algebraic [41,42], statistical [43,44], and mechanistic models [45] for breaking cryptographic systems [36] inspired a "critical mass" of scientifically influential minds of the twentieth century with the idea of equivalence between mechanical devices and strings of symbols (texts).

All the foregoing factors appear to have had a powerful effect on today's science. On the one hand, they led to the creation of symbol-manipulating devices both abstract (such as automata in the 1940s and 1950s [46,47] and formal grammars in the 1950s [48,49]) and real (such as electronic computers in the second half of the 1940s). On the other hand, they also led to the creation of molecular biology as a methodological blend of mechanistic modeling of chemical phenomena and semantics (essence of meaning).*

Biopolymers such as nucleic acids and proteins are chemical entities (represented by molecules), but they can also be seen as carriers of information. This is a reflection of the methodological duality between the material and symbolic aspects of observable phenomena [50,56–58]. Different fields of science explore this duality between material and symbolic properties in different ways and to different extents.

In molecular biology the material aspects of nucleic acids and proteins are reflected in their measurable physicochemical properties that have been traditionally studied by biochemists, structural biologists, biophysicists, and biological chemists. On the other hand there seem to be two (not just one!) symbolic aspects of nucleic acids and proteins:

1. The structural aspects (such as sequences and structures)
2. The mechanistic aspects (such as mechanisms of processes that may employ sequences and structures)

---

*The early extramechanistic rationale for biology given by (physicist) Schroedinger in 1944 [50–52] has been effectively modified to a kind of symbolic/informational approach to mechanistic models of biological phenomena by his followers (see Refs. 53–55 for an annotated historical account).

Most sound explanations in molecular biology indeed consist of plausible matching of sequences or structures with mechanisms. The process of such matching is complex. It may or may not explore the material aspects of chemicals that participate in the processes to be explained. However, with no known exceptions, it follows a scheme of encoding a structure with a mechanism or vice versa. In other words, a biologically relevant process is seen as a conversion of its one symbolic representation into another. In this way it is similar or identical to a general scheme of telecommunication or a cryptographic device that converts one symbolic representation of some real objects into another. From this perspective, molecular biology of DNA and proteins is indeed very similar to telecommunications engineering and cryptology (which in turn are similar to each other in this very respect).

From this perspective the field of molecular biology is literally founded on two assumptions that evoke both of the foregoing symbolic aspects of nucleic acids and proteins: sequence hypothesis and the central dogma of molecular biology.*

2.1. Sequence Hypothesis and the Central Dogma as Inspirations for Sequence Analysis

"Sequence hypothesis" implies that the biological specificity of DNA is encoded solely in its nucleotide sequence[†] but explicitly refers only to the DNA regions that undergo transcription (protein coding genes). It also refers (less directly) to genes for functional RNA such as species of transfer RNA (tRNA) or ribosomal RNA (rRNA). In the light of the sequence hypothesis, the fact that the functional essence of chromosomal DNA is to store encoded messages is implicitly obvious. The fact that the actual chemical structure and properties are less essential for the same functional essence appears clear as well.

The central dogma of molecular biology further stabilizes the importance of the symbolic aspects of nucleic acids and proteins for understanding

---

*The potentially pejorative term "dogma" in the phrase "central dogma" is probably a result of a frivolous joke motivated by a half-serious value judgment about biology served by the same renegade physicists of the early 1950s who originated molecular biology. The entire field of the life sciences was probably considered (by them) analogous to scholastic science of the Dark Ages when dogmas were taken more seriously than scientific methods. It is not impossible that the original central dogma was considered to be just a temporary working hypothesis that would be modified and renamed later on. Should that be the case, a joke or jargon in the name of hypothesis would be easy to explain.

[†] "Specificity of a piece of nucleic acid is expressed solely by the sequence of its bases, and this sequence is a (simple) code for the amino acid sequence of a particular protein" [59].

their general functional roles in the living cell. It implicitly states that the cellular protein biosynthesis can be viewed as an asymmetrical communications system consisting of a message storage box (memory), the "package" containing the encoded message (mRNA), and the receiver-decoder (polypeptide produced on the ribosome).* The dogma then states that the process is irreversible because the polypeptide cannot be reverse-translated into the mRNA that encoded it and therefore cannot be used to produce the same piece of DNA that encoded its own biosynthesis[†].

A focus on symbolic aspects of cellular processes led to profound generalizations of observations (such as the universal genetic code) as well as to opening new ways of thinking about the relationship between genotype and phenotype. It also made it possible to manipulate genetic systems in a way that leads to experimentally testable phenotypic variations. Thus predictions about the functions of genes and their control could be made at least in principle and in some cases in fact. On the other hand, the rise of computer-based technology along with developments in computer-oriented formal linguistics made it possible to effectively employ computers to study and analyze strings of symbols. Because nucleotide and protein sequences are, in fact, strings of symbols, the employment of computers for their analysis was not surprising to any one in either biology or computer science. In fact, computers are ideal tools for molecular biology in all these respects that explore symbolic aspects of living things at the context-free level.

An important contribution of the sequence hypothesis and central dogma to the field of computer-assisted sequence research is the exposition of mechanisms of cellular processes as symbol-manipulating devices (automata) in which nucleic acid and protein sequences serve as carriers of symbols rather than as chemical compounds. Because of this conceptual connection between sequences of symbols (polynucleotides and polypeptides) and mechanisms of cellular processes (such as the biosynthesis of proteins), biomolecular sequence analysis is literally inspired by language and machine

---

*The obvious analogy to Shannon's (1948) three-element model of a communication system [60] is probably not a coincidence. In Shannon's theory the general scheme for (tele)-communications is sender → channel → receiver, whereas the scheme for protein biosynthesis is DNA (coding region) → mRNA → polypeptide.

[†]The original formulation of the central dogma excessively evokes the concept of "information" being "passed," which again underscores the symbolic as opposed to the material (structural) aspects of representing biomolecules. "Once information has passed into protein, it cannot get out again. In more detail, the transfer of information from nucleic acid to nucleic acid, or from nucleic acid to protein may be possible, but transfer from protein to protein, or from protein to nucleic acid is impossible" [59].

metaphors. Sequences are produced by automata that resemble devices to print symbols in a sequential manner (such as old telegraph receivers, which punched paper tapes). Therefore there seems to be no reason why they (the sequences) should not be analyzed by methods similar to those used in the analysis of printed text and in cryptology.*

2.2. A Paradigm of Biomolecular Cryptology: Its Strengths and Limitations

The morphology of strings of symbols (texts) can be analyzed and assessed without regard to the meaning they convey. Strings of symbols can also be analyzed in a way that is independent of the reasons they were generated. Traditionally this exact approach was taken in cryptanalysis in an effort to uncover the plain text from a cryptogram by way of discovering the rules of the specific cryptosystem [36,43,44].

The idea of cryptosystem is important. In principle, a cryptosystem is a set of rules for encrypting messages before they are sent and decrypting them at their destination. One of the goals of cryptanalysis is to reconstruct the cryptosystem based on properties of cryptograms alone (as when messages are intercepted by a party not intended to receive them).

The idea of using methods of cryptanalysis in biology is at least as old as molecular biology [13,33,34,54,64–69]. An interesting difference between the art of cryptanalysis and biomolecular sequence analysis is that cryptanalysis ends once the meaning of a cryptogram is understood. Usually, once that happens we can decrypt all cryptograms generated by the same cryptosystem. In the case of nucleic acid sequences we have no idea about their "true meaning." Nor is it obvious that the "DNA as a text" metaphor entails any meaning in the linguistic sense. Therefore there is no natural end to sequence analysis unless we are lucky enough to break an entire "cryptosystem," i.e., recover a meaningful mechanism that leads to observed sequence patterns as its "signatures." However, even in this ideal case the mechanisms are only the (simple) models of real phenomena. Therefore there is no guarantee that recovering the mechanism of a sequence-dependent phenomenon will lead to knowledge of all the sequence patterns that could participate in processes explained by the same mechanistic model. Further

---

*There are plenty of reasons, however, why the power of these "text" analyses should not be overestimated [58,61–63], but this is an entirely different issue. In this chapter our main focus is the exposition of the aspects of text metaphor that lead to positive enrichment of our knowledge of biological facts or (usually mechanistic) plausible interpretations of ensembles of facts.

recoding and the reassigment of patterns to mechanisms appears to be the best way to gain more knowledge about the particular biological phenomenon at hand.*

2.2.1. Alphabets

Symbols (letters) used for writing down the primary structure (sequence) of nucleic acids or proteins constitute an initial elementary alphabet. For sequences of nucleic acids the initial elementary alphabet is the four-letter set $E_1 = \{A, C, G, T \text{ (or U)}\}$, where the letters in brackets stand for the nucleotides adenine, cytosine, guanine, and thymine (or uracil), respectively. (In the case of proteins, the initial elementary alphabet is most often the 20-letter set of symbols of amino acid residues.) Grouping nucleotide symbols together leads to other (two-letter) elementary alphabets frequently used in nucleic acid sequence analysis:

$E_2 = \{K, M\}$, where K is either guanine or thymine (uracil in RNA) and M is either adenine or cytosine

$E_3 = \{R, Y\}$, where R is a purine (adenine or guanine) and Y is pyrimidine (either cytosine or thymine)

$E_4 = \{S, W\}$, where S is either cytosine or guanine and W is either adenine or thymine (or uracil in RNA)

Once a sequence of nucleic acid or protein is given, we can impose any kind of annotations on it. In particular, we can label short oligonucleotides (or short oligopeptides in proteins) such that the symbols used for labeling will serve as letters of the new alphabet.

Among nonelementary alphabets are k-extensions of finite alphabets (where k is a fixed integer). The k-extension E^k of an elementary alphabet E that contains m letters is a k-gram alphabet over E defined as a set of all m^k k-grams (strings of length k; k-tuples) of symbols from E. The union of any number of extensions (between 1 and k extensions) of an alphabet E is a source of 1-grams through k-grams that is often used to find candidate sequence motifs. Any such set of strings of symbols is a subset of a generalized

---

*There seem to be only two notable exceptions to this sinister reality: (1) translation of the protein-coding part of mature mRNA into polypeptide according to the genetic code and (2) semiconservative replication of nucleic acids via complementary base pairing according to Watson–Crick–Chargaff rules (with or without additional RNA-world-specific extensions). That way the cryptogram and cryptosystem metaphors came full circle from their conceptual origins, through the hope that other "functional codes" exist, back to the genetic code and Watson–Crick base pairing (the original inspiration for the cryptosystem/code metaphor).

k-gram alphabet, which is the union of subsequent extensions of alphabet E (from 1 to a fixed k):

$$\{E\}^k = \bigcup_{i=1}^{k} E^i.$$

Examples

A 2-gram alphabet over E_1 is a set of 16 dinucleotides $E_1^2 = \{A, C, G, T\}^2 = \{AA, AC, AG, AT, CA, CC, CG, CT, GA, GC, GG, GT, TA, TC, TG, TT\}$.

Similarly, a 2-extension of E_2 is a set of only four dinucleotides, $E_2^2 = \{RR, RY, YR, YY\}$.

A 3-extension $E_2^3 = \{RRR, RRY, RYR, RYY, YRR, YRY, YYR, YYY\}$, whereas $E_2^3 = \{KKK, KKM, KMK, KMM, MKK, MKM, MMK, MMM\}$.

A generalized 3-gram alphabet over $E_2 = \{R, Y\}$ is $\{E_2^3\} = \{R, Y, RR, RY, YR, YY, RRR, RRY, RYR, RYY, YRR, YRY, YYR, YYY\}$ $= E_2 \cup E_2^2 \cup E_2^3$.

Other imaginable nonelementary alphabets include finite sets of symbols that represent secondary and tertiary structures of biopolymers, physicochemical properties of identifiable regions of nucleic acid structure, and even symbols that label (name) possible biological functions of such regions. Generally speaking, any set of annotations of regions in a sequence can be considered an alphabet. So can a set-theoretic union of various annotation alphabets with sequence-based alphabets, including the initial elementary alphabet.

2.2.2. Metaphor of Functional Codes is Too Esoteric to be Practical

By analogy to the art of cracking ciphers and cryptographic codes, it would be nice to think of sequence analysis as a search for alleged functional codes that would relate sequences (or structures) to biological functions somehow encoded in these sequences. It is unfortunate indeed that finding the actual functional code would mean removing the word "somehow" from the preceding sentence. In other words, a functional code word must signify a specific assignment of a well-defined function to a pattern of symbols in a sequence. The functional code would then be a complete set of all such code words. This means that each individual code word as well as the entire code would have to reflect real deterministic dependencies (not just a correlation of observations) within a mechanism in which a complete list of states and possible transitions between the states are known for a fact (i.e., experimentally verifiable fact).

The concept of a functional code becomes esoteric because it is difficult to define a biological function at the molecular—often "subfunctional"—level. It can be sensibly discussed only in specific cases of known mechanisms that can be represented (modeled) in a form of finite-state automata that take into account well-defined states and transitions between them. From this perspective we can sensibly say that "the function of the heart is to pump blood" or "the function of a bottle is to store liquids" because we assume knowledge of a larger system in which an organ or a tool is supposed to play a specific role (i.e., function). However, in the case of complex or ill-characterized systems, which cannot be adequately described by well-defined states and observables, the idea of specific functions becomes vague. Thus the concept of functional code has no clear meaning* and becomes impractical as far as sequence research is concerned.

2.2.3. Classification Codes and Functionally Equivalent Sequences

Is it possible to grasp the concept of function without defining it in terms of mechanisms? The answer is yes. Function can be defined implicitly either via an operational definition or by convention.

The operational definition evokes the concept of *functionally equivalent sequences* (FESs) [3,8,11,63,75–77]. One can, for instance, create a large set of sequences known to play similar or identical biological roles while the details of molecular mechanisms associated with this role are not required. A compiled collection of FESs (introns, exons, 5′-UTR, 3′-UTRs, transcription terminators, and so on) can be studied for the occurrence of conserved sequence patterns. Ideally such patterns should occur in every sequence from the collection (in which case, and only then, we could sensibly assume that the pattern could be a functional code word), but in reality they will occur in a significantly large number of (but not all) sequences. If a sequence pattern is annotated as significant, we will call it a classification code word or a motif. The convenience of this approach comes from the fact that the function is simply the name (or name-equivalent descriptor such as a pointer) of our collection of functionally equivalent sequences.

---

*The same is true of the general concept of "a function" (i.e., of *some* function in general). It does imply that some unknown parts of the ill-defined system play a role in moving it from one unknown state to another in an alleged pursuit of an unknown mechanism. Clearly there is no explanatory benefit from using such a general notion of function. Therefore, in this case also, the concept of functional code does not have a clear basis in factual reality. See Refs. 23 and 70–74 for more detailed discussions of biological function.

From this perspective, biomolecular sequence analysis is a search for classification code words. It relies entirely on the art of approximately evaluating (measuring) the functional significance of sequence and structural data without having been given a clear definition of function. Classification codes are clearly a reflection of our (the analysts') knowledge of the correlation between sequence patterns and a possible (even if unspecified) biological function shared by all sequences in a given FES. Individual motifs (classification code words) as well as attempts to generate entire alphabets of motifs are addressed by the methodology of pragmatic inference discussed later in this chapter.

3. VARIANTS AND CULTURES OF SEQUENCE ANALYSIS

3.1. Computational Biology vs. Bioinformatics

It may be of importance here to note that the culture of computational biology differs from the culture of bioinformatics. Sequence analysis plays important roles in both fields, but its methods and goals are understood differently by computational biologists and by bioinformaticians. Computational biology originally attracted a considerable number of practically minded theoretical biologists in the 1970s and early 1980s who were both curious about the phenomenon of life and mathematically literate. They wanted to study nucleic acid and protein sequences in order to better understand life itself. In contrast, bioinformatics has attracted a large number of skilled computer enthusiasts with knowledge of computer programs that could serve as tools for laboratory biologists (who in turn were assumed, probably wrongly, to be uneducated in mathematics and not particularly computer-literate). Typical bioinformatics professionals would give their undivided attention to producing and promoting computer tools (mostly software) that laboratory biologists would like better than the products advocated by competitors. Today's split between computational biology and bioinformatics appears to be a reflection of a profound cultural clash between curiosity-driven attitude of computational scientists and adversarial competitiveness of molecular biology software providers.

As far as computational biology is concerned we need to realize that a huge number of indispensable, must-know, methods of sequence analysis are not even mentioned in the vast majority of today's textbooks, monographs, and review papers in bioinformatics. In the rare cases when the methods of sequence analysis are mentioned at all, the coverage is generally misleading and pedagogically useless while its questionable factual and methodological correctness remains unchallenged.

Perhaps the most controversial attitude is exhibited by some bioinformaticians who deny the methodological value of all direct methods of alphabet acquisition (usually based on oligonucleotide or oligopeptide frequency counts) and at the same time overrate all methods of database searches with regular expressions over a known alphabet. (Some of these methods also advocate poorly verifiable—but impressive for science administrators who allocate funds—artificial intelligence techniques.) It appears that such controversial stands are a result of confusion caused by the marketing activities of bioinformatics software developers who have stakes in preventing new methods from being created because considerable financial investments have already been made in current (often unverified) methodologies.*

Genome sequencing projects initially held great promise for computational biologists [61,62,78,79]. Unfortunately, they quickly became corrupted and dysfunctional as far as the science of computational biology was concerned. Because of administrative pressures to rapidly annotate new genomes emerging from several genome projects, the research on precise, "sequence-only"-based methods for segmenting nucleic acid sequences were driven to stagnation by the early 1990s. They were replaced by costly, difficult to understand (at least for laboratory biologists), and methodologically confusing artificial learning tools such as artificial neural networks (ANNs) or hidden Markov models[†] (HMMs) that explore laboratory knowledge from

---

*The financial stakes of bioinformaticians in preventing progress in computational biology research are very high because of a very small market for bioinformatics software and because of enormous influx of free software available over the Internet as well as provided by government-subsidized commercial-like entities such as NCBI, NSF supercomputer centers, or EBI.

[†] Perhaps the most devastating blow for sequence research was the creation of large, government-subsidized centers of excellence for bioinformatics in the late 1980s and the 1990s. Not only did the majority of these centers eliminate the inventiveness of sequence researchers of 1980s, but they also misguided masses of computer-illiterate molecular biologists toward believing in the false premises of misrepresented computer usage paradigms. Brutal and often irresponsible hiring practices of the centers combined with their excellent funding situation led to profound deterioration of inventiveness and creativity in the computational biology community at large. Many of the best and the brightest of early periods of computational biology were forced to do things that massive numbers of others can do as well. Many others became disillusioned by the apparent lack of interest in their research and abandoned science altogether. Yet, despite these events, individuals associated with the bioinformatics establishment launched a decade-long popular press campaign advertising bioinformatics as a brand new science that goes together with genomics quite in the same way as Lenin's name went together with that of Stalin in (former) Soviet propaganda shows. It would be good if future historians of science could explain why political manipulators of bioinformatics centers of excellence were able to silence and cripple (or in some cases annihilate) a worldwide community of curious, skilled, and inventive sequence analysis researchers.

outside sequences together with known facts about sequences. Clearly, genome projects have favored (and financed) bioinformatics while at the same time neglecting computational biology. As a result, the confused scientific community of the 1990s withdrew into rating tools of bioinformatics for possible usefulness to laboratory workers instead of focusing on new and much-needed basic methods of sequence analysis. Fortunately (albeit almost by accident), the interest in sequence analysis proper has survived among individual scientists, and now is a perfect time to resume the tasks so unwisely abandoned in the 1990s.

3.2. Nucleic Acid vs. Protein Sequence Analysis

Other noticeable methodological premises that need to be mentioned are differences between nucleic acid and protein sequence analyses. On the surface, differences appear to be negligible and reducible to the size of the initial alphabet in which sequences are represented (four nucleotide symbols in nucleic acids and 20 amino acid residue symbols for proteins). However, as far as the practice of sequence analysis is concerned the differences are much more profound than that.

The first difference arises from the fact that Watson–Crick–Chargaff base-pairing rules and the knowledge of a general mechanism for protein biosynthesis conceptually bind sequences of nucleic acids to their structural specificity. DNA sequences can undergo semiconservative replication to produce double-stranded entities identical to themselves. They can also undergo transcription into RNA sequences complementary to one of the strands of double-stranded DNA. Similarly, RNA three-dimensional structures appear to be completely determined by their sequences via the use of rules of complementarity (at least in the case of small RNAs such as 5sRNA, the evidence that this is indeed the case is voluminous). In the case of proteins, the relation between sequences and structures is not so obvious, and very much depends on the definition of "function." Even protein folding (once believed to be entirely sequence-dependent) requires the involvement of chaperones, and generally the details of the sequence–structure relationship are difficult to demonstrate directly.

The second difference comes from the fact that the sequence hypothesis, the central dogma, and the universal genetic code provide the conceptual ambience for defining the function of genome fragments in mechanistic terms. However, for protein sequence fragments, mechanistic descriptions of function are much more difficult (if not impossible) to formulate. For instance, we can say that the function of a protein-coding region in DNA is simply to carry a "message" about a protein sequence to succeeding generations (via DNA replication) or to mRNA (via transcription). The concept of "function" in this case is naturally pertinent to our well-verified mechanistic models of cellular processes such as protein biosynthesis or DNA replication. There is no such clarity in the case of proteins. Although it is well known that replacement of even one amino acid by another in a sequence of protein can mean a difference between health and disease [80,81], there is no clear definition of "function" for a given sequence fragment at the molecular level. That is to say, it is easier to see the effect of mutations in proteins at the phenotype level of an entire organism than to relate mutation to a change in the "local" chemistry of the mutated protein.

The foregoing differences between fundamental assumptions imposed on nucleic acids and protein sequences shape the kind of questions that can be asked before sequence analysis begins. A nucleic acid sequence analyzer can sensibly ask, "What is the function-associated alphabet in which this particular sequence should be represented?," whereas the protein analyst would tend to ask, "What subsequences or groups thereof are most conserved in a given collection of sequences?" The nucleic acid analyst would perhaps want to know in what molecular level function a given sequence pattern could possibly participate, whereas the protein analyst would rather want to know how unlikely finding a given sequence pattern at random is according to a given model of chance. We can easily see that these are genuinely different questions. Not only do they address different problems and pertain to different kinds of knowledge but they also require different skills from researchers who do sequence analysis. That is to say, sequence analysis research should, in principle, be able to address all these questions, but the adequacy of answers and interpretations will generally be different for nucleic acids than for proteins.

3.3. Sequence Annotation "By Signal" vs. Annotation "By Content"

The legacy of thinking in terms of alleged "functional codes" (see Sec. 2.2.2) leads sequence analysis toward a search for patterns that appear to be a "signature" of a biological function. This procedure has been genuinely successful only in finding stop codons of the genetic code in nucleic acid sequences. Any stretch of sequence between two stop codons that does not contain such a codon is a potential protein-coding region (unidentified reading frame, so to speak). One can narrow the search by identifying sequences between putative start codons (which unfortunately often code also for the amino acid methionine) and stop codons. This could, in principle, be a method for identifying potential (intronless) protein coding regions. Unfortunately, but not entirely unexpectedly, the search for other potential "signals" often leads to great terminological and methodological complexity of annotation tools but does not help much with understanding a relationship between sequence patterns and biologically sound cellular mechanisms.

Annotation "by content" is methodologically very different than the foregoing search "by signal" [9,82–84]. Instead of looking for specific sequence fragments (say, probable "words"), we search for regions of sequence in which a sequence-associated parameter (usually numerical) assumes values from a range specified by the analyst before the analysis (or otherwise determined during the analysis). The analyzed sequence can be segmented according to the value of the parameter. The hope is that sequence segments defined by the parameter's value will be collinear with biologically important regions that are defined by their biological function. For instance, biologically speaking, introns are the regions in protein encoding sequences that are removed from pre-mRNA after transcription. On the other hand, statistically speaking, repeats of mono- and dinucleotides in introns happen much more often than in neighboring exons. Therefore, segmenting nucleotide sequences by using local compositional complexity (or other measures of repetitiveness) as a numerical parameter can (and often does) lead to finding introns in protein encoding genes.

Of course, computers can memorize properly scripted biological facts and their associations with sequences without us, the analysts, understanding the reasons for these correlations. This is a main reason why search "by signal" has been used in genome informatics to search for protein coding regions and possible regulatory domains in newly sequenced DNA. On the other hand, computers and analysts together can invent and test a large number of different content-associated parameters in a short time. Properly scripted parameter-based segmenting techniques can then be used to search for functionally important regions in newly sequenced chromosomes or even entire genomes.

As far as science is concerned, search by content appears more attractive because it leads to finding putative functionally important regions without knowing them in advance. In contrast, search "by signal" is based on knowing a "function" of the pattern in advance and therefore appears to be more boring than search by content. Naturally, the level of excitement of an analyst with either methodology is to a great extent a function of his or her interests and as such does not need to affect the process of analysis itself.

3.4. Set of Many Sequences vs. Single-Sequence Analysis

Yet another methodological intricacy arises when we want to select appropriate methods for sequence analysis: The analysis of a single sequence requires a different methodology than analysis of a finite set of many sequences.

In the case of multiple sequences the results of analysis are sensitive to the lengths of individual sequences. It is also practical to prealign multiple sequences according to a position (such as leftmost nucleotide) or a characteristic pattern that occurs in every sequence of the ensemble.

4. MODELS OF PRINTED TEXT AND OTHER LESSONS FROM LINGUISTICS

Sequence analysis draws on the metaphor of printed text because it deals with strings of symbols that could be seen as texts generated by a printer. The printed text metaphor in turn represents a variant of both the language and the machine metaphor (machine-generated language encoded in an alphabetic script, to be precise). It is therefore not at all surprising that models of printed texts have proven profitable for devising methods of analyzing nucleic acid and protein sequences. The most well-known of these models draw on an assumption of a strict relationship between a text-generating device and the text itself. More specifically it is assumed that strings of symbols can be printed, read, and composed by clockwork-like devices that are called by different names in different fields. In historical perspective, in the last half a century, linguists, physicists, and computer scientists most often referred to such devices as "automata," wheras in telecommunications engineering the term "source" (of sequences of symbols; messages) was preferred. Cryptologists appear to be more straightforward in their terminology, most commonly referring to the concept of "cryptographic device."

4.1. Syntax-Generating (Generative) Models of Language

Generative models of language are based on the assumption that a given language L is nothing more than a set of sentences that in turn are finite strings of words chosen from a finite vocabulary. (Sometimes words are called letters, and vocabulary is called an alphabet.) A *grammar* is then a finite set of rules for specifying the sentences of L and only those sentences.

The most common formalisms (i.e., metalanguages) to describe grammars are *generative grammars* (sometimes referred to as *rewriting systems*) that consist of four kinds of mathematical objects:

1. Finite set of variables (non-terminals) VN
2. Finite set of terminals (actual "words" of modeled language) T

3. Finite set of rules of production P
4. Start symbol S

In set-theoretic notation we can represent a rewriting system G as a quadruple of sets:

$$G = \langle VN, T, P, S \rangle$$

Rules of production are always in the form $x \rightarrow y$ (read "x can be rewritten as y"), where x and y are strings of symbols drawn from the union of VN and T, i.e.,

$$x, y \in VN \cup T$$

Alternative formalisms to describe formal languages are *automata*, abstract models of hypothetical mechanical devices that accept strings that belong to L.

It turns out that the two groups of formalisms are equivalent to each other in the sense that a language generated by a given class of rewriting systems is acceptable by a given class of automata [46,85–87]. One can distinguish four major classes of rewriting systems that differ from one another in the details of allowed rules of production (and by automata to which they correspond). Their properties are summarized in Table 1.

As far as sequence analysis is concerned, only regular grammars are being systematically employed thus far to design various (formal) languages of regular expressions for pattern matching.* Some aspects of context-free grammars may also be employed for studying expressions that involve self-complementary (in terms of Watson–Crick–Chargaff rules of base pairing) oligonucleotides. This by itself is helpful in writing computer programs and queries within bioinformatics tools but does not have any bearing on understanding biological properties of the studied systems and phenomena.

Biologists have understood the complementarity of messages and mechanisms since the beginning of molecular biology.[†] However, the lack

---

One reason for the focus on languages to describe patterns is computer scientists' desire to use standards similar to those of the UNIX operating system, where regular expressions are implemented as a tool of alphabetic string matching. Anyone who searches online dictionaries or databases for the occurrence of strings containing "wild" characters, such as like Bio or c**j *lina or *omics, is in fact using a regular expression within a given search engine.

[†] For one thing, the graphs ("cartoons") illustrating pathways of cellular mechanisms are terrific working examples of formal equivalence between regular grammars and finite-state automata.

TABLE 1   Types of Grammars and Their Properties

Grammar of Type n^a	Productions	Automaton	References
Type 0: Unrestricted rewriting system	$x \rightarrow y$, where x and y stand for any string of terminal and non-terminal symbols	Turing machine	28, 48, 86
Type 1: Context-sensitive	$xAy \rightarrow xzy$, where A is a non-terminal symbol and x, y, z are arbitrary strings of terminal and non-terminal symbols	Linear bounded automaton	87
Type 2: Context-free	$A \rightarrow x$, where A is a non-terminal symbol and x is an arbitrary but nonempty string of terminal and non-terminal symbols	Pushdown automaton	85
Type 3: Regular	$A \rightarrow aB$ or $A \rightarrow a$, where A and B are single non-terminal symbols and a is an arbitrary string of terminal symbols	Finite-state automaton	46, 47, 88

[a] Language of type n (n = 0, 1, 2, or 3) is defined as a set of strings consisting of terminal symbols generated by rewriting system of type n. It can be proved that for $n > 1$ every type n language is also a type $n-1$ language but not vice versa. For example, every regular language is context-free, but there exist context-free languages that are not regular. This property of generative models is sometimes called a Chomsky hierarchy to underline Chomsky's invention of formal grammars [48,49].

of specific knowledge of experimentally verified function- or structure-associated vocabularies makes the concept of grammar too general to be useful for understanding biological reality. The problem here is almost identical to the case of the metaphor of functional code (see Sec. 2.1.1): The grammar is a nice metaphor that can be used for picturing sequence research to beginners but does not have explanatory power for detailed understanding in biology. In fact, in order to understand biological phenomena, we do not need instructions about syntax and automata. Instead, the practice of sequence analysis is an active search for a listing (vocabulary; alphabet) of sequence motifs (i.e., analogs of words and letters) that pertain to our specific problem that involves nucleic acid or protein sequences.

4.2. Statistical Models of Printed Texts

The distribution of relative frequencies of letters appears to be the same in any sufficiently long text formulated in a given language and recorded in an alphabetic script. Although the reasons for language-specific letter frequency conservation are complex [60,89], it reportedly makes sense [43,90] to use this observation as a rationale for statistical modeling of the morphology of printed texts.

Figure 1 shows an example of the ad hoc letter frequency data in average printed texts in English and in Polish. It can be seen that the space bar (word divider) is the most frequent character in both texts. It can also be seen that the mean distance separating two nearest occurrences of the same letter is a decreasing function of letter frequency. Much less obvious findings are that the average word length in English is shorter than in Polish (distance between nearest space bars), and that the letters Z, J, and K, which occur relatively rarely in English are quite frequent in Polish. Similar statistics (data not shown) indicate that A is the first letter of English words three times as often as it is the last letter, but in Polish A occurs three times as often at the ends of words than at the beginnings of words (similar but opposite relations hold for O). If we had chosen other sufficiently long sample texts in these two languages, most of the foregoing observations would remain unchanged. This in turn means that statistics alone could indicate whether or not two strings of symbols are written in the same language.

An example of a simple but involved frequency count analysis for nucleotide sequences is shown in Figure 2. Simple ad hoc analyses of this kind have only a preliminary character and cannot be used as "hard" criteria for segmenting sequences. However, despite their approximate character they are extremely valuable for shaping intuition about new ways in which sequence data should be represented in order to detect significant patterns that can be used as criteria for reliable string segmentation.

Another example of useful frequency count is shown in Figure 3. It concerns distance analysis between selected simple patterns (motifs) in nucleotide sequences. Here the actual pattern consists of two motifs separated by a gap (punctuation) of variable length. (In Fig. 3 we show only a recurrence of the same motif at variable distances, but the same analysis can be done for a more general case of two different motifs.)

A host of significant defining patterns (descriptors) can be inferred from studying Figures 3A–3D. For instance, the fact that dinucleotides in introns occur at preferred distances of 0, 2, 4, ..., and generally $2n$ ($n = 1,2,3, ...$)

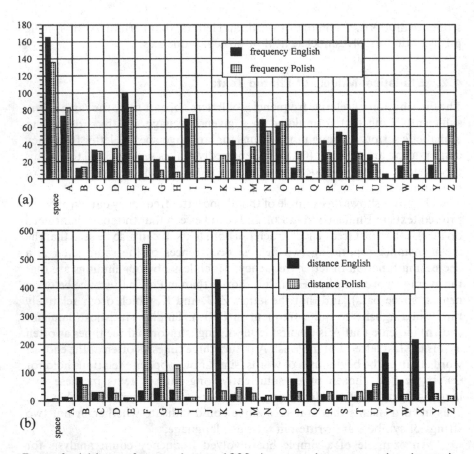

(a)

(b)

FIGURE 1 (a) Letter frequencies per 1000 characters in average printed texts in English and in Polish with the alphabets adjusted to 26 letters and a word divider (space bar). The sample texts chosen for this example were probably too short from the point of view of rigorous sampling, but the most conspicuous similarities and differences between the two frequency distributions conform to data known from all other studies. (b) Average distances between instances of the same character in English and Polish. The most important is the distance between word dividers because it indicates the average length of a word. In English, this appears to be five letters, whereas in Polish it is 6.4 (i.e., between 6 and 7, but most likely 6). As expected, rare letters are separated by large gaps, whereas instances of frequent letters are separated by short gaps.

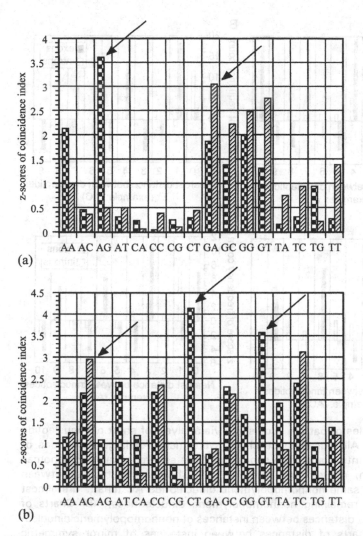

FIGURE 2 Example of using frequency counts to discriminate between introns (lighter bars) and exons (darker bars) from large nonredundant collections (FESs) of vertebrate intron and exon sequences. Figures show z-scores of the distribution of variance of (a) mononucleotide prefix and (b) mononucleotide suffix frequencies of each of 16 dinucleotides. As indicated by the arrows, some variances of prefix and suffix frequency distributions are markedly different in introns and exons. Sequence data were taken from the VIEW database in the NCBI repository [8,9].

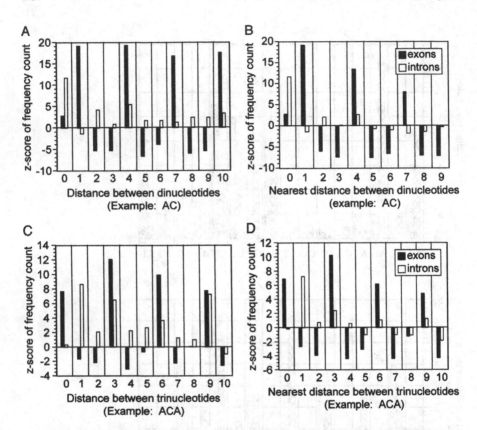

FIGURE 3 Examples of pattern acquisition via analysis of short oligonucleotide distance charts. All four distance charts were made for large collections of sufficiently long intron and exon sequences from several eukaryotic genomes (primarily human, mouse, and fruit fly.) (A) Charts of distances between instances of the same nonhomopolymeric dinucleotide AC (charts are almost identical for 11 remaining nonhomopolymeric dinucleotides). (B) Charts of nearest (shortest) distances between instances of nonhomopolymeric dinucleotide AC. (C) Charts of distances between instances of mirror symmetric trinucleotide ACA [it begins with dinucleotide AC whose charts are presented in the figures (3A) and (3B).] (D) Charts of nearest distance between instances of the same mirror-symmetric trinucleotide ACA.

whereas the preferred shortest distance is 0 indicates that tandem repeats of nonhomopolymeric dinucleotides is a predominant, defining, pattern in introns. The fact that mirror-symmetrical, nonhomopolymeric trinucleotides display 3-base quasi-periodicity in exons but not in introns is a clear confirmation of the triplet nature of the genetic code (avoidance of stop codons in protein-encoding reading frame). Even if the genetic code were not known, we could easily conclude that trinucleotides and their 3-base quasi-periodicity constitute defining, function-associated, patterns in exons but not in introns.

From the perspective of biomolecular sequence analysis, the appealing aspect of letter frequency count is its potential independence of either knowledge of language or actual meanings encoded in the analyzed strings. We do not need to know what words, sentences, and paragraphs mean in order to determine statistical regularities in them and then classify (segment) them according to criteria derived from such regularities. Ideally, our segmenting will reflect some functional property of the entire text (such as being formulated in the same language or being written by the same author) without knowledge of the essence of this underlying function.

An important factor in making quantitative models of texts is the assumption about a general principle of operation of the text-generating devices (sources). From the statistical point of view this assumption is equivalent to defining a model of chance. In the most general terms we can distinguish three models of chance that have been used in biomolecular sequence analysis:

1. Bernoulli text source
2. Markov text source (sometimes called a hidden Markov model)
3. General stochastic (non-Markov) source

Because of their simplicity and elegance, the concepts and terminology originated by Claude Shannon in the 1940s and 1950s [60,89,90] appear to be most appropriate for describing statistical methods of sequence analysis. However, in addition to Shannon H-function, other measures of deviation from equiprobability or from statistical independence can be, and often are, used in sequence analysis protocols.

4.2.1. Discrete Uniform Distribution of Letter Frequencies and Bernoulli Text

A set of events $A_1, A_2, \ldots, A_n$ such that one and only one of them must occur at each trial is called a *complete system of events*. If the events $A_1, A_2, \ldots, A_n$ of a complete system of events are given along with their probabilities $p_1, p_2, \ldots, p_n$

$(p_i > 0, p_1 + p_2 + \ldots + p_n = 1)$, we say that we have a *finite scheme* [91]:

$$A = \begin{pmatrix} A_1 & A_2 & \cdots & A_n \\ P_1 & P_1 & \cdots & P_n \end{pmatrix} \tag{1}$$

The probability distribution of letters from an alphabet of size n in a given text is an example of a finite scheme. For most purposes we only need the data on probabilities without regard to the events to which they are assigned. Therefore we will refer to the list of numbers (probabilities) in the second row of finite scheme (1) as a *probability vector* and denote it by $[Pn]$ (i.e., $[Pn] = [p_1, p_2, \ldots, p_n]$). It should be noted that because of the normalization condition (sum of probabilities in the finite scheme equals 1), the mean value of every list $[Pn]$ equals $1/n$, i.e.,

$$\langle [Pn] \rangle = \frac{1}{n} \sum_{i=1}^{n} P_i = \frac{1}{n} \cdot 1 = \frac{1}{n} \tag{1a}$$

The variance corresponding to this list is by definition

$$\mathrm{Var}([Pn]) = \frac{1}{n} \sum_{i=1}^{n} \left(p_i - \frac{1}{n} \right)^2 \tag{1b}$$

The alternative expression for the variance of list $[Pn]$ that can be derived from (1b) is

$$\mathrm{Var}([Pn]) = \frac{1}{n} \sum_{i=1}^{n} p_i^2 - \left(\frac{1}{n} \right)^2 \tag{1c}$$

The hypothetical long string of symbols (text) over a finite alphabet of a fixed size n is referred to as a *Bernoulli text* if the letters occur in it independently of each other and with equal probabilities, (i.e., $p_i = 1/n$). A Bernoulli text over an alphabet of size n has several properties of interest for sequence analysis and can be used as a model of chance (reference sequence). For one thing it constitutes a perfect example of a situation that should be modeled by a *discrete uniform distribution* (DUD) given by the finite scheme

$$\mathrm{DUD} = \begin{pmatrix} A_1 & A_2 & \cdots & A_n \\ 1/n & 1/n & \cdots & 1/n \end{pmatrix} \tag{2}$$

Here the symbols $A_1, A_2, \ldots, A_n$ stand for letters of an alphabet of size n, but they can represent any kind of statistically independent events belonging to a complete set of finite size (cardinality) n. The list of probabilities in the second row can be abbreviated by the symbol $[1/n]$.

It is easy to note that the mean probability of the list $[1/n]$ must equal $1/n$:

$$\langle [1/n] \rangle = \frac{1}{n} \sum_{i=1}^{n} \frac{1}{n} = \frac{1}{n} \cdot 1 = \frac{1}{n} \tag{2a}$$

By definition, the variance of this list equals 0 because every number in the list is equal to the mean:

$$\mathrm{Var}([1/n]) = \frac{1}{n} \sum_{i=1}^{n} \left(\frac{1}{n} - \frac{1}{n} \right)^2 = \frac{1}{n} \cdot 0 = 0 \tag{2b}$$

Because the mean value of the list $[1/n]$ equals $1/n$, the variance of $[Pn]$ given by either (1b) or (1c) is a measure of how much a given list $[Pn]$ differs from the DUD.

Another such measure can be Shannon entropy [60]:

$$H([Pn]) = -\frac{1}{n} \sum_{i=1}^{n} P_i \log_n P_i \tag{3}$$

Because we took the size of the alphabet n as the basis of the log function, Shannon entropy (3) reaches its maximum (equal to one n-ary unit per symbol) for the DUD and its minimum (equal to 0) for all n lists $[Pn]$ in which one of the n probabilities equals 1 and all others equal 0. The convention $0 * \log 0 = 0$ needs to be adopted.

Yet another measure of deviation of a given list of probabilities $[Pn]$ from the DUD is the so-called index of coincidence I:

$$I = \frac{\kappa_p}{\kappa_r} \tag{4}$$

where κ_p [43] is a sum of squares of probabilities in the list $[Pn]$:

$$\kappa_p([Pn]) = \sum_{i=1}^{n} P_i^2 \tag{4a}$$

The value of κ_r equals simply $1/n$:

$$\kappa_r([Pn]) = \sum_{i=1}^{n} \left(\frac{1}{n} \right)^2 = \frac{1}{n} \tag{4b}$$

It can be easily seen that

$$I([Pn]) = n\kappa_p([Pn]) \tag{5}$$

The relation between the variance of $[Pn]$ and the index of coincidence is also clear:

$$I([Pn]) = 1 + n^2 \, \text{Var}([Pn]) \tag{6}$$

It is clear from the foregoing discussion that all indices that indicate the deviation of a finite list of probabilities from the discrete uniform distribution case should be equivalent to each other and derivable from each other [11,44,63,92,93].

5. PRAGMATIC INFERENCE

The idea of context dependence of motifs and patterns is at the core of *pragmatic inference,* the art of determining sequence motifs from their instances and the knowledge context to which they pertain. Figure 4 shows the general scheme for a protocol of pragmatic inference as used by most practitioners of sequence analysis.

From a practical point of view, pragmatic inference is an organized ensemble (a system) of three protocols:

1. Sequence alignment and *string matching* (database searches)
2. *Pattern acquisition* (primarily statistical analysis of sequences)*
3. Redesigning and remaking database entries and annotations thereof

Each protocol addresses different questions. Sequence alignment can be done only when the alphabet in which the sequences are written is known. It can be the initial alphabet to represent sequences (A, C, G, T/U for nucleotide sequences and 20 symbols for amino acid residues in the case of proteins). It can also be a more sophisticated alphabet of regular expressions (i.e., flexibly defined patterns to match against sequences).

The goal of pattern acquisition methods (mostly statistical) is to find patterns that are meaningful indicators of function. Ideally we would want to find all such meaningful patterns (motifs) because then we would have the complete alphabet of motifs. Sequences could be rewritten in this alphabet

---

*Sometimes (particularly in older literature) the concept of pragmatic inference is used as an equivalent of pattern acquisition only.

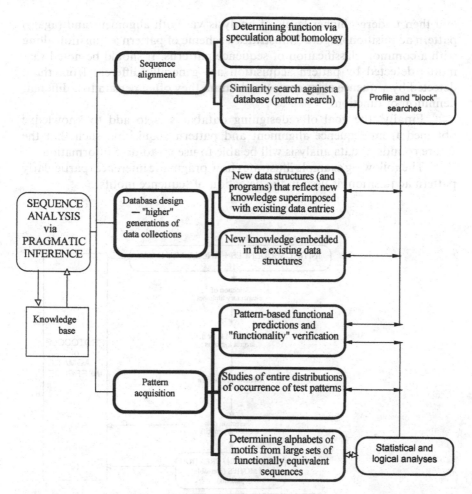

FIGURE 4 General scheme of pragmatic inference as perceived by most practitioners of sequence analysis can be divided into three methodologically distinct classes of protocols: (1) pattern matching (sequence alignment and database searches), (2) enrichment of databases, and (3) pattern acquisition. Pattern acquisition in the case of nucleotide sequences is generally based on frequency counts, visual inspection of sequences, and invention of new data representations.

and then undergo higher levels of analysis via both alignment and (again) pattern acquisition. Figure 5 illustrates a scheme of pattern acquisition along with a common classification of sequence patterns. It should be noted that motifs detected by pattern acquisition are generally different from those detected by sequence alignment. In addition, they often pertain to a different definition of function.

Finally, the goal of redesigning databases is to add to knowledge obtained from sequence alignment and pattern acquisition such that the future rounds of data analysis will be able to use up-to-date information.

The following example illustrates how pragmatic inference (particularly pattern acquisition) is used in the detection of sequence motifs.

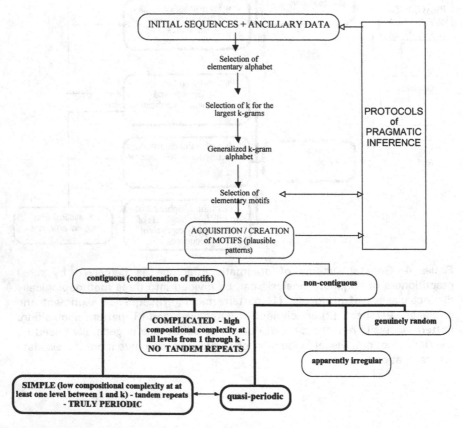

FIGURE 5   Pattern acquisition and classification of sequence patterns.

Example. Let us consider the following fragments (two single strands) in DNA sequences:

$$3'-\text{ATATATATATATATATAT}\cdots5'$$
$$3'-\text{GCATGCGCACGCGCACGC}\cdots5' \qquad (I)$$

Alignment result (II), albeit potentially useful, misses the principle of constructing these very nonrandom sequences:

$$\cdots\text{AT}\cdots\text{A}\cdots\text{A}\cdots$$
$$\cdots\text{AT}\cdots\text{A}\cdots\text{A}\cdots \qquad (II)$$

Pragmatic inference via pattern acquisition requires some experimentation with alphabets. We can, for instance, rewrite our sequences as an alternation of purines (R) and pyrimidines (Y), which leads to the observation that our sequences are tandem repeats of the motif RY (dinucleotide in the R/Y alphabet):

$$\text{RYRYRYRYRYRYRYRYRY}\cdots$$
$$\text{RYRYRYRYRYRYRYRYRY}\cdots \qquad (III)$$

We can also rewrite our sequences as tandem repeats of a hexanucleotide:

$$(\text{ATATAT})\,(\text{ATATAT})\,(\text{ATATAT})\cdots$$
$$(\text{GCATGC})\,(\text{GCATGC})\,(\text{GCATGC})\cdots \qquad (IV)$$

This in a hexanucleotide alphabet could be rewritten as

$$(\text{Hex-a})_3\cdots$$
$$(\text{Hex-b})_3\cdots \qquad (IVa)$$

where Hex-a and Hex-b are hexanucleotides ATATAT and GCATGC, respectively.

Another way of rewriting sequences (I) can be assisted by the observation that ATA and TAT are complementary to each other in terms of Watson–Crick base-pairing rules. ATA and TAT are also related by the relation of reverse complementarity, i.e., they would be strictly complementary to each other if they faced each other in two antiparallel strands of

DNA molecule. The same relation of reverse complementarity is seen in the case of trinucleotides GCA and TGC, which are components of Hex-b in (IVa). This leads us to the observation that both sequences (I) consist of tandem repeats of hexanucleotides that are composed of a trinucleotide followed by its reverse complement. We can therefore rewrite our sequences in the form

$$\beta\beta'\beta\beta'\beta\beta'\cdots$$
$$\partial\partial'\partial\partial'\partial\partial'\cdots \tag{V}$$

Taking into account the fact that hexanucleotide blocks are tandemly repeated (three times), we can rewrite (V) in the compressed form that conveys this information:

$$(\beta\beta')_3\cdots$$
$$(\partial\partial')_3\cdots \tag{Va}$$

where β and ∂ signify trinucleotides ATA and GCA, respectively, while β' and ∂' indicate their reverse complements TAT and TGC.

It should be noted that each of the patterns described by expressions (III)–(Va) can be considered a motif because it is abundant in original sequences in a logically and statistically significant way. On the other hand, expressions such as (III)–(Va) are regular expressions and therefore could be used for pattern matching in properly rewritten data collections. The results of such matching (alignment or database search) could then be used to supplement the knowledge base to assist in the next round (if needed) of pattern acquisition.

6. PRAGMATIC SEQUENCE ANALYSIS

6.1. Overlapping and Nonoverlapping Counts of k-Gram Frequencies

A k-gram alphabet over an alphabet A of size n (i.e., a k-extension of A) contains n^k elements. In a Bernoulli text of length L, the expected frequency of each k-gram is therefore given by the formula

$$F_0 = L/n^k \tag{5}$$

For the *nonoverlapping count*, variance is a function of n, k, and L given by the equation [17,94,95]

$$V = \frac{L}{n^k}\left(1 - \frac{2k-1}{n}\right) \tag{6}$$

For an *overlapping count*, variance is no longer a simple function of numbers n, k, and L but also depends on the specific sequence of the k-gram. Therefore it needs to be calculated for each individual k-gram separately from other k-grams. The reason for this complication is the variance's dependence on the self-overlap capacity of individual k-grams.

The self-overlap capacity of a given k-gram S can be determined from an autocorrelation polynomial:

$$K_S = \sum_{i=1}^{n} a_i x^i \tag{7}$$

Coefficients a_i equal 1 if the first and last $(k-i)$-grams in S are identical and equal 0 otherwise. For example, the polynomial for trinucleotide CCC is $K_{CCC} = 1 + x + x^2$. For trinucleotides ACA we have $K_{ACA} = 1 + x^2$, but for GAT there is no self-overlap, and therefore $K_{GAT} = 1$.

Finding coefficients of the polynomial K_S is more difficult in the case of tetranucleotides and longer k-grams. It generally requires more advanced combinatorics than we wish to describe in this chapter. Interested readers should consult, for instance, Ref. 17 and also get acquaint themselves with the original statement of the problem [94].

Let us consider an overlapping count of k-grams in a Bernoulli text of length L over an elementary alphabet that contains n letters. The variance of frequency distribution for a given k-gram S in such case is a function of $K_S(1/n)$ given by the formula

$$V_S = \frac{L}{n^k}\left[2K_S\left(\frac{1}{n}\right) - 1 - \frac{2k-1}{n^k}\right] \tag{8}$$

Failure to consider self-overlap capacity in overlapping counts can lead to huge errors (hundreds of percent) in both probability and variance calculation. (This is particularly evident in cases of small elementary alphabets and large values of k.) On the other hand, methods to calculate variance are often computationally expensive and difficult to implement [18,19]. That is why in the case of biomolecular sequence analysis it is practical to use frequency counts of nonoverlapping k-grams if at all possible.

6.2. Some Statistical Practices of Particular Importance in Sequence Analysis

In order to determine which sequence patterns can be considered classification code words we need to perform an oligonucleotide frequency count based on the following general assumptions [8,11,43,60,90,96–98] :

A1. The relative frequency of a given "true" code word should be approximately the same in a vast majority of sequences from a given sufficiently large set of sufficiently long functionally equivalent sequences (such as, transcription terminators, introns, or RNA polymerase promoter regions).

A2. The probabilities of at least some individual classification code words should differ markedly from one another in all—or at least a vast majority of—entries from a given collection of functionally equivalent sequences (functional class).

A3. The entire probability distribution of "true" code words should be approximately the same in a vast majority of sequences from a given functional class.

A4. Selected "true" code words should occupy distinct relative positions in a vast majority of sequences from a given functional class.

It is important at this point to determine what a "sufficiently long" sequence is. Given an alphabet A of size n, a sequence is sufficiently long if every symbol of the alphabet occurs in it at least once. In other words, the number of letters absent from the sequence should be zero. The minimal lengths of sequences that can be considered sufficiently long for letter frequency count to make statistical sense over different sizes of alphabets are shown in the graph in Figure 6 and listed in the table at the bottom of the figure.

The lengths of the sequences shown in Figure 6 can be evaluated from several statistics. For the purpose of this chapter I have used the so-called statistics of blanks [43] for Bernoulli texts. According to this statistics, the expected number of "blanks" (absent symbols) in a text of length L over an alphabet of size n can be approximated by the binomial distribution and is given by the formula

$$B(L) = \sum_{i=1}^{n} (1 - p_i)^L \tag{9}$$

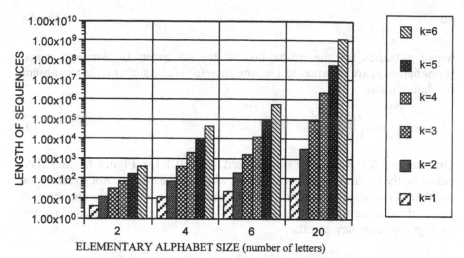

	2	4	6	20
$k=1$	4.32×10^0	1.28×10^1	2.25×10^1	1.03×10^2
$k=2$	1.28×10^1	7.86×10^1	2.09×10^2	3.31×10^3
$k=3$	3.28×10^1	4.10×10^2	1.65×10^3	9.03×10^4
$k=4$	7.86×10^1	2.01×10^3	1.23×10^4	2.29×10^6
$k=5$	1.82×10^2	9.45×10^3	8.76×10^4	5.53×10^7
$k=6$	4.10×10^2	4.35×10^4	6.09×10^5	1.30×10^9

FIGURE 6 Minimum lengths of sequences (L_0) that are required for statistical analyses of k-grams ($k = 1, 2, \ldots, 6$) over four elementary alphabets of sizes 2, 4, 6, and 20 letters, respectively. The alphabets of sizes 2 and 4 correspond to typical representations of nucleotide sequences, whereas the alphabet of size 20 can represent sequences of proteins. It can be seen that to talk sensibly of the frequency count of mononucleotides in a two-letter alphabet—such as {R,Y},{K,M}, or {S,W} representations of nucleic acids—we can do well with sequences as short as five nucleotides in length. However, for dinucleotides we need stretches of sequence of at least 13 nucleotides, for trinucleotides our minimum length L_0 is 33, and for tetranucleotides in a binary alphabet the minimum length is 79 nucleotides. When we consider the standard four-letter initial elementary alphabet {A, C, G, T/U} of nucleic acid sequences, we need a sequence at least 410 nucleotides in length to be able to sensibly study trinucleotide frequency counts and nothing shorter than 2010 nucleotides to study tetranucleotide frequency distributions. Taking into account the fact that many protein-coding genes (particularly in bacteria and viruses) are as short as 100–200 nucleotides, we should not be surprised that our conclusions from frequency counts of even trinucleotides cannot be very reliable for sequence segmenting purposes. (Still they can give us an intuitive idea about a more effective data representation in terms of finding function-associated patterns.)

where p_i stands for the probability of the ith letter. In Bernoulli texts, probabilities p_i are all equal to $1/n$, and therefore the number of letters absent by chance alone is

$$B'(L) = n\left(1 - \frac{1}{n}\right)^L \tag{10}$$

One can set $B'(L)$ close to 0—we used value 0.1 for Figure 6—and then calculate the minimum value of L_0 that will be large enough not to affect the number of missing letters. Only the sequences whose lengths are greater than or equal to L_0 are sufficiently long to be suitable for statistical analysis based on k-gram frequency count.

7. CONCLUDING REMARKS

Computer-assisted sequence analysis attracts the best and the brightest of modern science because it is fun to study sequences and because it is challenging to find methods to analyze strings of symbols with the prospect of almost instant laboratory experimental verification of results and interpretations. Progress in developing new methods of search by content will probably continue through the remainder of this century (which is almost an entire century) because of the enormous amount of genome sequence data accumulated thus far and because of remarkable emerging techniques for the study of cell biochemistry. I sincerely hope that this chapter along with the others in this handbook will help current and future readers to develop intellectual discipline, intuition, and sensibilities that will guide them through their computational biology research adventure. I hope it will be at least as fulfilling and joyous an experience for them as it has been and continues to be for my coworkers and myself.

REFERENCES

1. Konopka AK. Is the information content of DNA evolutionarily significant? J Theor Biol 1984; 107:697–704.
2. Konopka AK. Theory of degenerate coding and informational parameters of protein coding genes. Biochimie 1985; 67:455–468.
3. Konopka AK, Smythers GW. DISTAN—a program which detects significant distances between short oligonucleotides. Comput Appl Biosci 1987; 3:193–201.
4. Konopka AK, Chatterjee D. Distance analysis and sequence properties of functional domains in nucleic acids and proteins. Gene Anal Tech 1988; 5:87–93.
5. Konopka AK, Owens J. Non-contiguous patterns and compositional complexity of nucleic acid sequences. In: Bell G, Marr T, eds. Computers and DNA. Redwood City, CA: Addison-Wesley Longman, 1990:147–155.

6. Konopka AK, Owens J. Complexity charts can be used to map functional domains in DNA. Gene Anal Tech Appl 1990; 7:35–38.
7. Konopka AK. Towards mapping functional domains in indiscriminately sequenced nucleic acids: a computational approach. In: Sarma RH, Sarma MH, eds. Human Genome Initiative and DNA Recombination. Vol. 1. Guilderland NY: Adenine Press, 1990:113–125.
8. Konopka AK. Plausible classification codes and local compositional complexity of nucleotide sequences. In: Lim HA, Fickett JW, Cantor CR, Robbins RJ, eds. The Second International Conference on Bioinformatics, Supercomputing, and Complex Genome Analysis. New York: World Scientific, 1993:69–87.
9. Konopka AK. Sequences and codes: fundamentals of biomolecular cryptology. In: Smith D, ed. Biocomputing: Informatics and Genome Projects. San Diego: Academic Press, 1994:119–174.
10. Konopka AK. Theoretical molecular biology. In: Meyers RA, ed. Molecular Biology and Biotechnology. Weinheim: VCH, 1995:888–896.
11. Konopka AK. Theoretical molecular biology. In: Meyers RA, ed. Encyclopedia of Molecular Biology and Molecular Medicine. Vol. 6. Weinheim: VCH, 1997: 37–53.
12. Konopka AK. Lecture Notes on Foundations of Computational Molecular Biology. San Diego: SDSU-Interdisciplinary Research Center, 1991. Unpublished.
13. Shulman MJ, Steinberg CM, Westmoreland N. The coding function of nucleotide sequences can be discerned by statistical analysis. J Theor Biol 1981; 88:409–420.
14. Fickett JW. The gene identification problem: an overview for developers. Comput Chem 1996; 20:103–118.
15. Waterman MS. Introduction to Computational Biology: Maps, Sequences and Genomes. London: Chapman & Hall/CRC-Press, 1995.
16. Pevzner PA. Computational Molecular Biology: An Algorithmic Approach-Cambridge, MA: The MIT Press, 2000.
17. Gentleman JF, Mullin RC. The distribution of the frequency of occurrence of nucleotide subsequences, based on their overlap capability. Biometrics 1989; 45:35–52.
18. Régnier M. A unified approach to word statistics. Proc RECOMB 98, New York, 1998: 207–213.
19. Klaerr-Blanchard M, Chiapello H, Coward E. Detecting localized repeats in genomic sequences: a new strategy and its application to *Bacillus subtilis* and *Arabidopsis thaliana* sequences. Comput Chem 2000; 24:57–70.
20. Rosen R. Essays on Life Itself. New York: Columbia Univ Press, 2000.
21. Pagels HR. The Dreams of Reason: The Computer and the Rise of the Sciences of Complexity. New York: Bantam, 1989.
22. Konopka AK, Crabbe JC. Practical aspects of practicing interdisciplinary science. Comput Biol Chem 2003; 27:163–164.
23. Mikulecky DC. Robert Rosen (1934–1998): a snapshot of biology's Newton. Comput Chem 2001; 25:317–327.
24. Davis M. The Universal Computer. New York: Norton, 2000.
25. Kleene SC. A note on recursive functions. Bull Am Math Soc 1936; 42:544–546.

26. Tarski A. The concept of truth in formalized languages. In: Corcoran J, ed. Logic, Semantics and Metamathematics. Indianapolis: Hackett, 1933:152–278.
27. Tarski A. The semantic conception of truth and the foundation of semantics. J Phil Phenom Res 1944; 4:341–375.
28. Turing AM. On computable numbers with an application to the Entscheidungsproblem. Proc Lond Math Soc 1936; 42(ser 2):230–265.
29. Turing AM. Computing machinery and intelligence. Mind 1950; LIX.
30. Harshey AD, Chase M. Independent functions of viral proteins and of nucleic acids in the growth of the bacteriophage. J Gen Physiol 1952; 36:39–56.
31. Avery OT, MacLeod CM, McCarty M. Studies of the chemical nature of the substance inducing transformation of pneumococcal types. I. Induction of transformation by a deoxyribonucleic acid fraction isolated from *Pneumococcus* Type III. J Exp Med 1944; 79:137–158.
32. Beadle GW, Tatum EL. Genetic control of biochemical reactions in *Neurospora*. Proc Natl Acad Sci USA 1941; 27:499–506.
33. Gatlin LL. Information Theory and the Living System. New York: Columbia Univ Press, 1972.
34. Campbell J. Grammatical Man: Information, Entropy, Language, and Life. New York: Simon and Schuster, 1982.
35. Davies N. Heart of Europe. Oxford: Oxford Univ Press, 2001.
36. Kahn D. The Codebreakers: The Story of Secret Writing. New York: Macmillan, 1967.
37. Hodges A. Alan Turing: The Enigma. New York: Simon & Schuster, 1983.
38. Kahn D. Interviews with cryptologists. Cryptologia 1980; 4:65–70.
39. Kasparek C, Woytack R. In memoriam: Marian Rajewski. Cryptologia 1982; 6:19–25.
40. Woytack R. Conversations with Marian Rejewski. Cryptologia 1982; 6:50–60.
41. Hill LS. Concerning certain linear transformation apparatus of cryptography. Am Math Monthly 1931; 38:135–154.
42. Hill LS. Cryptography in an algebraic alphabet. Am Math Monthly 1929; 36:306–312.
43. Kullback S. Statistical Methods in Cryptanalysis. Laguna Hills, CA: Aegean Park Press, 1976.
44. Sinkov A. Elementary Cryptanalysis, a Mathematical Approach. New York: Random House, 1968.
45. Vernam GS. Cipher printing telegraph systems for secret wire and radio telegraphic communications. J Am Inst Elec Eng 1926; XLV:109–115.
46. Kleene SC. Representation of events in nerve nets and finite automata. In: Shannon C, McCarthy J, eds. Automata Studies. (Ann Math Studies, No. 34) Princeton, NJ: Princeton Univ Press, 1956:3–41.
47. McCullough WS, Pitts E. A logical calculus of the ideas immanent in nervous activity. Bull Math Biophys 1943; 5:115–133.
48. Chomsky N. Three models for the description of language. IRE Trans Inf Theory 1956; 2:3:113–124.
49. Chomsky N. Syntactic Structures. The Hague: Mouton, 1957.

50. Schrödinger E. What is Life? Cambridge: Cambridge Univ Press, 1944.

51. Rosen R. What is biology? Comput Chem 1994; 18:347–352.

52. Rosen R. Biology and the measurement problem. Comput Chem 1996; 20:95–100.

53. Watson J. Genes, Girls and Gamov. Oxford: Oxford Univr Press, 2001.

54. Crick F. What Mad Pursuit. New York: Basic Books, 1988.

55. Morange M. A History of Molecular Biology. Cambridge, MA: Harvard Univ Press, 2000.

56. Pattee HH. How does a molecule become a message? In: Lang A, ed. 28th Symposium of the Society of Developmental Biology. New York: Academic Press, 1969:1–16.

57. Pattee HH. The physics of symbols: bridging the epistemic cut. BioSystems 2001; 60:5–21.

58. Konopka AK. Grand metaphors of biology in the genome era. Comput Chem 2002; 26:397–401.

59. Crick FHC. On protein synthesis. Symp. Soc Exp Biol 1958; 12:138–163.

60. Shannon CE. A mathematical theory of communication. Bell Syst Tech J 1948; 27:379–423 (623–656).

61. Konopka AK, and Salamon, P. Workshop on Computational Molecular Biology. Human Genome News, 1992; January 1992, 3.

62. Konopka AK, and Salamon, P. Second International Workshop on Open Problems in Computational Molecular Biology. Human Genome News 1993; May 1993, 5.

63. Konopka AK. Biomolecular sequence analysis: pattern acquisition and frequency counts. In: Cooper DN, ed. Nature Encyclopedia of the Human Genome. Vol. 1. London: Nature Publishing Group Reference, 2003:311–322.

64. Gamov G. Possible relation between deoxyribonucleic acid and protein structures. Nature 1954; 173:318.

65. Gamov G. Possible mathematical relation between deoxyribonucleic acid and proteins. Kgl Danske Videnskab Selskab, Biol Medd 1954; 22:1–13.

66. Gamov G, Ycas M. Statistical correlation of protein and ribonucleic acid compositions. Proc Natl Acad Sci USA 1955; 41:1011–1019.

67. Golomb SW. Cryptographic reflections on the genetic code. Cryptologia 1980; 4:15–19.

68. Kay LE. Cybernetics, information, life: the emergence of scriptural representations of heredity. Configurations 1997; 5:23–97.

69. Shepherd JCW. Method to determine the reading frame of a protein from the purine/pyrimidine genome sequence and its possible evolutionary justification. Proc Natl Acad Sci USA 1981; 78:1596–1600.

70. Cummins R. Functional analysis. In: Sober E, ed. Conceptual Issues in Evolutionary Biology. Cambridge, MA: MIT Press, 1994:49–69.

71. Wright L. Functions. In: Sober E, ed. Conceptual Issues in Evolutionary Biology. Cambridge, MA: MIT Press, 1994:27–47.

72. Rosen R. Anticipatory Systems. New York: Pergamon Press, 1985.

73. Mayr E. This is Biology: The Science of The Living World. Cambridge, MA: Belknap Press / Harvard Univ Press, 1997.

74. Konopka AK. This is biology: The science of the living world, by Ernst Mayr. Comput Chem 2002; 26:543–545. Book review.
75. Mengeritzky G, Smith TF. Recognition of characteristic patterns in sets of functionally equivalent DNA sequences. Comput Appl Biosci 1987; 3:223–227.
76. Konopka AK, Smythers GW, Owens J, Maizel JV Jr. Distance analysis helps to establish characteristic motifs in intron sequences. Gene Anal Tech 1987; 4:63–74.
77. Konopka AK. Compilation of DNA strand exchange sites for non-homologous recombination in somatic cells. Nucleic Acids Res 1988; 16:1739–1758.
78. Konopka AK. Towards understanding life itself. Comput Chem 2001; 25:313–315.
79. Konopka AK, Salamon P. Computational Molecular Biology Workshop. Human Genome News May 1994; 6.
80. Zuckerkandl E, Pauling L. Molecules as Documents of Evolutionary History. Gates and Crellin Laboratories of Chemistry—California Institute of Technology. 1962; 1–9. Contribution No. 3041.
81. Pauling LC. Current opinion: molecular disease. Pfizer Spectrum 1958; 6:234–235.
82. Nakata K, Kanehisa M, DeLisi C. Prediction of splice junctions in mRNA sequences. Nucleic Acids Res 1985; 13:5327–5340.
83. Staden R. Measurement of the effects that coding for a protein has on a DNA sequence and their use for finding genes. Nucleic Acids Res 1984; 12:551–567.
84. Staden R. Computer methods to locate signals in nucleic acid sequences. Nucleic Acids Res 1984; 12:505–519.
85. Chomsky N. Context-free grammars and pushdown storage. MIT Res Lab Electron, Quart Prog Rep No 65 1962; 187–194.
86. Chomsky N. On certain formal properties of grammars. Inform Control 1959; 2:137–167.
87. Hartmanis J, Hunt HB. The LBA problem and its importance in the theory of computing. SIAM-AMS Proc 1973; 8:27–42.
88. Rabin MO, Scott D. Finite automata and their decision problems. IBM J Res Dev 1959; 3:114–125.
89. Shannon CE. Communication theory of secrecy systems. Bell Syst Tech J 1949; 28:657–715.
90. Shannon CE. Prediction and entropy of printed English. Bell Syst Tech J 1951; 30:50–64.
91. Khinchin AI. The entropy concept in probability theory. Usp Mat Nauk 1953; VIII:3–20. In Russian.
92. Miller GA. The magical number seven, plus or minus two: some limits on our capacity for processing information. Psych Rev 1956; 63:81–97.
93. Konopka AK. Sequence complexity and composition. In: Cooper DN, ed. Nature Encyclopedia of the Human Genome Vol. 5. London: Nature Publishing Group Reference, 2003:217–224.
94. Guibas LJ, Odlyzko AM. String overlaps, pattern matching and nontransitive games. J Comb Theory (A) 1981; 30:183–208.

95. Pevzner PA, Borodovsky MY, Mironov AA. Linguistics of nucleotide sequences I: The significance of deviations from mean statistical characteristics and prediction of the frequencies of occurrence of words. J Biomol Struct Dynam 1989; 6:1013–1026.

96. Akhmanova OS, Mel'chuk IA, Frumkina RM, Paducheva EV. Exact Methods in Linguistic Research. Rand Project R-397-PR. Santa Monica, CA: Rand Corp 1963:300.

97. Harris ZS. Distributional structure. Word 1954; 10:775–793.

98. Harris ZS. From phoneme to morpheme. Language 1955; 31:190–222.

3

Motifs in Sequences: Localization and Extraction

Maxime Crochemore
Institut Gaspard-Monge, University of Marne-la-Vallée, Marne-la-Vallée, France

Marie-France Sagot
INRIA, Laboratoire de Biométrie et Biologie Évolutive,
University Claude Bernard, Lyon, France

1. MOTIFS IN SEQUENCES

Conserved patterns of any kind are of great interest in biology, because they are likely to represent objects upon which strong constraints are potentially acting and may therefore perform a biological function. Among the objects that may model biological entities, we shall consider only strings in this chapter. As is by now well known, biological sequences, whether DNA, RNA or proteins, may be represented as strings over an alphabet of four letters (DNA/RNA) or 20 letters (proteins). Some of the basic problems encountered in classical text analysis have their counterpart when the texts are biological sequences; among them is pattern matching. However, this problem comes

*Maxime.Crochemore@univ-mlv.fr, http://www-igm.univ-mlv.fr/~mac
†Marie-France.Sagot@inria.fr, http://www.inrialpes.fr/helix/people/sagot/

with a twist once we are in the realm of biology; exact patterns hardly make sense in this case. By exact, we mean identical, and there are, in fact, at least two types of "nonidentical" patterns one must consider in biology. One comes from looking at what "hides" behind each letter of the DNA/RNA or protein alphabet, and the other corresponds to the more familiar notion of "errors." The errors concern mutational events that may affect a molecule during DNA replication. Those of interest to us are point mutations, that is, mutations operating each time on single letters of a biological sequence: substitution, insertion, or deletion. Considering substitutions only is sometimes enough for dealing with some problems.

There are basically two questions that may be addressed when trying to search for known or predicted patterns in any text. Both are discussed in general computational biology books such as Durbin et al.'s [1], Gusfield's [2], Meidanis and Setubal's [3], or Waterman's [4]. One, rather ancillary, question has to do with position: Where are these patterns localized (pattern localization prediction)? The second question, more conceptual, concerns identifying and modeling the patterns ab initio: What would be a consensual motif for them (pattern consensus prediction)? In biology, it is often the second question that is the most interesting, although the first is far from being either trivial or solved. Indeed, in general what is interesting to discover is which patterns, unknown at the beginning, match the string(s) more often than "expected" and therefore have a "chance" of representing an interesting biological entity. This entity may correspond to a binding site, i.e., to a (in general small) part of a molecule that will interact with another, or it may represent an element that is repeated in a dispersed or periodic fashion (for instance, tandemly). The role played by a repetition of whatever type is often unknown. Some repeats, in particular small tandem ones, have been implicated in a number of genetic diseases and are also interesting for the purposes of studying polymorphism; other types of repeats, such as short inverted ones, seem to be hot spots for recombination.

We will address both kinds of problems (pattern localization prediction and pattern consensus prediction) after we have discussed some notions of "nonidentity," that is, of similarity, that we shall be considering. These are presented in Sec. 2. We start with the identity, both because it may sometimes be of interest and because this allows us to introduce some notations that are used throughout the chapter. Such notations are based on those adopted by Karp et al. [5] in a pioneering paper on finding dispersed exact repeats in a string. From there, it is easy to derive a definition of similarity based not on the identity but on any relation between the letters of the alphabet for the strings. In particular, this relation can be, and in general is, nontransitive (contrary to equality). This was introduced by Soldano et al. [6]. Finally, definitions of similarity taking errors (substitutions, insertions, and deletions) into account are discussed, and the idea of models is presented. This idea was initially formally defined by Sagot et al. [7].

We review the pattern localization prediction question in Section 3. Because many methods used to locate patterns are inspired from algorithms developed for matching fixed patterns with equality, we state the main results concerning this problem. Complexity bounds have been intensively studied and are known with good accuracy. This is the background for broader methods aimed at locating approximate patterns. The most widely used approximation is based on the three alignment operations recalled in Section 2. The general method designed to match an approximate pattern is an extension of the dynamic programming method used for aligning strings. Improving this method has also been intensively investigated because of the multitude of applications it generates. The fastest known algorithms are for a specialization of the problem with weak but extra conditions on the scores of edit operations.

For fixed texts, pattern matching is more efficiently solved by using some kind of index. Indexes are classical data structures aimed at providing fast access to textual databases. As such, they can be considered as abstract data types or objects. They consist of both data structures to store useful information and operations on the data (see Salton [8], or Baeza-Yates and Ribero-Neto [9]). The structures often memorize a set of keys, as in the case of an index at the end of a technical book. Selecting keys is a difficult question that sometimes requires human action. In this chapter, we consider full indexes, which contain all possible factors (segments) of the original text, and we refer to these structures as factor or suffix structures. These structures help find repetitions in strings, search for other regularities, solve approximate matchings, or even match two-dimensional patterns, to mention a few applications. Additional or deeper analysis of pattern matching problems may be found in books by Apostolico and Galil [10], Crochemore and Rytter [11], Gusfield [2], and Stephen [12].

Section 4 deals with the problem of finding repeats, exact or approximate, dispersed or appearing in a regular fashion along a string. Perhaps the most interesting work in this area is that of Karp et al. [5] for identifying exact, dispersed repeats. This is discussed in some detail. Combinatorial algorithms also exist for finding tandem repeats. The most interesting are those of Landau and Schmidt [13] and Kannan and Myers [14], which allow for any error scoring system, and that of Kurtz et al. [15], which uses a suffix tree for locating such repeats and comes with a very convenient visualization tool. In biology, so-called satellites constitute another important type of repetition. Satellites are tandem arrays of approximate repeats varying in the number of occurrences between two and a few million and in length between two and a few hundred, sometimes thousands, of letters. To date, only one combinatorial formulation of the problem has been given [16], which we describe at some length.

Finally, motif extraction is considered in Section 5. A lot of the initial work done in this area used a definition of similarity that is based on the relative entropy of the occurrences of a motif in the considered set of strings. This often produces good results for relatively small data sets, and the method has therefore continuously improved. Such a definition, however, leads to exact algorithms that are exponential in the number of strings, so heuristics have to be employed. These do not guarantee optimality, that is, they do not guarantee that the set of occurrences given as a final solution is the one with maximal relative entropy. We do not treat such methods in this chapter. The reader is referred to Ref. 17 for a survey of these and other methods from the point of view of biology.

A definition of similarity based on the ideas of models (which are objects that are external to the strings) and a maximum error rate between such models and their occurrences in strings can lead to combinatorial algorithms. Some algorithms in this category are efficient enough to be used for more complex models. An algorithm for extracting simple models as well as more complex ones, called structured models, elaborated by Marsan and Sagot [18] is treated in some detail.

2. NOTIONS OF SIMILARITY

2.1. Preliminary Definitions

If s is a string of length $|s| = n$ over an alphabet Σ, that is, $s \in \Sigma^n$, then its individual elements are noted s_i for $1 \le i \le n$, so that we have $s = s_1 s_2 \cdots s_n$. A nonempty word $u \in \Sigma^*$ is a factor of s if $u = s_i s_{i+1} \cdots s_j$ for a given pair (i, j) such that $1 \le i \le j \le n$. The empty word, denoted by λ, is also a factor of s.

2.2. Identity

Although identity is seldom an appropriate notion of similarity to consider when working with biological objects, it may sometimes be of interest. This is a straightforward notion we nevertheless define properly, because it allows us to introduce some notations that are used throughout the chapter.

The identity concerns words in a string, and we therefore adopt Karp et al.'s [5] identification of such words by their start position in the string. To facilitate exposition, this and all other notions of similarity are given for words inside a single string. It is straightforward to adapt them to the case of more than one string (for instance, by considering the string resulting from the concatenation of the initial strings with a distinct forbidden symbol separating any two adjacent strings). Let us use E to denote the identity relation on the alphabet Σ (the E stands for "equivalence").

Relation E between elements of Σ may then be extended to a relation E_k between factors of length k in a string s in the following way:

Definition 1. Given a string $s \in \Sigma^n$ and i, j two positions in s such that $i, j \leq n - k + 1$, then

$$i \; E_k \; j \Leftrightarrow s_{i+l} \; E \; s_{j+l} \text{ for all } l \text{ such that } 0 \leq l \leq k - 1$$

In other words, $i \; E_k \; j$ if and only if $s_i s_{i+1} \cdots s_{i+k-1} = s_j s_{j+1} \cdots s_{j+k-1}$. For each $k \geq 1$, E_k establishes an equivalence relation that corresponds to 1a relation between occurrences of words of length k in s. This provides a first definition of similarity between such occurrences. Indeed, each equivalence class of E_k having cardinality greater than 1 is the witness of a repetition in s.

2.3. Nontransitive Relation

When dealing with biological strings, one has to consider that the "letters" represented by such strings are complex biological objects with physicochemical properties, as, for instance, electric charge, polarity, size, different levels of acidity, etc. Some, but seldom all, of these properties may be shared by two or more objects. This applies more to proteins than to DNA/RNA but is true to some extent for both.

A more realistic relation to establish between the letters of the protein or DNA/RNA alphabet (respectively called amino acids and nucleotides) would therefore be reflexive, symmetric, but nontransitive [6]. An example of such a relation, noted R, is given below.

Example 1. Let $\Sigma = \{A,C,D,E,F,G,H,I,K,L,M,N,P,Q,R,S,T,V,W,Y\}$ be the alphabet of amino acids and R be the relation of similarity between these amino acids given by the graph in Figure 1. The maximal cliques of R are

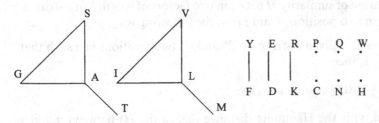

FIGURE 1 Example of a relation of similarity between the letters of the protein alphabet (called amino acids).

the sets $\{A,S,G\}, \{A,T\}, \{I,L,V\}, \{L,M\}, \{F,Y\}, \{D,E\}, \{K,R\}, \{C\}, \{P\}, \{N\},$ $\{Q\}, \{H\}, \{W\}$.

It may be represented by a graph whose nodes are the elements of Σ and where an edge links two nodes if the elements of Σ labeling the nodes correspond to biological objects sharing enough physicochemical properties to be considered similar.

As previously, the relation R between elements of Σ may easily be extended to a relation R_k between factors of length k in a string s.

Definition 2. Given a string $s \in \Sigma^n$ and i, j two positions in s such that $i, j \le n - k + 1$, then

$$i \; R_k \; j \; \Leftrightarrow \; s_i + l \; R \; s_{j+1} \text{ for all } l \text{ such that } 0 \le l \le (k-1)$$

For each $k \ge 1$, R_k establishes a relation that is no longer an equivalence between positions (factors of length k) in a string s. The concept that is important here is that of a (maximal) clique.

Definition 3. Given an alphabet Σ and a nontransitive relation on Σ, a set C of elements of Σ is a (maximal) clique of relation R if for all $\alpha, \beta \in C$, $\alpha \; R \; \beta$ and for all $\gamma \in \Sigma \setminus C$, $C \cup \{\gamma\}$ is not a clique.

Definition 4. Given a string $s \in \Sigma^n$, a set C_k of positions in s is a clique of relation R_k if for all $i, j \in C_k$, $i \; R_k \; j$ and for all $l \; [1..n] \setminus C_k$, $C_k \cup \{l\}$ is not a clique.

Cliques of R_k give us, then, a second way of establishing a definition of similarity between factors of length k in a string.

2.4. Allowing for Errors

2.4.1. Introducing the Idea of a Model

Let us initially assume that the only authorized errors are substitutions. In view of the definitions established in previous sections, one would be tempted to define a relation of similarity H between two factors of length k in a string s, that is, between two positions i and j in s, the following way.

Definition 5. Given a string $s \in \Sigma^n$ and i, j two positions in s such that $i, j \le n - k + 1$, then

$$i \; H_k \; j \; \Leftrightarrow \; \text{dist}_H(s_i \cdots s_{i+k-1}, s_j \cdots s_{j+k-1}) \le e$$

where $\text{dist}_H(u, v)$ is the Hamming distance (hence the H) between u and v (that is, the minimum number of substitutions one has to operate on u in order to obtain v) and e is a nonnegative integer that is fixed.

Parameter e corresponds to the maximum number of substitutions that are tolerated. In the same way as in Section 2.3, cliques of H_k provide us with another possible definition of similarity between factors of length k in a string.

Even before trying to consider how to adapt the above definition to the case of a Levenshtein (or any other type of) distance where insertions and deletions are permitted as well as substitutions (this is not completely trivial; indeed, given two words u and v respectively starting at positions i and j in s and such that $i\ L_k\ j$, what is the meaning of k?), one may intuitively note that calculating H_k (and, a fortiori, L_k) is no longer as easy as computing E_k or R_k.

The reason is that, although the definitions given in Sections 2.2 and 2.3 involve pairs of positions in a string s, it is possible to rewrite them in such a way that, given a position i in s and a length k, it is immediate to determine to which class or clique(s) i belongs in the sense that the class or clique(s) can be uniquely identified just by "reading" $s_i \cdots s_{i+k-1}$. Let us consider first the simpler case of an identity. Straightforwardly, position i belongs to the class whose label is $s_i \cdots s_{i+k-1}$. In the case of a nontransitive relation R between letters of Σ, let us name C the set of (maximal) cliques of R and define clique$_R$ (α) as the cliques of R to which a letter α belongs. Then, position i belongs to all the sets of R_k whose labels may be spelled from the (regular) expression clique$_R(s_i) \cdots$clique$_R(s_{i+k-1})$ and that are maximal under R_k. Note the small difference here from the identity relation: Maximality of a validly labeled set has to be checked [6].

No such easy rewriting and verification are possible in the case of the definition of H_k (or L_k had we already written it) if we wish to build the notion of similarity between factors in a string upon that of the cliques of H_k. Indeed, to obtain such cliques we need to compare (a possibly great number of) pairs of positions between themselves. This is expensive.

One may, however, rewrite the definition of H_k in a way that refers to labels as we did above for E_k and R_k, although such labels are no longer as immediately identifiable. A possible definition (still for the case where only substitutions are considered) is the following.

Definition 6. Given a string $s \in \Sigma^n$ and i, j two different positions in s such that $i, j \leq n - k + 1$, then

$$i\ H_k\ j \iff \exists m \in \Sigma^k \text{ such that } \text{dist}_H(m,\ s_i \ldots s_{i+k-1}) \leq e$$
$$\text{and } \text{dist}_H(m,\ s_j \ldots s_{j+k-1}) \leq e$$

where $\text{dist}_H(u, v)$ and e are as before.

Generalizing this gives the following definition.

Definition 7.   A set S_k of positions in s represents a set of factors in s of length k that are all similar between themselves if and only if there exists (at least) a string $m \in \Sigma^k$ such that, for all elements i in S_k, $\text{dist}_H (m, s_i \cdots s_{i + k - 1})$ $\le e$ and, for all $j \in [1..n] \setminus S_k$, $\text{dist}_H(m, s_j \cdots s_{j + k - 1}) > e$.

Observe that extension of both definitions to a Levenshtein distance now becomes straightforward. We reproduce below, after modification, just the last definition.

Definition 8.   A set S_k of positions in s represents a set of factors of length k that are similar if and only if there exists (at least) a string $m \in \Sigma^k$ such that for all elements i in S_k, $\text{dist}_L(m, s_i \cdots) \le e$, and for all $j \in [1..n] \setminus S_k$, $\text{dist}_L(m, s_j \cdots) > e$.

Because the length of an occurrence of a model m may now be different from that of m itself (it varies between $|m| - e$ and $|m| + e$), we denote the occurrence by $(s_i \cdots)$ leaving indefinite its right endpoint.

Observe also that it remains possible, given a position i in s and a length k, to obtain the label of the group(s) of the relation H_k (or L_k) to which i belongs. Such labels are represented by all strings $m \in \Sigma^k$ such that dist_H (or $\text{dist}_L)(m, s_i \cdots) \le e$, that is, such that their distance from the word starting at position i in s is no more than e.

We call models such group labels. Positions in s indicating the start of a factor of length k are e-occurrences (or simply occurrences where there is no ambiguity) of a model m if $\text{dist}(m, s_i \cdots) \le e$, where "dist" is either the Hamming or Levenshtein distance. Observe that a model m may have no exact occurrence in s.

Finally, we have considered so far what is called a "unitary cost distance" (unitary because the cost of each operation—substitution, insertion, or deletion—is one unit). We could have used instead a "weighted cost distance," that is, we could have used any cost for each operation, in the range of integers or real numbers.

2.4.2.   Expanding on the Idea of Models—Two more Possible Definitions of Similarity

Nontransitive Relation and Errors.   Models allow us to considerably enrich the notion of conservation. For instance, they enable us to simultaneously consider a nonrelative transition between the letters of the alphabet (amino acids or nucleotides) and the possibility of errors. In order to do that, it suffices to permit the model to be written over an extended alphabet composed of a subset of the set of all subsets of Σ [noted $\mathcal{P}(\Sigma)$], where Σ is the alphabet of amino acids or nucleotides. Such an alphabet can be, for instance, one defined by the maximal cliques of the relation R given in Figure 1. Definition 8 then becomes

Definition 9. A set S_k of positions in s represents a set of factors of length k that are all similar between themselves if and only if there exists (at least) one element $M \in P^k$ with $P \subseteq \mathcal{P}(\Sigma)$ such that, for all elements i in S_k, setdist$(M, s_i \cdots) \leq e$ and, for all $j \in [1..n] \setminus S_k$, setdist$(M, s_j \cdots) > e$, where setdist$(M, v)$ for $v \in \Sigma^k$ is the minimum Hamming or Levenshtein distance between v and all $u \in M$.

The alphabet Σ itself may belong to P, the alphabet of models. It is then called a wild card. It is obvious that this may lead to trivial models. Alphabet P may then come with weights attached to each of its elements indicating how many times (possibly infinite) it may appear in an interesting model. Observe that another way of describing the alphabet P of models is as the set of edges of a (possibly weighted) hypergraph whose nodes are the elements of Σ.

When e is zero, we obtain a definition of similarity between factors in the string that closely resembles that given in Section 2.3. Note, however, that, given two models M_1 and M_2, we may well have that the set of occurrences of M_1 is included in that of M_2. The cliques of Definition 4 correspond to the sets of occurrences that are maximal.

A Word Instead of Symbol-Based Similarity. Errors between a group of similar words and the model of which they are occurrences can either be counted as unitary events (possibly with different weights), as was done in the previous sections, or they can be given a score. The main idea behind scoring a resemblance between two objects is that it allows us to average the differences that may exist between them. It may thus provide a more flexible function for measuring the similarity between words. A simple example illustrates this point.

Example 2. Let $\Sigma = \{A,B,C\}$ and
$$score \, (i,i) \; = 1 \qquad \forall i \in \Sigma$$
$$score(A,B) \; = \; score(B,A) \; = \; -1$$
$$score(A,C) \; = \; score(C,A) \; = \; -1$$
$$score(B,C) \; = \; score(C,B) \; = \; -1$$

If we say that two words are similar either if the number of substitutions between them is ≤ 1 or their score is ≥ 1, then by the first criterion the words AABAB and AACCB are not similar, whereas by the second criterion they are, the second substitution being allowed because the two words on average share enough resemblance.

In the example and in the definition of similarity introduced in this section, gaps are not allowed, only substitutions. This is done essentially for the sake of clarity. Gaps may, however, be authorized; the reader is referred to Ref. 19 for details.

Let a numerical matrix $\mathcal{M}$ of size $|\Sigma| \times |\Sigma|$ be given such that

$$\mathcal{M}\,(a,b) = \text{score between } a \text{ and } b \text{ for all } a, b \in \Sigma$$

If this score measures a similarity between a and b, we talk of a similarity matrix (two well-known examples of which in biology are PAM250 [20] and BLOSUM62 [21]), whereas if the score measures a dissimilarity between a and b we talk of a dissimilarity matrix. A special case of this latter matrix is when the dissimilarity measure is a metric, that is, when the scores obey, among other conditions, the triangular inequality. In that situation, we talk of a distance matrix (an example of which is the matrix proposed by Risler et al. [22]).

In what follows, we consider that $\mathcal{M}$ is a similarity matrix.

Definition 10. Given $u = u_1 u_2 \cdots u_k \in \Sigma^k$, $m = m_1 m_2 \cdots m_k \in \Sigma^k$ a model of length k, and $\mathcal{M}$ a matrix, we define

$$score_{\mathcal{M}}\,(u,m) = \sum_{i=1}^{k} \mathcal{M}\,(u_i, m_i)$$

Definition 11. A set S_k of positions in s represents a set of factors of length k that are similar if and only if given w a positive integer such that $w \leq k$ and t a threshold value

1. There exists (at least) one element $m \in \Sigma^k$ such that, for all elements i in S_k and for all $j \in \{1, \ldots, |m| - w + 1\}$, $score_{\mathcal{M}}\,(m_j \cdots m_{j + w - 1}, s_i \cdots s_{i + w - 1}) \geq t$.

2. For all $i \in [1..n] \setminus S_k$, there exists at least one $j \in \{1, \ldots, |m| - w + 1\}$ such that $score_{\mathcal{M}}\,(m_j \cdots m_{j + w - 1}, s_i \cdots s_{i + w - 1}) < t$.

An example is given below.

Example 3. Let $\Sigma = \{A, B, C\}$, $w = 3$, and $t = 6$. Let $\mathcal{M}$ be the following matrix:

	A	B	C
A	3	1	0
B	1	2	1
C	0	1	3

Given the three strings

$$s1 = \text{ABCBBABBBACABACBBBAB}$$
$$s2 = \text{CABACAACBACCABCACCACCC}$$
$$s3 = \text{BBBACACCABABACABACABA}$$

then the longest model that is present in all strings is CACACACC (at positions 9, 1, and 12, respectively in strings $s1$, $s2$, $s3$).

3. MOTIF LOCALIZATION

We review in this section the main results and combinatorial methods used to locate patterns in strings. The problem is of major importance for several reasons. From a theoretical point of view, it is a paradigm for the design of efficient algorithms. From a practical point of view, the algorithms developed in this chapter often serve as basic components in string facility software. In particular, some techniques are used for the extraction of unknown motifs.

We consider two instances of the question, depending on whether the motif is fixed or the string is fixed. In the first case, preprocessing the pattern accelerates the search for it in any string. Searching a fixed string is made faster if a kind of index on it is preprocessed. At the end of the section, we sketch how to search structural motifs for the identification of tRNA motifs in biological sequences.

3.1. Searching for a Fixed Motif

String searching or string matching is the problem of locating all the occurrences of a string x of length p, called the pattern, in another string s of length n, called the sequence or the text. The algorithmic complexity of the problem is analyzed by means of standard measures: running time and amount of memory space required by the computations. This section deals with solutions in which the pattern is assumed to be fixed. There are mainly three kinds of methods to solve the problem: sequential methods (simulating a finite automaton), practically fast methods, and time–space optimal methods. Methods that search for occurrences of approximate patterns are discussed in Section 3.2. Alternative solutions based on a preprocessing of the text are described in Section 3.3.

Efficient algorithms for the problem have a running time that is linear in the size of the input [i.e., $O(n + p)$]. Most algorithms require an additional amount of memory space that is linear in the size of the pattern [i.e., $O(p)$]. Information stored in this space is computed during the preprocessing phase and later used during the search phase. The time spent during the search phase is particularly important. The number of comparisons made and the number of inspections executed have therefore been evaluated with great care. For most algorithms, the maximum number of comparisons (or number of inspections) made during the execution of the search is less than $2n$. The minimum number of comparisons necessary is $\lfloor n/p \rfloor$, and some algorithms reach that bound in ideal situations.

The complexity of the string searching problem is given by the following theorem due to Galil and Seiferas [23]. The proof is based on space-economical methods that are outside the scope of this chapter (see Ref. 11, for example). Linear time, however, is met by many other algorithms. Note that in the $O(\cdot)$ notation, coefficients are independent of the alphabet size.

Theorem 1.   The string searching problem, locating all occurrences of a pattern x in a text s, can be solved in linear time, $O(|s| + |x|)$, with a constant amount of additional memory space.

The average running time of the search phase is sometimes considered more significant than the worst-case time complexity. Despite the fact that it is usually difficult to model the probability distribution of specific texts, results for a few algorithms (with a hypothesis on what "average" means) are known. Equiprobability of symbols and independence between their occurrences in texts represent a common hypothesis used in this context and gives the next result (Yao [24]). Althoug h the hypothesis is too strong, the result reflects the actual running time of algorithms based on the method described below. In addition, it is rather simple to design a string searching algorithm working in this time span.

Theorem 2.   Searching a text of length n for a preprocessed pattern of length p can be done in optimal expected time $O([(\log p)/p]n)$.

String searching algorithms can be divided into three classes. In the first class, the text is searched sequentially, one symbol at a time, from beginning to end. Thus all symbols of the text (except perhaps $p-1$ of them at the end) are inspected. Algorithms simulate a recognition process using a finite automaton. The second class contains algorithms that are practically fast. The time complexity of the search phase can even be sublinear, under the assumption that both the text and the pattern reside in main memory. Algorithms from the first two classes usually require $O(p)$ extra memory space to work. Algorithms from the third class show that the additional space can be reduced to a few integers stored in a constant amount of memory space. Their interest is mainly theoretical so far.

The above classification can be somehow refined by considering the way the search phases of algorithms are designed. It is convenient to consider that the text is examined through a window. The window is assimilated to the segment of the text it contains, and it usually has the length of the pattern. It runs along the text from beginning to end. This scheme is called the sliding window strategy and is described below. It uses a scan-and-shift mechanism.

1. put window at the beginning of text;
2. **while** window on text do
3.   *scan*: **if** window = pattern **then** report it;
4.   *shift*: shift window to the right and
5.   memorize some information for use during next scans and shifts;

During the search, the window on the text is periodically shifted to the right according to rules that are specific to each algorithm. When the window is placed at a certain position on the text, the algorithm checks whether the pattern occurs there, i.e., if the pattern equals the content of the window. This is the scan operation during which the algorithm acquires from the text information that is often used to determine the next shift of the window. Part of the information can also be kept in memory after the shift operation. This information is then used for two purposes: first, saving time during the next scan operations, and, second, increasing the length of further shifts. Thus, the algorithms operate a series of alternate scans and shifts.

A naive implementation of the scan-and-shift scheme (no memorization, and uniform shift of length 1) leads to a searching algorithm running in maximum time $O(p \times n)$; the expected number of comparisons is $4n/3$ on a four-letter alphabet. This performance is quite poor compared to preceding results.

3.1.1. Practically Fast Searches

We describe a string searching strategy that is considered the fastest in practice. Derived algorithms apply when both the text and the pattern reside in main memory. We thus do not take into account the time to read them. Under this assumption, some algorithms have a sublinear behavior. The common feature of these algorithms is that they scan the window in the reverse direction (from right to left).

The classical string searching algorithm that scans the window in reverse direction is the BM algorithm (Boyer and Moore [25]). At a given position in the text, the algorithm first identifies the longest common suffix u of the window and the pattern. A match is reported if it equals the pattern. After that, the algorithm shifts the window to the right. Shifts are done in such a way that the occurrence of u in the text remains aligned with an equal segment of the pattern; such shifts are often called match shifts. The length of the shift is determined by what is called the displacement of u inside x and is denoted by $d(u)$. A sketch of the BM algorithm is displayed below.

1. **while** window on text do
2. $u := $ longest common suffix of window and pattern;
3.   **if** $u = $ pattern **then** report a match;
4.   shift window $d(u)$ places to the right;

The function d depends only on the pattern x, so it can be precomputed before the search starts. In the BM algorithm, an additional heuristics on mismatch symbols of the text is also usually used. This yields another displacement function used in conjunction with d. It is a general method that may improve almost all algorithms in certain real situations.

The BM algorithm is memoryless in the sense that after a shift it starts scanning the window from scratch. No information about previous matches is kept in memory. When the algorithm is applied to find all occurrences of A^p inside A^n, the search time becomes proportional to $p \times n$. The reason for the quadratic behavior is that no memory is used at all. It is, however, very surprising that the BM algorithm turns out to be linear when the search is limited to the first occurrence of the pattern. By the way, the original algorithm was designed for that purpose. Only very periodic patterns may increase the search time to a quadratic quantity, as shown by the next theorem [26]. The bound it gives is the best possible. Only a modified version of the BM algorithm can therefore make less than $2n$ symbol comparisons at search time.

Theorem 3. Assume that pattern x satisfies $\text{period}(x) > |x|/2$. Then the BM searching algorithm performs at most $3|s| - |s|/|x|$ symbol comparisons.

Theorem 3 also suggests that only a little information about configurations encountered during the process has to be kept in memory in order to get a linear time search for any kind of pattern. This is achieved, for instance, if prefix memorization is performed each time an occurrence of the pattern is found. However, this is also achieved with a better bound by an algorithm called TURBO_BM. This modification of the BM algorithm forgets all the history of the search except for the most recent comparison. Analysis becomes simpler, and the maximum number of comparisons at search phase becomes less than $2n$.

Searching simultaneously for several (a finite number of) patterns can be done more efficiently than searching for them one at a time. The natural procedure takes an automaton as pattern. It is an extension of the single-pattern searching algorithms based on the simulation of an automaton. The standard solution is from Aho and Corasick [27].

3.2. Approximate Matchings

The search for approximate matchings of a fixed pattern produces the position in the text s of an approximation of the pattern x. A search for texts for approximate matchings is usually done with methods derived from the exact string searching problem described above. Either they include an exact string matching as an internal procedure or they transcribe a corresponding algorithm. The two classical ways to model approximate patterns consist in assuming that a special symbol can match any other symbol or that operations to transform one pattern into another are possible.

In the first instance we have, in addition to the symbols of the input alphabet Σ, a wild card (also called a don't care symbol) ϕ with the property that ϕ matches any other character in Σ. This gives rise to variants of the string searching problem where, in principle, ϕ appears (I) only in the pattern, (2) only in the text, or (3) in both the pattern and the text. Variant (1) is solved by an adaptation of the multiple string matching and of the pattern matching automaton of Aho and Corasick [27]. For other variants, a landmark solution is that of Fischer and Paterson [28]. They transpose the string searching problem into an integer multiplication problem, thereby obtaining a number of interesting algorithms. This observation brings string searching into the family of Boolean, polynomial, and integer multiplication problems and leads to an $O(n \log p \log \log p)$ time solution in the presence of wild cards (provided that the size of Σ is fixed).

The central notion for comparing strings is based on three basic edit operations on strings introduced in Section 2. It may be assumed that each edit operation has an associated nonnegative real number representing the cost of that operation, so that the cost of deleting from w an occurrence of symbol a is denoted by $D(a)$, the cost of inserting some symbol a between any two consecutive positions of w is denoted by $I(a)$, and the cost of substituting some occurrence of a in w with an occurrence of b is denoted by $S(a,b)$.

The string editing problem for input strings x and s consists in finding a sequence of edit operations, or edit script, Γ of minimum cost that transforms x into s. The cost of Γ is the edit distance between x and s (it is a mathematical distance under some extra hypotheses on operation costs). Edit distances where individual operations are assigned unit costs occupy a special place.

It is not difficult to see that the general problem of edit distance computation can be solved by an algorithm running in $O(p \times n)$ time and space through dynamic programming. Owing to the widespread application of the problem, however, such a solution and a few basic variants were discovered and published in an extensive literature. The reader can refer to Apostolico and Giancarlo [29], or to Ref. 10 for a deeper exposition of the question.

The computation of edit distances by dynamic programming is readily set up. For this, let $C(i, j)$ $(0 \le i \le |s|$ and $0 \le j \le |x|)$ be the minimum cost of transforming the prefix of s of length i into the prefix of x of length j. Then $C(0, 0) = 0$, $C(i, 0) = C(i - 1, 0) + D(s_i)$ $(i = 1, 2, \cdots, |s|)$, $C(0, j) = C(0, j - 1) + I(x_j)$ $(j = 1, 2, \cdots, |x|)$, and $C(i, j)$ equals

$$\min\{C(i - 1, j - 1) + S(s_i, x_j), \ C(i - 1, \ j) + D(s_i), \ C(i, \ j - 1) + I(x_j)\}$$

for all i, j $(1 \le i \le |s|, 1 \le j \le |x|)$. Observe that, of all entries of the C matrix, only the three entries $C(i - 1, j - 1)$, $C(i - 1, j)$, and $C(i, j - 1)$ are involved in the computation of the final value of $C(i, j)$. Hence $C(i, j)$ can be evaluated row by row or column by column in $\Theta(|s| \times |x|) = \Theta(p \times n)$ time. An optimal edit script can be retrieved at the end by backtracking through the local decisions made by the algorithm.

A few important problems are special cases of string editing, including the computation of a longest common subsequence, local alignment, i.e., the detection of local similarities in strings, and some important variants of string searching with errors, or searching for occurrences of approximate patterns in texts.

3.2.1. String Searching with Differences

Consider the problem of computing, for every position of the textstring s, the best edit distance achievable between x and a substring w of s ending at that position. Under the unit cost criterion, a solution is readily derived from the recurrence for string editing given above. The first obvious change consists in setting all costs to 1 except that $S(x_i, s_j) = 0$ for $x_i = s_j$. We thus have now, for all i, j $(1 \le i \le |x|, 1 \le j \le |s|)$,

$$R(i, j) = \min\{R(i - 1, \ j - 1) + 1, \ R(i - 1, j) + 1, \ R(i, j - 1) + 1\}$$

A second change affects the initial conditions, so that we have now $R(0,0) = 0$, $R(i, 0) = i$ $(i = 1, 2, \cdots, p)$, $R(0, j) = 0$ $(j = 1, 2, \cdots, n)$. This has the effect of setting to zero the cost of prefixing x by any prefix of s. In other words, any prefix of the text can be skipped at no cost in an optimum edit script.

The computation of R is then performed in much the same way as indicated for matrix C above, thus taking $\Theta(|x| \times |s|) = \Theta(p \times n)$ time. We are interested now in the entire last row of matrix R.

In practice, it is often more interesting to locate only those segments of s that present a high similarity with x under the adopted measure. Formally, given a pattern x, a text s, and an integer e, this restricted version of the problem consists in locating all terminal positions of substrings w of s such that the edit distance between w and x is at most e. The recurrence given above will clearly produce this information. However, there are more efficient methods to deal with this restricted case. In fact, a time complexity $O(e \times n)$ and even sublinear expected time are achievable. We refer to, e.g., Refs. 10 and 11 for detailed discussions. In the following, we review some of the basic principles behind an $O(e \times n)$ algorithm for string searching with e differences due to Landau and Vishkin [30]. Note that when e is a constant the corresponding time complexity then becomes linear.

It is essential here that edit operations have unitary costs. Matrix R has an interesting property that is intensively used to get the $O(e \times n)$ running time. Its values are in increasing order along diagonals, and consecutive values on the same line or the same column differ by at most one unit (see Fig. 2).

Because of the monotonicity property on diagonals and unitary costs, the interesting positions on diagonals are those corresponding to a strict incrementation. Computing these values only produces a fast computation in time $O(e \times n)$. This is possible if queries on longest common prefixes, as suggested in Figure 2, are answered in constant time. This, in turn, is possible because strings can be preprocessed in order to get this time bound.

To do so, we consider the suffix tree (see Sec. 3.3), $Ac(\mathrm{Suff}(z))$, of $z = x\$s$, where $\$ \notin \mathrm{alph}(s)$. String $w = \mathrm{LCP}(x[\ell + 1..p-1], s[d + \ell + 1..n-1])$ is also $\mathrm{LCP}(x[\ell + 1..p-1]\$s, s[d + \ell + 1..n-1])$ because $\$ \notin \mathrm{alph}(s)$. Let f and g be the nodes of $Ac(\mathrm{Suff}(z))$ associated with strings $x[\ell + 1..p-1]\$s$ and $s[d + \ell + 1..n-1]$. Their common prefix of maximal length is then the label of the path in the suffix tree starting at the root and ending at the lowest common ancestor of f and g. Longest common prefix queries are thus transformed into

R	-1	0	1	2	3	4	5	6	7	8	9	10	11
		C	A	G	A	T	A	A	G	A	G	A	A
-1	0	0	0	0	0	0	0	0	0				
0 G	1	1	1	0	1	1	1	1	0				
1 A			1	1	0	1	1	1	1	0			
2 T					1	0	1			1	1		
3 A						1	0	1				1	
4 A							1	0	1				1

FIGURE 2 Simulation of fast searching for approximate matchings. Searching y = CAGATAAGAGAA for x = GATAA with at most one difference. Pattern x occurs at right position 6 on y without errors ($R[4, 6] = 0$) and at right positions 5, 7, and 11 with one error ($R[4, 5] = R[4, 7] = R[4, 11] = 1$). After initialization, values are computed diagonalwise, value 0 during the first step and value 1 during the second step. Value $R[4, 6] = 0$ comes from the fact that GATAA is the longest common prefix of x and $y[2..11]$. And, as a second example, $R[4, 11] = 1$ because AA is the longest common prefix of $x[3..4]$ and $y[10..11]$. When queries related to longest common prefixes are answered in constant time, the running time is proportional to bold values in the table.

lowest common ancestor queries that are answered in constant time by an algorithm due to Harel and Tarjan [31], simplified later by Schieber and Vishkin [32]. The consequence of the above discussion is the next theorem.

Theorem 4. On a fixed alphabet, after preprocessing x and s, searching s for occurrences of x with at most e differences can be solved in time $O(e \times |s|)$.

In applications to massive data, even an $O(e \times n)$ time may be prohibitive. By using filtration methods, it is possible to set up sublinear expected time queries. One possibility is to first look for regions with exact replicas of some pattern segment and then scrutinize those regions. Another possibility is to look for segments of the text that are within a small distance of some fixed segments of the pattern. Some of the current top-performing software for molecular database searches is engineered around these ideas [33–36]. A survey can be found in Ref. 37.

3.3. Indexing

Full indexes are designed to solve the pattern matching problem, searching s for occurrences of x, when the text s is fixed. Having a static text allows us to build a data structure to which the queries are applied. Efficient solutions require a preprocessing time $O(|s|)$ and need $O(|x|)$ searching time for each query.

Full indexes store the set of factors of the text s. Because factors are beginnings of suffixes of s, this is equivalent to storing all suffixes of the text. Basic operations on the index are find whether pattern x occurs in s, give the number of occurrences of x in s, and list all positions of these occurrences. But many other operations admit fast solutions through the use of indexes.

Indexes are commonly implemented by suffix trees suffix automata [also called suffix DAWGs (directed acyclic word graphs)], or suffix arrays. The latter structure realizes a binary search in the ordered list of suffixes of the text. The former structures are described in the remainder of this section.

Suffixes of s can be stored in a digital tree called the suffix trie of s. It is an automaton whose underlying graph is a tree. Branches are labeled by all the suffixes of s. More precisely, the automaton accepts Suff(s), the set of suffixes of s. A terminal state outputs the position of its corresponding suffix. Figure 3 displays the suffix trie of $s = $ **ababbb**.

3.3.1. Compaction

The size of a suffix trie can be quadratic in the length of s, even if pending paths are pruned (it is the case with the word $a^k b^k a^k b^k$, $k \in \mathbf{N}$). To cope with this problem, another structure is considered. It is the compacted version of the

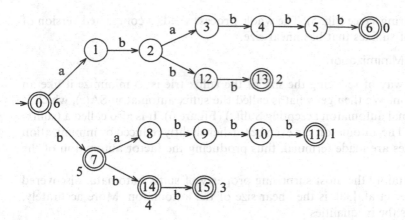

FIGURE 3 Suffix trie of ababbb.

trie, called the suffix tree and denoted ST(s). It keeps from the trie states that are either terminal states or forks (nodes with outdegree greater than 1). Removing other nodes leads to label arcs with words that are non-empty segments of s (see Fig. 4).

It is fairly straightforward to see that the number of nodes of ST(s) is no more than $2n$ (if $n > 0$), because nonterminal internal nodes have at least two children, and there are at most n external nodes. However, if the labels of arcs are stored explicitly, again the implementation can have quadratic size. The technical solution is to represent labels by pairs of integers in the form (position, length) and to keep in main memory both the tree ST(s) and

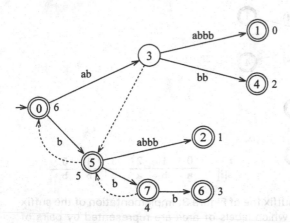

FIGURE 4 Suffix tree of **ababbb**.

the textstring s (see Fig. 5). The whole process yields a compacted version of the trie of suffixes that has linear size.

3.3.2. Minimization

Another way of reducing the size of the suffix trie is to minimize it like an automaton. We then get what is called the suffix automaton SA(s), which is the minimal automaton accepting Suff(s) (Figure 6). It is also called a (suffix) DAWG. The automaton can even be further slightly reduced by minimization if all states are made terminal, thus producing the factor automaton of the text.

Certainly the most surprising property of suffix automata, discovered by Blumer et al. [38], is the linear size of the automaton. More accurately, it satisfies the inequalities

$$|s| + 1 \leq \# states \leq 2|s| - 1$$
$$|s| \leq \# arcs \leq 3|s| - 4$$

3.3.3. Efficient Constructions

The construction of suffix structures can be carried on in linear time. Indeed, running times depend on the implementation of the structures, mainly on that of the transition function. If arcs are implemented by sets of successors, transitions are done by symbol comparisons, which leads to an O($|s|$ log card Σ) construction time within $O(|s|)$ memory space. This is the solution to choose for unbounded alphabets. If arcs are realized by a transition table that assumes that the alphabet is fixed, transitions are done by table lookups, and

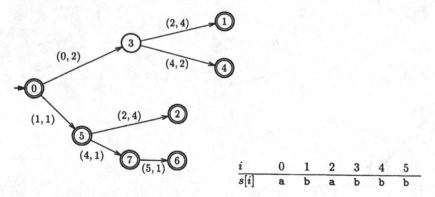

i	0	1	2	3	4	5
$s[i]$	a	b	a	b	b	b

FIGURE 5 Compaction of the suffix trie of Figure 3: Implementation of the suffix tree of **ababbb** of Figure 4 in which labels of arcs are represented by pairs of integers.

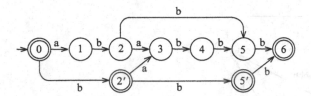

FIGURE 6 Suffix automaton of **ababbb**: minimal deterministic automaton accepting Suff (s).

the construction time becomes $O(|s|)$ using, however, $O(|s|$ card $\Sigma)$ memory space. These two techniques are referred to as the comparison model and the branching model, respectively.

Classical algorithms that build suffix trees are those of Weiner [39], McCreight [40], and Ukkonen [34]. The latter algorithm is the only one to process the text in a strictly on-line manner. DAWG construction was first designed by Blumer et al. and later extended to suffix and factor automata (see Refs. 41 and 42).

To complete this section, we compare the complexities of the above structures to the suffix array designed by Manber and Myers [43]. A preliminary version of the same idea appears in the PAT system of Gonnet et al. [44]. A suffix array is an alternative implementation of the set of suffixes of a text. It consists of both a table storing the permutation of suffixes in lexicographic order and a table storing the maximal lengths of common prefixes between pairs of suffixes (LCP table). Access to the set of suffixes is managed via a binary search with the help of the LCP table. Storage space is obviously $O(|s|)$; access time is only $O(p + \log |s|)$ to locate a pattern of length p [it would be $O(p \times \log |s|)$ without the LCP table]. Efficient preprocessing is the most difficult part of the entire implementation; it takes $O(|s| \log |s|)$ time, although the total size of suffixes is $O(|s|^2)$.

3.3.4. Efficient Storage

Among the many implementations of suffix structures, we can mention the notion of sparse suffix trees due to Kärkkäinen and Ukkonen [45], which considers a reduced set of suffixes; the suffix cactus due to Kärkkäinen [46], who degenerates the suffix tree structure without overly increasing the access time; and the version dedicated to external memory (SB-trees) by Ferragina and Grossi [47]; but several other variations exist (see Refs. 48 and 49, for example).

An excellent solution to save on the size of suffix structures is to simultaneously compact and minimize the suffix trie. Compaction and minimization are commutative operations, and when both are applied they yield the

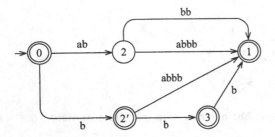

FIGURE 7 Compact suffix automaton of **ababbb** with explicit labels on arcs.

compact suffix automaton, denoted by CSA(s). Figures 7 and 8 display an example of a compact suffix automaton. The direct construction of the compact suffix automaton CSA(s) is possible without first building either the suffix automaton SA(s) or the suffix tree (see Ref. 50). It can be realized within the same time and space as that of other structures.

Table 1 gives an idea of the minimum and maximum sizes of suffix structures (in the comparison model). The average analysis of suffix automata, including their compact version, was started by Blumer et al. [51] and later completed by Raffinot [52].

The size of an implementation of the above structures is often evaluated by the average number of bytes necessary to store one letter of the original text. It is commonly admitted that these ratios are 4 for suffix arrays, 9–11 for suffix trees, and slightly more for suffix automata, provided the text is not too large (of the order of a few megabytes).

Kurtz [53] provides several implementations of suffix trees having this performance. Holub (personal communication, 1999) designed an implementation of compact suffix automata having a ratio of 5, a result that is extremely

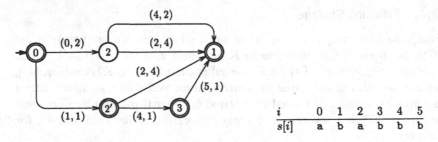

FIGURE 8 Compact suffix automaton of **ababbb**. It is the compacted version of SA(s) and the minimized version of ST(s). Labels of arcs are represented by pairs of integers as in the suffix tree (see Fig. 5).

TABLE 1 Compared Sizes of Suffix Structures

Text of length n	Number of states		Number of arcs	
	min	max	min	max
Suffix trie	$n + 1$	$O(n^2)$	$2n$	$O(n^2)$
Suffix tree	$n + 1$	$2n + 2$	n	$2n + 1$
Suffix automaton	$n + 1$	$2n - 1$	n	$3n - 4$
Compact SA	2	$n + 1$	n	$2n - 2$

good compared to the space for a suffix array. Recently, Balik [54] gave an implementation of another type of suffix DAWG whose ratio is only 4 and sometimes even less.

3.3.5. Indexing for Approximate Matchings

Though approximate pattern matching is much more important than exact string matching for treating real sequences, it is quite surprising that no specific data structure exists for this purpose. Therefore, indexing strategies for approximate pattern matching use the data structures presented above and adapt the search procedure. This one is then based on the next result.

Lemma 1. If x and s match with at most e differences, then x and s must have at least one identical substring of length $r = \lmax\{|x|, |s|\}/(e + 1)\rfloor$.

An original solution was proposed by Manber and Baeza-Yates [55], who considered the case where the pattern embeds a string of at most e wild cards, i.e., has the form $x = u\phi^i v$, where $i \le e$, $u, v \in \Sigma^*$, and $|u| \le p$ for some given e and m. Their algorithm is off-line (on the text) in the sense that the text s is preprocessed to build the suffix array associated with it. This operation costs $O(n \log |\Sigma|)$ time in the worst case. Once this is done, the problem reduces to one of efficient implementation of two-dimensional orthogonal range queries.

Some other solutions preprocess the text to extract its q-grams or q-samples. These, and possibly their neighbors up to some distance, are memorized in a straightforward data structure. This is the strategy used, for example, by the two famous programs FASTA and BLAST, which makes them run fairly fast.

There is a survey on this aspect of indexing techniques by Navarro [56].

3.4. Structural Motifs

Real motifs in biological sequences are often not just simple strings. They are sometimes composed of several strings that appear in organized fashion along the sequence at bounded distances from one another. Possible variations of bases can be synthesized by regular expressions. Efficient methods exist that make it possible to locate motifs described in this manner.

Motifs can also be repetitions of a single seed (tandem repeats) or (biological) palindromes, again with possible variations on individual bases. Palindromes, for instance, represent the basic elements of the secondary structures of RNA sequences. In contrast to the previous type of motifs, a regular expression cannot deal with repetitions and palindromes (at least if there is no assumption on their length).

A typical problem one may wish to address concerns the localization of tRNAs in DNA sequences. It is an instance of a wider problem that is related to the identification of functional regions in genomic sequences. The problem is to find all positions of potential tRNAs in a sequence, given a model obtained from an alignment of experimentally identified tRNAs.

There are basically two approaches to the solution: One consists of a general-purpose method that integrates searching and folding; the other consists of a self-contained method specifically designed for tRNAs. The latter produces more accurate results and faster programs. This is needed to explore complete genomes. We briefly describe the strategy implemented by the program FAStRNA of El-Mabrouk and Lisacek (see Ref. 57 for more information on other solutions), an algorithmic improvement on the tRNAscan algorithm by Fichant and Burks [58].

FAStRNA depends on two main characteristics of tRNAs (at least of the tRNAs in the training set used by the authors of the software): the relative invariance of some nucleotides in two highly conserved regions forming the $T\Psi C$ and D signals, and the cloverleaf structure composed of four stems and three loops (see Fig. 9).

In a preliminary step, the program analyzes the training set to build consensus matrices on nucleotides. This provides the invariant bases of the $T\Psi C$ and D regions used to localize the two signals. After discovering a signal, the program tries to fold the stem around it. Other foldings are performed to complete the test for the current position in the DNA sequence. Various parameters help tune the program to increase its accuracy, and an appropriate hierarchy of searching operations makes it possible to decrease the running time of the program.

The built-in strategy produces a very low rate of false positives and false negatives. Essentially, it fails for tRNAs containing a very long intron. Searching for signals is implemented by a fast approximate matching procedure of the type described above, and folding corresponds to doing an alignment as presented earlier. The program runs 500 times faster than previous tRNA searching programs.

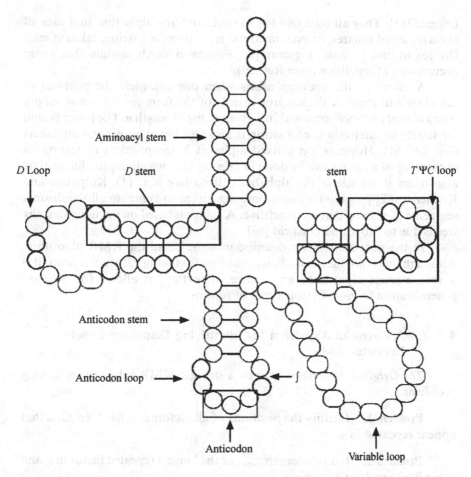

Aminoacyl stem ⟶

D Loop D stem   stem $T\Psi C$ loop

Anticodon stem ⟶

Anticodon loop ⟶   ⟵ ∫

Anticodon

Variable loop

FIGURE 9 Cloverleaf secondary structure of a tRNA.

4. REPEATED MOTIFS IDENTIFICATION

4.1. Exact Repetitions

4.1.1. General Algorithms

One of the first methods that made it possible to discover exact repetitions in strings was designed by Karp et al. [5]. Their algorithm (henceforward called KMR) runs in $O\ (n \log n)$ time on a string of length n but cannot find all repetitions. However, various solutions based on closely related ideas were proposed by Crochemore [59], Apostolico and Preparata [60], and Main and

Lorentz [61]. They all take $O(n \log n)$ time, and any algorithm that lists all occurrences of squares, or even maximal repetitions in a string, takes at least $\Omega(n \log n)$ time because, for example, Fibonacci words contain that many occurrences of repetitions (see Ref. 59).

A more specific question arises when one considers the problem of detecting and locating the squares (words of the form uu, for a non-empty string u) that possibly occur within a given string of length n. The lower bound for testing squarefreeness of a string is also $\Omega(n \log n)$ on general alphabets (see Ref. 61). However, on a fixed alphabet Σ the problem of testing an occurrence of a square can be done in $O(n \log |\Sigma|)$, which implies linear-time algorithms if the size of the alphabet is fixed (see Ref. 11). Kolpakov and Kucherov [62] proposed a linear-time algorithm to compute all the distinct segments of a string that are repetitive. A solution based on the use of a suffix tree is due to Stoye and Gusfield [63].

In the next section, we describe in some detail the KMR algorithm. Although this is not the most efficient method for finding all exact repeats, it is a very elegant algorithm, and, more important, it allows for an easy generalization to more flexible types of repeats.

4.1.2. A Powerful Algorithm for Identifying Dispersed Exact Repeats—KMR

The Original Algorithm.   Given a string s, KMR solves the following problems.

Problem 1.   Identify the positions of all factors of a fixed length k that appear repeated in s.

Problem 2.   Find the length $k_{\max}$ of the longest repeated factor in s, and solve Problem 1 for $k = k_{\max}$.

KMR rests on the definition of an equivalence relation given in Sec. 2.2. Problem 1 and the second part of Problem 2 can then be formulated as the problem of finding the partition associated with E_k. Problem 2 further requires finding the maximum value of k such that E_k is not an identity. The algorithm is based on an iterative construction of partitions E_l for $l \leq k$. The mechanism for performing such constructions rests on the following lemma.

Lemma 2.   Given $a, b \geq 1$ two integers with $b \leq a$, and i, j two different positions in s such that $i, j < n - (a + b) + 1$, then

$$i \; E_{a+b} \; j \;\; \Leftrightarrow \;\; i \; E_a \; j \quad \text{and} \quad (i+b) \; E_a \; (j+b)$$

The main idea behind the KMR algorithm is to use the lemma with $a = b$ for as long as possible. The lemma is consequently called the doubling lemma. This means finding repeats of length $2a$ by using previously acquired information on the repeats of length a that may become the prefixes and suffixes of those of length $2a$. If we are dealing with Problem 1, and if k is not a power of 2, we then use the lemma with $b < a$ in a last step in order to obtain E_k. If we are treating Problem 2, we may need more than one step to find the value of k_{max} such that E_{kmax} is not the identity but $E_{kmax + 1}$ is. The search for k_{max} from the smallest power of 2 that is bigger than k_{max}, let us say it is 2^p, can be done by applying the lemma with $b < a$ in a binary search fashion between $2^p - 1$ and 2^p.

Building the partitions E_a basically corresponds to performing a set intersection operation. The intersections can be implemented using, for instance, stacks. More precisely, we need an array V_a of size n that stores, for each position i in s, the label of the class of E_a to which the a-long factor starting at i belongs. Lemma 2 is applied by means of two arrays of stacks P and Q. Stacks in P are filled by traversing V_a. Such stacks are, in fact, a dual of V_a. Each one corresponds to a class c of E_a and contains the positions i in s belonging to c. Array P therefore serves to sort the prefixes of length a of the repeats of length $2a$ one is trying to identify. The content of each stack of P in turn is then poured into the appropriate stack of Q. A division separates, within the same stack of Q, elements coming from different stacks of P. Like P, array Q has as many stacks as there are classes in E_a. It serves to sort the suffixes of length a of the repeats of length $2a$. One then just needs to pour Q in an orderly fashion into V_{2a} the obtain the classes of E_{2a}, checking the quorum as one goes.

As mentioned, KMR time complexity is $O(n \log k)$. When solving Problem 2, this leads to an $O(n \log n)$ complexity because of possible degenerate cases (such as that of a string s composed of a single letter). KMR space complexity is $O(n)$.

Nontransitive Relations Without Errors.   KMR may be adapted to deal with a nontransitive relation R [6]. The problems solved are the same as for KMR.

Lemma 1 applies analogously, except that one needs to substitute R for relation E.

Lemma 3.   Given $a, b, \geq 1$ two integers with $b \leq a$, and i, j two different positions in s such that $i, j \leq n - (a + b) + 1$, then

$$i \; R_{a + b} \; j \Leftrightarrow i \; R_a \; j \qquad \text{and} \qquad (i + b) \; R_a \; (j + b)$$

Computing relations R_l for $l \leq k$ requires the same structures as for KMR except that, as we saw, a set of positions pairwise related by R_l is no longer an equivalence class but a clique. The algorithm was in consequence called KMRC (the "C" standing for clique) [6]. In particular, a position may belong to two or more distinct cliques of R_l. Array Vl must now therefore be an array of stacks, like P and Q. It indicates, for each cell i corresponding to a position in s, the cliques of relation R_l to which i belongs.

The construction itself follows the same schema as indicated for KMR. Some of the sets of similar factors obtained at the end of each step may not be maximal. A further operation is therefore needed to eliminate sets included in another one so as to get maximal cliques at the end.

To calculate the complexity of the KMRC algorithm, we need to define a quantity g that measures the "degree of nontransitiveness" of relation R.

Definition 12. Given R, a nontransitive relation on Σ, we call g the greatest number of cliques of R to which a symbol may belong, that is,

$$g = \text{Max} \{ g_a \mid a \in \Sigma, g_a = \text{number of cliques to which } a \text{ belongs}\}$$

We call $\bar{g}$ the average value of g_a for $a \in \Sigma$, that is,

$$\bar{g} = \frac{\sum_a g_a}{n_c}$$

where n_c is the number of cliques of R.

If one does not count the set inclusion operations to eliminate non-maximal cliques, KMRC has time complexity $O(ng^k \log k)$, because each position i in s may belong to at most g^k (or, on the average, $\bar{g}^k$) cliques of R_k. Inclusion tests based on comparing the positions contained in each set take $O(n^2 g^{2k})$ time at the end of step k. At least one other approach for testing set inclusion is possible and may result in a better theoretical (but not necessarily better in practice—this is discussed in Ref. 6) time complexity. Space complexity is $O(ng^k)$.

4.2. Inexact Repetitions—The Particular Case of Tandem Arrays (Satellites)

4.2.1. Model for Tandem Arrays (Satellites)

Tandem arrays (called tandem repeats when there are only two units) are a sequence of repeats that appear adjacent in a string. As concerns biology, such tandemly repeated units are divided into three categories depending on the length of the repeated element, the span of the repeat region, and its location within the chromosome [64]. Repeats occurring in or near the centromeres and telomeres are called simply satellites. Their span is large, up to a million bases, and the length of the repeated element varies greatly, anywhere from five to a few hundred base pairs. In the remaining, euchromatic, region of the chromosome, the kinds of tandem repeats found are classified as either micro- or minisatellites, according to the length of the repeated element. Micro-satellites are composed of short units, of two to five base pairs, in copy numbers in general around 100. Minisatellites, on the other hand, involve slightly longer repeats, typically around 15 base pairs, in clusters of variable sizes, comprising between 30 and 2000 elements.

Figure 10 shows an example of a tandem repeat starting at position 391131 on chromosome IX from yeast (in the sequence as recovered from the ftp site ftp://ftp.mips.embnet.org/yeast/). This repeat is composed of 41 full units, 16 of which contain a deletion of nine bases against the other elements. Apart from this, the repeat is well conserved overall (on average, one mutated base per element), except for the first six units and the last one. The repeat is located inside a coding region (in the other strand) corresponding to a glucoamylase s_1/s_2 precursor protein (SwissProt id: AMYH_YEAST).

Satellites of whatever type demand a more complex definition of models than that given in Section 2.4, requiring additional constraints.

We have, in fact, two definitions related to a satellite model, the prefix model and the consensus model. The latter concerns satellite models strictly speaking, whereas prefix models are, in fact, models for approximately periodic repetitions that are not necessarily tandem.

Formally, a prefix model of a satellite is a string $m \in \Sigma^*$ [or $\mathcal{P}(\Sigma)$] that approximately matches a train of wagons. A wagon of m is a factor u in s such that dist$(m, u) \leq e$. A train of a satellite model m is a collection of wagons $u_1, u_2, \ldots, u_p$ ordered by their starting positions in s and satisfying the following properties.

Property 1. $p \geq min\_repeat$, where min_repeat is a fixed parameter that indicates the minimum number of elements a repeating region must contain.

Property 2. $left_{u_{i+1}} - left_{u_i} \in JUMP$, where $left_u$ is the position of the left end of wagon u in s and

$$JUMP = \{y : y \in \cup_{x \in [1,\ max\_jump]} x \times [min\_range,\ max\_range]\}$$

with the three parameters min_range, max_range, and max_jump fixed.

A prefix model m is said to be valid if there is at least one train of m in the string s. Similarly, a train, when viewed simply as a sequence of substrings of s, is valid if it is the train for some model m. A prefix model represents the invariant that must be true as we progressively search for our final goal, which is to arrive at a consensus model. This is a prefix model that further satisfies the following property.

```
GTTGCTAGAGGAAGATGGGGTTGGTACTGGTGCTACAGAGCTTTC
AGTGGTGGAGCTGGAT              ACTGGAGCAGAAGAGCTTTC
AGTAGTAGAGCTTGATCGGGTTCGTACTGGAACAGAAGAGCTTTC
AGTGCTAGAGCTGAATGGCGTTGAAGATGGAGCCGAGGAAGTGAT
GTTGCTAGAGGAAGATGGGGTTGGTACTGGTGCTACAGAGCTTTC
AGTAGTAGAGCTTGATGGGGTTGGTACTGGAGCAGAAGAGCTTTC
GGTAGTAGAGCTGGATGGAGTTGGCACTGGAGCAGAAGAGCTTTC
AGTAGTAGAGCTGGATGGAGTTGGTACTGGAGCAGAAGAGCTTTC
AGTGGTAGAGCTGG        TT    ACTGGAGCAGAAGAGCTTTC
AGTAGTAGAGCTGCATGCACTTCGTACTGGACCAGAAGAGCTTTC
AGTAGTAGAGCTGCATGCAGTTCGTACTGGAGCAGAAGAGCTTTC
AGTAGTAGAGCTGGATGGAGTTGGTACTGGAGCAGAAGAGCTTTC
AGTGGTAGAGCTGG        TT    ACTGGAGCAGAAGAGCTTTC
AGTGGTAGAGCTGG        TT    ACTGGAGCAGAAGAGCTTTC
AGTAGTAGAGCTGGATGGAGTTGGTACTGGAGCAGAAGAGCTTTC
AGTAGTAGAGCTGGATGGGGTTGGTACTGGAGCAGAAGAGCTTTC
AGTAGTAGAGCTGGATGGAGTTGGTACTGGAGCAGAAGAGCTTTC
AGTGGTAGAGCTGG        TT    ACTGGAGCAGAAGAGCTTTC
AGTACTACAGCTTCATCGGCTTGGAGCTGGAGCAGAAGAGCTTTC
AGTAGTAGAGCTTGATGGAGTTCGTACTCGACCAGAAGAGCTTTC
AGTAGTAGAGCTTGATGGAGTTGGCACTGGAGCAGAAGAGCTTTC
AGTGGTGGAGCTGG        TT    ACTGGAGTAGAAGAGCTTTC
AGTAGTAGAGCTGGATGGAGTTGGTACTGGAGCAGAAGAGCTTTC
AGTGGTAGAGCTGG        TT    ACTGGAGCAGAAGAGCTTTC
AGTGGTGGAGCTTGATGGGGTTGGAGCTGGAGCAGAAGAGCTTTC
AGTAGTAGAGCTGGATGGAGTTGGTACTGGAGCAGAAGAGCTTTC
AGTGGTAGAGCTGG        TT    ACTGGAGCAGAAGAGCTTTC
AGTAGTAGAGCTGCATGCACTTCGTACTCGACCAGAAGAGCTTTC
AGTGGTACAGCTGG        TT    ACTGGAGCAGAAGAGCTTTC
AGTAGTAGAGCTGGATGGAGTTGGTACTGGAGCAGAAGAGCTTTC
AGTGGTACAGCTGG        TT    ACTGGAGCAGAAGAGCTTTC
AGTGGTAGAGCTGG        TT    ACTGGAGCAGAAGAGCTTTC
AGTGGTAGAGCTGG        TT    ACTGGAGCAGAAGAGCTTTC
AGTGGTAGAGCTGG        TT    ACTGGAGCAGAAGAGCTTTC
AGTAGTAGAGCTGGATGGAGTTGGTACTGGAGCAGAAGAGCTTTC
AGTGGTAGAGCTGG        TT    ACTGGAGCAGAAGAGCTTTC
AGTGGTAGAGCTGG        TT    ACTGGAGCAGAAGAGCTTTC
AGTAGTACAGCTTGATCGGGTTCGTACTGGACCAGAAGAGCTTTC
AGTGGTACAGCTGG        TT    ACTGGAGCAGAAGAGCTTTC
AGTAGTAGAGCTGGATGGAGTTGGTACTGGAGCAGAAGAACTTTC
AGTAGTAGAGCTTGATGGGGTTGGTACTGGAGTAGTAGTCTTCTT
```

FIGURE 10 An example of a tandem repeat in chromosome IX of yeast *Saccharomyces cerevisiae*, starting at position 391131.

Property 3. $left_{u_{i+1}} - right_{u_i} \in$ GAP, where $right_u$ is the position of the right end of wagon u, and

$$\text{GAP} = \{y : y \in \cup_{x \in [0, \; max\_jump-1]} x \times [min\_range, \; max\_range]\}$$

Parameter *max_jump* allows us to deal with very badly conserved elements inside a satellite (by actually not counting them) although we require that the satellite be relatively well conserved globally. Fixing *max_jump* at a value strictly greater than 1 means that we allow some wagons (the badly conserved ones) to be "jumped over." This may be seen as "meta-errors," that is, as errors involving not a single letter inside a wagon but a wagon inside a train. Note that $0 \in$ GAP. This guarantees that when jumps are not authorized, the repeats found are effectively tandem.

Because mutations affecting a unit concern indels (that is, insertions and deletions) as well as substitutions, it is sometimes interesting to work with a variant of the above properties where *JUMP* and GAP are defined as

$$JUMP = \left\{ y : \begin{array}{l} y \in [min\_range, \; max\_range] \quad \text{or} \\ y \in \cup_{x \in [2, \; max\_jump]} x \times [min\_range - g, max\_range + g] \end{array} \right\}$$

$$\text{GAP} = \left\{ y : \begin{array}{l} y \in [min\_range, \; max\_range] \quad \text{or} \\ y \in \cup_{x \in [1, \; max\_jump]} x \times [min\_range - g, max\_range + g] \end{array} \right\}$$

and $g \geq e$ is a fixed value. The idea is to allow the length of the badly conserved elements to vary in a larger interval than is permitted for the detection of "good" wagons.

The satellite problem we propose to solve is the following.

Problem 3. Given a string s and parameters *min_repeat*, *min_range*, *max_range*, *max_jump*, and e (possibly also g), find all consensus models m that are valid for s and for each such m.

In fact, the original papers [16,65] report a set of disjoint "fittest" trains realizing each model, given a measure of "fitness."

The algorithm presented in Sec. 4.2.3 is the only combinatorial, non-heuristic developed so far for identifying tandem arrays. Other exact approaches either treat the case of tandem repeats only [13,14], do not allow for errors [50,59,66,67], or require the generation of all possible (not just valid) models of a given length [68–70].

4.2.2. Building Prefix Satellite Models

As with all previous cases considered in this chapter, satellite models are constructed by increasing lengths. To determine if a model is valid, we must have some representation of the train or wagons that make it so. There are two possibilities:

1. We can keep track of each valid train and its associated wagons.
2. We can keep track of individual wagons and, on the fly, determine if they can be combined into valid trains.

The first possibility is appealing because model extension is straight-forward. We would just have to verify, for each wagon of each train, whether it can be extended according to the extended model and then count how many wagons remain to check to determine whether the train it belonged to is still a valid train. However, there are generally many overlapping trains involving many of the same wagons for a given model. Common wagons may be present more than once in the list of occurrences of m if this is kept as a list of trains. This approach entails redundancies that lead to an inefficient algorithm. We therefore adopt the second approach, of keeping track of wagons and determining if they can be assembled into trains as needed.

The rules of prefix-model extension are given in Lemma 4. A wagon is identified by a triple (i, j, d) indicating that it is the substring $s_i s_{i+1} \cdots s_j$ of s and that it is $d \le e$ differences away from its model. Position i indicates the left end of the wagon and j its right end. Unlike the other algorithms presented in this chapter, models and their occurrences (the wagons) will be extended to the left. This is just to facilitate verifying Property 2. Strictly speaking, we should then speak of suffix models instead of prefix models. Right ends of occurrences are calculated but are used only for checking Property 3.

Lemma 4.   The triple (i, j, d) encodes a wagon of $m' = \alpha m$ with $\alpha \in \Sigma$ and $m \in \Sigma^k$ if and only if at least one of the following conditions is true:

(match)	$(i + 1, j, d)$ is a wagon of m and $s_i = \alpha$,
(substitution)	$(i + 1, j, d - 1)$ is a wagon of m and $s_i \ne \alpha$,
(deletion)	$(i, j, d - 1)$ is a wagon of m,
(insertion)	$(i + 1, j, d - 1)$ is a wagon of αm, and, furthermore, $d \le e$.

For each prefix model m, we keep a list of wagons of m that are in at least one train validating m. We describe such wagons as being valid with respect to m. When we extend a model (to the left) to $m' = \alpha m$, we perform two tasks:

First, determine which valid wagons of m can be extended as above to become wagons of m'.

Second, of these newly determined wagons of m', keep only those that are valid with respect to m'. This requires effectively assembling wagons into trains, something that is not needed in an approach that would keep track of trains directly.

Note that we need not actually enumerate the trains in the second step, we must simply determine if a wagon is part of a train. This will allow us to perform an extension step in time linear with respect to the string length.

As a final insight, consider the directed graph $G = (V,E)$, where V is the set of all valid wagons and there is an edge from wagon u to v if $left_v - left_u \in JUMP$. Then a wagon u is valid if it is part of a path of length min_repeat or more in G. Determining this property is quite simple, because the graph is clearly acyclic. In the computation that follows, we effectively compute both the length of the longest path to u in $Lcnt_u$ and the length of the longest path from u in $Rcnt_u$. If $Lcnt_u + Rcnt_u > min\_repeat$, then u is valid.

4.2.3. Consensus Satellite Models

We encode the collection of all wagons of m in a set $L_m \subseteq \{1, \ldots,n\}$ and an $(n + 1) \times (2e + 1)$-element array D_m as follows:

1. $i \in L_m$ if and only if i is the left end of at least one wagon valid with respect to m.
2. For each $i \in L_m$, the value $D_m[i, \delta]$ for $\delta [-e,e]$ is the edit distance of m from wagon $s_i s_{i + 1} \cdots s_{i + |m| - 1 + \delta}$.

Intuitively, L_m gives the left ends of all valid wagons, which is all we need to verify Properties 1 and 2. D_m gives us the distances we need for extending models, together with the right ends needed for verifying Property 3. Formally, $(i, i + |m| - 1 + \delta, d)$ is a valid wagon of m if and only if $i \in L_m$ and $d = D_m[i, \delta] \le e$.

The complete algorithm is given below. When Extend(αm) is called, it is assumed that L_m is known along with the relevant D_m values. The routine computes these items for the extension αm and recursively for the extensions thereof. Lines 1–6 compute the set of left ends of wagons for αm derivable from wagons of m that are valid. While Lemma 4 gives us a way to do this computation, recall that we are using dynamic programming to compute all extensions simultaneously. This corresponds to adding the last row to the dynamic programming matrix of s versus αm. At start, L_m gives all the positions in row $|m|$ that have value e or less (and are valid) and D_m gives their values. From these we compute the positions in row $|m| + 1$ in the obviously sparse fashion to arrive at the values $L_{\alpha m}$ and $D_{\alpha m}$.

procedure Extend(αm)
1. $L_{\alpha m} \leftarrow \emptyset$
2.   **for** $i + 1 \in L_m$ (in decreasing order) **do**
3.   **for** $\delta \in [-e,e]$ **do**
4. $D_{\alpha m}[i,\delta] \leftarrow min \left\{ \begin{array}{l} D_m[i + 1, \ \delta] + (\textbf{if } s_i = \alpha \textbf{ then } 0 \textbf{ else } 1), \\ \textbf{if } i \in L_m \textbf{ then } D_m [1, \ \delta + 1] + 1, \\ \textbf{if } i + 1 \in L_{\alpha m} \textbf{ then } D_{\alpha m} [i + 1, \ \delta - 1] + 1 \end{array} \right\}$

5.   **if** $min_\delta\{D_{\alpha m}[i,\delta]\} \le e$ **then**
6. $L_{\alpha m} \leftarrow L_{\alpha m} \cup \{i\}$
7.   **for** $i \in L_{\alpha m}$ (in decreasing order) **do**
8. $Rcnt\,[i] \leftarrow max_{k\in(i\,+\,JUMP)\cap\,L_{\alpha m}}\{Rcnt[k]\} + 1$
9.   **for** $i \in L_{\alpha m}$ (in increasing order) **do**
10. $Lcnt[i] \leftarrow max_{k\in(i\,-JUMP)\cap\,L_{\alpha m}}\{Lcnt[k]\} + 1$
11.   **for** $i \in L_{\alpha m}$ **do**
12.   **if** $Lcnt[i] + Rcnt[i] \le min\_repeat$ **then** $L_{\alpha m} \leftarrow L_{\alpha m}-\{i\}$
13.   **if** $L_{\alpha m} \ne 0$ **then**
14.   **if** $|\alpha m| \in [min\_range,max\_range]$ **then**
15.   Record(αm)
16.   **if** $|\alpha m| < max\_range$ **then**
17.   **for** $\beta\epsilon\Sigma$ **do**
18.   Extend ($\beta\alpha m$)

After wagons have been extended whenever possible, we have to elim-inate those that are no longer valid. This is performed by lines 7–12. We compute, for each position $i \in L_{\alpha m}$, the maximum number of wagons in a train starting with a wagon whose left end is at i in $Rcnt[i]$ (including itself), and the maximum number of wagons in a train ending with a wagon whose left end is at i in $Lcnt[i]$. The necessary recurrences are given in lines 8 and 10 of the algorithm, where we recall that $JUMP = \{y : y \in \bigcup_{x\in[1,max\_jump]} x \times [min\_range,max\_range]\}$ and $i + JUMP$ denotes adding i to each element of $JUMP$. Observe that $Rcnt[i] + Lcnt[i]-1$ is the length of the longest train containing a wagon whose left end is at position i.

Clearly lines 7–10 take $O(|L_{\alpha m}||JUMP|)$ time. However, when $L_{\alpha m}$ is a very large fraction of n, one can maintain an Rcnt(Lcnt)-prioritized queue of the positions in $(i + JUMP) \cap L_{\alpha m}$, to obtain an $O(n\,max\_jump\,\log|JUMP|)$ bound.

Finally, in the remaining steps, lines 13–18, the algorithm calls Record to record potential models and then recursively tries to extend the model if possible. Routine Record confirms that the model is a consensus model by verifying Property 3 and recording the intervals spanned by trains that are valid for the consensus model, if any.

The total time taken by the algorithm is $O(n\,(|JUMP| + e)\,max\_range\,\mathcal{N}(e, max\_range)) = O(n\,max\_range^2\,max\_jump\,\mathcal{N}(e, max\_range)$ as $e < max\_range$. The term $\mathcal{N}(e, max\_range)$ corresponds to the number of words in the e-neighborhood of a word w of length max_range, that is, words that are at a Levenshtein distance of at most e from w. This number is bounded over by k^e.

The space requirement is that of keeping all the information concerning at most max_range models at a time (a model m and all its prefixes). It is therefore $O(n\,max\_range\,e)$, because only $O(n\,e)$ storage is required to record the left-end positions and edit-distance at each possible right end.

5. MOTIF EXTRACTION

5.1. Spelling Simple Models

We now present increasingly sophisticated models and algorithms for extracting models that occur in a set of strings (possibly not all). Such models correspond in general to binding sites—that is, to sites in a biological molecule that will come into contact with a site in another molecule, thus permitting some biological process to start (for instance, transcription or translation). We start by considering simple models.

The problem we wish to solve is the following.

Problem 4. Given a set of N strings $S = s_1, \cdots, s_N$, an integer $e \geq 0$, and a quorum $q \leq N$, find all models m such that m is valid, that is, occurs with at most e errors in at least q strings of set S.

The spelling of models is done using a suffix tree. The idea comes from the observation that long strings, especially when they are defined over a small alphabet, may contain many exact repetitions. We do not want to compare such repeated parts more than once with the potentially valid models. One way of doing that is to use a representation of the strings that allows us to put together some of the repetitions, that is, to use an index of the strings such as a suffix tree.

Trees for representing all the suffixes of a set of strings $\{s_i, 1 \leq i \leq N$ for some $N \geq 2\}$ are called generalized suffix trees and are constructed in a way very similar to the construction of the suffix tree for a single string [71,72]. We denote such generalized trees by $\mathcal{GT}$. They share all the properties of a suffix tree given in Section 3.3 with string s substituted by strings $s_1, \cdots, s_N$.

In particular, a generalized suffix tree $\mathcal{GT}$ satisfies the fact that every suffix of every string s_i in the set leads to a distinct leaf. When p strings, $p \geq 2$, have the same suffix, the generalized tree has p leaves corresponding to this suffix, each associated with a different string. To achieve this property during construction, we just need to concatenate to each string s_i of the set a symbol that is not in Σ and is specific to that string.

To be able to spell valid models (i.e., models satisfying the quorum constraint), we need to add some information to the nodes of the suffix tree.

In the case where we are looking for repeats in a single string s, we just need to know, for each node x of $\mathcal{T}$, how many leaves are contained in the subtree rooted at x. Let us use the term leaves$_x$ to denote this number for each node x. Such information can be added to the tree by a simple traversal of it.

If we are dealing with $N \geq 2$ strings, and therefore a generalized suffix tree $\mathcal{GT}$, it is no longer enough to know the value of leaves$_x$ for each node x in $\mathcal{GT}$ in order to be able to verify whether a model remains valid. Indeed, this time, for each node x, we need to know not only the number of leaves in the subtree of $\mathcal{GT}$ having x as root but also that number for each different string to which the leaves refer.

In order to do that, we must associate to each node x in $\mathcal{GT}$ an array, denoted $colors_x$, of dimension N that is defined as

$$colors_x[i] = \begin{cases} 1 & \text{if at least one leaf in the subtree} \\ & \quad \text{rooted at } x \text{ represents a suffix of } s_i \\ 0 & \text{otherwise} \end{cases}$$

for $1 \le i \le N$.

The array $colors_x$ for all x can also be obtained by a simple traversal of the tree in which each visit to a node takes $O(N)$ time. The additional space required is $O(N)$ per node.

One must observe that occurrences are now grouped into classes, and "real" ones, that is, occurrences considered as individual words in the strings, are never manipulated directly. Present case occurrences of a model are thus, in fact, nodes of the generalized suffix tree (we denote them by the term "node-occurrences") and are extended in the tree instead of in the string. Once the process of model spelling has ended, the start positions of the "real" occurrences of the valid models can be recovered by traversing the subtrees of the nodes reached so far and reading the labels of their leaves.

The algorithm is a development of the recurrence formula given in Lemma 5, below, where x denotes a node of the tree, father (x) its father, and d the number of errors between the label of the path going from the root to x as against a model m.

Lemma 5. (x,d) is a node-occurrence of $m' = m\alpha$ with $m \in \Sigma^k$ and $\alpha \in \Sigma$ if and only if one of the following two conditions is satisfied:

(match) $(father(\mathrm{x}),d)$ is a node-occurrence of m, and the label of the
                     arc from father (x) to x is α.

(substitution) $(father(x),d-1)$ is a node-occurrence of m, and the label
                     of the arc from father(x) to x is $\beta \neq \alpha$.

(deletion) $(x,d-1)$ is a node-occurrence of m.

(insertion) $(father(x),d-1)$ is a node-occurrence of $m\alpha$.

Furthermore, $d < e$.

The algorithm time complexity is $O(nN^2\mathcal{N}(e,k))$.

5.2. Structured Models

5.2.1. Introducing Structured Models

Although the objects defined in the previous section can be reasonable, algorithmically tractable models for single binding sites, they do not take into account the fact that such sites are often not alone (in the case of eukaryotes, they may even come in clusters) and, in particular, that the relative positions of such sites, when more than one participates in a biological process, are in general not random. This is particularly true for some DNA binding sites such as those involved in the transcription of DNA into RNA (e.g., the so-called promoter sequences).

There is therefore a need to define biological models as objects that take such characteristics into account. This has the motivation just mentioned and also presents interesting algorithmic aspects; exploiting such characteristics could lead to algorithms that are both more sensitive and more efficient. Models that incorporate such characteristics are called structured models. They are related to the structured motifs of Section 3.

Formally, a structured model is a pair (m,d) where

> m is a p-tuple of simple models $(m_1, \cdots, m_p)$ representing the p parts of a structured model (we shall call these parts boxes)
>
> d is a $(p-1)$-tuple $((d_{\min_1}, d_{\max_1}, \delta_1), \cdots, (d_{\min_{p-1}}, d_{\max_{p-1}}, \delta_{p-1}))$ of triplets representing the $p-1$ intervals of distance between two successive boxes in the structured model

with p a positive integer, $m_i \in \Sigma^+$, and $d_{\min_i}$, $d_{\max_i}$ ($d_{\max_i} \geq d_{\min_i}$), δ_i are non-negative integers.

Given a set of N strings $s_1, \cdots, s_N$ and an integer q, $1 \leq q \leq N$, a model (m,d) is said to be valid if, for all i, $1 \leq i \leq (p-1)$ and for all occurrences u_i of m_i there exist occurrences $u_1, \cdots, u_{i-1}, u_{i+1}, \cdots, u_p$ of $m_1, \cdots, m_{i-1}, m_{i+1}, \cdots, m_p$ such that

> $u_1, \cdots, u_{i-1}, u_i, u_{i+1}, \cdots, u_p$ belong to the same string of the set.
>
> There exists d_i, with $d_{\min_i} + \delta_i \leq d_i \leq d_{\max_i} - \delta_i$, such that the distance between the end position of u_i and the start position of u_{i+1} in the string is equal to $d_i \pm \delta_i$.
>
> d_i is the same for p-tuples of occurrences present in at least q distinct strings.

The term d_i represents a distance and $\pm \delta_i$ an allowed interval around that distance. When $\delta_i = (d_{\max_i} - d_{\min_i} + 1)/2$, then δ_i is omitted, and d in a structured model (m,d) is denoted by a pair $(d_{\min_i}, d_{\max_i})$. An example of a model with $p = 2$ is given in Figure 11.

Observe that simple models are indeed only a special case of structured models.

5.2.2. Statement of the Structured Model Problem

Concerning structured models, solutions to variants of increasing generality of the same basic problem are proposed. Suffix trees are used in all cases.

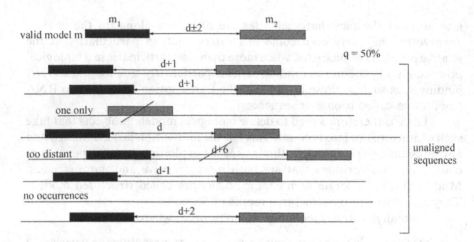

FIGURE 11   Example of a model with two boxes ($p = 2$).

These variants may be stated as follows, given a set of N strings $s_1, \cdots, s_N$, a nonnegative integer e, and a positive integer q.

Problem 5.   Find all models of the form $((m_1, m_2),(d_{\min_i},d_{\max_i}))$ that are valid.

Problem 6.   Find all models of the form $((m_1, \cdots, m_p), ((d_{\min_i},d_{\max_i}), \cdots, (d_{\min_{p-1}}, d_{\max_{p-1}})))$ that are valid, where $p \geq 2$.

Problem 7.   Find all models of the form $((m_1, m_2), (d_{\min_i},d_{\max_i}, \delta_1))$ that are valid.

Problem 8.   Find all models of the form $((m_1, \cdots, m_p), ((d_{\min_i},d_{\max_i},\delta_1), \cdots, (d_{\min_{p-1}}, d_{\max_{p-1}},\delta_{p-1})))$ that are valid, where $p \geq 2$.

Problems 5 and 6 represent situations where the exact intervals of distances separating the parts of a structured site are unknown, the only known fact being that these intervals cover a restricted range of values. How restricted is indicated by the δ_i parameters. We present below algorithms for the first two problems only. Further details on the other two may be found in Ref. 18.

To simplify matters, we shall consider that for $1 \leq i \leq p, m_i \in \Sigma^k$, where k is a positive integer, i.e., that each single model m_i of a structured model (m,d) is of fixed, unique length k. In a likewise manner, we shall assume that each part mi has the same error rate e and, when dealing with models composed of more than two boxes, that the $d_{\min_i}, d_{\max_i}$, and possibly δ_i for $1 \leq i \leq p-1$ have identical values. We denote these values by d_{min}, d_{max}, and δ. Problem 6 is then formulated as finding all models $((m_1, \cdots, m_p), (d_{min}, d_{max}))$ that are valid, and Problem 8 as finding all valid models $((m_1, \cdots, m_p), (d_{min}, d_{max}, \delta))$.

Besides fixing a maximum error rate for each part in a structured model, one can also establish a maximum error rate for the whole model. Such a global error rate allows us to consider in a limited way possible correlations between boxes in a model.

Another possible global, or local, constraint one may wish to consider for some applications concerns the composition of boxes. One may, for instance, determine that the frequency of one or more nucleotide in a box (or among all boxes) is below or above a certain threshold. For structured models composed of more than p boxes, one may also establish that a box i is palindromic in relation to a box j for $1 \leq i \leq j \leq p$. In algorithmical terms, the two types of constraints just mentioned are not equivalent. The first type, box composition whether local or global, can in general be verified only a posteriori, whereas the second type (palindromic boxes) will result in a, sometimes substantial, pruning of the virtual trie of models.

Introducing such additional constraints may in some cases require changes to the basic algorithms described below. The interested reader can find details concerning such changes in the original papers [18,73].

We present, in the next section, first a naive approach and then two algorithms that are efficient enough to tackle structured model extraction (Problem 5) from big data sets. The second algorithm has a better time complexity than the first but needs more space. The first is easier to understand and implement. Both are described in more detail than previous algorithms because structured models in some ways incorporate almost all other kinds of motifs we are considering. The most notable exception concerns satellites, which are discussed in Section 4.2. We then show how to extend these to treat Problem 16. Details on the algorithms for solving Problems 7 and 8 can be found in Ref. 18.

Other combinatorial approaches have been developed for treating somewhat similar kinds of structured motifs. They either enumerate all possible (not just valid) motifs [74], do not allow for errors [75,76], or are heuristics [77,78].

5.2.3. Algorithms for the Special Case of a Known Interval of Distance

Naive Approach. A naive way of solving Problem 5 consists in extracting and storing all valid single models of length k (given q and e) and then, once this is finished, verifying which pairs of such models could represent valid structured models (given an interval of distance $[d_{min}, d_{max}]$).

The lemma used for building valid single models is similar to Lemma 5 except that in practice, for most biological problems we wish to address [17,79], in general only substitutions are allowed. The lemma therefore becomes as stated below.

Lemma 6. (x,d) is a node-occurrence of $m' = m\alpha$ with $m \in \Sigma^k$ and $\alpha \in \Sigma$ if and only if one of the following two conditions is satisfied:

(match)   (*father* $(x),d$) is a node-occurrence of m, and the label of the arc from *father* (x) to x is α.

(substitution)   (*father*$(x),d - 1$) is a node-occurrence of m, and the label of the arc from *father*(x) to x is $\beta \neq \alpha$.

Furthermore, $d \leq e$.

One way of doing the verification profits from the simple observation that two single models m_1 and m_2 may form a structured model if and only if at least one occurrence of m_1 is at the right distance of at least one occurrence of m_2. Building an array of size nN where cell i contains the list of models having an occurrence starting at that position in $s = s_1 \cdots s_N$ allows us to compare models in cell i to models in cells $i + d_{min}, \cdots, i + d_{max}$ only. If the sets of occurrences of models are ordered, this comparison can be done in an efficient way (in time proportional to the size of the sets of node-occurrences, which is upper-bounded by nN).

First Algorithm: Jumping in the Suffix Tree. A first non-naive approach to solving Problem 5 starts by extracting single models of length k. Because we are traversing the trie of models in depth-first fashion (also in lexicographic order), models are recursively extracted one by one. At each step, a single model m (and its prefixes) is considered. Once a valid model m_1 of length k is obtained together with its set of T node-occurrences V_1 (which are nodes located at level k in GT), the extraction of all single models m_2 with which m_1 could form a structured model $((m_1, m_2), (d_{min},d_{max}))$ starts. This is done with m_2 representing the empty word and having as node-occurrences the set V_2 given by

$$V_2 = (w, \ e_w = e_v) \mid \exists v \in V_1 \text{ with } d_{min} \leq level \ (w) - level(v) \leq d_{max}$$

where $level(v)$ indicates the level of node v in GT. From a node-occurrence v in V_1, a jump is therefore made in GT to all potential start node-occurrences w of m_2. These nodes are the d_{min} to d_{max} generation, descendants of v in GT. Exactly the same recurrence formula as that given in Lemma 6 can be applied to the nodes w in V_2 to extract all single models m_2 that, together with m_1, could form a structured model verifying the conditions of the problem for all valid m_1. An illustration is given in Figure 12, and a pseudocode is presented below. The procedure ExtractModels is called with arguments

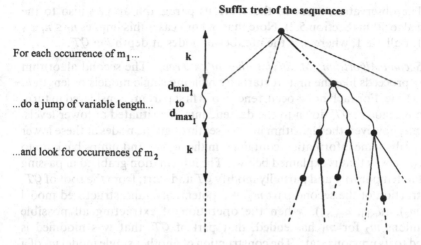

Suffix tree of the sequences

For each occurrence of m_1...

...do a jump of variable length...

...and look for occurrences of m_2

k

d_{min_1}
to
d_{max_1}

k

FIGURE 12 Extracting structured models (in the context of Problem 5) with a suffix tree—an illustration of the first algorithm.

m equal to the empty word having as sole node-occurrence the root of $\mathcal{GT}$ and i equal to 1.

procedure ExtractModels(Model m, Block i)
1.   **for** each node-occurrence v of m **do**
2.   **if** $i = 2$ **then**
3.   put in *PotentialStarts* the children w of v at levels $k + d_{min}$ to $k + d_{max}$
4.   **else**
5.   put v (i.e., the root) in *PotentialStarts*
6.   **for** each model m_i (and its occurrences) obtained by doing a recursive depth-first traversal from the root of the virtual model *tree* $\mathcal{M}$ while simultaneously traversing $\mathcal{GT}$ from the node-occurrences in *PotentialStarts* (Lemma 6 and quorum constraint) **do**
7.   **if** $i = 1$ **then**
8.   ExtractModels($m = m_1, i + 1$)
9.   **else**
10.   report the complete model $m = ((m_1, m_2), (d_{min}, d_{max}))$ as valid

Because the minimum and maximum lengths of a structured model (m,d) that may be considered are, respectively, $2k + d_{min}$ and $2k + d_{max}$, we need only build the tree of suffixes of length $2k + d_{min}$ or more and for each such suffix consider at most the first $2k + d_{max}$ symbols.

The observation made in the previous paragraph applies also to the second algorithm (Section 5.2). Note that in both cases this implies $n_i \leq n_{i+1} \leq Nn$ for all $i \geq 1$, where n_i is the number of nodes at depth i in $\mathcal{G}T$.

Second Algorithm: Modifying the Suffix Tree. The second algorithm initially proceeds like the first. It starts by building single models of length k, one at a time. For each node-occurrence v of a first part m_1 considered in turn, a jump is made in $\mathcal{G}T$ down to the descendants of v situated at lower levels. This time, however, the algorithm just passes through the nodes at these lower levels, grabs some information contained in the nodes, and jumps back up to level k (in a way that is explained below). The information grabbed in passing is used to temporarily and partially modify $\mathcal{G}T$ and start, from the root of $\mathcal{G}T$, the extraction of the second part m_2 of a potentially valid structured model $((m_1, m_2), (d_{\min}, d_{\max}))$. When the operation of extracting all possible companions m_2 for m_1 has ended, that part of $\mathcal{G}T$ that was modified is restored to its previous state. The construction of another single model m_1 of a structured model $((m_1, m_2), (d_{\min}, d_{\max}))$ then follows, and the whole process unwinds in a recursive way until all structured models that satisfy the initial conditions are extracted.

More precisely, the operation performed between the spelling of models m_1 and m_2 locally alters $\mathcal{G}T$ up to level k to a tree $\mathcal{G}T'$ that contains only the k-long prefixes of suffixes of $\{s_1, \cdots, s_N\}$ starting at a position between $d_{\min}$ and $d_{\max}$ from the end position in s_i of an occurrence of m_1. Tree $\mathcal{G}T'$ is, in a sense, the union of all the subtrees t of depth at most k rooted at nodes that represent start occurrences of a potential companion m_2 for m_1.

For each model m_1 obtained, before spelling all possible companions m_2 for m_1, the contents of colors$_z$ for all nodes z at level k in $\mathcal{G}T$ are stored in an array L of dimension n_k (this is for later restoration of $\mathcal{G}T$). Tree $\mathcal{G}T'$ is then obtained from $\mathcal{G}T$ by considering all nodes w in $\mathcal{G}T$ that may be reached during a descent of, this time, $k + d_{\min}$ to $k + d_{\max}$ arcs down from the node-occurrences (v, e_v) of m_1. These correspond to all end node-occurrences (instead of start as in the first algorithm) of potentially valid models having m_1 as first part. The Boolean arrays colors$_w$ or all w indicate the input strings to which these occurrences belong. This is the information we grab in passing and take along the only path of suffix links in $\mathcal{G}T$ that leads back to a node z at level k in $\mathcal{G}T$. If it is the first time z is reached, colors$_z$ is assigned colors$_w$, otherwise colors$_w$ is added (Boolean "or" operation) to colors$_z$. When all nodes v and w have been treated, the information contained in the nodes z that were reached during this operation are propagated up the tree from level k to the root (using normal tree arcs) in the following way: If $\bar{z}$ and $\hat{z}$ have the same parent z, then colors$_z$ = colors$_{\bar{z}}$ $\cup$ colors$_{\hat{z}}$. Any arc from the root that is not visited at least once in such a traversal up the tree is not part of $\mathcal{G}T'$, nor are the subtrees rooted at its end node.

The extraction of all second parts m_2 of a structured model (m,d) follows, as for single models in the initial algorithm (Lemma 6 in Section 5.2).

Restoring the tree $\mathcal{GT}$ as it was before the operations described above requires restoring the value of colors preserved in L for all nodes z at level k and propagating the information (state of Boolean arrays) from z up to the root.

Because nodes w at level between $2k + d_{min}$ to $2k + d_{max}$ will be solicited for the same operation over and over again, which consists in following the unique suffix-link path from w to a node z at level k in $\mathcal{GT}$, $\mathcal{GT}$ is pretreated so that one single link has to be followed from z. Going from w to z therefore takes constant time.

An illustration is given in Figure 13. A pseudocode of the algorithm is as follows. The procedure ExtractModels is called, as for the first algorithm,

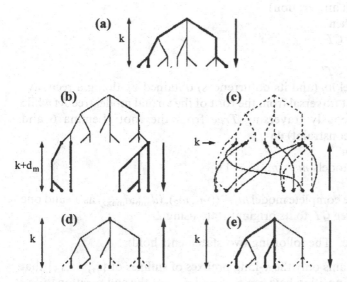

FIGURE 13 Extracting structured models (in the context of Problem 5) with a suffix tree—an illustration of the second algorithm. Part (a) corresponds to the extraction of the first single models m_1 of structure models (m,d). Part (b) corresponds the jump of $k + d_{min}$ to $k + d_{max}$ down normal tree arcs to grab some information (to lighten the figure, we made here $d_{min} = d_{max} = d_m$). Part (c) shows the jump back up to level k following suffix links with the information grabbed in passing. (d) represents the propagation of the information received at level k up to the root. Finally, (e) illustrates the search for second single models m_2 of structure models (m, d) in tree T'.

with both arguments m equal to the empty word having as sole node-occurrence the root of $\mathcal{GT}$ and i equal to 1.

procedure Extract Models (Model m, Block i)
  1.   **for** each node-occurrence v of m **do**
  2.   **if** $i = 2$ **then**
  3.   put in *PotentialEnds* the children w at levels $2k + d_{\min}$ to $2k + d_{\max}$
  4.   **for** each node-occurrence w in *PotentialEnds* **do**
  5.   follow fast suffix-link to node z at level k
  6.   put z in L
  7.   **if** first time z is reached **then**
  8.   initialize *colors$_z$* with zero
  9.   put z in NextEnds
 10.   add *colors$_w$* to *colors$_z$*
 11.   do a depth-first traversal of $\mathcal{GT}$ to update the Boolean arrays from the root to all z in *NextEnds* (let $\mathcal{GT}'$ be the k-deep tree obtained by such an peration)
 12.   **if** $i = 1$ **then**
 13.   Tree $= \mathcal{GT}$
 14.   else
 15.   Tree $= \mathcal{GT}'$
 16.   **for** each model m_i (and its occurrences) obtained by doing a recursive depth first traversal from the root of the virtual model tree $\mathcal{M}$ while simultaneously traversing *Tree* from the root (Lemma 6 and quorum constraint) **do**
 17.   **if** $i = 1$ **then**
 18.   ExtractModels($m = m_1, i + 1$)
 19.   **else**
 20.   report the complete model $m = ((m_1, m_2), (d_{\min}, d_{\max}))$ as a valid one
 21.   restore tree $\mathcal{GT}$ to its original state using L

Proposition 1.   The following two statements hold:

  1. $\mathcal{GT}'$ contains only the k-long prefixes of suffixes of $\{s_1, \cdots, s_N\}$ that start at a position between $d_{\min}$ and $d_{\max}$ of the end position in $\{s_1, \cdots, s_N\}$ of an occurrence of m_1.
  2.   The above algorithm solves Problem 5.

The proof is straightforward and can be found in the original papers [18,73].

Complexity.   The naive approach to solving Problem 5 requires $nN^2 \, \mathcal{N}(e,k)$ time to find single models that could correspond to either part of a structured model [and $nN\mathcal{N}(e,k)$ space to store all potential parts]. If we denote by Δ the value $d_{\max} - d_{\min} + 1$, finding which pair of single models can be put together to produce a structured model could then be done in time proportional to

$$\underbrace{\mathcal{N}(e,k)}_{(1)} \; \underbrace{\Delta\mathcal{N}(e,k)}_{(2)} \; \underbrace{nN}_{(3)} \; \underbrace{nN}_{(4)}$$

where (1) is the maximum number of single models to which a position may belong, (2) is the maximum number of models to which a position at a distance between $k + d_{min}$ and $k + d_{max}$ from the first may belong, (3) is the maximum number of comparisons that must be done to check whether two single models may form a structured one, and (4) is the number of starting positions to consider.

The total time complexity of the second algorithm is $O(Nn_k\mathcal{N}^2(e,k) + Nn_{2k + d_{max}}\mathcal{N}(e,k))$. Space complexity is slightly higher than for the first algorithm: $O(N^2n + Nn_k)$, where $n_k \leq Nn$. The second term is for array L.

In either case, the complexity obtained is better in terms of both time and space than the one given by a naive solution to Problem 5.

5.2.4. Extending the Algorithms to Extract Structured Models Having $p > 2$ Parts

First Algorithm: Jumping in the Suffix Tree. Extending the first algorithm to extract structured models composed of $p > 2$ parts, that is, solving Problem 6, is immediate. After extracting the first i parts of a structured model $((m_1, \cdots, m_p), (d_{min}, d_{max}))$ for $1 \leq i < p-1$, one jumps down in the tree $\mathcal{GT}$ (following normal tree arcs) to get to the d_{min} to d_{max} descendants of every node-occurrence of $((m_1, \cdots, m_i), (d_{min}, d_{max}))$, then continues the extraction from there using Lemma 6.

A pseudocode is given below.

procedure ExtractModels(Model m, Block i)
1.   **for** each node-occurrence v of m **do**
2.   **if** $i = 2$ **then**
3.   put in *PotentialStarts* the children w of v at levels $(i-1)k + (i-1)d_{min}$ to $(i-1)k + (i-1)d_{max}$
4.   **else**
5.   put v (the root) in *PotentialStarts*
6.   **for** each model m_i (and its occurrences) obtained by doing a recursive depth-first traversal from the root of the virtual model tree $\mathcal{M}$ while simultaneously traversing $\mathcal{GT}$ from the node-occurrences in *PotentialStarts* (Lemma 6 and quorum constraint) **do**
7.   **if** $i = 1$ **then**
8.   ExtractModels($m = m_1 \cdots m_i, i + 1$)
9.   **else**
10.   report the complete model $m = ((m_1, m_2), (d_{min}, d_{max}))$ as valid one

Second Algorithm: Modifying the Suffix Tree. Extending the second algorithm to solve Problem 6 is slightly more complex and thus calls for a few remarks. The operations done to modify the tree between building m_i and m_{i+1}, $i \geq 1$, are almost the same as those described in Section 5.2 except for two facts. One is that up to $p-1$ arrays L are now needed to restore the tree after each modification it undergoes. The second difference, more important, is that we need to keep, for each node v_k at level k reached from an ascent up $\mathcal{GT}$'s suffix links, a list, denoted $Lptr_{v_k}$, of pointers to those nodes, at lower levels, that affected the content of v_k. The reason for this is that tree $\mathcal{GT}$ is modified up to level k only (resulting in tree $\mathcal{GT}'$), because these are the only levels concerned by the search for occurrences of each box of a structured model. Lower levels of $\mathcal{GT}$ remain unchanged, in particular the Boolean arrays at each node below level k. To obtain the correct information concerning the potential end node-occurrences of boxes i for $i > 2$ (i.e., the strings to which such occurrences belong), we therefore cannot move down $\mathcal{GT}$ from the ends of node-occurrences in $\mathcal{GT}'$ of box $(i-1)$. If we did, we would not miss any occurrence but we could get more occurrences, e.g., the ones that did not have an occurrence of a previous box in the model. We might thus overcount some strings and consider as valid a model that, in fact, no longer satisfied the quorum. We have to go down $\mathcal{GT}$ from the ends of node-occurrences in $\mathcal{GT}$, that is, from the original ends of node-occurrences in $\mathcal{GT}$ of the boxes built so far. These are reached from the list of pointers $Lptr_{v_k}$ for the nodes v_k that are identified as occurrences of the box just treated. For models composed of p boxes, we need at most $p-1$ lists $Lptr_{v_k}$ for each node v_k at level k.

A pseudocode for the algorithm is as follows.

procedure ExtractModels(Model m, Block i)
1.   **for** each node-occurrence v of m **do**
2.   **if** $i > 2$ **then**
3.   put in *PotentialEnds* the children w at levels $ik + (i-1)d_{min}$ to $ik + (i-1)d_{max}$
4.   **for** each node-occurrence w in *PotentialEnds* **do**
5.   follow fast suffix-link to node z at level k
6.   put z in $L(i)$
7.   **if** first time z is reached **then**
8.   initialize *colors$_z$* with zero
9.   put z in *NextEnds*
10.   add *colors$_w$* to *colors$_z$*
11.   do a depth-first traversal of $\mathcal{GT}$ to update the Boolean arrays from the root to all z in NextEnds (let $\mathcal{GT}'$ be the k-deep tree obtained by such an operation)
12.   **if** $i = 1$ **then**

13. $Tree = \mathcal{GT}$
14.   **else**
15. $Tree = \mathcal{GT}'$
16.   **for** each model m_i (and its occurrences) obtained by doing a recursive depth-first traversal from the root of the virtual model tree $\mathcal{M}$ while simultaneously traversing $Tree$ from the root (Lemma 6 and quorum constraint) **do**
17.   **if** $i < p$ **then**
18. $ExtractModels(m = m_1 \cdots m_i, i + 1)$
19.   **else**
20.   report the complete model $m = ((m_1, \cdots, m_p), (d_{min}, d_{max}))$ as a valid one
21.   **if** $i > 1$ **then**
22.   restore tree $\mathcal{GT}$ to its original state using $L(i)$

Complexity. The first algorithm requires $O(Nn_{pk + (p-1)d_{max}} \mathcal{N}^p(e,k))$ time, where $\mathcal{N}^p(e,k) \le k^{pe}|\Sigma|^{pe}$. The space complexity remains the same as for solving Problem 3, that is, $O(N^2 n)$.

The total time complexity of the second algorithm is $O(Nn_k \mathcal{N}^p(e,k) + Nn_{pk + (p-1)d_{max}} \mathcal{N}^{p-1}(e,k))$. The space complexity is $O(N^2 n + N(p-1)n_k)$.

ACKNOWLEDGMENTS

We are grateful to N. El Mabrouk for helpful discussions and to our research partners J. Allali, C. Allauzen, A. Vanet, L. Marsan, N. Pisanti, M. Raffinot, and A. Viari.

REFERENCES

1. Durbin R, Eddy SR, Krogh A, Mitchison G. Biological Sequence Analysis. Probabilistic Models of Proteins and Nucleic Acids. London: Cambridge Univ Press, 1998.
2. Gusfield D. Algorithms on Strings, Trees and Sequences: Computer Science and Computational Biology. London: Cambridge Univ Press, 1997.
3. Setubal JC, Meidanis J. Introduction to Computational Molecular Biology. Florence (KY): Thomson Learning, 1996.
4. Waterman MS. Introduction to Computational Biology. Maps, Sequences and Genomes. New York: Chapman and Hall, 1995.
5. Karp RM, Miller RE, Rosenberg AL. Rapid identification of repeated patterns in strings, trees and arrays. Proc 4th Annu ACM Symp Theory of Computing, 1972:125–136.
6. Soldano H, Viari A, Champesme M. Searching for flexible repeated patterns using a non transitive similarity relation. Pattern Recog Let 1995; 16:233–246.

7. Sagot M-F, Escalier V, Viari A, Soldano H. Searching for repeated words in a text allowing for mismatches and gaps. In: Baeza-Yates R, Manber U, eds. Second South American Workshop on String Processing. Viñas del Mar, Chili: University of Chili, 1995:87–100.

8. Salton G. Automatic Text Processing. Reading, MA: Addison-Wesley, 1989.

9. Baeza-Yates R, Ribero-Neto B. Modern Information Retrieval. Reading, MA: Addison-Wesley, 1999.

10. Apostolico A, Galil Z, eds. Pattern Matching Algorithms. New York: Oxford Univ Press, 1997.

11. Crochemore M, Rytter W. Text Algorithms. New York: Oxford Univ Press, 1994.

12. Stephen GA. String Searching Algorithms. Singapore: World Scientific, 1994.

13. Landau G, Schmidt J. An algorithm for approximate tandem repeats. In: Galil Z, Apostolico A, Crochemore M, Manber U, eds. Combinatorial Pattern Matching. Lecture Notes in Computer Science. Vol 684. New York: Springer-Verlag, 1993:120–133.

14. Kannans K, Myers EW. An algorithm for locating non-overlapping regions of maximum alignment score. In: Galil Z, Apostolico A, Crochemore M, Manber U, eds. Combinatorial Pattern Matching. Lecture Notes in Computer Science. Vol. 684. New York: Springer-Verlag, 1993:7486.

15. Kurtz S, Ohlebusch E, Schleiermacher C, Stoye J, Giegerich R. Computation and visualization of degenerate repeats in complete genomes. In: Eighth International Symposium on Intelligent Systems for Molecular Biology. Menlo Park, CA: AAAI Press, 2000.

16. Sagot M-F, Myers EW. Identifying satellites and periodic repetitions in biological sequences. J Comput Biol 1998; 10:10–20.

17. Vanet A, Marsan L, Sagot M-F. Promoter sequences and algorithmical methods for identifying them. Res Microbiol 1999; 150:779–799.

18. Marsan L, Sagot M-F. Algorithms for extracting structured motifs using a suffix tree with an application to promoter and regulatory site consensus identification. J Comput Biol 2000; 7:345–362.

19. Sagot M-F, Viari A, Soldano H. A distance-based block searching algorithm. In: Rawlings C, Clark D, Altman R, Hunter L, Lengauer T, Wodak S, eds. Third International Symposium on Intelligent Systems for Molecular Biology. Menlo Park, CA: AAAI Press, 1995:322–331.

20. Dayhoff MO, Schwartz RM, Orcutt BC. A model of evolutionary change in proteins. In: Dayhoff MO, ed. Atlas of Protein Sequence and Structure. Washington, DC: Natl Biomed Res Found, 1978; 5(suppl 3):345–352.

21. Henikoff S, Henikoff JG. Amino acid substitution matrices from protein blocks. Proc Natl Acad Sci USA 1992; 89:10915–10919.

22. Risler JL, Delorme MO, Delacroix H, Hénaut A. Amino acid substitutions in structurally related proteins: a pattern recognition approach. J Mol Biol 1988; 204:1019–1029.

23. Galil Z, Seiferas J. Time-space optimal string matching. J Comput Syst Sci 1983; 26(3):280–294.

24. Yao AC. The complexity of pattern matching for a random string. SIAM J Comput 1979; 8(3):368–387.

25. Boyer RS, Moore JS. A fast string searching algorithm. Commun ACM 1977; 20(10):762–772.

26. Cole R. Tight bounds on the complexity of the Boyer-Moore string matching algorithm. SIAM J Comput 1994; 23(5):1075–1091.

27. Aho AV, Corasick MJ. Efficient string matching: an aid to bibliographic search. Commun ACM 1975; 18(6):333–340.

28. Fischer MJ, Paterson M. String matching and other products. In: Karp RM, ed. Complexity of Computation. Providence, RI, SIAM-AMS, 1974:113–125.

29. Apostolico A, Giancarlo R. Sequence alignment in molecular biology. J Comput Biol 1998; 5(2):173–196.

30. Landau GM, Vishkin U. Fast string matching with k differences. J Comput Syst Sci 1988; 37(1):63–78.

31. Harel D, Tarjan RE. Fast algorithms for finding nearest common ancestors. SIAM J Comput 1984; 13(2):338–355.

32. Schieber B, Vishkin U. On finding lowest common ancestors: simplification and parallelization. SIAM J Comput 1988; 17(6):1253–1262.

33. Altschul SF, Gish W, Miller W, Myers EW, Lipman DJ. A basic local alignment search tool. J Mol Biol 1990; 215:403–410.

34. Ukkonen E. On-line construction of suffix trees. Algorithmica 1995; 14(3):249–260.

35. Baeza-Yates RA, Perleberg C. Fast and practical aproximate string matching. In: Galil Z, Apostolico A, Crochemore M, Manber U, eds. Combinatorial Pattern Matching. Lecture Notes in Computer Science. Vol. 644. New York: Springer-Verlag, 1992:185–192.

36. Chang SC, Lawler EL. Sublinear expected time approximate string matching and biological applications. Algorithmica 1994; 12:327–344.

37. Apostolico A, Crochemore M. String pattern matching for a deluge survival kit. In: Abello J, Pardalos PM, Resende MGC, eds. Handbook of Massive Data Sets. Boston: Kluwer Academic, 2002:151–194.

38. Blumer A, Blumer J, Ehrenfeucht A, Haussler D, Chen MT, Seiferas J. The smallest automaton recognizing the subwords of a text. Theor Comput Sci 1985; 40(1):31–55.

39. Weiner P. Linear pattern matching algorithm. Proc 14th Annu IEEE Symp on Switching and Automata Theory, Washington, DC, 1973:1–11.

40. McCreight EM. A space-economical suffix tree construction algorithm. J Algorithms 1976; 23(2):262–272.

41. Blumer A, Blumer J, Ehrenfeucht A, Haussler D, Chen MT, Seiferas J. The smallest automaton recognizing the subwords of a text. Theoret Comput Sci 1985; 40:31–55.

42. Crochemore M. Transducers and repetitions. Theoret Comput Sci 1986; 45(1): 63–86.

43. Manber U, Myers G. Suffix arrays: a new method for on-line string searches. SIAM J Comput 1993; 22(5):935–948.

44. Gonnet G, Baeza-Yates R, Snider T. Lexicographical indices for text: inverted files vs. PAT trees. Tech Rep OED-91–01. Centre for the New OED, University of Waterloo. Toronto, ON, Canada, 1999.

45. Kärkkäinen J, Ukkonen E. Sparse suffix trees. Proceedings of the 2nd Annual International Computing and Combinatorics Conference. Lecture Notes Comput Sci 1090. New York: Springer-Verlag, 1996:219–230.

46. Kärkkäinen J. Suffix cactus: a cross between suffix tree and suffix array. In: Galil Z, Ukkonen E, eds. Proceedings of the 6th Annual Symposium on Combinatorial Pattern Matching. Lecture Notes Comput Sci 937. New York: Springer-Verlag, 1995:191–204.

47. Ferragina P, Grossi R. A fully-dynamic data structure for external substring search. Proceedings of the 27th ACM Symposium on the Theory of Computing. New York: ACM Press, 1995.

48. Anderson A, Nilson S. Efficient implementation of suffix trees. Softw Pract Exp 1995; 25:129–141.

49. Irving RW. Suffix binary search trees. Tech Rep 1995-7. University of Glasgow, 1995.

50. Crochemore M, Vérin R. On compact directed acyclic word graphs. In: Mycielski J, Rozenberg G, Salomaa A, eds. Structures in Logic and Computer Science. Lecture Notes Comput Sci 1261. New York: Springer-Verlag, 1997:192–211.

51. Blumer A, Ehrenfeucht A, Haussler D. Average sizes of suffix trees and DAWGS. Discrete Appl Math 1989; 24:37–45.

52. Raffinot M. Asymptotic estimation of the average number of terminal states in dawgs. Proceedings of the 4th South American Workshop on String Processing. Ottawa, Canada: Carleton Univ Press, 1997:140–148.

53. Kurtz S. Reducing the space requirement of suffix trees. Tech Rep 98-03. University of Bielefeld: Germany, 1998.

54. Balík M. Searching substrings. Tech Rep DC-PSR-2000-02. Czech Tech Univ, Prague 2000.

55. Manber U, Baeza-Yates R. An algorithm for string matching with a sequence of don't cares. Inform Process Lett 1991; 37(3):133–136.

56. Navarro G. Indexing methods for approximate string matching. Tech Rep Univ Chile, Santiago, 2000.

57. El-Mabrouk N, Lisacek F. Very fast identification of RNA motifs in genomic DNA. Application to tRNA search in the yeast genome. J Mol Biol 1996; 264:46–55.

58. Fichant GA, Burks C. Identifying potential tRNA genes in genomic DNA sequences. J Mol Biol 1991; 220:659–671.

59. Crochemore M. An optimal algorithm for computing the repetitions in a word. Inform Proc Lett 1981; 12:244–250.

60. Apostolico A, Preparata FP. Optimal off-line detection of repetitions in a string. Theoret Comput Sci 1983; 22(3):297–315.

61. Main MG, Lorentz RJ. An O(n log n) algorithm for finding all repetitions in a string. J Algorithms 1984; 5(3):422–432.

62. Kolpakov R, Kucherov G. Finding maximal repetitions in a word in linear time. Symposium on Foundations of Computer Science (FOCS), New York (USA). Los Alamitos, CA: IEEE Comp Soc, 1999:596–604.

63. Stoye J, Gusfield D. Simple and flexible detection of contiguous repeats using a suffix tree. Theor Comput Sci 2002; 270(1–2):843–856.

64. Charlesworth B, Sniegowski P, Stephan W. The evolutionary dynamics of repetitive DNA in eukaryotes. Nature 1994; 371:215–220.

65. Sagot M-F, Myers EW. Identifying satellites in nucleic acid sequences. In: Istrail S, Pevzner P, Waterman M, eds. RECOMB'98. Proceedings of Second Annual

International Conference on Computational Molecular Biology. New York: ACM Press, 1998:234–242.

66. Clift B, Haussler D, McConnell R, Schneider TD, Stormo D. Sequence landscapes. Nucleic Acids Res 1986; 14:141–158.

67. Milosavljevic A, Jurka J. Discovering simple DNA sequences by the algorithmic significance method. Comput Appl Biosci 1993; 9:407–411.

68. Delgrange, O. Un algorithme rapide pour une compression modulaire optimale. Application á l'analyse de séquences génétiques. PhD thesis, 1997. Thése de doctorat - Université de Lille I.

69. Fischetti V, Landau G, Schmidt J, Sellers P. Identifying periodic occurrences of a template with applications to protein structure. In: Galil Z, Apostolico A, Crochemore M, Manber U, eds. Combinatorial Pattern Matching. Lecture Notes Comput Sci. Vol 644. New York: Springer-Verlag, 1992:111–120.

70. Rivals E, Delgrange O. A first step toward chromosome analysis by compression algorithms. In: Bourbakis NG, ed. First International IEEE Symposium on Intelligence in Neural and Biological Systems. Los Alamitos, CA: IEEE Comp Soc Press, 1995:233–239.

71. Bieganski P, Riedl J, Carlis JV, Retzel EM. Generalized suffix trees for biological sequence data: applications and implementations. Proceedings of the 27th Hawaii International Conference on Systems Science. Los Alamitos, CA: IEEE Comp Soc Press, 1994:35–44.

72. Hui LCK. Color set size problem with applications to string matching. In: Apostolico A, Crochemore M, Galil Z, Manber U, eds. Combinatorial Pattern Matching. Lecture Notes Comput Sci. Vol 644. New York: Springer-Verlag, 1992:230–243.

73. Marsan L, Sagot M-F. Extracting structured motifs using a suffix tree—algorithms and application to promoter consensus identification. In: Istrail S, Pevzner P, Waterman M, eds. RECOMB'00. Proceedings of Fourth Annual International Conference on Computational Molecular Biology. New York: ACM Press, 2000.

74. van Helden J, Rios AF, Collado-Vides J. Discovering regulatory elements in non-coding sequences by analysis of spaced dyads. Nucleic Acids Res 2000; 28:1808–1818.

75. Jonassen I, Collins JF, Higgins DG. Finding flexible patterns in unaligned protein sequences. Protein Sci 1995; 4:1587–1595.

76. Jonassen I. Efficient discovery of conserved patterns using a pattern graph. Comput Appl Biosci 1997; 13:509–522.

77. Fraenkel YM, Mandel Y, Friedberg D, Margalit H. Identification of common motifs in unaligned DNA sequences: application to *Escherichia coli lrp* regulon. Comput Appl Biosci 1995; 11:379–387.

78. Klingenhoff A, Frech K, Quandt K, Werner T. Functional promoter modules can be detected by formal models independent of overall nucleotide sequence similarity. Bioinformatics 1 1999; 15:180–186.

79. Vanet A, Marsan L, Labigne A, Sagot M-F. Inferring regulatory elements from a whole genome. An analysis of the σ^{80} family of promoter signals. J Mol Biol 2000; 297:335–353.

4

Protein Sequence Analysis and Prediction of Secondary Structural Features

Jaap Heringa

Centre for Integrative Bioinformatics VU (IBIVU), Faculty of Sciences
and Faculty of Earth and Life Sciences, Free University,
Amsterdam, The Netherlands, and MRC National
Institute for Medical Research, London, England

1. AMINO ACID SEQUENCE COMPARISON

1.1. Pairwise Comparison

As soon as two protein sequences need to be compared, a task known as alignment arises. Alignment involves matching the amino acids of the two sequences in such a way that their similarity can be determined best. For example, if the two sequences are CDFG and CDEFS, it is clear that the alignment

C D – F G

C D E F S

provides the best assessment of their similarity. However, in comparing two considerably different protein sequences of, for instance, 300 amino acids each, the real problem begins. Ideally, the alignment of two sequences should be in agreement with their evolution, i.e., the patterns of descent as well as the molecular structural and functional development. Unfortunately, the evolutionary traces are often very difficult to detect. For example, in divergent evolution of two protein sequences from a common ancestor, amino acid mutations, insertions, and deletions of residues, gene doubling, transposed gene segments, repeats, domain structures, and the like can blur the ancestral tie beyond recognition.

1.1.1. Dot Matrix Analysis

One way to represent all possible alignments of two sequences is to compare them in a two-dimensional matrix, where one sequence is written out vertically with its amino acids forming the matrix rows while the other sequence forms the columns. Each intersection of a matrix row with a matrix column represents the comparison of associated amino acids in the two sequences such that all possible local alignments can be found along diagonals parallel to the major matrix diagonal. The simplest way to express the similarity of matched amino acid pairs is to place a dot in a matrix cell whenever the two are identical. Such matrices are therefore often referred to as dot matrices and were used early on to visualize the relationship of two sequences [1,2]. Their use remains because the human eye is a very strong device to perform pattern recognition in spaces of up to three dimensions. For example, overall similarity is discernible by piecing together local subdiagonals through insertions and deletions.

More biological insight is normally obtained by using more varying amino acid similarity values than the identity values mentioned above (see Sec. 1.3). Most protein sequence comparison methods use amino acid exchange values for each possible exchange, normally incorporated in a symmetrical 20×20 matrix in which each value approximates the evolutionary likelihood of a mutation from one amino acid into another; the matrix diagonal contains the odds for self-conservation (see Sec. 1.3).

To increase the signal-to-noise ratio for dot plots, McLachlan [3–5] was the first to develop filtering techniques. He devised "double matching probabilities" to estimate the significance of regions showing high similarity. Such regions were identified by using windows of fixed length that are effectively slid over the two sequences to compare all possible stretches of, for example, five matched residue pairs. To establish a level of background noise, McLachlan took a large number of randomly shuffled sequences and calculated the mean value and standard deviation over all randomized windows of a given length and compared these numbers with those from the two query sequences (using the Z-score, i.e., the number of standard deviations above the random mean). Often, the output values are filtered on the basis of a cutoff value for placing dots in the comparison matrix. Argos [6] used nonredundant real protein sequences instead of randomized ones to take into account natural protein sequence biases arising from structure, such as amphipathic helices or strands found in nearly all proteins. McLachlan's initial program CMPSEQ [3] was elaborated by Staden [7], who devised the widely used dot matrix program DIAGON. He added output filtering and also allowed different amino acid similarity scoring systems. Following McLachlan, Pustell and Kafatos [8,9] devised further filtering methods to improve the signal-to-noise ratio and used letters instead of dots in the output matrices to allow scaling of the window scores.

Important biological issues include the choice of amino acid similarity scores to use as well as the adopted length of the windows in the dot matrix analysis. Argos [6] altered the classical alignment-based amino acid similarity scores by including physicochemical amino acid parameters in the calculation of the similarity values. He used these together with windows of different lengths that were tested simultaneously in order to be less dependent on the actual choice of an individual window length. Other researchers have suggested the use of biologically meaningful lengths of windows, such as the length of an average protein secondary or supersecondary structure [10] or the peptide length resulting from an average exon size [11].

1.1.2. Dynamic Programming Methods

Although dot matrix methods can show multiple regional similarities between amino acid sequences, there has been only a single evolutionary pathway from one sequence to another. In the absence, however, of observed evolutionary traces, the matching of two sequences is regarded as mimicking evolution best when the minimum number of mutations are used to arrive at one sequence from the other. An approximation of this is finding the highest similarity value determined from summing substitution scores along matched residue pairs minus any insertion/deletion penalties (see below). Such an alignment is generally called the optimal alignment. Unfortunately, testing all possible alignments including the insertion of a gap at each position of each sequence, is infeasible. For example, there are about 10^{88} possible alignments of two sequences of 300 amino acids [12], a number clearly beyond all computing capabilities. However, when introductions of gaps are also assigned scoring values such that they can be treated in the same manner as the mutation of one residue into another, the number of calculations is greatly reduced and becomes readily computable. The technique to calculate the highest scoring or optimal alignment is generally known as the dynamic programming (DP) technique. Although the physicist Richard Bellman first conceived DP and published a number of papers on the topic between 1955 and 1975, Needleman and Wunsch [13] introduced the technique to the biological community, and their paper remains among the most cited in the area.

A DP algorithm operates in two steps. First a search matrix is set up in the same way as a dot matrix (see Sec. 1.1.1), with one sequence displayed horizontally and the other vertically (Fig. 1). The matrix is traversed from the upper left to the lower right. Each cell [i,j] in the matrix receives as a score the value composed of the maximum value of the scores in row $i-1$ and column $j-1$ (with subtraction of the proper gap penalty values) added

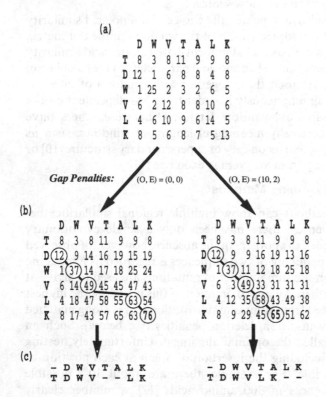

FIGURE 1 Two alignments of sequences DWVTALK and TDWVLK by dynamic programming. The Dayhoff PAM 250 [14] exchange weights (see Fig. 5) were added to a constant of 8 to make all values nonnegative. (a) Needleman–Wunsch search matrix after initialization with the appropriate PAM250 values. (b) The states of two exchange matrices after execution of the forward pass. The left matrix is traversed without gap penalties, whereas for the right-hand matrix penalty values of 10 and 2 are used for gap opening and extension, respectively. Optimal alignment paths through both search matrices are indicated. (c) The alignments resulting from tracing back from the highest scoring cell for each of the search matrices. It can be seen that the two different gap penalty regimes lead to different alignments and alignment scores.

to the exchange value of the associated matched residue pair of cell [*i,j*]. Cell[*i,j*] therefore contains the maximum score of all possible alignments of the two subsequences up to cell[*i,j*]. Writing the above in a more explicit form:

$$S[i,j] = s[i,j] + \text{Max} \left\{ \begin{array}{l} S[i-1,j-1] \\ \max_{1<x<i}(S[i-x,j-1] - P(x-1)) \\ \max_{1<y<j}(S[i-1,j-y] - P(y-1)) \end{array} \right\} \quad (1)$$

where $S[i,j]$ is the alignment score for the first subsequence from 1 to i and the second subsequence from 1 to j, Max denotes the maximum value of the three arguments between brackets, $s[i,j]$ is the substitution value for the amino acid exchange associated with cell[*i,j*], $P(x-1)$ is the nonnegative penalty value for a gap of length $x-1$, and $\max_{1<x<i}$ represents the maximum value of all argument values over the range 1 to i.

Needleman and Wunsch [13] used a fixed penalty value for the inclusion of a gap of any length, whereas Sellers [15] added a penalty value for each inserted gap position. Most present alignment routines take an intermediate approach by using the formula $P(x) = P_o + P_e x$, where P_o is the penalty used upon the opening of a gap of length x and P_e is the value for each extension of the gap. Many researchers use P_o 10–30 times larger than P_e. The choice of proper gap penalties is also closely connected to the residue exchange values used in the analysis. When the search matrix is traversed, the highest scoring matrix cell is selected from the bottom row or the rightmost column, and its score is guaranteed to be the optimal alignment score. The second step of a dynamic programming algorithm is usually called the traceback step; the actual optimal alignment is reconstructed from the matrix cell containing the highest alignment score. The path then follows successively lower scores but each time selects the highest available score in the preceding row and column up to the current matrix cell (Fig. 1).

Because classical Needleman–Wunsch-type dynamic programming algorithms use a two-dimensional search matrix, so that the algorithmic speed and storage requirements are both of the order $N*M$, when two sequences consisting of N and M amino acids in length are matched. The large computer memory requirements of Needleman–Wunsch-type algorithms are due to the traceback step, where the matches of the optimal alignment are reconstructed. Furthermore, the amount of computation required makes the dynamic programming technique infeasible for a query sequence search against a large sequence database on personal computers.

Gotoh [16,17] devised a dynamic programming algorithm that dramatically decreased the storage requirements from order N^2 to order N (assuming that two sequences each N amino acids in length are matched) while keeping speed on the order of N^2. Myers and Miller [18] constructed an even more memory-efficient linear space algorithm, based on the Gotoh approach and on a traceback strategy proposed by Hirschberg [19]; that is only slightly slower.

To speed up DP algorithms, calculation of regions in the search matrix unlikely to be selected in the final alignments can be avoided, saving valuable computer time. Sankoff and Kruskal [20] computed only a band of a certain width along the main diagonal of the search matrix; however, this is useful only when the sequences are fairly similar with no extensive gaps. Ukkonen [21] and Ficket [22] iteratively increased the bandwidth until the optimal alignment was found. Another possibility involves finding regions of high similarity and then applying dynamic programming over only the remaining part of the search matrix.

A problem with global dynamic programming methods that match complete sequences can arise when highly dissimilar sequences are compared. In such cases global alignment techniques might fail to recognize highly similar internal regions because such regions may be overshadowed by dissimilar regions and strong gap penalties required to achieve their proper matching. Moreover, many biological sequences are modular and show shuffled domains [23], which can render a global alignment of two complete sequences meaningless. The occurrence of varying numbers of internal sequence repeats [24] can also severely limit the applicability of global methods. In general, when there is a large difference in the lengths of two sequences to be compared, global alignment routines become tricky. To address these problems, Smith and Waterman [25] developed a so-called local alignment technique in which the most similar regions in two sequences are selected and aligned first. To get from global to local dynamic programming, an important prerequisite is that the amino acid exchange values that are used must include negative values. Any score in the search matrix lower than zero is to be set to zero. Formula (1) is then changed to

$$S[i,j] = \text{Max} \begin{cases} s[i,j] + S[i-1,j-1] \\ \\ s[i,j] + \max_{1<x<i}(S[i-x,j-1] - P(x-1)) \\ \\ s[i,j] + \max_{1<y<j}(S[i-1,j-y] - P(y-1)) \\ \\ 0 \end{cases} \qquad (2)$$

where Max is now the maximum of four terms. A consequence of this scenario is that the final highest alignment score value does not have to be in the last row or column as in global alignment routines but can be anywhere in the search matrix. The local alignment algorithm relies on dissimilar subsequences producing negative scores that are subsequently discarded by placing zero values in the associated submatrix cells. An arbitrary issue in using the algorithm is deciding the zero cutoff relative to the 20 × 20 residue exchange weights matrix.

Waterman and Eggert [26] generalized the local alignment routine by devising an alignment routine that allows the calculation of a user-defined number of top-scoring local alignments instead of only the optimal local alignment. The obtained local alignments do not intersect; i.e., they have no matched amino acid pair in common. If during the procedure an alignment is encountered that intersects with any of the top-scoring alignments listed thus far, then the highest scoring of the conflicting pair is retained in the top list. Huang et al. [27] developed an implementation (SIM) of the technique in which they reduced the memory requirements from order N^2 to order N, thereby allowing very long sequences to be searched at the expense of only a small increase in computation time. Another popular version of the same technique is LALIGN [28], which is part of the popular FASTA package [29](see Sec. 1.5).

It is not always clear whether the top-scoring (local or global) alignment of two sequences is biologically the most meaningful. There may be biologically plausible alignments that score close to the top value. For example, a multiple alignment of reasonably diverged sequences normally shows the sequences matched in such a way that most if not all alignments between any two sequences in the multiple alignment are suboptimal compared to their optimal pairwise alignment (for multiple sequence alignment, see Sec. 1.2). Vingron and Argos [30] and Zuker [31] approached this issue by constructing an algorithm that determines all optimal and suboptimal alignments and depicts them in a dot plot. Reliably aligned regions can be defined as those for which alternative local alignments do not exist. Mott [32] derived estimates for the statistical significance of alignment scores using maximum likelihood methods, thus avoiding computationally intensive comparisons with randomized sequences.

Bucher and Hofmann [33] put local alignments in a new perspective. They interpreted each cell$[i, j]$ in the DP search matrix as the total probability that a local alignment would go through it, which is equivalent to summing the scores of all local alignments intersecting cell$[i,j]$. Using this approach, Bucher and Hofmann [33] found an increase in pairwise sequence search capabilities.

1.2. Multiple Alignment

Multiple sequence comparison involves the search for similarity in more than two sequences. The alignment of a set of sequences can provide important information about structure–function relationships within the proteins, such as the evolutionary conservation of functional amino acids at certain sequence positions or conserved hydrophobicity patterns in particular regions. Further, there are many techniques for sequence analysis that rely on a multiple alignment, such as phylogenetic analysis, secondary structure prediction, and sequence–structure comparison (see below). These observations and derived techniques, however, depend crucially on the quality of the multiple alignment.

As for pairwise alignment (see Sec. 1.1), two basic classes of multiple alignment programs have been developed: global and local methods. Global alignment programs attempt to align the sequences over their whole length, whereas local programs search only for the most conserved regions and leave the other parts of the sequences unaligned. The most effective alignment algorithm depends on the nature of the sequences to be aligned. Global algorithms produce the most accurate and reliable alignments when all the sequences in the data set are of similar length. However, when the sequences differ greatly in length, local alignment programs are often more successful at identifying the conserved regions.

The two most explored computational techniques for multiple sequence alignment are the DP technique [13,25] and, more recently, hidden Markov modelling (HMM) [34,35]. Whereas the DP technique is deterministic (see above), HMM is a stochastic approach, which has proven powerful if applied to sequence database searches. Krogh et al. [35] described an HMM procedure for multiple alignment in which the alignment process is modeled in a finite automata fashion with three basic alignment steps considered as states: match, insert, and delete. Probabilities are attached to the state transitions by an expectation maximization algorithm trained during the alignment such that position-dependent amino acid substitution, insertion, and deletion probabilities are generated. Although HMM methods incorporate more detail than the classical DP methods, such as the mentioned position-specific scoring schemes, HMM approaches to multiple sequence alignment generally perform poorly compared with other methods [36,37]. This is mainly due to the inherently complex parameterization of the technique. As a consequence, state-of-the-art methods for multiple alignment are all based on the DP technique.

1.2.1. Global Multiple Alignment Methods

The problem of finding an optimal or highest scoring global alignment of two sequences was solved three decades ago with the DP technique [13], which guarantees the finding of the highest scoring or optimal alignment based on an amino acid substitution scoring scheme and insertion/deletion penalties (see Sec. 1.3). Unfortunately, the calculation of the optimal alignment generally becomes computationally infeasible for four or more sequences. Murata et al. [38] extended the Needleman–Wunsch procedure for the optimal alignment of three sequences, using a three-dimensional search matrix, but the application is limited to sequences of about 200 amino acids and uses a gap penalty independent of the gap length. Gotoh [16] devised a similar algorithm but devised a linear gap weighting function. In general, an algorithm for optimal alignment needs a number of computational steps and an amount of memory of at least the order of the product of the sequence lengths. Rigorous methods for the simultaneous alignment of four or more sequences thus cannot evaluate all possible matches but attempt to find the optimal alignment by considering only a small fraction of all possible comparisons. Over the years, various heuristic approaches have been developed, leading to a large number of multiple alignment programs that adopt different strategies.

Carillo and Lipman [39] showed that the optimal alignment path of N sequences is limited to a small region in the N-dimensional search matrix of which the upper bounds can be inferred from pairwise comparisons of the sequences. The algorithm MSA of Lipman et al. [40] is based on the Carillo and Lipman approach and generalizes the pairwise diagonal strip method of Ficket [22] to N dimensions, where N is the number of sequences to be aligned. Up to 10 sequences of 200–300 residues in length can be aligned with the Lipman et al. method. Furthermore, the algorithm addresses an additional problem in the comparison of multiple sequences, which is the weighting of the aligned sequences because similar sequences should not dominate the multiple sequence alignment. Lipman et al. [40] used the weighting scheme suggested by Altschul et al. [41] based on phylogenetic trees. More recently, the MSA method was extended to larger data sets using a divide-and-conquer strategy [42] implemented in the method DCA [43]. However, the approach remains extremely CPU- and memory-intensive and is thus applicable to only small data sets.

Other approaches attempt to limit the number of comparisons by considering only subsequences in the original sequence set that can be used as anchor points across the sequences. Johnson and Doolittle [44] used sliding windows to pinpoint residue positions in each of the sequences that should be aligned. Their method is feasible for the alignment of up to about 10 protein sequences. Sobel and Martinez [45] based their routine MALIGN upon the occurrence of identical words in each of the sequences, which makes the method less suitable for proteins than for nucleic acid sequences. However, their method early on provided the generation of suboptimal alignments as well as tests for the statistical significance of the final alignment. Waterman [46] extended the word search to nonidentical but similar words that should occur in at least a preset fraction of the initial sequences. A list of such best words is compiled using statistical criteria that depend partly on the word length. The alignment is then obtained by matching the associated regions in the sequences with the best words. Waterman and Jones [47] devised an algorithm for this strategy and extended it by allowing the occurrence of gaps in the words. Vingron and Argos [48] used a graph-theoretical method based on the rapid pairwise sequence comparison algorithm by Wilbur and Lipman [49] to find identical dipeptides that string consecutively across the sequences and serve as anchor points for extending the alignment. Regions in between two successive anchor points are ordered from long to short and are successively aligned using the standard Needleman–Wunsch algorithm coupled with the profile technique where amino acid frequencies in the already aligned block of intervening sequences are taken into account as subsequences are added. A problem with all of these anchoring techniques is that they do not display sustained accuracy and might be very good or bad depending on the actual sequence set.

In general, the most popular and successful approach has been the progressive alignment method [50,51], with which a multiple alignment is built up gradually by aligning the closest sequences first and successively adding in the more distant ones. Most widely used methods thus work in an agglomerative way by aligning sequences, following a heuristically determined order, until all sequences are joined in a final multiple alignment. The sequences are either aligned according to a previously determined order that, for instance, can be derived from a known phylogenetic tree or are reassessed at each step during the progressive alignment. Most present day methods use a dendrogram constructed from the pairwise sequence similarities as a guide tree and then invoke a DP algorithm to compare the sequence pairs, a block of aligned sequences with a single sequence, or two blocks of aligned sequences. The initial step thus involves performing all pairwise comparisons between the sequences, which has a time complexity of N^2L^2, where N is the number of sequences and L the average sequence length. This complexity is often the bottleneck for computing large alignments. However, using a guide tree is a good heuristic, because the sequences are progressively aligned from similar to divergent, which results in less error because alignment of similar sequences is more accurate than that of distant sequences and error propagation is minimized. A consequence of this scenario is that whenever a gap is introduced in any sequence during an alignment step, the gap will remain in further steps. Figure 2 shows an example of a guide tree and the resulting multiple alignment of repeating motifs within transcription factor IIIA (TFIIIA).

Hogeweg and Hesper [50] were the first to devise an integrated agglomerative algorithm. In their method, a dendrogram is constructed based on all

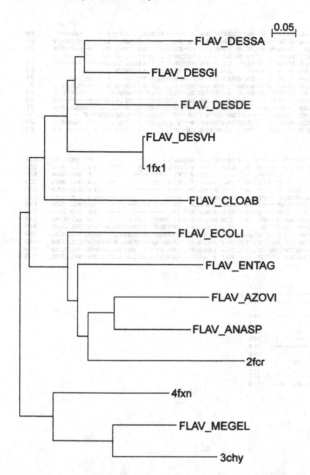

FIGURE 2 Guide tree and alignment by the program ClustalX [52] of the signal transduction protein cheY (PDB code 3chy) and 13 flavodoxin sequences, all showing the basic flavodoxin topology. (a) Guide tree based on pairwise sequence identity values. (b) Multiple alignment of the 14 sequences. Polar amino acids are dark, hydrophobic residues are light gray.

pairwise similarities of sequences matched by dynamic programming. The method is flexible in the clustering procedure used, and, for example, Unweighted Pair-Group Mean Average (UPGMA) [53], the present-day ancestor method [54], or the neighbor-joining method [55] can be alternatively chosen. The sequences are aligned progressively following the branch order of the dendrogram. During the alignment, internode or ancestral sequences are

```
      3chy  --ADKEL KFL VVDDF ST RRRI VRNL KEL G----- Y RNVEKAEDGVDAL N------ K. QAGGY GFV--I------ SDWN MP    60
 FLAV_MEGEL  --MVE IVYWSGT GNTEAMANE IEAAVKAAG-A-BVESVRIEDTNVDDVAS-KDVILL GCPAMGSE--E------LEBSVV    67
      4fxn  ---MKIVYWSGT GNTEKMAEL IAKGI IKSG-K-DVNT INVSBVNIDELLN-EDIL IL GCSAMGBE--V------LEESEF    66
 FLAV_DESGI  MPKAL IVYGSTG GNTE GVAEA IAKFLNSEG-M-BTTVVNVADVTAPGL AEGVDVVLL GCSTVGDDE IE------LQEDEY    72
 FLAV_DESSA  MSKSL IVYGSTG GNTE TAAEYVAEAI ENKE-I-BVELKNVTDVSVADL GNGYDIVLF GCSTVGEEE IE------LQBDFI    72
 FLAV_DESDE  MSKVL IVFGSST GNTE SIAQKLEEL IAAGG-H-EVTLLNAADASAENLADGYDAVLF GCSAVGMEDLE-----MQBDFL    72
      1fx1  -PKAL IVYGSTG GNTE YTAEI IARQL ANAG-Y-EVBSRDAASVEAGGLFEGFBLVLL GCSTVGDDS IE------LQBDFI    71
 FLAV_DESVH  MPKAL IVYGSTG GNTE YTAEI IARELADAG-Y-EVBSRDAASVEAGGLFEGFBLVLL GCSTVGDDS IE------LQBDFI    72
 FLAV_CLOAB  -MKI SIL YSSKTGKTERVAKL IEEGVKRSGNI -EVKTMVLBAVDKKFL QE-SEGI YF GTP TYYAN--------ISVEMK    68
 FLAV_ANASP  SKKI GLFYGT QT GKTE SVAEI IRDEF GNDVVF----L HDVSQAEVIDL ND-YQY L.II GCPTWNIGELQ--SD------ VK    67
 FLAV_AZOVI  -RKI GLFFGSNT GKIRKVAKS IKKRF DDETMSD---AL NVKRVSAEDFAQ-YQFL.II GTPTL GEGELPGLS SDGENESVK    75
      2fcr  --KIGIFFST STGNTTE VADE I GKTL GAKADAP---I DVDDVTDPQALKD-YBLL FL GAPTWNTGABTERSGT----SVD    70
 FLAV_ENTAG  MATIGIFFGSDTGQTRKVAKL IHQKL DGIADAP---LDVERATREQFL S--Y P VLLL GTPTL GDGELPGVEAGSQYDSVQ    75
 FLAV_ECOLI  -AITGIFFGSDTGNTENIAKI QKQL GKDVAD----VHDIAKSSKEDLEA-YDIL.LL GIP TWYYGEAQ-CD-------VD    66
```

```
      3chy  NMDG-LELLKTIBAD-----GAMSALFVL------MVTAEAKNENIIARAQAGAS----------------GYV-VKNFTA   112
 FLAV_MEGEL  EPFF-TDLAPKLKGRKVGL FGSYGWGSGE-----VMDAWKQNTEDTGATVIGTA--------------IVN-EMVDNA   124
      4fxn  EPFI-EEISTKI SGKKVALFGSYGWGDGK-----VMRDFEERGNGYGCVVVETT--------------LIVQKEDEA   124
 FLAV_DESGI  PLYE-DLDRAGLKDKVGVFGCGDSSYTYF--CGAVDVIEKNAEEL GATLVASS----------------LKIDGLVDSA   133
 FLAV_DESSA  FLYD-SLENADLKGKKVSVFGCGDSDYTYF--CGAVDAIEENLEKMGAVVIGDS----------------LKIDGDPERD   133
 FLAV_DESDE  SLFE-EFNRFGLAGRKVAAFASGDQEYEHF--CGAVPAIEERAKEL GATIIAEG-----------------LKMEGDASND   133
      1fx1  PLFD-SLEET GAQGRKVACFGCGDSSYEYF--CGAVDAIEEKLKNL GAEIVQDG---------------LRIDGDPRAA   132
 FLAV_DESVH  PLFD-SLEET GAQGRKVACFGCGDSSYEYF--CGAVDAIEEKLKNL GAEIVQDG---------------LRIDGDPRAA   133
 FLAV_CLOAB  XVID-ESSEYNLEGKL GAAFSTANSIAGGS--DIALLTILNGMLMVKGMLVYSGGVA----FGKPRTML GYVHIMEIQENE   141
 FLAV_ANASP  GLYS-ELDDVBFNGKLVAYFGTGDQIGYADNEQDAIGILLEEKISQRGGKIVGYWSTDGYDFNDSKALR-NGKIVGLALDE   145
 FLAV_AZOVI  EFLP-XIEGLBFSGKTVALFGLGDQVGYPENYLDALGELYSFFKDRGAKIVGSWSTDGYEIESSEAVV-DGKFVGLALDL   153
      2fcr  EFLYDKL FEVDMKDLPVALFGLGDAEGYPDNFCDAIEEINDCFAKGGAKPVVGFSNPDBYDYEESKSVR-DGKFLGLFLDM   149
 FLAV_ENTAG  XFTN-TLSEABL TGKTVALFGLGDQLHYSKNF VSAMRIL YDLVIARGACVVGNWPREGYKFSFSAALLENNKFVGLVLDQ   154
 FLAV_ECOLI  DFFP-TLEEIDFNGKLVALFGCGDQEDYAEYFCDAL GTIRDIIEPRGATIVGHWPTAGYHFEASKGLADDDHFVGLAIDE   145
```

```
      3chy  ATLEEKINKIFEKL GM----------------   128
 FLAV_MEGEL  FECKE-L GEARAKA----------------   137
      4fxn  EQDCIEFGKKIANI----------------   138
 FLAV_DESGI  E--VLBWARE VLARV----------------   146
 FLAV_DESSA  E--IVSWGGGIADKI----------------   146
 FLAV_DESDE  FEAVASFAEDVLKQL----------------   148
      1fx1  RDDIVGWAHDVRGAI----------------   147
 FLAV_DESVH  RDDIVGWAHDVRGAI----------------   148
 FLAV_CLOAB  DENARIFGERIAPKVKQIF----------   160
 FLAV_ANASP  DNQSDLTHDRIKSWVAQLKSEFGL-------   163
 FLAV_AZOVI  DNQSGKTBERVAAWLAQIAPEFGLSL----   179
      2fcr  VHDQIPMEKRVAGVEAAVVSETGV-------   173
 FLAV_ENTAG  ENQYDLTKERIDSWLEKIKPAVL--------   177
 FLAV_ECOLI  DRQVELTAERVEKWVKQISEELHLDEILHA   175
```

(b)

FIGURE 2   Continued.

constructed to represent already aligned groups of sequences. Hogeweg and Hesper [50] argued that a multiple alignment and the associated phylogenetic tree cannot be separated. They showed this using an iterative procedure for the first time (see Sec. 1.2.3). Thus, from the initial tree based on pairwise alignments carrying no information yet of related groups of sequences, a multiple alignment is generated and the associated pairwise similarities are inferred. Then, a new tree is iteratively constructed from which a succeeding alignment is created, based on the increased information.

The method MULTALIGN of Barton and Sternberg [56] establishes a simple chain order in which the individual sequences are aligned one by one. Initially, all pairwise alignment scores are determined and the two most similar sequences are matched first. During further iterations, the sequence showing the highest alignment score when matched with the prealigned sequence block is added to that block. To determine the alignment score, each sequence position i of the kth sequence matched with position j of a prealigned block of $k-1$ sequences receives a score per matched position averaged over the corresponding residue substitution values:

$$S_{i,j} = \frac{\sum_{p=1}^{k-1} D(A_{k,i}, A_{p,j})}{k-1} \tag{3}$$

where $D(A_{k,i}, A_{p,j})$ is the amino acid exchange weight. The PAM250 substitution matrix [14] (see Sec. 3.3) is used with a constant of 8 added to remove all negative matrix elements. Matched gaps are evaluated by the lowest exchange weight of zero. The resulting multiple alignment can be progressively refined by realigning each sequence with the previous alignment from which that sequence is deleted; i.e., sequence A1 is matched with aligned sequences A2,...,AN, sequence A2 is then realigned with the alignment of A1, A3,...,AN, and so forth. This process is repeated until all N sequences are realigned. Barton and Sternberg [56] recommend two such complete refinement cycles. The quality of a multiple alignment is assessed by comparisons with alignment scores over randomized sequences if the sequence groups are not too large; otherwise, normalized alignment scores (NASs) are used. An NAS is the alignment score divided by either the length of the shorter matched sequence or the number of residues not aligned with gaps.

Feng and Doolittle [51] devised a method for the construction of a phylogenetic tree through progressive alignment of the sequences. The algorithm works using only strictly pairwise sequence comparisons; no consensus sequences or averaging of similarities to compare blocks of sequences are used. Gaps in already aligned sequences are fixed by inserting special gap characters at gap positions, according to the credo "once a gap, always a gap" [51]. First, a rough branching order is determined using the phylogenetic tree-building method of Fitch and Margoliash [57]. This tree order is basically followed, but the alignment order of nearest neighbors in each obtained subgroup of sequences is reversed and the highest scoring alignment is selected for further comparison. For example, if the initial branch order is ABCD, then A and B are aligned first. The alignment orders ABC and BAC (alignment taking place successively from left to right) are then checked, and the best alignment, for example, BAC, is taken. Then BACD and BADC are examined. Only nearest neighbors are swapped, to keep computation manageable; other possible permutations are not considered.

The method MULTALIN of Corpet [58] follows the Hogeweg and Hesper approach in that it also uses hierarchical clustering and iteration. It is different in that it uses for the alignment of two sets of sequences the average similarity score between a pair of alignment columns i and j, one from each set, which is the average over the amino acid exchange values associated with all pairwise intercolumn residue comparisons:

$$S_{i,j} = \frac{\sum_{m=1}^{M}\sum_{n=1}^{N} D(A_{i,m}, A_{j,n})}{M * N} \tag{4}$$

where $A_{i,m}$ is the amino acid type in sequence m of alignment column i, $A_{j,n}$ is the amino acid type in sequence n of alignment column j, D is the amino acid exchange weight, and M and N denote the number of sequences in the two aligned sequence blocks. This way of scoring alignment positions is effectively profile comparison [59] (see Sec. 1.5).

Higgins and Sharp [60] constructed a fast and widely used method, Clustal, which was especially designed for use on small workstations. Speed was obtained during the pairwise alignments of the sequences through the Wilbur and Lipman [49,61] algorithm (see Sec. 3.1). From the pairwise similarities, a tree is constructed using the UPGMA clustering criterion. Then the sequences are aligned following the branching order of the tree. For the comparison of groups of sequences, Higgins and Sharp [60] used consensus sequences to represent aligned subgroups of sequences and also employed the Wilbur–Lipman technique to match these. Clustal does not provide the possibility to iterate the procedure as do the Hogeweg and Hesper [50] and Corpet [58] approaches. The Clustal package has been subjected to a number of revision cycles. Higgins et al. [62] implemented an updated version ClustalV in which the memory-efficient dynamic programming routine of Myers and Miller [18] is used, enabling the alignment of large sets of sequences using little memory. Further, two alignment positions, each from a different alignment, are compared in ClustalV using the average alignment similarity score of Corpet [58]. The largely extended version ClustalW [63] uses the alternative Neighbor-Joining (NJ) algorithm [55], which is widely used in phylogenetic analysis, to construct a guide tree. Sequence blocks are represented by a profile in which the individual sequences are additionally weighted according to the branch lengths in the NJ tree. An integrated user interface has been implemented in ClustalX [52], which is downloadable from http://www-igbmc.u-strasbg.fr/pub/ClustalX/ and comes with accessory programs for tree depiction. The Clustal versions W and X have generally become the most popular methods for multiple sequence alignment.

The PILEUP routine from the GCG package (Genetics Computer Group [64]) closely follows the earlier V version of Clustal. It generates an UPGMA-based tree and, for the alignment of two sets of matched sequences, uses the average alignment similarity score of Corpet [58].

The method MULTAL of Taylor [65] is very fast and constructs a tree during the progressive alignment as in the method of Feng and Doolittle [51]. It uses a fast sequential branching method to align the closest pairs of sequences first and subsequently align the next closest sequences to those already aligned. The order in which the sequences are aligned is based largely on the global amino acid composition of the sequences, which saves the overhead of performing all-against-all pairwise alignments. Scoring blocks of aligned sequences is done by dynamic programming akin to Corpet [58], but the similarity of two alignment columns is additionally normalized by the minimum number of sequences in either of two compared alignment blocks,

$$
S_{i,j} = \frac{\sum_{m=1}^{M}\sum_{n=1}^{N} D(A_{i,m}, A_{j,n})}{M * N * \min(M, N)}
\tag{5}
$$

where the variables are as under Eq. (4).

The method PRALINE [66] does not use a precalculated search tree but at each alignment step performs a full profile search (see Sec. 1.5) with the most recently aligned sequence block. It thus reevaluates at each alignment step which sequences or blocks of sequences should be aligned and hence determines the alignment order during progressive alignment. The technique offers a number of strategies to optimize the quality of multiple alignment, including global or local profile preprocessing and secondary structure prediction-based alignment (see Sec. 2.4.8). The profile preprocessing strategy is aimed at incorporating into each sequence trusted information from other sequences, through either global or local alignments. For each sequence, a preprocessed alignment is created by stacking other sequences (*master–slave* alignment) that score beyond a user-specified threshold when aligned pairwise with the sequence considered. A low threshold would result in a preprocessed alignment for each sequence comprising all other sequences, whereas higher thresholds would allow fewer and fewer sequences into the alignment. For each of the thus-formed preprocessed alignments, a profile is constructed (see Sec. 1.5). PRALINE then performs progressive multiple alignment using the preprocessed profiles to represent each of the original sequences. Because the preprocessed profiles for each of the sequences incorporate knowledge about other sequences (in particular, similar sequences) and comprise position-specific gap penalties, they enable increased matching of distant sequences and likely placement of gaps outside ungapped core regions during progressive alignment. The multiple alignment of the preprocessed profiles can also be used to derive consistency scores for each amino acid in the alignment, which for each sequence reflect the consistency among the pairwise alignments used that include the sequence. The second strategy of exploiting secondary structure prediction to optimize alignments will be described in Sec. 1.2.3.

Incorporating a priori knowledge into a multiple sequence alignment can greatly enhance its reliability and reduce computation time. For example, when two or more three-dimensional structures are known in a set of sequences to be matched, conserved structural regions can be identified and used to guide the alignment. The method of Taylor [67] allows the specification of one or more templates as consensus subsequences that, for example, can be associated with secondary structural elements. Based on these templates the sequences are included in a multiple alignment one by one. The templates are progressively updated to include the new residue variabilities.

1.2.2. Local Multiple Alignment Methods

Local multiple alignment methods focus on the comparison of conserved motifs in a set of protein sequences. Motif-based methods generally align shorter and ungapped sequence fragments and succeed when these fragments are recognizable as motifs in all or most of the sequences. Most motif-based methods, therefore, are not particularly suitable for the alignment of distant sequence groups. Generally, because these methods are not designed to yield full alignments, they should be attempted on sets of sequences of varying lengths if there is suspicion of shared motifs or domain structures.

The method MATCH-BOX [68] aims to find ungapped sequence regions with a high degree of similarity across a set of input sequences. This is done by comparing the frequency distribution of all pairwise-aligned sequence fragments (which are gathered from global alignments) with that derived from shuffled sequences. Using a set of the most similar nine-residue fragments, local alignments are created for each fragment if similarity beyond a threshold is found with segments across all other sequences. Boxes of ungapped regions are then delineated from these local alignments and assembled in a final alignment with unaligned amino acids and gaps between the boxes. The method also generates a reliability index for the aligned positions within the boxes, which relies on statistics derived from analyzing a relatively small number of known family alignments.

The Boguski et al. [69] semimanual program suite incorporates the space-efficient local alignment routine SIM of Huang et al. [27] (see Sec. 1.4.3) as well as the MSA method for multiple sequence alignment [40] (see Sec. 1.2). For each pair of sequences, the highest scoring local alignments containing gaps are determined and ungapped regions occurring in each of the sequences are extracted from them. Whenever meaningful, neighboring blocks of such motifs with the intervening sequence fragments are aligned using the MSA method, thus allowing gaps. The method of Boguski et al.

[69] also provides a user interface through which parts of the alignment can be manually edited.

Schuler et al. [70] introduced the Multiple Alignment Construction and Analysis Workbench (MACAW), which allows the user to lock or shift regions in an alignment while nonlocked subsequences are aligned automatically. The method is semiautomatic and produces blocks of alignments shared by all or a subset of the sequences. It is possible to iteratively define conserved regions such that the fraction of poorly defined segments that must be aligned automatically becomes smaller at each iteration cycle. The GIBBS method of Lawrence et al. [71] (see Sec. 1.4.2) has been incorporated in the MACAW procedure to detect the local fragments.

The method DIALIGN of Morgenstern and coworkers [72,73] constructs a multiple alignment by assembling a collection of high-scoring ungapped segments in a sequence-independent progressive manner. It is based on segment–segment comparisons rather than the residue–residue comparisons used in other programs. The segments are incorporated into a multiple alignment by using an iterative mathematical procedure that aims to find the optimal order for building in the segments. Only sequence fragments for which matched segments are found are aligned; regions between blocks of similar segments are left unaligned.

The program ITERALIGN (Brocchieri and Karlin [74]) aims to optimize the consistency between local pairwise alignments and their embedment in a multiple alignment across all input sequences. The authors went so far as to edit individual sequences by replacing amino acids by those that are preponderant at a corresponding position in a multiple alignment to achieve better recognition of crucial alignment regions.

These local programs perform well when there is a clear block of ungapped alignment shared by all of the sequences or when there are blocks of alignment separated by long insertions or deletions. They perform poorly, however, on general sets of test cases compared with global methods (see Sec. 1.2.4).

1.2.3.   Iterative Multiple Alignment Optimization and Alternative Techniques

Several new alignment algorithms have recently been developed, where a common point of interest has been the application of iterative strategies to refine and improve the initial multiple alignment. Iterative strategies, first proposed by Hogeweg and Hesper [50], provide an interesting alternative to simultaneous alignment (see above), because they are applicable to relatively large data sets. Although they do not provide any guarantees about finding an optimal solution, they are reasonably robust and certainly much less sensitive to the number of sequences than, for example, the aforementioned deterministic method MSA [40].

The PRRP program (Gotoh [75]) optimizes a progressive global alignment by iteratively dividing the sequences into two groups that are subsequently realigned using a global group-to-group alignment algorithm. Pairwise sequence weights are derived from a tree constructed with the UPGMA cluster criterion and used to calculate the alignment scores when sequence blocks are matched. Gotoh reported better accuracy [75] than that of ClustalW [63].

The PRALINE method [66] can perform iteration based on the consistency of preprocessed profiles (see Sec. 1.2.1) to optimize the alignment. Each of the preprocessed profiles can contain information about all the sequences, so each sequence in the final alignment can be assessed in terms of the degree of consistency reached across the profiles, which is translated into a consistency score for each amino acid in the multiple alignment. Iteration is then guided by the thus obtained scores, which are used as weights in the construction of alignments (using the dynamic programming protocol) during the next multiple alignment step [76]. From the resulting set of iterative alignments, the one with the highest cumulative score over all pair-wise matched amino acids in the alignment (sum-of-pairs score) can be selected as a safeguard to prevent alignments from wandering away to less optimal areas in the alignment space [76].

The PRALINE method can also use secondary structure prediction (see Sec. 2) to optimize the alignments in an iterative fashion. Most reliable secondary structure prediction methods use sequence information in multiple alignments (see Sec. 2.4), and their prediction accuracy relies crucially on the quality of the multiple alignment used (see Sec. 2.5). In the PRALINE approach, the multiple alignment is guided by predicted secondary structure, so that an iterative scheme arises that optimizes both the quality of the multiple alignment and that of the secondary structure prediction. An initial multiple alignment is constructed without information about the corresponding secondary structure (Fig. 3). Then, for each sequence the secondary structure is predicted by the PREDATOR [77,78], or PHD method [79] (see Sec. 2.4), although in principle this could be done by any available method, and a new alignment is iteratively constructed, now using the predicted secondary structure. The initial alignment is constructed using a default residue exchange matrix (the BLOSUM62 matrix) and gap penalty values. After secondary structure prediction, PRALINE uses the thus obtained secondary structure information as illustrated in Figure 4. During progressive alignment, pairs of sequences (and/or profiles representing already aligned sequence blocks) are matched using three secondary structure–specific residue exchange matrices [80], each with associated gap penalties. As shown in Figure 4, the residue exchange weights for matched sequence positions with identical secondary structure states are taken from the

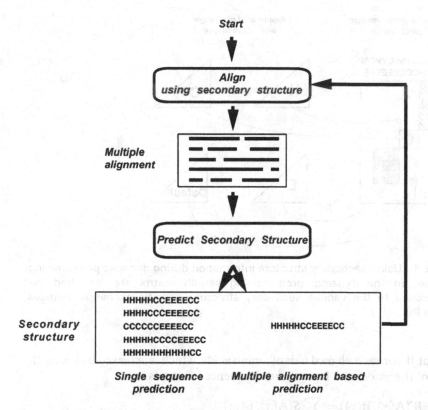

Start

Align
using secondary structure

Multiple
alignment

Predict Secondary Structure

Secondary
structure

HHHHHCCEEEECC
HHHHCCCEEEECC
CCCCCCEEEECC
HHHHHCCCCEEECC
HHHHHHHHHHHCC

HHHHHCCEEEECC

Single sequence
prediction

Multiple alignment based
prediction

FIGURE 3 Iterative multiple alignment and secondary structure prediction. Information from secondary structure prediction programs predicting for single sequences as well as multiple sequences can be used.

corresponding residue exchange matrix, whereas matched sequence positions with nonidentical secondary structure states are assigned the corresponding value from the default exchange matrix. The secondary structure information is thus used in a conservative manner based upon the assumption that consistent secondary structure predictions are indicative of their reliability (Fig. 3). In this iterative scenario, multiple alignment guides secondary structure prediction, which in turn guides alignment.

The multiple alignment method T-Coffee [81] combines information from global and local pairwise alignments. For each sequence pair, a single global alignment and 10 top-scoring nonintersecting local alignments are generated by the programs ClustalW [63] and LALIGN [28], respectively. The global and local alignment scores are then combined to yield a synthetic

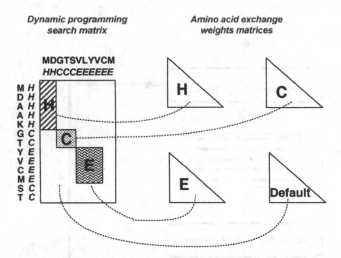

FIGURE 4 Using secondary structure information during dynamic programming. Regions in the dynamic programming search matrix are identified that correspond to the various secondary structure–specific exchange matrices. (From Ref. 66.)

weight W for each aligned pair of amino acids, which is achieved by taking the sum of the associated basic scores (sequence identities):

$$W(A(x), B(y)) = \sum S(A(x), B(y))$$

where $A(x)$ is residue x in sequence A, and summation is over the scores S of the global and local alignments containing the residue pair $(A(x), B(y))$, while for S each time the sequence identity percentage of the associated alignment is taken. This scenario results in a library of weights for each nonredundant residue pair. The information in the library is then further enhanced by a procedure called matrix extension [81]. Each library weight $W(A(x), B(y))$ is recalculated to reflect the degree to which residues $A(x)$ and $B(y)$ align consistently, as judged by all other library weights involving either $A(x)$ or $B(y)$. This is done by using a triplet approach aimed at calculating the contribution of third sequences I onto the direct alignment of sequences A and B, based on the notion that a triplet alignment A-I-B effectively provides an alternative alignment of A and B. Each extended score W' is then calculated as

$$W'(A(x), B(y)) = W(A(x), B(y)) + \sum_{I \neq A,B} Min(W(A(x), I(z)),$$

$$W(I(z), B(y)))$$

where x, y, and z are sequence positions in sequences A and B and the intermediate sequence I, respectively, and summation is done over all third sequences I other than A or B. The minimum of $W(A(x), I(z))$ and $W(I(z), B(y))$ is taken to use information from third sequences conservatively. The more intermediate sequences support the alignment of the pair, the higher its extended weight becomes. The extended library weights W' for the individual matched amino acid pairs are then used to fill the DP search matrix and align the associated input sequences. Library extension is performed at each step during the progressive alignment, which is carried out basically following the ClustalW protocol [63]. The dramatic increase in sensitivity of the T-Coffee method is mainly a result of its matrix extension scenario, which combines local and global alignment, where an incorrect direct alignment of sequences A and B can effectively be overridden by consistent alignments of other sequences acting as intermediates in the above triplet alignments.

In addition to the DP technique, other computational strategies have been explored. The program SAGA [82] uses a genetic algorithm (GA) to select from an evolving alignment population the alignment that optimizes, as an objective function (OF), the weighted sum of pairs used in the MSA program. More recently, a measure of consistency between the considered multiple alignment and a corresponding library of Clustal pairwise alignments was taken. This OF was developed for the COFFEE algorithm [36]. As mentioned above, hidden Markov models (HMMs) have also been tried out as statistical models of the primary structure consensus for a sequence family [34,35]. The program HMMT [83] uses a simulated-annealing method to maximize the probability that an HMM represents the sequences to be aligned.

1.2.4. Evaluation of Multiple Alignment Methods

Some recent evaluations of available multiple alignment techniques were carried out [37] using the database of benchmark alignments BAliBASE [84]. These studies showed the method PRRP [75] to be marginally the most accurate, closely followed by ClustalW, which is a much faster program. Notredame et al. [81] showed, also using the BAliBASE set of reference alignments [84] as the standard of truth, that the T-Coffee algorithm generates significantly improved alignments compared to ClustalW [63], PRRP [75], and DIALIGN2 [73]. The overall relative improvements measured using the column score (see below) were 8.6%, 8.6%, and 17.2%, respectively. Other methods included in the assessment tests, such as the local alignment method DIALIGN [72], the HMM-based method HMMT [83], and the Gibbs sampling method GIBBS [71], generally fell behind. It must be stressed, however, that DIALIGN was relatively successful in aligning sequences with very large insertions or deletions. The high degree of accuracy of the T-Coffee method was confirmed in benchmark tests by [76], who, in addition, found that the various PRALINE alignment strategies lead to reliable alignments of very divergent sequences. Virtually the same accuracy as that achieved with ClustalW was attained by the PRALINE method with default parameters, not using strategies such as profile preprocessing or predicted secondary-structure-induced alignment [76].

1.3. Scoring Amino Acid Substitutions

Dynamic programming techniques of the Needleman–Wunsch type rely on an amino acid exchange weight matrix and gap penalty values. The optimal or highest scoring alignment of two sequences is evaluated by the pairwise amino acid substitution scores summed over all matched positions less the penalties arising from each gap in the alignment. For alignment of similar sequences (>35% in sequence identity), the scoring system used is not critical [85]. In more divergent comparisons with residue identity fractions in the so-called twilight zone (15–25%) [86], different scoring regimes can lead to dramatically deviating alignments.

Many different substitution matrices have been devised over more than three decades, each trying to optimize the signal-to-noise ratio in the detection of homologies among sequences. A combination of physicochemical characteristics of amino acids can be used to derive a substitution matrix that basically contains pairwise amino acid similarity values. Other data from which residue exchange matrices have been computed include sequence alignments, structure-based alignments, and common sequence motifs. The remainder of this section will deal with these approaches.

Fitch [1] constructed the first nonidentity residue exchange weights matrix. He used the minimum number of nucleotide base changes for each amino acid substitution. Values of 0, 1, 2, and 3, required to substitute one residue with another, were converted to similarity values 4, 2, 1, and 0, respectively.

Because the aim of a sensitive scoring system is proper alignment, a number of widely used scoring matrices have been inferred from multiple sequence alignments. Dayhoff et al. [14,87] derived the classical PAM250 matrix from an evolutionary model for residue substitutions (Fig. 5). Sequences from 72 protein families were compared for which similarities were high enough (a fraction of 85% or more identical residues) to yield accurate multiple alignments by eye. The amino acid substitutions observed in these matches were then tabulated and converted to mutational probabilities according to 1% accepted point mutations (one amino acid changed out of 100). This so-called PAM1 matrix is converted into a PAM250 matrix by 250 self-multiplications, but this number can be varied to yield matrices associated with greater or smaller genetic distances. The most widely used substitution table is the PAM250 log-odds matrix where each PAM250 matrix

```
A    2
R   -2   6
N    0   0   2
D    0  -1   2   4
C   -2  -4  -4  -5  12
Q    0   1   1   2  -5   4
E    0  -1   1   3  -5   2   4
G    1  -3   0   1  -3  -1   0   5
H   -1   2   2   1  -3   3   1  -2   6
I   -1  -2  -2  -2  -2  -2  -2  -3  -2   5
L   -2  -3  -3  -4  -6  -2  -3  -4  -2   2   6
K   -1   3   1   0  -5   1   0  -2   0  -2  -3   5
M   -1   0  -2  -3  -5  -1  -2  -3  -2   2   4   0   6
F   -4  -4  -4  -6  -4  -5  -5  -5  -2   1   2  -5   0   9
P    1   0  -1  -1  -3   0  -1  -1   0  -2  -3  -1  -2  -5   6
S    1   0   1   0   0  -1   0   1  -1  -1  -3   0  -2  -3   1   2
T    1  -1   0   0  -2  -1   0   0  -1   0  -2   0  -1  -3   0   1   3
W   -6   2  -4  -7  -8  -5  -7  -7  -3  -5  -2  -3  -4   0  -6  -2  -5  17
Y   -3  -4  -2  -4   0  -4  -4  -5   0  -1  -1  -4  -2   7  -5  -3  -3   0  10
V    0  -2  -2  -2  -2  -2  -2  -1  -2   4   2  -2  -1  -1   0  -6  -2   2  -2   4
     A   R   N   D   C   Q   E   G   H   I   L   K   M   F   P   S   T   W   Y   V
```

FIGURE 5   Dayhoff PAM250 log odds residue exchange matrix. Single-letter codes have been used for the 20 amino acids. The most negative matrix value of -8 is for the tryptophan–cysteine (WC) exchange, and the highest value is for tryptophan conservation (WW). Note that the tryptophan–arginine (WR) exchange value is unreasonably high, probably due to paucity of data used for deriving the log odd values. (From Ref. 14.)

element C is converted by $10 \times \log(C)$. Jones et al. [88] repeated the work of Dayhoff et al. [87] and constructed a PAM250 matrix over a database about 18 times as large (23,000 sequences compared to about 1300 in Dayhoff et al. [87]). Gonnet et al. [89] performed an exhaustive matching of a database of 1.7×10^6 residues in which sequences were pregrouped using a special tree formalism. They also derived a PAM250-based substitution matrix and suggested a gap penalty regime exponentially related to the gap length. Henikoff and Henikoff [90] used the PROSITE sequence motif database [91] to construct about 2000 multiple subsequence alignments associated with the conserved PROSITE motifs. From this database, Henikoff and Henikoff constructed the BLOSUM series of exchange matrices. For example, the alignment blocks showing pairwise sequence identities of ≤62% were used to construct the scoring matrix BLOSUM62. Very similar sequence groups were downweighted by taking the average value of their contributions to the matrix.

Structure-based alignments of protein sequences have also been used for the derivation of scoring matrices. Risler et al. [92] used 2860 residue substitutions derived from rigid-body superpositions of 32 tertiary structures from 11 families. Only those substitutions were recorded that came from well-superposed parts within each equivalenced structural pair. The exchange weights were calculated from the observed substitutions using statistical χ^2 values and therefore suffer from the scarcity of data. Johnson and Overington [93] repeated the approach, but now equivalenced 235 tertiary structures from 65 families using the superposition method of Šali and Blundell [94]. They collected about 200,000 residue exchanges, two orders of magnitude more than Risler et al. [92]. Their odds matrix was calculated using substitution preferences where the log-odds ($O_{i,j}$) for each substitution of residue type i into j was calculated as

$$O_{i,j} = 10 \log_{10} \left(\frac{S_{i,j} \bigg/ \sum_{n=1}^{20} S_{i,n}}{\sum_{m=1}^{20} S_{m,j} \bigg/ \sum_{m=1}^{20} \sum_{n=1}^{20} S_{m,n}} \right) \qquad (6)$$

where $S_{i,j}$ is the observed frequency of the substitution from residue i into j and summation is over the 20 residue types. Johnson and Overington [93] also compared their exchange matrix with 12 others. Various gap penalty values were tested for each matrix, and the Needleman–Wunsch comparison technique was used with a large set of structurally aligned protein sequences as a benchmark database. Another evaluation is that of Henikoff and Henikoff [95], which circumvented gap penalty assignments by using the BLAST comparison method [96] (see Sec. 1.5). Vogt et al. [97] also assessed a large set of exchange matrices. They optimized the gap penalties individually for each substitution matrix using the 3D-ALI database [98] as the standard of truth. Vogt et al. found that the Gonnet et al. [89] matrix leads to the most accurate alignments. These comparative studies taken together show the best overall performance for the scoring systems of Henikoff and Henikoff [90] (BLOSUM62) and Gonnet et al. [89].

Probably the currently most widely used matrix is BLOSUM62, which has relatively high diagonal values compared to the PAM250 matrix and so is a more conservative matrix. This has an appreciable effect on the pairwise identity values, the most popular way of identifying sequence relationships, because the BLOSUM62-based identity scores are dramatically higher than those of the PAM250 matrix, particularly between divergent sequences (Fig. 6). However, "softer" matrices than the BLOSUM62 matrix, such as the higher PAM series for the Dayhoff or Gonnet matrices or the BLOSUM50

	1fx1	2fcr	4fxn	FLAV_ANASP	FLAV_AZOVI	FLAV_CLOAB	FLAV_DESDE	FLAV_DESGI	FLAV_DESSA	FLAV_DESVH	FLAV_ECOLI	FLAV_ENTAG	FLAV_MEGEL	3chy
1fx1	*	29	30	29	18	17	44	54	53	95	35	21	26	13
2fcr	35	*	23	27	35	16	24	28	26	24	33	34	18	16
4fxn	33	25	*	20	21	25	25	30	30	28	26	25	41	14
FLAV_ANASP	30	40	30	*	49	23	28	29	33	31	46	39	21	16
FLAV_AZOVI	25	38	28	50	*	22	22	28	26	26	45	47	23	18
FLAV_CLOAB	24	22	30	27	24	*	18	18	18	21	26	20	20	14
FLAV_DESDE	44	29	26	28	31	25	*	47	44	47	35	25	30	18
FLAV_DESGI	54	31	33	33	33	23	47	*	55	58	32	28	34	18
FLAV_DESSA	53	31	34	38	33	25	45	55	*	56	33	27	35	12
FLAV_DESVH	95	31	30	33	29	27	47	58	56	*	37	26	28	16
FLAV_ECOLI	37	36	31	46	46	26	38	40	39	40	*	41	26	19
FLAV_ENTAG	27	35	32	42	49	25	31	36	28	30	42	*	24	20
FLAV_MEGEL	28	26	43	24	26	28	34	36	36	32	34	31	*	8
3chy	23	27	16	21	23	22	23	22	23	23	24	25	20	*

FIGURE 6 Pairwise sequence identity percentages of 13 flavodoxin sequences and the signaling protein cheY (PDB code 3chy) using the PAM250 matrix (upper right triangle) with gap (open, extend) penalties of 10 and 1 as well as the BLOSUM62 matrix (lower left triangle) with penalties of 12 (gap opening) and 1 (gap extension). Proteins identified by four-letter codes are designated using their PDB identifiers, whereas for the others the Swiss-Prot identifiers are given.

matrix, are particularly useful in global alignment, because they are more suitable for aligning divergent sequences. They can also be useful in local alignment searches, aiding the recognition of distant familial relationships.

It is evident that constructing a single amino acid exchange matrix to represent all localized exchange patterns in protein structures is a gross oversimplification and might lead to error in specific protein families. Attempts to group local characteristics of protein structure and to construct specific exchange matrices for each of those groups have included secondary structure [80,99] and solvent accessibility [100]. The latter was carried out by identifying amino acid substitution classes (a more binary approach using sets of amino acid types) that were maximally indicative of local protein structure. Recent attempts have included adapting the amino acid exchange matrix to a local family in order to optimize the representation of the family and thus enhance searching for distant members [101].

1.4. Sequence Patterns

A collection of related sequences contains more information than a single sequence. Taking advantage of this multiple sequence information for feature extraction and databank searching is quite natural and has been explored by many researchers. Two closely related problems have to be addressed: (1) how to represent the collective information contained in many sequences and (2) how to quantify the similarity between the multiple representation and each individual sequence in the databank to be searched. In this section the recognition of segments conserved among a group of protein sequences will be considered. Section 1.5 will concentrate on how to optimally use the available sequence information for protein families.

With the growth in the amount of structural information available, it has become clear that certain regions of a protein molecular structure are less liable to change than others. Knowledge of this kind can be used to elucidate certain characteristics of a protein's architecture such as buried versus exposed location of a segment or the presence of specific secondary structural elements [102,103]. The most salient aspect, however, involves the proteins functionality. The most conserved protein region very often serves as a ligand binding site, a target for posttranslational modification, the enzyme catalytic pocket, and the like. Detecting such sites in newly determined protein sequences could save immense experimental effort and help characterize functional properties. Moreover, using only the conserved regions of a protein rather than its whole sequence for databank searching can reduce background noise and help considerably to establish distant relationships.

1.4.1. Classifying Sequence Motifs

In principle, the purpose of any method of sequence comparison, such as the alignment of two sequences, is to find primary structure conservation and establish reliable regions of local homology. Sequence motifs are usually considered in the context of multiple sequence comparison, when certain contiguous sequence spans are shared by a substantial number of proteins. Staden [104] gave the following classification for protein sequence motifs:

Exact match to a short defined sequence
Percentage match to a defined short sequence
Match to a defined sequence, using a residue exchange matrix and a
    cutoff score
Match to a weight matrix with cutoff score
Direct repeat of fixed length
A list of allowed amino acids for each position in the motif

Although sequence patterns are often referred to as a combination of several elementary motifs [104] separated by intervening sequence stretches, here patterns and motifs will be used interchangeably.

Protein sequence patterns are often derived from a multiple sequence alignment, so that the quality of the alignment determines the correctness of the patterns. Defining a consensus sequence from a multiple alignment can be approached as in social choice theory, using majority and plurality rules of voting [105]. A consensus sequence can generally be written as a set of heterogeneous rules of some kind for each alignment position. Taylor [106] used Venn diagrams to make subdivisions of the amino acid residue types, which were placed into a number of partially intersecting classes, such as CHARGED, SMALL, HYDROPHOBIC, POSITIVE, etc. Any position within a sequence pattern can thus be described by a logical rule of the type "TINY.or.POLAR_non-AROMATIC.or.PROLINE." The stringency of the restraint imposed on a particular position should depend on how crucial a given feature is for the protein family under scrutiny. Allowed elements can range from an absolutely required residue type to a gap.

1.4.2.   Sequence Pattern Detection Methods

In the Taylor [67] approach, an initial alignment of sequences is generated on the basis of structural information available (e.g., superposition of C_α atoms) to ensure reliability. After creating the template from the initial alignment, it can then be extended to include additional related proteins. This process is repeated iteratively until no other protein sequence can be added without giving up on essential features. In an attempt to make the pattern less dependent on the quality of the alignment, Patthy [107] adopted the following iterative approach. The sequences are first pairwise aligned and the most similar of them grouped. For each group, the alignments are inspected to identify residues conserved in most of the sequences, and an initial pattern is formulated. Then every sequence within the group is optimally aligned with the pattern, resulting in the generation of a multiple alignment. As a next step, the consensus sequences derived for the different groups are amalgamated, each individual sequence realigned with the pattern, an extended multiple alignment generated, and so on. While producing the consensus sequences, the algorithm relies on user-specified thresholds, for example, the fraction of residues deemed similar or identical according to the Dayhoff PAM250 residue exchange matrix (see Sec. 1.3) at a given position for it to be included in the consensus. In a similar approach, Smith and Smith [108] also used pairwise comparisons as a starting point for determining sequence patterns.

Vihinen [109] derived conserved motifs by superimposing dot matrices from pairwise comparisons that involve a given reference sequence. This approach was generalized by Vingron and Argos [110], who included all pairwise dot matrices for a given set of sequences. They elucidated consistent and related regions in all matrices through matrix multiplication. Other, often indirect, measures of conservation have also been attempted as criteria for motif delineation, such as the existence of homogeneous regions in a protein's physical property profiles (see, e.g., Chappey and Hazout [111]).

Frequently a sequence motif is proven experimentally to be responsible for a certain function. A collection of functionally related sequences should then yield a discriminating pattern that occurs in all the functional sequences and does not occur in other unrelated sequences. Such patterns often consist of several sequentially separated elementary motifs, which, for example, join in the three-dimensional structure to form a functional pocket. Such discriminating motifs were studied extensively early on for particular cases such as helix–turn–helix motifs [112] or G-protein-coupled receptor fingerprints [113].

A consistent semimanual methodology for finding characteristic protein patterns was developed by Bairoch [114]. The aim of his approach was to make the derived patterns as short as possible but still sensitive enough to recognize the maximum number of related sequences and also sufficiently specific to reject, in a perfect case, all unrelated sequences. A large collection of motifs gathered in this way is available in the PROSITE databank [91]. Associated with each motif is an estimate of its discriminative power. For many PROSITE entries, published functional motifs serve as a first approximation of the pattern. In other cases, a careful analysis of a multiple alignment for the protein family under study is performed. Then the initial motif is searched against the Swiss-Prot sequence databank (see Sec. 2), and search hits are analyzed. The motif is then extended or shrunk in sequence length to achieve the minimal amount of false positives and maximal amount of true positives. For example, the putative AMP-binding pattern ([LIVMFY]-x(2)-[STG]-[STAG]-G-[ST]-[STEI]-[SG]-x-[PASLIVM]-[KR]) given in PROSITE is found in 150 motifs over 112 sequences from the current Swiss-Prot databank (Release 38); 137 hits in 99 sequences are true positives, and 13 hits in 13 sequences are known false positives. This gives a precision of $137/(137+13) = 91.3\%$. Four motifs in four proteins known to be AMP-binding could not be identified using the motif, so that the so-called recall fraction $= 137/(137+4) = 99.3\%$. The PROSITE databank and related software are an invaluable and generally available tool for detecting the function of newly sequenced and uncharacterized proteins. In addition to specifying regular expressions (such as that for the putative AMP-binding pattern above) for many protein sequence motifs, the PROSITE database also represents entries using the extended profile formalism of Bucher et al. [115], an example of which is included in Figure 7.

A description of sequence motifs in a regular expression such as in the PROSITE database involves enumeration of all residue types allowed at particular alignment positions or by less stringent rules previously described. This necessarily leads to some loss of information, because particular sequences might not be considered. Also, the formalism is not readily applicable to all protein families. Increased information about a conserved sequence span can be stored in the form of an ungapped sequence block. Henikoff and Henikoff [116] derived a comprehensive collection of such blocks (known as the BLOCKS databank) from groups of related proteins as specified in the PROSITE databank [91] using the sensitive technique of Smith et al. [117]. In the latter method, the search for conserved motifs begins with listing all common three-residue combinations with the maximal length of allowed spacers between these three residues set at 24 amino acids. The most frequent occurrences among the group of specific combinations are found and joined into blocks, and a mean score for each column of the block is calculated using the PAM250 residue exchange matrix. Then the best matching subsequences from the rest of the proteins in the group are found. Searching protein sequences against the library of conserved sequence blocks or an individual block against the whole sequence library [118] provides a sensitive way of establishing distant evolutionary relationships between proteins. Recently, the BLOCKS database was extended [119] by using, in addition to PROSITE, nonredundant information from four more databases: PRINTS [120], Pfam-A [121], ProDom [122], and Domo [123].

As faster and more reliable sequence comparison tools became available (see Secs. 1.1 and 1.2), several other attempts were made early on to compile automatically as many patterns as possible from the full protein sequence databank (e.g., Refs. 124–126). For example, Sheridan and Venkataraghavan used the BLAST3 [127] searching tool to generate all pairwise local alignments resulting from the full sequence databank self-comparison and then clustered these into multiple alignments that they subsequently applied to find additional representatives of the corresponding protein family in the databank. More recent programs that attempt to detect domains from multiple alignment information are DOMAINER [126], MKDOM [128], and DIVCLUS [129].

Pattern matching methods based upon machine learning procedures have also been attempted. The formalism of neural networks (see Sec. 2.4.1) has been particularly explored, because they have the ability to learn an internal nonlinear representation from presented examples. They also are well suited for recognition of ill-defined, fuzzy objects (for reviews see Refs. 130 and 131).

The method of Frishman and Argos [132] exploits neural networks to delineate conserved sequence blocks and can then use these blocks to flexibly search sequence databanks. First, a neural network is used to elucidate unknown patterns from a multiple alignment of N sequences. One segment of width W in each position of the alignment is tested and the net is trained on the alignment of the segment including $N-1$ sequences, after which the excluded sequence segment is submitted for recognition and the network output is recorded. This is repeated with each of the N segments removed.

ID   KRINGLE_2; MATRIX.
AC   PS50070;
DT   NOV-1997 (CREATED); NOV-1997 (DATA UPDATE); JUL-1998 (INFO UPDATE).
DE   Kringle domain profile.
MA   /GENERAL_SPEC: ALPHABET='ABCDEFGHIKLMNPQRSTVWYZ'; LENGTH=79;
MA   /DISJOINT: DEFINITION=PROTECT; N1=6; N2=74;
MA   /NORMALIZATION: MODE=1; FUNCTION=LINEAR; R1=.7529; R2=.00952475; TEXT='NScore';
MA   /CUT_OFF: LEVEL=0; SCORE=813; N_SCORE=8.5; MODE=1;
MA   /CUT_OFF: LEVEL=-1; SCORE=603; N_SCORE=6.5; MODE=1;
MA   /DEFAULT: D= -20; I= -20; B1= -50; E1= -50; MI= -105; MD= -105; IM= -105; DM= -105;
MA   /I: B1=0; BI= -105; BD= -105;
MA   /M: SY='D'; M=-15,29,-30,44,37,-36,-15,1,-34,5,-25,-24,10,-6,13,-4,0,-10,-30,-34,-19,25;
MA   /M: SY='C'; M=-10,-20,120,-30,-30,-20,-30,-30,-30,-30,-20,-20,-20,-40,-30,-30,-10,-10,-10,-50,-30,-30;

MA   /M: SY='C'; M=-10,-20,120,-30,-30,-20,-30,-30,-30,-30,-20,-20,-20,-40,-30,-30,-10,-10,-10,-50,-30,-30;
MA   /I: I=-8; MI=-5; IM=-5; DM=-15; MD=-15;
MA   /M: SY='D'; M=-7,23,-24,24,6,-28,-7,-1,-24,0,-24,-15,18,-10,3,-5,5,-3,-22,-34,-18,4;
MA   /M: SY='T'; M=-8,-29,-22,-34,-26,3,-34,-26,34,-27,26,18,-24,-25,-21,-24,-20,-6,26,-22,-2,-26;
MA   /M: SY='P'; M=-7,-9,-30,-6,1,-26,-18,-10,-20,0,-25,-15,-7,39,-2,-6,-2,-1,-23,-29,-20,-3;
MA   /M: SY='R'; M=-7,-2,-25,-2,5,-22,-17,-3,-18,8,-17,-8,-1,-12,13,14,0,-3,-15,-23,-9,7;
MA   /M: SY='C'; M=-10,-20,120,-30,-30,-20,-30,-30,-30,-30,-20,-20,-20,-40,-30,-30,-10,-10,-10,-50,-30,-30;
MA   /I: E1=0; IE=-105; DE=-105;
NR   /RELEASE=38,80000;
NR   /TOTAL=136(41); /POSITIVE=136(41); /UNKNOWN=0(0); /FALSE_POS=0(0);
NR   /FALSE_NEG=0; /PARTIAL=0;
CC   /TAXO-RANGE=??E??; /MAX-REPEAT=38;
DR   P98140, FA12_BOVIN, T; Q04962, FA12_CAVPO, T; P00748, FA12_HUMAN, T;
DR   Q04756, HGFA_HUMAN, T; P80009, PLMN_CANFA, T; P80010, PLMN_HORSE, T;

DR   Q29485, PLMN_ERIEU, T; P00747, PLMN_HUMAN, T; P12545, PLMN_MACMU, T;
DR   P20918, PLMN_MOUSE, T; P06867, PLMN_PIG , T;
3D   1AOH; 1BHT; 1CEA; 11KRN; 1KDU; 1PK2; 1PK4; 1PKR; 2PK4; 1PMK; 1PML;;
3D   1TPK;;
DO   PDOC00020;
//

FIGURE 7 An example entry of the PROSITE databank presenting the Kringle domain profile. Identifiers in the first column are line codes for easy computer access. After general information such as the entry name (ID), accession number (AC), update information (DT) and the full name of the pattern (DE), the profile description (MA) follows with the following identifiers: /GENERAL SPEC: provides the basic alphabet (one letter amino acid codes) and the length of the profile; /DISJOINT: indicates here that any two sequence segments cannot overlap with the profile from position 6 (N1) to 74 (N2); /NORMALIZATION: identifies the parameters for the normalized score $Y = R1 + R2*X$, where X is the raw score; /CUT_OFF: designates the cutoff level for score significance— two levels are given here (0 and −1) with corresponding raw and normalized scores; /DEFAULT: specifies default scores for deletions (D), insertions (I), internal initiation (B1), internal termination (E1), transition from match to insertion (MI), transition from match to deletion (MD), transition from insertion to match (IM), and transition from deletion to match (DM); /I: specifies local

Lawrence et al. [71] described a method called GIBBS aimed at detecting conserved regions with a residual degree of similarity from unaligned sequences. The technique is based on iterative sampling of individual sequence segments of given length W from a set of N sequences. In the first step the segments are taken from random positions of $N-1$ sequences, one randomly selected sequence being excluded. A tentative "conserved" region of length L is constructed from these segments, and observed and statistically expected residue frequencies are calculated for each of the L positions. Then all possible segments from the excluded sequence are tested, one by one, for their consistency with the amino acid probabilities of the generated subalignment. If at least a small fraction of the randomly selected segments are actually related, thus providing a weak information signal, the probability of successful extension of the nascent pattern by related segments from other sequences in the set will be slightly higher than could be expected for a completely random situation. The procedure is repeated iteratively, and the pattern probabilities are recalculated at each step, with the discriminative power of the pattern possibly increasing with the inclusion of each new related member.

The program MEME [133,134] is a tool for unsupervised motif searching within DNA and protein sequences that which operates using an expectation maximization (EM) algorithm. It finds occurrences of motifs by comparing the residue composition at each position of a putative motif against the general composition of background sequence regions that do not display the motif. Regions showing the most discriminating compositions are then selected as motifs. A limitation of the MEME motifs is that they are

---

values for insertions/deletions into the profile—three such sites are included in the figure; /M: gives the position-specific scores for the amino acids in the order specified under /GENERAL_SPEC. The NR lines show the numerical results of the Swiss-Prot databank searches (here for release 38 with 80,000 sequences) indicating that there are 136 occurrences of the pattern, all of which are true positives. Because a sequence can contain multiple copies of a pattern, the numbers of individual sequences containing the pattern are given in parentheses. Line CC represents the taxonomic range of the pattern, where the letter E stands for eukaryotes and the question marks indicate that it is not known whether this pattern is present in archebacteria, bacteriophages, prokaryotes, and eukaryotic viruses, respectively. Also, the maximum number of the pattern's repeats in a single sequence is specified. Then follows information on all the sequences associated with the pattern (DR lines). For each sequence, its accession number, Swiss-Prot databank identifier, and the type of the hit (T for true, or F for false) are specified.

ungapped, but the program can find multiple occurrences in individual sequences, which, however, do not need to be encountered within each input sequence. Another useful feature of the MEME method is that it is geared to finding DNA palindrome sequences, which are often implicated as DNA binding sites for proteins. To increase the chance of finding palindrome sets, the nucleotide probabilities of corresponding motif columns (columns 1 and W, 2 and $W-1$, and so forth, with W the width of the motif) are constrained to be the same.

The evolution of pattern derivation methods and resulting motifs in turn has triggered the development of tools of varying degrees of complexity to search individual sequences and whole sequence databanks with user-specified motifs [135–141]. Techniques to evaluate the discriminative power of a given pattern are also available [142]. In particular, many programs are available for searching sequences against the PROSITE databank; a full list of publicly available and commercial programs for this purpose is supplied with PROSITE [91,114].

The method of Frishman and Argos [132] mentioned earlier exploits neural networks to delineate conserved sequence blocks and can then use these blocks to flexibly search sequence databanks. This is done by combining the trained neural nets to detect additional representatives of the "trained" family. The average net recognition is used as a measure to select the most conserved alignment regions. In the database search step of the algorithm, the M most conserved protein blocks are used to extensively train M corresponding neural networks, which are then used to scan the protein sequence databank. Variable constraints can be imposed on the distances between the blocks, although the M blocks must be in the same sequential order as in the multiple alignment.

A further extension of the motif-searching techniques is provided by methods that compare sequence templates with target sequences while allowing gaps for insertions and deletions (see, e.g., Rhode and Bork, [143]). These methods are similar to the profile-like methods described in Sec. 1.6. Alternatively, it is possible to search a databank with multiple ungapped motifs independently and then merge the hit lists for the motifs to find consistent occurrences of certain databank sequences [113].

1.4.3. Internal Sequence Repeat Detection

An important characteristic of genomes, particularly those of eukaryotes, is the high frequency of internal sequence repeats. For example, the human genome is estimated to consist of more than 50% reiterated sequences [24]. Genomic repeats have been implicated in a number of cellular processes. For example, there are known cases where repeat sequences are used by bacteria to increase their colonization and infection of human individuals [144]. Moreover, palindromic (i.e., reverse complemented) repeats often form DNA hairpin structures, which are associated with replication or structural mechanisms [145]. To analyze repeats at the genomic level, a fast method named REPUTER was developed that is able to detect palindromic and near-perfect repeats [146]. This method owes its speed to the implementation of suffix trees to organize the occurrence of repeat types.

Given the widespread duplication and rearrangement of genomic DNA and subsequent gene fusion events, also at the protein level, internal sequence repeats are abundant and are found in numerous proteins. Gene duplication may enhance the expression of an associated protein or result in a pseudogene where less stringent selection of mutations can quickly lead to divergence that results in an improved protein. An advantage of duplication followed by gene fusion at the protein level is that the protein resulting from the new single-gene complex shows a more complex and often symmetrical architecture, conferring the advantages of multiple, regulated, and spatially localized functionality. Repeat proteins often fulfill important cellular roles, such as zinc-finger proteins that bind DNA, the β-propeller domain of integrin α-subunits implicated in cell–cell and cell–extracellular matrix interactions, or titin in muscle contraction, which consists of many repeated Ig and Fn3 domains. The similarities found within sets of internal repeats can be 100% in the case of identical repeats, down to the level where any discernible sequence similarity has been lost as a result of mutation and insertion/deletion events. A classical example of this is chymotrypsin, where fusion of two duplicated genes, each coding for a separate β-barrel domain, has resulted in a two-domain enzyme. The active site consists of amino acids of both domains and shows greatly enhanced activity compared to a suspected ancestral active center within an individual ancestral barrel [147]. The amino acid sequences of the two barrels have diverged so much that the duplication event had to be inferred from the structural similarity [148].

The problem of recognizing internal sequence repeats in proteins has been tackled by many researchers. One of the pioneers in the automatic detection of repeats was McLachlan, who devised the first methods over three decades ago [3]. The first methods relied on Fourier analysis [149,150], and this technique remained popular [151]. Although Fourier transforms are designed to detect periodic behavior, their application to protein sequence signals is compromised by the fact that many repeats are distant through mutations and insertions/deletions and different irregular sequence stretches can intervene. Moreover, proteins can contain multiple repeat types, all with different basic periodicities, which decreases the periodic signal for any one type. Finally, Fourier techniques require a relatively large number of repetitions, whereas many proteins contain only a few repeats.

Another approach to delineate repeats in protein sequences was made by exploring dynamic programming (DP). The first attempts were made by McLachlan [5], who used the DP technique over fixed window lengths on myosin rod repeats. Boswell and McLachlan [152] elaborated the method by incorporating dampening factors and allowing the occurrence of gaps. Argos [6] (see Sec. 1.1) also adopted the window technique but exploited physico-chemical properties of amino acids in addition to the PAM250 residue exchange matrix [14] and used the technique to detect repeats in, for example, frog transcription factor IIIA (TFIIIA), human hemopexin, and chick tropoelastin. Huang et al. [27] used local alignments [25] to find the repeats in rabbit globin genes. Their method SIM is a memory-optimized implementation of the approach introduced by Waterman and Eggert [26], which calculates a list of top-scoring nonintersecting local alignments, meaning that no two alignments have a given matched amino acid pair in common.

Heringa and Argos [153] adapted the basic Waterman–Eggert algorithm to repeat situations within a single protein by demanding, in addition to top-scoring alignments being nonintersecting, that locally aligned fragments not overlap. They introduced a graph-based iterative clustering mechanism that takes the list of top-scoring nonoverlapping local alignments thus produced for a single query sequence, declares the N-terminal matched amino acid pair in each top alignment as the start site of a repeats pair, and then attempts to delineate associated start sites within the top alignments that match the repeat type based on alignment consistency with already clustered members of the repeat type. If such new repeats are found, the cluster procedure is iterated. The cluster consistency criterion assesses the number of established repeats that align with a putative repeat and accepts a new repeat only if three or more such top-scoring alignments can be found and if at least one of these associated alignments has already contributed one or more repeat members to the current repeat type and therefore can be trusted to be "in phase" with that repeat type. After the clustering phase, the repeats can be multiply aligned and turned into a profile (see Sec. 1.5), which can then be slid over the query sequence to verify the repeats already found and possibly detect new incarnations missed by the preceding algorithmic steps [153]. If new repeats are found, the profile can be updated and the procedure iterated. The REPRO algorithm is able to detect multiple repeat types independently and is a very sensitive but slow technique. A webserver for the REPRO algorithm is available at http://mathbio.nimr.mrc.ac.uk [154].

A quick algorithm for calculating the length and copy number of internal repeat sets was devised by Pellegrini et al. [155]. The method uses the Waterman–Eggert algorithm and converts the scores of the selected top alignments to probabilities. An $N \times N$ path matrix, where N is the length of the protein sequence, is then filled with 1s for matrix cells corresponding to local nonintersecting alignments that score above a preset threshold value for the probabilities, and 0s elsewhere. Two very simple summing protocols are then applied to this matrix to obtain an approximate notion of the repeat length and copy number, albeit the repeat boundaries are not determined. Marcotte et al. [156] used the algorithm to derive a general census of repeats in proteins using the Swiss-Prot protein sequence database.

The method Radar [157] basically follows the algorithmic steps of the REPRO method [153]. It calculates nonintersecting local alignments and then uses these in an iterative procedure to determine the shortest nonreducible repeat unit and the associated boundaries. A profile is constructed from a multiple alignment of a found repeat set and is slid over the query sequence to capture more repeats. The whole procedure is then iterated in an attempt to find multiple repeat types. The Radar step to find the shortest possible repeat unit includes (1) an iterative wraparound DP algorithm that mimics sliding a repeat profile over the query sequence to find more incarnations followed by updating and sliding the profile in the next iteration and (2) a recursive procedure to detect the smallest repeat unit within a potentially reducible set of repeats delineated by the wraparound DP scenario. The Radar method is sensitive and sufficiently fast for genomic application.

1.5. Profile Analysis and Homology Searching

1.5.1. Modes for Calculating Profiles

A natural extension of the motif-searching techniques is provided by methods that use information over an entire sequence alignment of a certain protein family to find additional related family members. The earliest conceptually clear technique of this kind of sequence searching was called profile analysis [59]; it combines a full representation of a sequence alignment with a sensitive searching algorithm. The procedure takes as input a multiple alignment of N sequences. First, a profile is constructed from the alignment; i.e., an alignment-specific scoring table that comprises the likelihood of each residue type to occur in each position of the multiple alignment. A typical profile has $L(20 + 2)$ elements, where L is the total length of the alignment, 20 is the number of amino acid types, and the last two columns contain gap penalties (see below). As a measure of similarity between different types of residues, one of the residue exchange matrices described in Sec. 1.3 is used. Then each element of the profile is calculated for each alignment position r and residue type c as

$$\text{Profile}(r, c) = \sum_{d=1}^{M} W_{d,r} \, \text{Comp}(\text{Residue}_d, \text{Residue}_c) \tag{7}$$

where $M = 20$ is the number of amino acid types. Comp is the comparison value or substitution weight between the residue type c and each possible type of residues d. $W_{d,r}$ is the weight, which depends on the number of times each residue type occurs in position r of the alignment. The two commonly used weighting schemes are linear,

$$W_{d,r} = \frac{\sum_{i=1}^{N} W_i * \delta_d}{\sum_{i=1}^{N} W_i}, \qquad \delta_d = \begin{cases} 1 & \text{if Residue}_{i,r} = \text{Residue}_d \\ 0 & \text{if Residue}_{i,r} \neq \text{Residue}_d \end{cases} \qquad (8)$$

and logarithmic,

$$W_{d,r} = \frac{\ln\left[\dfrac{1 - \sum_{i=1}^{N} w_i * \delta_d}{1 + \sum_{i=1}^{N} w_i}\right]}{\ln\left[\left(1 + \sum_{i=1}^{N} w_i\right)^{-1}\right]} \qquad (9)$$

where w_i is a weight assigned to each sequence in the alignment, usually 1.0.

Linear weighting simply reflects the fraction of each residue type at position r, whereas logarithmic weighting upweights the most frequent residue types. For any weighting scheme, $W_{d,r} = 0$ or 1 if a certain type of residue d does not occur or occurs exclusively in position r, respectively. An example of a profile after Gribskov et al. with linear weighting is provided in Figure 8. Gribskov et al. [59] used a single extra column in the profile to describe the local weight for both the gap opening and the gap extension penalty. For alignment positions not containing gaps, $P_{open} = P_{extend} = 100$, whereas for positions with insertions/deletions these values are multiplied by the weighting factor

$$G_{\max}/(1.0 + G_{\text{inc}} L_{\text{gap}}) \qquad (10)$$

where $G_{\max}$ is the maximum possible multiplier for an alignment position containing gaps, G_{inc} scales the decrease of this quantity as the observed gap length grows, and L_{gap} is the length of the gap crossing a given alignment position. The advantage of such positional gap penalties is that multiple alignment regions with gaps (loop regions) will be assigned lower gap penalties, and hence will be more likely than core regions to attract gaps in a target sequence during profile searching, consistent with structural considerations. However, the implementation by Gribskov et al. [59] does not take the frequency of gaps at each alignment position into account for the estimation of gap opening and/or extension penalties. This does not correspond with the expectation that a position rich in gaps would correspond to a

Pos	Alignment	Cons	A	C	D	E	F	G	H	I	K	L	M	N	P	Q	R	S	T	V	W	Y	Gap
1	GM..VVV...	V	0	-28	-16	-16	-16	-2	-20	22	-16	12	18	-16	-12	-16	-18	-8	-2	26	-58	-26	22
2	VGG.HNHVG.	G	2	-46	6	-2	-50	18	8	-14	-18	-30	-22	4	-12	0	-18	0	-4	-2	-86	-42	22
3	LFL.FFLLLL	L	-18	-96	-84	-66	78	-78	-36	30	-66	84	48	-60	-66	-54	-60	-54	-42	18	-24	30	22
4	TTS.TTTSTS	T	18	-24	0	0	-54	6	-18	-6	0	-42	-24	6	6	-18	-12	24	42	-4	-72	-54	22
5	DAD.AAPPTA	A	22	-52	12	8	-82	8	-4	-24	-12	-48	-30	4	28	6	-22	14	14	-12	-110	-66	22
6	VDGHEREGAD	E	6	-76	36	54	-88	20	8	-30	-14	-54	-38	14	-14	18	-26	2	0	-20	-128	-70	100
7	QQEDEBEDQQ	E	0	-100	56	60	-104	-12	34	-40	8	-56	-36	24	-12	56	-4	-8	-8	-40	-134	-80	100
8	VEWAKKKKKI	K	-20	-82	-20	-18	-56	-12	-18	-4	20	-26	-2	-10	-30	-10	10	-8	-8	-8	-48	-54	100
9	AAQEATSSKS	A	18	-46	10	12	-78	6	-8	-24	6	-30	-26	12	8	6	-8	18	16	-18	-66	-66	100
10	LLLLALANAT	L	-2	-84	-36	-28	-18	-54	-24	10	-34	34	28	-26	-26	-20	-44	-20	-8	16	-74	-38	100
11	VVVVIVIVIV	V	-112	-88	-40	-40	-4	-36	-36	88	-40	40	40	-4	-28	-40	-40	-20	0	80	-112	-32	100
12	KNLLINTKQQ	K	-8	-88	0	0	-60	-28	8	-16	16	-16	0	-4	-20	12	-14	-8	0	-16	-56	-56	100
13	SSNKSGAADA	S	4	-46	14	8	-82	20	-10	-30	2	-58	-34	16	4	0	-14	22	12	-20	-90	-66	100
14	SSVCTLLAHS	S	-8	-18	-28	-24	-58	-14	-14	-14	-28	-14	-10	-14	-12	-24	-26	0	0	2	-76	-38	100
15	FWWNWWWWWF	W	-20	-164	-136	-132	56	-132	-56	-76	-68	-24	-64	-80	-116	-100	16	-44	-92	-100	272	28	100
16	........F.	F	-2	-8	-12	-10	18	-10	-4	2	-10	-4	-28	-8	-10	-4	-8	-6	-6	-2	0	14	22
17	EEGGRSGGL.	G	16	-66	12	10	-72	30	-18	-34	-12	-44	-28	2	-18	-4	-28	6	-2	-18	-98	-72	22
18	ESKGLKKKN.	K	-6	-76	4	4	-50	-12	-4	-28	32	-34	-10	10	-18	6	10	2	-2	-24	-68	-62	22
19	FFVVVVVII.	V	-16	-44	-52	-48	50	-42	-36	64	-48	36	28	-44	-38	-48	-44	-26	-12	52	-80	4	22
20	NKEEDNNGK.	N	10	-80	54	28	-86	4	14	-38	22	-58	-32	24	-18	20	0	-8	0	-42	-92	-42	22

FIGURE 8 Example of a profile for the first 20 positions of a multiple alignment of 10 globin sequences. The first column contains the multiple alignment position, the second and third columns, respectively, the aligned amino acids and the consensus amino acid for each alignment position. Then follow 20 columns with the propensities for each amino acid obtained by using linear weighting [Eq. (8)] and amino acid exchange values from the PAM250 matrix rescaled by multiplying each value by 2 [Eq. (7)]. For example, the propensity for glycine (G) at alignment position 1 is 1*Comp(G, G) + 1*Comp(G, M) + 3*Comp(G, V) = 1*10 + 1*-6 + 3*-2 = -2. A single position-dependent gap score is given, which is calculated using Eq. (10) with parameters $G_{max} = 33.3$ and $G_{inc} = 0.1$. For example, alignment positions 1–5 are spanned by a single gap, so that for these positions $P_{open} = P_{extend} = 33.3/(1.0 + 0.1 \times 5) = 22$. (From Ref. 59.)

loop site such that gaps in a sequence matched with the profile should be accommodated more easily at this position. Many alternative profile implementations therefore reserve the two last columns of the profile for positional gap opening (P_{open}) and gap extension (P_{extend}) penalties, which can be individually determined using protocols that take the above considerations into account. Another issue with the Gribskov et al. implementation is that an alignment column with, for example, a single glycine and gaps for all other sequences would show the same score for glycine as a column consisting of identically conserved glycine residues.

1.5.2. Profile Search Methods

In the Gribskov et al. [59] approach, a profile is aligned with each sequence in the databank by means of the Smith and Waterman [51] dynamic programming procedure (described in Sec. 1.1.2), which finds the best local alignment path in a search matrix where appropriate profile values are placed into a comparison matrix cell corresponding to each residue in the database sequence and each alignment (or profile) position. Each match between a databank residue and a profile position receives the profile propensity for the databank residue type as a score. For each database sequence, the alignment score corresponding to the best local alignment quantifies the degree of similarity of this sequence with the probe profile. The scores are then corrected for sequence length, represented in the form of Z-scores, and ranked to create the final list of databank search hits. Top-scoring sequences with scores above some threshold level are then likely to be related to the multiply aligned sequences used to build the profile. In addition to aligning a single sequence to a profile, it is also possible to align two profiles. In this case two matched profile positions receive a score by summing over the 20 residue types the products of the corresponding propensities from the two profiles.

Another sequence searching technique based on flexible protein sequence patterns was introduced by Barton and Sternberg [158]. Significant residue positions are selected on the basis of sequence conservation, functional importance, or the presence of secondary structure. These residues, constituting the pattern, can be separated by gaps that serve to exclude variable regions from the analysis. For each gap, minimal and maximal possible lengths are derived from the initial sequence set. A lookup table similar to a profile is then calculated, which results in scores to compare each element of the pattern with each residue type. This feature distinguishes the method from regular expression pattern matching algorithms based on positional match sets, which essentially use a binary exchange matrix. The flexible pattern is subsequently compared to every databank sequence using a modified Needleman–Wunsch technique [13] for alignment (see Sec. 1.1.2). Because only partial information contained in the protein family is used, which represents the most essential structural features, the method has high discriminating power. It is especially recommended for sequence alignments in which crucial elements are separated by long noisy stretches.

A technique with somewhat inverted logic compared to that of the profile-related methods was described by Altschul and Lipman [127]. Rather than searching individual databank sequences with the highest degree of similarity to a target set of sequences, their algorithm uses a single query sequence and attempts to find consistent sets of similar sequences in the databank. Similarity to the query sequence is determined by means of alignment of ungapped sequence segments with high individual scores.

1.5.3. Sequence Weighting

Altschul et al. [41] and Vingron and Argos [48] proposed sequence weighting as a means to deal with the fact that the complete set of natural sequences belonging to a certain protein family or superfamily is normally unequally represented in a given set of input sequences. This will lead to an amplification of the overrepresented sequences in profile searching, which, if such newly found sequences are added to the profile for further searching, will cause a more unbalanced representation. To address this problem, in sequence weighting a weight is assigned for each sequence before any average value is calculated. This means that if the alignment contains sequences that are redundant according to a preconceived criterion, then information derived from those sequences will be downweighted in the profile. The Altschul et al. weighting scheme is integrated in the multiple alignment method MSA of Lipman et al. [40] and derives the sequence weights from a rooted tree. The idea of this scheme is to weight the influence of each external node on the root distribution of the tree. The Vingron–Argos method simply weights each sequence on the basis of its average distance from all the other sequences and therefore does not need any tree-based information. Interestingly, the Altschul et al. method upweights sequences close to the root of the phylogenetic tree, because those would contribute more information to the central point of interest, whereas the Vingron–Argos method (as well as the other published techniques) upweight distant sequences. Sibbald and Argos [159] also upweighted outlier sequences by employing Voronoi cell volumes to assign individual weights to a set of sequences. Their technique does not rely on a phylogenetic tree but is dependent on sampling the sequence space in order to obtain the weights.

Thompson et al. [63] derived the sequence weights of sequences in a profile directly from the branch lengths of a phylogenetic tree constructed with the neighbor-joining technique of Saitou and Nei [55]. They used these weights during progressive alignment in the construction of profiles representing prealigned sequence groups. Independently, Lüthy et al. [160] mod-

ified the profile search technique by employing Voronoi-based weighting
[159]. Both techniques make use of the BLOSUM62 matrix, constructed from
ungapped alignment blocks [90]. In line with using the BLOSUM62 matrix
(derived from blocks of ungapped alignments), Thompson et al. exclude from
analysis all alignment positions with a percentage of gaps higher than a
certain specified threshold. Such regions would be expected to consitute loop
regions in the associated protein structures showing less consistent amino acid
conservation patterns.

Although in principle it is a good idea to upweight more distant
sequences because they should carry more information for each alignment
position, this is warranted only if the alignment is evaluated for correctness.
When sequence weighting is used in progressive multiple alignment, the
increased chance of mistakes in aligning distant sequences often leads to the
amplification of misinformation. To quantify this effect, Vogt et al. [97]
compared local and global alignments of pairwise sequences with a databank
of structure-based alignments [98] and included a large set of substitution
matrices with optimized gap penalties. The highest-scoring combination of
global alignment with the Gonnet residue exchange matrix [89], with added
constant to make all values nonnegative, on average showed 15% incorrect
residue matching when sequences with 30% residue identity were aligned; this
error rate quickly increased to 45% incorrect matches at 20% residue identity
of the aligned sequences and to 73% error at 15% sequence identity. These
statistics clearly show the risk of upweighting the importance of more distant
sequence alignments, particularly given that incorrect alignments resulting
from the multiple minima problem typically yield a low score, so that the
sequences involved would appear to be more distant than they actually are.

In addition to global sequence weighting, weighting of individual
alignment positions was also proposed. The deletion of gapped alignment
positions in the aforementioned techniques of Barton and Sternberg [158] and
Thompson et al. [63] (see preceding section) is a first approach to positional
sequence weighting. Sunyaev et al. [161,162] developed a weighting scenario
reminiscent to phylogenetic parsimony methods. They weighted the amino
acid likelihoods at each alignment position according to the probability that
identical amino acids occur in more than one sequence at the alignment
position. The idea is that if more alignment positions show identical conservation for a given subset of sequences (not necessarily the same conserved
amino acid type over the alignment positions involved), the occurrence of the
amino acids at those positions becomes more expected, which is corrected for
by appropriately lowering the weight for the considered position. This
approach leads to position-specific sequence weights that are calculated in
the absence of phylogenetic trees and can be implemented in the position-
specific probabilities for each type of amino acid.

Other avenues to deal with the underrepresentation of amino acids at alignment positions include pseudo-count approaches [163–166], all effectively extrapolating the numbers of amino acids at each alignment position based on amino acid exchange probabilities. Henikoff and Henikoff [167] proposed an embedment strategy to add extra information from a multiple alignment to local regions within a representative sequence. They represented reliably aligned regions by a position-specific scoring matrix (PSSM), while single sequence information was retained for uncertain alignment regions. The approach is adopted in the popular PSI-BLAST search engine [168]. Following Henikoff and Henikoff [167], Brocchieri and Karlin [74] went so far as to edit individual sequences by replacing amino acids by those that are preponderant at a corresponding position in a multiple alignment. Although aimed at achieving better recognition of crucial alignment regions during alignment, the latter approach can lead to error progression in the case of misalignment.

1.5.4. HMM-Based Profile Searching

Baldi et al. [34] and Krogh et al. [35] used hidden Markov models (HMMs) to represent an aligned block of sequences, thus extending the definition of profiles with position specific amino acid exchange values. HMMs address the problems described above for the classical profile approach of Gribskov et al. [59]. A typical profile HMM consists of a chain of match, insert, and delete nodes, with all transitions between nodes and all character costs in the insert and match nodes trained to specific probabilities. A query sequence can then be aligned to the model, usually by using a standard dynamic programming algorithm so that the sum of the probabilities is maximized. For a comprehensive account on HMM methods in sequence analysis, see Durbin et al. [169]. The difference between the methods of Baldi et al. [34] and Krogh et al. [35] lies in the estimation of the probability parameters, which in the Baldi et al. method is done using gradient descent, whereas Krogh et al. used expectation maximization (EM) techniques. Extensive libraries of HMMs for protein domains are deposited in the PFAM database [121] and PROSITE profiles database.

Profile searching using HMMs is currently one of the most sensitive search techniques. In contrast to HMMs applied to multiple sequence alignment, estimating the parameters for an HMM of an established multiple alignment is normally successful. The chance to end up in a local trap in parameter space is limited compared to that in HMM-based multiple sequence alignment.

Bucher et al. [33] unified the profile, motif, and HMM approaches through extension of the profile definition with regular expression-like patterns, weight matrices, and HMMs. They proved that their generalized profiles are equivalent to certain types of HMMs. The generalized profiles have been used to extend the PROSITE protein motif database [33], which in its basic form is a library of regular expressions. The profile syntax enables the emulation of most common motif search techniques, such as direct searching for PROSITE patterns, searching for patterns without gaps [138], searching using the profile definition of Gribskov et al. [59], flexible pattern searches [158], searching using the Viterbi algorithm for HMMs [35], and domain and fragment searches using the HMMER method [170]. Owing to advances in computational performance, procedures for sequence database homology searching have been developed, such as the HMM tools SAM-T98 [171] and HMMER2 [172] (see Sec. 1.6.4).

1.6. Sequence Databank Searching

Until recently, Needleman–Wunsch–type dynamic programming techniques (see Sec. 1.1.2) were computationally too expensive for the comparison of a large number of protein sequences, such as those contained in the PIR or Swiss-Prot databases. The total number of annotated sequences deposited in these databases is presently greater than 80,000. However, for any biologist who has a new protein sequence of unknown functionality, the comparison with all known and annotated sequences is paramount. Therefore, fast routines have been devised that enable database searches on even small computers with only a small loss of sensitivity compared to searches using full dynamic programming. With the recent advent of parallel multiprocessor computers at central sites, researchers can routinely perform multiple sequence searches over complete sequence databases.

1.6.1. FASTA

Until recently, the most widely used quick routine was FASTA [29]. The FASTA program compares a given query sequence with a library of sequences and calculates for each pair the highest scoring local alignment. The speed of the algorithm is obtained by delaying application of the dynamic programming technique to the moment where the most similar segments are already identified by faster and less sensitive techniques. To accomplish this, the FASTA routine operates in four steps. The first step searches for identical words of a user-specified length occurring in the query sequence and the target sequence(s). The technique is based on that of Wilbur and Lipman [49,61] and involves searching for identical words (k-tuples) of a certain size within a specified bandwidth along search matrix diagonals. For not-too-distant sequences (> 35% residue identity), little sensitivity is lost while speed is greatly increased. The search is performed by hashing techniques, where a lookup table is constructed for all words in the query sequence, which is then used to compare all words encountered in the target sequence(s). Generally, for proteins a word length of two residues is sufficient ($ktup$ = 2). Searching with higher $ktup$ values increases the speed but also the risk that similar regions will be missed. For each target sequence the 10 regions with the highest density of ungapped common words are determined. In the second step, these 10 regions are rescored using the Dayhoff PAM250 residue exchange matrix [14], and the best scoring region of the 10 is reported under *init1* in the FASTA output. In the third step, regions scoring higher than a threshold value and sufficiently near each other in the sequence are joined, now allowing gaps. The highest score of these new fragments can be found under *initn* in the FASTA output. The fourth and final step performs a full dynamic programming alignment [172] over the final region, which is widened by 32 residues at either side, and the score is written under *opt* in the FASTA output.

1.6.2. BLAST

Another speed-optimized and widely used algorithm that maintains significant sensitivity is the program BLAST (Basic Local Alignment Search Tool) [96], which is based on an exhaustive statistical analysis of ungapped alignments [173]. Basically, BLAST generates a list of all tripeptides from a query sequence and for each of those derives a table of tripeptides that are deemed similar; the number of similar tripeptides is only a fraction of the total number possible. The BLAST program quickly scans a database of protein sequences for regions showing high similarity by using the tables of similar peptides. The BLAST algorithm also provides a rigorous statistical framework, based on the extreme value theorem, to estimate the statistical significance of tentative homologs. The E value given for each sequence found indicates the expected number of sequences with an alignment score equal to or greater than that of the sequence considered. The original BLAST program could detect only local alignments without gaps and therefore might miss some significant similarities. A more recent version of the BLAST algorithm is able to insert gaps in the alignments, which leads to greater sensitivity [168]. The original statistical framework for ungapped alignments is used to assess the significance of the gapped alignments, although no mathematical proof for this is available yet [174]. However, computer simulations have indicated that the theory probably applies to gapped alignments as well. The most recent development for the BLAST engine is position-specific iterated BLAST (PSI-BLAST) [168], which exploits the increase in sensitivity offered by multiple alignments and derived profiles in an iterative fashion. The program initially operates on a single query sequence by performing a gapped BLAST search. Then it takes the significant local alignments found, constructs a multiple alignment, and abstracts a position-specific scoring matrix (PSSM) from this alignment. This is a type of profile (see Sec. 1.5) that is used to rescan the database in a second round aimed at finding more sequences. The scenario is iterated until the user decides to stop or the search has converged, i.e., no more significantly scoring sequences can be found in subsequent iterations. An example of a PSI-BLAST run is provided in Figure 9. The web server for PSI-BLAST, located at http://www.ncbi.nlm.nih.gov/BLAST, enables the user to specify at each iteration round which sequences should be included in the profile; by default, all sequences are included that score beyond a user-set E value (Figs. 9b and 9d). However, the user needs to activate every

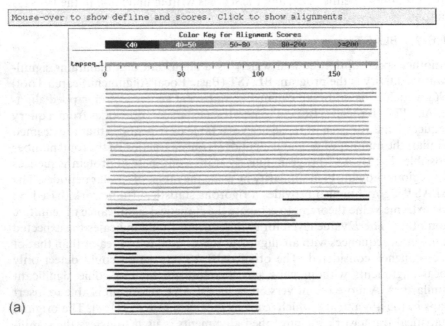

(a)

FIGURE 9 An example of a PSI-BLAST database query and output when the PSI-BLAST web server is used. As a query sequence, the primary structure of the flavodoxin with Swiss-Prot code FLAV_ENTAG was taken. (a) Stacking of the 69 significantly scoring local alignments onto the query sequence. It can be seen that some of the local alignments correspond only with a short query sequence stretch. (b) Listing of the top-scoring sequences with associated random probabilities (E scores). (c) Local alignment of the bottom five sequences in (b). (d) Lowest significantly scoring sequences after three complete PSI_BLAST iterations (convergence of the search has not yet occurred). Newly found sequences after each iteration are marked.

Sequences with E-value BETTER than threshold

Sequences producing significant alignments:		Score (bits)	E Value
sp|P28579|FLAV_ENTAG	FLAVODOXIN >gi|95560|pir||A39414 flavodo...	360	4e-99
sp|P71169|FLAW_ENTAG	FLAVODOXIN >gi|1497840|emb|CAA57484| (X8...	352	1e-96
sp|P04668|FLAW_KLEPN	FLAVODOXIN >gi|78953|pir||S02511 flavodo...	232	2e-60
sp|P52967|FLAV_RHOCA	FLAVODOXIN 2 >gi|987626 (L42290) niff [R...	178	3e-44
sp|P00324|FLAV_AZOVI	FLAVODOXIN 2 >gi|625194|pir||FXAVEP flav...	171	2e-42
sp|P23001|FLAV_AZOCH	FLAVODOXIN B (FLDB) >gi|398014 (M73019) ...	171	3e-42
prf||752055A	flavodoxin [Azotobacter vinelandii]	168	2e-41
sp|O07026|FLAV_KLEPN	FLAVODOXIN >gi|2289914 (U67169) flavodox...	143	1e-33
sp|P23243|FLAV_ECOLI	FLAVODOXIN 1 >gi|95841|pir||A37319 flavo...	143	1e-33
pdb|1AG9|A	Chain A, Flavodoxins That Are Required For Enzyme ...	141	5e-33
sp|P44562|FLAV_HAEIN	FLAVODOXIN >gi|1074088|pir||C64053 flavo...	138	3e-32
sp|P27319|FLAV_SYNY3	FLAVODOXIN >gi|481443|pir||S38632 flavod...	133	1e-30
sp|P10340|FLAV_SYNP7	FLAVODOXIN >gi|79632|pir||A28670 flavodo...	129	1e-29
pdb|1OFV|	Flavodoxin (Oxidized Form)	127	6e-29
pir||S18374	flavodoxin - Anabaena sp. (PCC 7119) (fragment) >...	125	2e-28
pdb|1FLV|	Flavodoxin >gi|999876|pdb|1RCF| Flavodoxin Compl...	125	2e-28
sp|P11241|FLAV_ANASP	FLAVODOXIN >gi|79771|pir||S04600 flavodo...	125	2e-28
sp|O25776|FLAV_HELPY	FLAVODOXIN >gi|2314319|gb|AAD08207.1| (A...	123	9e-28
sp|P31158|FLAV_SYNP2	FLAVODOXIN >gi|477721|pir||B47673 flavod...	123	9e-28
gi|4155685	(AE001536) Flavodoxin [Helicobacter pylori J99]	122	2e-27
gi|887845	(U28375) ORF_o173 [Escherichia coli] >gi|1789262 (A...	114	4e-25
sp|P41050|FLAW_ECOLI	FLAVODOXIN 2 >gi|1073381|pir||S52315 fla...	114	4e-25
gi|2935176	(AF021093) flavodoxin A [Helicobacter pylori]	113	7e-25
sp|P14070|FLAV_CHOCR	FLAVODOXIN >gi|81145|pir||S06648 flavodo...	99	2e-20
sp|P56268|FLAW_KLEOX	FLAVODOXIN 2 >gi|216725|dbj|BAA00243| (D00...	95	4e-19
sp|O52659|FLAV_TRIER	FLAVODOXIN 2 >gi|2865512 (AF044318) flavod...	93	1e-18
prf||0903247A	promoter nifLA [Klebsiella pneumoniae]	54	8e-07
gi|149271	(J01743) nifF [Klebsiella pneumoniae]	53	2e-06
gi|149251	(M24106) nifF [Klebsiella pneumoniae]	52	4e-06
sp|O34589|FLAW_BACSU	PROBABLE FLAVODOXIN 2 >gi|2632237|emb|CA...	49	3e-05
emb|CAA55301.1|	(X78558) nifF [Pantoea agglomerans]	45	4e-04

(b)

FIGURE 9   Continued.

subsequent iteration. An alternative to the PSI-BLAST web server is a stand-alone version of the program, downloadable from the aforementioned WWW address, that allows the user to specify beforehand the desired number of iterations. Although a consistent and very powerful tool, a limitation of the PSI-BLAST engine is that the statistics do not take compositional biases into account. Biased amino acid compositions might well confuse the algorithm and lead to a buildup of error in subsequent iterations. This is significant for cross-genome comparison, because it is becoming increasingly clear from genome sequencing efforts that large overall compositional differences exist between different genomes. Another matter of debate is the way in which the PSSM is generated, because this is essentially a simple stacking of the found local regions onto the query sequence used as a template (*N*-to-1 alignment), without keeping track of cross-similarities between the added regions. Because there are also no safeguards to control the number of sequences added to the PSSM at each iterative step, where all sequences having an

```
prf||0903247A promoter nifLA [Klebsiella pneumoniae]
             Length = 81

Score = 53.9 bits (127), Expect = 8e-07
Identities = 23/36 (63%), Positives = 32/36 (88%)

Query: 1   MATIGIFFGSDTGQTRKVAKLIHQKLDGIADAPLDV 36
           MA IGIFF +DTG+TRK+AK+IH++L  +ADAPL++
Sbjct: 1   MANIGIPFATDTGKTRKIAKMIHKQLGELADAPLNM 36


gi|149271 (J01743) niff [Klebsiella pneumoniae]
          Length = 34

Score = 52.7 bits (124), Expect = 2e-06
Identities = 22/34 (64%), Positives = 31/34 (90%)

Query: 2   ATIGIFFGSDTGQTRKVAKLIHQKLDGIADAPLD 35
           A IGIFFG+DTG+TRK+AK+IH++L  +ADAP++
Sbjct: 1   ANIGIFFGTDTGKTRKIAKMIHKQLGELADAPVN 34


gi|149251 (M24106) niff [Klebsiella pneumoniae]
          Length = 43

Score = 51.5 bits (121), Expect = 4e-06
Identities = 26/44 (59%), Positives = 34/44 (77%), Gaps = 1/44 (2%)

Query: 134 KFSFSAALLENNEFVGLPLDQENQYDLTEERIDSWLEKLKPAVL 177
           +FS S+AL E + FVGL LDQ+NQ+D TE R+ SWLE++K  VL
Sbjct: 1   EFSASSAL-EGDRFVGLVLDQDNQFDQTEARLASWLEEIKRTVL 43


sp|O34589|FLAW_BACSU PROBABLE FLAVODOXIN 2 >gi|2632237|emb|CAA10879| (AJ222587) YkuP
                     protein [Bacillus subtilis] >gi|2633788|emb|CAB13290|
                     (Z99111) similar to sulfite reductase [Bacillus
                     subtilis]
                     Length = 178

Score = 48.8 bits (114), Expect = 3e-05
Identities = 33/104 (31%), Positives = 52/104 (49%), Gaps = 17/104 (16%)

Query: 1   MATIGIFFGSDTGQTRKVAKLIHQKLDGIADAPLDVRRATR------EQFLSYPVLLLGT 54
           MA I + + +G T +A LI +  G+ +A +V R        + F Y +++GT
Sbjct: 1   MAKILLVYATMSGNTEAMADLIEK---GLQEALAEVDRFEAMDIDDAQLFTDYDHVIMGT 57

Query: 55  PTLGDGELPGVEAGSQYDSWQEFTNTLSEADLTGKTVALFGLGD 98
            T GDG+LP        D + +  + E D +GKT A+FG GD
Sbjct: 58  YTWGDGDLP--------DEFLDLVEDMEEIDFSGKTCAVFGSGD 93


emb|CAA55301.1| (X78558) niff [Pantoea agglomerans]
               Length = 21

Score = 44.9 bits (104), Expect = 4e-04
Identities = 21/21 (100%), Positives = 21/21 (100%)

Query: 1   MATIGIFFGSDTGQTRKVAKL 21
           MATIGIFFGSDTGQTRKVAKL
Sbjct: 1   MATIGIFFGSDTGQTRKVAKL 21
```

(c)

FIGURE 9 Continued.

expectation value (or E-value) lower than a preset threshold are selected, it is clear that erroneous alignments are likely to drive the engine to include false positives. However, Jones [175] found that PSSMs made by PSI-BLAST are well suited to serve as a basis for accurate secondary structure prediction (see Sec. 2.1).

1.6.3. Fast Smith–Waterman Local Alignment Searches

Collins and Coulson [176] devised a computer protocol to perform database searches based on an implementation of the full Smith and Waterman

sp|P00322|FLAV_CLOBE	FLAVODOXIN >gi|65882|pir||FXCLEK flavodoxin...	_87_	1e-16
sp|P00321|FLAV_MEGEL	FLAVODOXIN >gi|65881|pir||FXMB flavodoxin -...	_86_	2e-16
gi	3822405	(AF088999) inducible nitric oxide synthase; iNOS [Sal...	_86_	2e-16
pdb|1FVX|	Clostridium Beijerinckii Flavodoxin Mutant: G57n Oxi...	_85_	3e-16
pdb|5NUL|	Clostridium Beijerinckii Flavodoxin Mutant: G57t Sem...	_84_	5e-16
pdb|4NLL|	Clostridium Beijerinckii Flavodoxin Mutant: G57d Oxi...	_84_	7e-16
pdb|4NUL|	Clostridium Beijerinckii Flavodoxin Mutant: D58p Oxi...	_84_	9e-16
pdb|3NLL|	Clostridium Beijerinckii Flavodoxin Mutant: G57a Oxi...	_84_	9e-16
prf||711671A	flavodoxin [Peptostreptococcus elsdenii]	_80_	1e-14
sp|Q57746|Y298_METJA	HYPOTHETICAL PROTEIN MJ0298 >gi|2128223|pir...	_77_	9e-14
sp|P56268|FLAW_KLROX	FLAVODOXIN >gi|216725|dbj|BAA00243| (D00339...	_76_	2e-13
emb|CAB36512.2|	(AL035439) putative NADPH cytochrome reductase [...	_71_	5e-12
sp|P55887|FLAW_SALTY	FLAVODOXIN 2 >gi|1916334 (U92524) putative ...	_70_	2e-11
dbj|BAA87084.1|	(AB027780) NADPH-cytochrome p450 reductase [Schi...	_66_	2e-10
gi	2621938	(AE000861) unknown [Methanobacterium thermoautotrophi...	_66_	2e-10
gb|AAD39340.1|	(AF068681) endothelial nitric oxide synthase [Can...	_64_	9e-10
gi|882603	(U29579) ORF_o479 [Escherichia coli] >gi|1789064 (AE00...	_62_	3e-09
dbj|BAA05933|	(D28595) unknown [Escherichia coli]	_62_	3e-09
sp|Q62600|NOS3_RAT	NITRIC-OXIDE SYNTHASE, ENDOTHELIAL (EC-NOS) (...	_59_	2e-08
pir||S37156	NADPH--ferrihemoprotein reductase (EC 1.6.2.4) - Jer...	_58_	7e-08
gb|AAD42410.1|AF157493_18	(AF157493) trp repressor binding prote...	_56_	2e-07
gb|AAC09468.1|	(AF002698) putative NADPH-cytochrome P450 reducta...	_53_	1e-06
gi	149251	(M24106) nifF [Klebsiella pneumoniae]	_52_	2e-06
pir||A61338	flavodoxin - Clostridium pasteurianum (fragment)	_52_	3e-06
sp|Q57954|Y534_METJA	HYPOTHETICAL PROTEIN MJ0534 >gi|2826294 (U6...	_49_	3e-05
pir||F64366	flavoprotein - Methanococcus jannaschii	_49_	3e-05
prf||0903247A	promoter nifLA [Klebsiella pneumoniae]	_48_	4e-05
gb|AAF12546.1|AE001826_15	(AE001826) ribonucleotide reductase, N...	_47_	7e-05
gi	2621267	(AE000809) flavoprotein A homolog (II) [Methanobacter...	_47_	1e-04
gi	1914879	(L23871) flavodoxin [Klebsiella pneumoniae]	_46_	2e-04
sp|P30849|WRBA_ECOLI	TRP REPRESSOR BINDING PROTEIN >gi|1787239 (...	_46_	2e-04
pir||I59246	trp repressor binding protein - Escherichia coli >gi...	_46_	2e-04
gi	687617	(U17835) flavoprotein [Methanobacterium thermoautotrop...	_45_	4e-04
sp|P27863|HEMG_ECOLI	PROTOPORPHYRINOGEN OXIDASE (PPO) >gi|107358...	_45_	4e-04
gi	148250	(M87049) o181 [Escherichia coli]	_45_	4e-04
gi	2622931	(AE000934) unknown [Methanobacterium thermoautotrophi...	_45_	5e-04
dbj|BAA16728|	(D90900) potential FMN-protein [Synechocystis sp.]	_44_	6e-04

(d)

FIGURE 9 Continued.

[25] local alignment technique (see Sec. 1.1). They implemented the MPsrch protocol [176] on massively parallel computers with SIMD (single instruction multiple data) type processors, named the BLITZ server. Following Collins and Coulsen [176], a number of implementations that enabled fast Smith–Waterman-based local searches have arisen. One of the central computer sites where such programs are running is the European Bioinformatics Institute (EBI) outstation of the European Molecular Biology Laboratory, where they are integrated in a web server. Available are

Compugen's implementation of the BLITZ server http://www2.ebi.ac.uk/ bic_s) and a fast heuristic implementation of the true Smith–Waterman algorithm (Scanps) of Barton [177], both allowing users to perform database queries via the worldwide web. The output of a query is a list of top-scoring local alignments (one per protein) where statistical significance measures are also given based on the mean value and standard deviation of the distribution of scores over the entire database [176]. The speed of the techniques allows several PAM exchange weight matrices (based on different evolutionary distances; see Sec. 1.3) to be used in searching the databanks with the same query sequence. A widely used implementation of the full Smith–Waterman algorithm is the SSEARCH method [29], which comes with the same rigid statistical procedures as those employed in the PSI-BLAST engine.

1.6.4. HMM-Based Database Search Engines

Owing to advances in computational performance, procedures for sequence database homology searching have been developed that are based on more computationally intensive formalisms such as the HMM tools SAM-T98 [171] and HMMER2 [172] (see Sec. 1.5.4). Although SAM-T98 was assessed by its authors as being superior to FASTA [29], PSI-BLAST [168], and BLAST [96], other assessments are less conclusive (e.g., Shi et al. [178]). The SAM-T98 method incorporates reversed-model score adjustment to correct for length, composition bias, and other effects related to secondary structure, amphipathicity, and the like.

2. PREDICTING SECONDARY STRUCTURAL FEATURES FROM PROTEIN SEQUENCES

Protein structure is hierarchical in its internal organization. At higher levels within this hierarchy, especially for structural domains or higher order structures, the connectivity of the polypeptide backbone between substructures is flexibly maintained. This means, for example, that a protein can form a stable structure irrespective of the sequential arrangement of its constituent domains [23].

For protein secondary structure, however, the elements are crucially context-dependent; i.e., they rely critically on other secondary structure elements in their environment. In general, about 50% of the amino acids within known protein structures fold into α-helices or β-strands, so that roughly half of the protein structures are regularly shaped. The reason for the regularity observed for helices and strands is the inherent polarity of the protein backbone, which contributes a polar nitrogen and oxygen atom for each amino acid. To satisfy energy constraints, main-chain regions buried in the internal protein core need to form hydrogen bonds between those polar atoms. The α-helix and β-strand conformations are optimal, because each main-chain nitrogen atom within these conformations can in principle associate with an oxygen partner (and *vice versa*). In order to satisfy the hydrogen-bonding constraints, β-strands need to interact with other β-strands, which they can do in a parallel or antiparallel fashion, to form a β-pleated sheet. β-Strands thus rely on crucial long-range interactions between residues remote in sequence. They are therefore more context-dependent than α-helices, which are more able to fold "on their own." The fact that prediction methods typically have the greatest difficulty in delineating β-strands correctly is believed to be due to this context dependency.

2.1. Properties of Secondary Structure

In addition to differing context dependencies, there are also general compositional differences between helix, strand, and coil conformations, and this is the signal used in many of the prediction methods for single sequences. Methods that rely on multiple alignments can also exploit the fact that the amino acid exchange patterns are different for the three secondary structure states [99]. A number of more specific observations for secondary structures, as are found in the large collection of protein structures deposited in the Protein Data Bank (PDB) [179], can be used to recognize secondary structures. For each of the secondary structures α-helix, β-strand, and coil (loop) these can be summarized as follows.

The number of residues per turn is 3.6 in the ideal case, and helices are often positioned to shield a buried protein core, so they often have one phase contacting hydrophobic amino acids while the other phase interacts with the solvent. Such helices are therefore amphipathic [180] and have a hydrophobic phase and a hydrophilic phase. They thus show a periodicity of three to four residues in hydrophobicity along the associated sequence stretch (Fig. 10).

Proline residues are not expected to occur in middle segments, because they disrupt the α-helical turn. However, they are seen in the first two positions of α-helices.

β-Strands fold into so-called β-pleated sheets, which have two solvent-exposed strands at either edge. The hydrophobic nature of such edge strands is different from that of buried strands within a β-sheet. The side-chains along a β-strand protrude in an alternating direction, and strands at the edge of β-sheets typically show an alternating pattern of hydrophobic–hydrophilic residues whereas buried strands tend to contain merely hydrophobic residues (Fig. 10). Because the β-strand is the most extended conformation (i.e., consecutive C_α atoms are farthest apart), it takes relatively few residues to

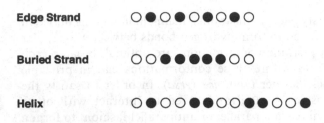

Edge Strand

Buried Strand

Helix

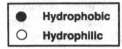

● **Hydrophobic**
○ **Hydrophilic**

FIGURE 10 Hydrophobicity patterns in α-helices and β-strands. Different patterns are observed for edge and buried strands and helices.

cross the protein core with a strand. Therefore, the number of residues in a β-strand is usually limited and can be as little as two or three amino acids, whereas helices shielding such strands from solvent comprise more residues. β-Strands can be disrupted by single residues that induce a kink in the extended conformation of the main chain. Such so-called β-bulges often comprise relatively hydrophobic residues.

Multiple alignments of protein sequences often display gapped and/or highly variable regions, which would be expected to correspond to loop regions rather than the two other basic secondary structures. Loop regions contain a high proportion of small polar residues such as alanine, glycine, serine, and threonine. Glycine residues are often seen in loop regions because of their inherent flexibility. Proline residues are often seen in loops as well. They are generally not observed in helices (but see above) and strands because they kink the main chain, although they can occur in the N-terminal two positions of α-helices as mentioned above.

2.2. Assessing Prediction Accuracy

The most widely used way to assess the quality of an alignment is by calculating the overall per residue three-state accuracy, called the Q_3:

$$Q_3 = \frac{P_H + P_E + P_C}{N} \times 100\% \tag{11}$$

where N is the total number of residues predicted and P_S is the number of correctly predicted residues in state S ($S = H, E,$ or C). Some researchers use the so-called Matthews correlation coefficient, because it more stringently estimates the prediction for each structural state:

$$C_S = \frac{(TP_S \times TN_S) - (FP_S \times FN_S)}{(TP_S + FP_S) \times (TP_S + FN_S) \times (TN_S + FP_S) \times (TN_S + FN_S)^{1/2}} \quad (12)$$

where TP_S and TN_S are, respectively, the numbers of positive and negative cases correctly predicted for the structural state S, and FP_S are and FN_S are the numbers of false positives and negatives, respectively. Matthews correlations can be used to measure over- or underprediction for any of the structural states. Because Q_3 is the most intuitive measure and leads to a single percentage, it is most frequently used in the literature to report prediction accuracy.

An important issue in assessing performance is the notion of sustained accuracy. Knowledge about the average accuracy of a given method over a set of predicted proteins is not meaningful if the variance of those predictions is not known. It is important also to know what worst-case predictions can be expected from a method even if the average accuracy is quite high. A standard scenario to assess sustained accuracy of prediction is the jackknife test carried out over a large set of test proteins. It involves the following four steps: (1) taking one protein out of the complete set of N proteins, (2) training the method on the remaining $N-1$ proteins (the training set), (3) predicting the secondary structure for the protein taken out, and (4) repeating steps 1–3 for all N proteins and calculating the average accuracy and variance. This scenario provides insight into the influence of different training sets on the sustained accuracy of single-protein prediction. It also ensures that no information about a query sequence or multiple alignment is used in training the method. It is possible, in principle, to test the method by averaging the predictions over all combinations of x proteins ($1 < x < N$), each time using the method trained on the remaining $N-x$ proteins being predicted. However, as the number of combinations grows rapidly with x, the training phase of most methods is too slow for extensive testing using this mode. It can, however, also be used to save computation time if the database is split evenly into test groups of sequences containing multiple sequences (e.g., sevenfold cross-validation in Ref. 79), because each sequence within a test group is associated with a single training set, thus saving training overhead. Nonetheless, unnoticed but systematic tuning of the method to the database might still occur, so that the most rigorous test of any method is the prediction of test cases that have no homologs in the database and were not seen during the development of the method.

Notwithstanding the importance of the measures for accuracy listed above, in practice the success of a secondary structure prediction depends on how the information is being used. An example is fold recognition, where accurate delineation of the edges of secondary structural elements is not essential, but missing a structure that is crucial for the basic topology is costly. However, both the Q_3 and Matthews correlation coefficients evaluate equally, for example, missing two residues at either side of a seven-amino-acid strand or missing a complete topologically essential strand of four residues.

The accuracy of prediction methods is normally assessed by using known tertiary structures from the Protein Data Bank (PDB) [179] as a benchmark, with secondary structural elements assigned using the DSSP method of Kabsch and Sander [181]. A number of evaluation studies have been carried out including DSSP and other secondary structure assignment methods. These showed that assignment methods can yield significantly different assignments, such that the agreement of the methods can be as low as 65% [182–184]. Further, in structurally aligned sets of homologous proteins with known tertiary structure, the corresponding secondary structural elements can vary in length and/or show shifts of one to a few residues. A realistic maximum prediction accuracy per residue has therefore been estimated to be in the range of 80–100% [185]. Many researchers have suggested that prediction evaluation should be based on the overlap of predicted and observed segments rather than just counting the individual positions [186–191].

The most widely used benchmark assignments for secondary structures are those produced by the DSSP method [181]. The assignments are classified into eight different states: H, α-helix; G, 3/10 helix; I, π-helix; E, β-strand, B, β-bulge; T, hydrogen-bonded turn; S, bend; ' ', coil. For the evaluation of three-state predictions, the eight states must be grouped to yield the three states helix, strand, and coil. Differences among prediction evaluations in the literature concern the DSSP states grouped as helix (e.g., H, G, and I, or only H taken as helix) or strand (E and B, or only E taken as strand). Different grouping schemes can lead to differences as high as 3% in the apparent accuracy attained by a considered method [192].

A more recent secondary structure assignment program that combines many of the features of earlier methods, such as considering hydrogen bonding patterns and stereochemical characteristics, is the knowledge-based method STRIDE [193]. This method generally yields assignments in close agreement with those made by crystallographic experts.

The program XTLSSTR [194] employs pseudovisual characteristics for assigning secondary structure elements to mimic what an expert would do by eye. The algorithm combines two angles and three distances, which should fall within specified ranges for each defined secondary structure symbol. Some further priority rules lead to the finally assigned secondary structure for each sequence position in the case of more than one possible assignment as a result of partially overlapping ranges for the five parameters. An example of a priority rule for assigning a β-strand is that it has a lower priority than an α-helix, 3_{10}-helix, or hydrogen-bonded turn.

Taylor [195] developed a method to optimally dissect a protein structure into secondary structure elements, following ellipsoid approximation. In this approach, each stretch of amino acids is modeled as an ellipsoid and a score is calculated using the ellipsoid length and the ratio of the length of longest axis to that of the axis perpendicular to it. Dynamic programming (see Sec. 1) is then applied to the score matrix of all possible segments to obtain the optimal segmentation for the structure.

2.3. The Early Prediction Methods

Attempts to predict protein secondary structure emerged more than four decades ago (e.g., Szent-Györgyi and Cohen [196] and Periti et al. [197]). The first computer algorithms for secondary structure prediction followed about 15 years later [198–200]. The algorithms of Nagano [198] and Chou and Fasman [200] were based on statistical information, while Lim's method [199] was stereochemically oriented and relied on conserved hydrophobic patterns in secondary structures such as amphipathicity in helices [180]. Secondary structure prediction has generally been formulated for three states: helix, strand, and coil. Although the early and popular GOR method [201,202] used four-state prediction (with turn as an extra class), recent versions of the program predict secondary structure using three states [203]. The early methods of Lim [199] and Chou and Fasman [200] as well as the GOR method [201,202] will be described in more detail.

Lim [199] developed a set of complicated stereochemical prediction rules for α-helices and β-sheets based on their packing as observed in globular proteins. Apart from being the most successful early method (see below), Lim's stereochemical rules have contributed to the understanding of protein folding. For instance, Lim's hydrophobicity rules are important for α-helix formation. They state that terminal hydrophobic pairs are at sequence positions i and $i + 1$, hydrophobic pairs in middle helical segments are positioned at $(i, i + 4)$, and middle hydrophobic triplets are at $(i, i + 1, i + 4)$ or $(i, i + 3, i + 4)$. Nonetheless, the Lim method never gained widespread popularity because a computer implementation was not available until recently, with more accurate techniques now available.

The most popular pioneering method is that of Chou and Fasman [200]. In this method, predictions are based on differences in residue composition for three states of secondary structure: α-helix, β-strand, and turn (i.e., neither α-helix nor β-strand). Using a number of crystallographically determined protein tertiary structures, Chou and Fasman calculated the frequency of each amino acid type in four states. The position of turn residues was included as a fourth state in the frequency calculations because there are significant differences in residue type occurrences at different positions within turn sites. The frequencies were normalized to yield amino acid preferences for each of the structural states. For helix and strand, effects of neighboring residues in the protein sequence were taken into account by averaging the preferences over three residues for α-helix predictions and over two for β-strands. Secondary structures were then predicted for each position according to the highest preference values of the structural states. Extensions were made as long as preferences remained high enough and certain disruptive residues were not encountered (such as proline, which breaks an α-helix).

The GOR method quickly became the standard and remained so for about a decade after its first appearance [201]. Although the first versions GOR I and II, as mentioned above, predicted four states by discriminating between coil and turn secondary structures, GOR III [202] and the most recent version, GOR IV [203], perform the common three-state prediction. Like the Chou–Fasman method, the GOR method relies on the amino acid frequencies observed. However, it uses a 17-residue window (i.e., eight residues N-terminal and eight C-terminal of the central window position) for each of the three structural states. The amino acid frequencies are exploited using an information function based on conditional probabilities defined as

$$I(S; R) = \log\left[\frac{P(S/R)}{P(S)}\right] \tag{13}$$

where S is one of the structural states H, E, or C, and R is one of the 20 residue types. The factor $P(S|R)$ denotes the conditional probability of a secondary structural state for a sequence position given that it is occupied by residue type R. Rewriting the formula for frequencies gives

$$I(S; R) = \log\left[\frac{f_{s,r}/f_R}{f_s/N}\right] \tag{14}$$

where $f_{S,R}$ is the frequency of residue type R in state S, f_R is the general frequency of residue type R, and f_S/N is that of structural state S. Important in this formula is that the information of a particular residue type in one of the structural states is based not only on the normalized frequency but also on the inverse fraction of all residues in that state. The information difference between the various states defined as $I(\Delta S; R) = I(S; R) - I(!S; R)$ with $!S$ denoting all other states (*not S*). Using Eq. (14), the formula then becomes

$$I(S; R) = \log[f_{S,R}/f_S] - \log[f_{!S,R}/f_{!S}] \tag{15}$$

An important issue is that it is not feasible to sample all possible 17-residue fragments directly from the PDB, because there are 17^{20} possibilities. Subsequent versions of the GOR method over the years have therefore explored increasingly detailed approximations of this sampling problem, along with the growth of data available in the PDB:

> GOR I and GOR II just treated the 17 positions in the window independently, so single-position information could be summed over the 17-residue window.
>
> GOR III refined the earlier scheme by including pair frequencies derived from 16 pairs between each noncentral residue and the central residue in the 17-residue window. Because the PDB at the time was not large enough to provide sufficient data, dummy frequencies were calculated [202].
>
> The current version, GOR IV [203], uses pairwise information over all possible paired positions in a window (there are $17 \times 16/2$ possibilities), although relatively small weights are used compared with the GOR I single-position information, which is also included.

The theoretical principles used in the GOR method are statistically sound and no ad hoc rules or artificial variables are invoked, which makes it one of the most elegant methods with a high accuracy given its single-sequences prediction. As with many other methods (see below), however, a postprocessing step was introduced for the GOR IV method to refine the predictions: Helices are required to be at least four residues in length, and strands should consist of two or more residues. If a shorter helix or strand fragment is initially predicted, the method assesses the probabilities of extending the fragment to the minimum associated length or deleting it (i.e., changing it to coil).

The Chou–Fasman, GOR III, and Lim methods were assessed early on and showed accuracies of 50%, 53%, and 56%, respectively [204]. Version IV of the GOR method, however, is reported to raise the single-sequence prediction accuracy to 64.4% [203], as assessed through jackknife testing over a database of 267 proteins with known structure. These accuracies should be compared with random predictions, which would yield about 40% correctness given the observed distribution of the three states in globular proteins (with roughly 30% helix, 20% strand and 50% coil). Although significantly higher than random, these single-sequence prediction accuracies are generally not sufficient to allow the successful prediction of protein topology.

2.4. Modern Methods

The early Chou–Fasman and GOR prediction methods exploit general compositional biases exhibited by the three types of secondary structures and predict the secondary structure of single sequences. Zvelebil et al. [205] were the first to exploit multiple alignments in automatic prediction. They extended the GOR method and reported that predictions were improved by 9% compared to single sequence prediction. Levin et al. [206] quantified the effect and observed an 8% increase in accuracy when multiple alignments of homologous sequences (with sequence identities of ≥25%) were used. As a result, current state-of-the-art methods all use input information from multiple sequence alignments. Early on, sequence pattern matching techniques have also been attempted [103,187,207–209], albeit not with a dramatically increased success rate. More recently, however, researchers have used novel computational concepts to optimize the implementation of the observed patterns in mapping the primary onto the secondary structure, with the aim of enhancing prediction. These concepts include neural network applications [79,175,210], nearest-neighbor methods [77,78,211, 212], linear discriminant analysis [213], and inductive logic programming (ILP) [214]. The www addresses of some popular secondary structure prediction servers are given in Table 1.

TABLE 1 Web Sites of Various Secondary Structure Prediction Methods and Related Services

Service	Reference	URL
GOR4	Garnier et al. [203]	http://absalpha.dcrt.nih.gov:8008/gor.html
PHD[a]	Rost and Sander [79]	http://www.embl-heidelberg.de/predictprotein/predictprotein.html[b]
Pred2ary	Chandonia and Karplus [210]	http://yuri.harvard.edu/~jmc/2ary.html
NNSSP[a]	Salamov and Solovyev [212]	http://dot.imgen.bcm.tmc.edu:9331/pssprediction/pssp.html
PREDATOR[a]	Frishman and Argos [77,78]	http://www.embl-heidelberg.de/cgi/predator_serv.pl
DSC[a]	King and Sternberg [213]	http://bonsai.lif.icnet.uk/bmm/dsc/dsc_read_align.html
Zpred[a]	Zvelebil et al. [205]	http://kestrel.ludwig.ucl.ac.uk/zpred.html
Jpred	Cuff and Barton [192]	http://jpred.ebi.ac.uk/
PSIPRED	Jones [175]	http://www.cs.ucl.ac.uk/staff/d.jones/psipred/
COILS2	Lupas et al. [262]	http://www.isrec.isb-sib.ch/coils/COILS_doc.html

[a] Method can also be run using the Jpred server.
[b] Mirror websites for PHD can be found here as well.

2.4.1. Neural Network Methods

Neural networks are organized as interconnected layers of input and output units and can also contain intermediate (or "hidden") unit layers (for a review, see Ref. 215). Each unit in a layer receives information from one or more other connected units and determines its output signal based on the weights of the input signals. A neural network can be regarded as a black box that is trained to optimize the grouping of a set of input patterns into a set of output patterns by adjusting the weights of the internal connections. Therefore, neural nets are learning systems based upon complex nonlinear statistics.

The PHD method (Profile network from Heidelberg) [79] combines the added information from multiple sequence information with the optimization potential of the neural network formalism. PHD makes use of three consecutive complete neural networks. The first network produces the initial raw three-state prediction for each alignment position. It takes as input the fractions of the amino acids in a 13-residue window that is slid along the sequence. For each window position, the secondary structure of the central residue in the window is predicted. The output of the first network for each alignment position comprises three probabilities for the three states (helix, strand, and coil).

A second network refines the raw predictions of the first-level network. The three-state probabilities are processed now using a 17-residue window. The output of the second network comprises for each alignment position the three adjusted state probabilities. This refinement step in the second network for the raw predictions of the first network is aimed at correcting infeasible predictions and would, for example, change (EEEHHEE) into (EEEEEEE).

The first two networks perform the basic prediction of the secondary structure associated with a query multiple alignment. However, because neural nets can be trained in various ways, PHD employs a number of separately trained consecutive network pairs (networks 1 and 2) and feeds the outputs (refined three-state probabilities) of all those nets into a third network for a so-called jury decision.

The predictions made by the jury network undergo a final filtering step to delete predicted helices of one or two residues and change those into coils. The PHD method was trained [79] on a nonredundant set of 130 alignments from the HSSP database [216], each alignment containing one template sequence with a known structure and aligned homologous sequences. The PHD method showed an overall prediction accuracy of 70.8% in a jackknife test over 126 alignments (four of 130 alignments were transmembrane protein families), which for computational reasons were divided in seven groups for jackknife cross-validation. Although this is not the highest accuracy reported, the PHD method shows sustained performance and is therefore likely to come up with useful information in a wide variety of test cases. The PHD method is available via web servers at various sites. If given a single sequence for prediction, the server performs a BLAST search to find a set of homologous sequences and aligns those using the MAXHOM alignment program [216]. The resulting alignment is then fed into the actual PHD neural net algorithm. It is recommended that one produce an alignment using alternative multiple alignment methods as well.

An algorithm that follows the principles of PHD rather closely is the Pred2ary method [210], which was assessed with an accuracy of 74.8% and balanced prediction over the three structural states. The method employs a second neural net to filter the raw predictions of the first net and a third net for a jury decision, as does the PHD method [79]. A recent extended version, which combines the outputs of a massive number of 120 networks individually trained, is claimed to predict with an accuracy (Q_3) of 75.9% $\pm$ 7.9%. This is achieved by converting all combinations of network output weights for helix and strand, using a fine-grained two-dimensional matrix, into a priori probabilities of their correct prediction of the true structural state. These probabilities are then used for a final state prediction corresponding to the highest of the a priori probabilities for each of the three states.

A recent method that incorporates multiple sequence information and neural nets is PSIPRED [175]. This method exploits position-specific scoring matrices as generated by the PSI-BLAST algorithm [168] and feeds them to a two-layered neural network. Jones evaluated the method as at least 76.5% accurate. This top accuracy was confirmed by blind tests at the CASP3 meeting, where assessments of the state-of-the-art prediction methods were made. Moreover, the method is fast and can be easily ported to any common computer system.

2.4.2. *k*-Nearest-Neighbor Methods

As with neural network methods, the application of *k*-nearest-neighbor methods requires an initial training phase in which a pool of so-called exemplars is established. This pool consists of sequence fragments of a certain length derived from a database of known structures, so that the central residue of such fragments (exemplars) can be assigned the true secondary structural state as a label. Then a window of the same length is slid over the query sequence, and for each window the *k* most similar fragments are determined using a certain similarity criterion, after which the distribution of the secondary structure labels is used to derive propensities for the three states. In the methods covered below, *k* is in the range 25–100.

In the method of Yi and Lander [211], a database of 110 proteins with known tertiary structure was used to derive a large collection of 19-residue fragments (exemplars), of which the environmental states were noted. For each 19-residue window slid over the query protein, 50 nearest-neighbor exemplars were identified using the amino acid environmental scoring system of Bowie et al. [217], which includes as parameters the secondary structure state, accessible surface area, and polarity and scores the likelihood of a query residue type being in a particular state (or range) over these three environmental parameters. As a score, the average was taken of 19 residues of a query window matched with the 19-position exemplar considered. For each exemplar, a cutoff score was determined that should be met by a query fragment compared with it to count the exemplar as a neighbor. The cutoff score can be viewed as a reliability check for the predictive value of the exemplars. The 50 thus obtained nearest neighbors for each query window showed a distribution of the associated secondary structure labels, from which probability estimates for the three structural states were derived. Yi and Lander explored various scoring systems and found that the best protocol included 15 environmental classes (three secondary structures combined with five different accessibility/polarity classes) in conjunction with an amino acid exchange score taken from the Gonnet et al. [89] amino acid exchange matrix (see Sec. 1.3). This scenario resulted in a prediction accuracy of 67.1%. Using a neural network for a jury decision over six different scoring systems led to the final accuracy of 68%, as assessed through jackknife testing.

The NNSSP (Nearest Neighbor Secondary Structure Prediction) [212] method uses the nearest-neighbor approach of Yi and Lander [211] with the following differences:

1. N- and C-terminal positions of helices and strands and β-turns are explicitly taken as additional secondary structure types.
2. When predicting, the database of exemplars (see above) is restricted to sequences similar to the query sequence. This reduces computation and leads to more closely biologically related nearest neighbors.
3. Predictions are made for multiple alignments.
4. Alignment regions with insertions/deletions are explicitly taken into account.

Salamov and Solovyev [212] explored various window lengths and finally chose predictors combining window sizes of 11, 17, or 23; nearest-neighbor numbers of 50 or 100; and balanced or nonbalanced training (i.e., $3 \times 2 \times 2 = 12$ predictors). A simple majority rule over the 12 predictors increased the accuracy by 0.9%. A few simple filters were effected to refine the predictions thus obtained:

1. Helices consisting of one or two residues are deleted (changed to coil), but (EHE) becomes (EEE).

2. Strands of length 1 or 2 are deleted, but (HEEH) becomes (HHHH).
3. Helices of length 4 or less are deleted. The latter rule is applied after a full cycle of rules (1) and (2).

The overall accuracy of the method is 72.2% [212], which results from a jackknife test over the database of 126 proteins by Rost and Sander [79].

The PREDATOR method of Frishman and Argos [77,78] owes its accuracy mostly to the incorporation of long-range interactions for β-strand prediction and attains 68% prediction accuracy for single sequences. Using a k-nearest-neighbor approach (with $k = 25$ and 13-residue windows), propensities for the three general states (P^H, P^E, and P^C) are determined for each residue. Two more propensities for β-strands are determined by assessing the likelihood for all pairwise five-residue fragments (separated by more than six amino acids) to form a parallel or antiparallel β-bridge. The two propensities, one for antiparallel and another for parallel bridges, are based on summing residue hydrogen bonding propensities obtained from known structures. For each residue, the parallel and antiparallel β-strand propensities (P^{Par} and $P^{Antipar}$) correspond to the maximum scoring window pair with the considered residue at the N-terminal position in one of the windows. Pairwise hydrogen bonding potentials are also determined for α-helical residues at a sequence separation of four residues. The sum for each residue pair is calculated over a seven-residue window, which gives an extra helix propensity for the residue N-terminal in the window (P^{Helix}). The last additional propensity for residues in the β-turn conformation(P^{Turn}) is obtained by summing single-residue propensities in classic β-turn positions 1–4 [218] using a four-residue window. For each of the seven independent propensities thus obtained, threshold values (T) are calculated and used in the following five rules applied consecutively to arrive at a three-state prediction for each residue:

1. If ($P^{Par} > T^{Par}$ or $P^{Antipar} > T^{Antipar}$) and $P^{Helix} < T^{Helix}$, then predict β-strand; otherwise, if $P^{Helix} > T^{Helix}$, then predict α-helix, otherwise predict coil.
2. If $P^C > T^C$, then predict coil.
3. If $P^E > T^E$, then predict β-strand.
4. If $P^H > T^H$, then predict α-helix.
5. If $P^{Turn} > T^{Turn}$, then predict coil.

Apart from the novel scheme to predict strands using long-range pairwise strand potentials, the method can also use information from multiple sequences. However, PREDATOR does not use a multiple alignment but compares the sequences through pairwise local alignments [25]. For a single base sequence within a set of sequences, a set of highest scoring local alignments is compiled through matching the base sequence with each of the other sequences. A weight is then calculated for each matched local fragment, based on the fragment's alignment score and length. Using these weights, a final propensity is calculated for each of the seven states by calculating the weighted sum over all included local fragments (stacked as shown in Fig. 9a, but here with possibly more than one local alignment per sequence). The resulting seven propensities are then subjected to the above five rules to yield the final predicted state. The accuracy of the PREDATOR method is 74.8% [78], as assessed using jackknife testing over the *RS* protein set. As for the Pred2ary method (see above) with identical accuracy, this Q_3 is the second best reported in the literature [after PSIPRED with 76.5% accuracy (see above)].

2.4.3. Discriminant Analysis

The DSC method [213] combines the compositional propensities from multiple alignments with empirical rules important for secondary structure prediction (see Sec. 2.1). The information is processed using linear statistics. The following rules and concepts are used:

> N-terminal and C-terminal sequence fragments are normally coil.
> Periodicity in positions of hydrophobic residues.
> Alignment positions comprising gaps are indicative for coil regions.
> Periodicity in positions of conserved residues.
> Autocorrelation.
> Residue ratios in the alignment.
> Feedback of predicted secondary structure information.
> Simple filtering.

These concepts are applied in five steps:

1. The GOR method is used on each of the aligned sequences, and the average GOR score for each of the three states is computed for each alignment position.
2. For each position in the query multiple alignment, a so-called attribute vector is compiled, consisting of 10 attributes: three averaged GOR scores for *H*, *E*, and *C* (step 1); distance to alignment edge; hydrophobic moment assuming helix; hydrophobic moment assuming strand; number of insertions; number of deletions; conservation moment assuming helix; and conservation moment assuming strand.
3. The positional vectors are doubled in number of attributes by adding the same 10 attributes in a smoothed fashion (running average).

4. Seven more attributes are added to the 20 attributes of the preceding step: weights for predicted α-helix and β-strand based on the 20-attribute vectors of step 3, and the fractions of the five most discriminating residue types, His, Glu, Gln, Asp, and Arg. To convert the attribute vectors to three-state propensities, a linear discrimination function is used. This is effective a set of weights for the attributes in the positional vector corresponding to each of the secondary structure states. The weights used in the DSC method were obtained by using a training set of known 3D structures. For each alignment position, the secondary structure associated with the highest scoring discrimination function is then taken.

5. A set of 11 simple filter rules are used for a final prediction, such as, ($[E/C]CE[H/E/C][H/C]$)→C. These filter rules have been constructed automatically using machine learning techniques.

The accuracy (Q_3) of DSC, as assessed for each of the five steps based on the Rost–Sander protein set, comprises 63.5%, 67.8%, 68.3%, 69.4%, and 70.1% (actual DSC method) [213], respectively. The DSC method shows the best performance for moderately sized proteins in the range of 90–170 residues. As an additional option, the method can also refine a prediction by the PHD algorithm (see above) using the above concepts. The average Q_3 of this PHD-DSC combinatorial procedure is 72.4% [213], which is 0.6% higher than the accuracy of PHD alone.

2.4.4. Inductive Logic Programming

Inductive logic programming (ILP) is designed for learning structural relationships between objects. Muggleton et al. [214] used the ILP computer program Golem to automatically derive qualitative rules for residues in the α-helix conformation and central in a nine-residue window. The rules made use of the physicochemical amino acid characterizations of Taylor [106] and were established during iterative training steps over a small set of only 12 known α/α protein structures. The predictive ability of the knowledge base thus constructed by the Golem algorithm was assessed using four α/α protein structures [214]; the knowledge base has limited use because it is able to predict only helices in all-helical proteins.

2.4.5. Exploring Secondary Structure-Specific Amino Acid Exchanges

The SSPRED method [99] exploits an alternative aspect of the positional information provided by multiple alignment: It uses the evolutionary information within a multiple sequence alignment by considering the pairwise amino acid exchanges observed for each multiple alignment position. Using the 3D-ALI database [98] of combined structure and sequence alignments of distantly homologous proteins, three amino acid exchange matrices are compiled for helix, strand, and coil, respectively. Each matrix contains preference values for amino acid exchanges associated with its structural state as observed in the 3D-ALI database. The matrices are then used to predict the secondary structure of a query alignment by listing the observed residue exchanges for each alignment position and summing the corresponding preference values over each of the three exchange matrices. The observed residue exchanges are counted only once at each alignment position, which provides an implicit weighting of the sequence information, avoiding predominance of related sequences in the input alignment. The secondary structure associated with the matrix showing the highest sum is then assigned to the alignment position. Following these raw predictions, three simple cleaning rules are applied and completed in three successive cycles:

1. If a sequence site is predicted in one structural state and the two flanking positions in another, the position is changed into that of the consistent flanking sites, for example, $(H[E/C]H)$ becomes (HHH), where $[E/C]$ indicates E or C.
2. If in five consecutive positions, two middle sites are of a type other than the three flanking sites, the middle positions are changed to the flanking types. For instance, $(HH[E/C][E/C]H)$ or $(H[E/C][E/C]HH)$ becomes $(HHHHH)$.
3. Helices predicted less than or equal to 4 and strands less than or equal to 2 in length are changed into coil predictions.

The accuracy of the method was assessed at 72% using a relatively small test set of 38 protein families and jackknife testing. This is a high accuracy, given that the method is fast and conceptually simple.

2.5. Consensus Prediction and Alignment Quality

The JPRED server at the EMBL-European Bioinformatics Institute (Hinxton, U.K.) [192] conveniently runs state-of-the-art prediction methods such as PHD [79], PREDATOR [77,78], DSC [213], and NNSSP [212] and also includes ZPRED [205] and MULPRED (Barton, unpublished). The NNSSP method has to be activated explicitly, because it is the slowest of the ensemble. The server accepts a multiple alignment and predicts the secondary structure of the sequence on top of the alignment; alignment positions showing a gap for the top sequence are deleted. A single sequence can also be given to the server. In the latter case, a BLAST search is performed to find homologous sequences, which are subsequently multiply aligned using ClustalX and then

PRALINE:

```
3chy       ADKELKFLIVDDFSTMRRIVRNLLKELGFNNVEEAEDGVDALNKLQAGGYGFVISDWNMPNMDGLELLKTIRADGAMSALPVLMVTAEAKKENIIAAAQAGASGYVVKPFTAATLEEKLNKIFEKLGM
1fx1       ..PKALIVFGSSTGNTEYTAETIARQLANAGYEDSRDAASVEAGGLPEGFDLVLSTWGDDS.DFIPLFDSLEETGAQGRKVACFGCGDS.SEYFCGAKNLGAEGLRIDRAARDDIVGWAHDVRGAI..
FLAV_DESDB MSKVLIVFGSSTGNTESIAQKLEELIAAGGHETLLNAADASAENLADGTDAVLSAWGMED.DFLSLPEEFNRFGLAGRKVAAFASGDQ.EEHFCGAKELGATGLKMESNDPEAVASFAEDVLKQL..
FLAV_DRSVH MPKALIVFGSSTGNTEYTAETIARELADAGYEDSRDAASVEAGGLPEGFDLVLSTWGDDS.DFIPLFDSLEETGAQGRKVACFGCGDS.SEYFCGAKNLGAEGLRIDRAARDDIVGWAHDVRGAI..
FLAV_DRSGI MPKALIVFGSSTGNTEYTAETIARQLANAGYEDSRDAASVEAGGLPEGFDLVLSTWGDDE.DFVPLYEDLDRAGLKDKKVGVFGCGDS.STYFCGABELGATSLKIDD..SABVLDWAREVLARV.
FLAV_DESSA MSKSLIVYGSSTGNTETRAEYVAEAFENKEIDELKNVTDVSVADLGNGYDIVLSTWGEBE.DFIPLYDSLENADLKGKKVSVFGCGDS.DTYFCGAERMGAVSLKIDE..RDEIVSWGSGIADKI.
4fxn       MK..IVYWSGTGNTEKMAELIAKGIIESGKDVNTIDVNEINELDGMGEDEVLN.DILISAMGDEV.EFPFPIBREIS.TKISGKKVALFG.....SGWGSGEEDTGAT.AIVNDNA.PECKELGEAAAKA..
FLAV_MEGEL MVB..IVYWSGTGNTERAMANEIEAAVKAAGADESVRFEDTNVDDVASK.DVILPAMGSBE.VVEPFFTDLA.PKLKGKKVGLFG......SGWGSGEEDTGAT.AIVNDNA.PECKELGEAAAKA...
2fcr       .KIGIFFSTSTGNTTEVADFIGKTLGAK.ADFIDVD.DVTDPQALKDYDLLFPTWNTG.WD.EFLVDKLPEVDMKDLPVAIFGLGDABGDNFCDAAKQAKGLPLDIPMEKRVAGWVEAVVSETGV
FLAV_ANASP SKKIGLFYGTQTGKTESVAEIIRDEFGND.VVLHDVS.QAE.VTDLNDYQYLIPTWNIGEWE.GLY.SELDDVDFNGKLVAYFGTDQIGDNFQDASQRGGKLALDDLTDDRIKSWVAQLKSEFGL
FLAV_ECOLI .AITGIFFGSDTGNTENIAKMIQKQLGKD.VAVHDIA.KSS.KEDLRAYDILLPTWYYGEWD.DFF.PTLEEIDFNGKLVALFGCGDQRDEYFCDAEPRGATGLAIDELTAERVKEKWVKQISEELHL
FLAV_AZOVI .AKIGLFFGSNTGKTRKVAKSIKKRFDDETMSALNVN.RVS.AEDFAQYQFLIPTLGEGEWE.EFL.PKIEGLDFSGKTVALFGLGDQVGENYLDAKDRGAKGLALDGKTDERVAAWLAQIAPEFGL
FLAV_ENTAG MATIGIFFGSDTQTRKVALLHQKLDGI.ADPLDVR.RAT.REQFLSYVLLPTLGDGEWQ.EFT.NTLSEADLTGKTVALFGLGDQLKNFVSAIARGACGLPLDDLTEERIDSWLEKLKPAV.L
FLAV_CLOAB .MKISILYSSKTGKTRRVAKLLEEGVKRSGNIVKTMNLDAVDKKFLQESEGIIPTTYAN.WEMKKWIDESSEPNLEGKLGAAFSTANSIASDI.AMVKGMLEIQENRIFGRRIANKVKQIF..
dsc        ------EEEEEE-----------HHHHH---------------HHHHH-----HH-------HHHHHHHHHHHH
mul        --EEEEE-------------HHHHHHHHHH-----------HHH------H-HH-----------HHHHHHHHHHHHHH-H
phd        --EEEEEE--------HHHHHHHHHHHH-------------HHHH----BB--BB----------B----HHHHHHHHHHHHH
pred       ---EEEEE-----------HHHHHHHHHH-------------HHHH------EE--BB-------EEEE---HHHHHHHHHHHH
zpred      ---EEEEE------HHHHHHHHHHHHHHHHH--------HHHHHHHHHHHHHEEEEEE-----BEEHHHHHHHHHHHHHH
```

CLUSTALX:

```
3chy       ADKELKFLIVDDFSTMRRIVRNLLKELGFNNVEEAEDGVDALNKLQAGGYGFVISDWNMPNMDGLELLKTIRADGAMSALPVLMVTAEAKKENIIAAAQAGASGYVVKPFTAATLEEKLNKIFEKLGM
FLAV_MEGEL MVEIVYWSGTGNTRAMANEIEAAVKAAGVESVRFEDTNVDDVALLGCPAMGSERLEDSVVEPFFTDLAPKLKGKSYGWGSGEWMDAWKQRTEDTGATVIGTAIVNEMPDNAPECKE.LGEAAAKA.
4fxn       .MKIVYWSGTGNTEKMAELIAKGIIESGVNTINVSDVNIDELLILGCSAMGDEVLBESBFPFEEISTKISGKGSYGWGDGKWMRDFEERMNGYGCVVVETPLIVNEPDEAEQCIEFGKKIANI..
FLAV_DESGI KALIVFGSSTGNTEGVABAIAKTLNSEGTTVVNVADVTAPGLALLGCSTWGDDBLQBDFVPLYEDLDRAGLKDKGCGDSSYTYAVDVIEKKABELGATIVASSLKIGEPDSAB..VLDWAREVLARV
FLAV_DESSA KSLIVFGSSTGNTEETRAEYVAEAFENKVELKNVTDVSVADLGLFGCSTWGEBELQDDFIPLYDSLENADLKGKKVSVFGCGDSDTYYAVDAIBEKLBEKMGAVVIGDSLKIGDPERDE..IVSWGSGIADKI
KVLIVFGSSTGNTESIAQKLEELIAAGGVTLLNAADASAENLALFGCSAWGMEEMQDDFLSLPFEFNRFGLAGRASGDQEYEHAVPAIEERAKELGATIIAEGLKMGDASNDPEAVASFAEDVLKQL.
1fx1       KALIVFGSSTGNTEYTAETIARQLANAGVDSRDAASVEAGGLFLLGCSTWGDDELQDDFIPLFDSLEETGAQGRQCGDSSYEYAVDAIEEKLKNLGABIVQDGLRIGDPRAARDDIVGWAHDVRGAI
FLAV_DRSVH KISILYSSKTGKTRRVAKLLEEGVKRSGVKTMNLDAVDKKFLQIFGTPTYANISWEMKWKIDESSEPNLEGKSTANSIAGGALLTILNHLMVKGMLVYSGGVHIEIQENEDNARIFGERIANKVK
FLAV_CLOAB KIGLFYGTQTGKTESVAEIIRDEPGNDVLHDVSQABVTDLNIIGCPTWNIQQWEGLYSELDDVDFNGKGTDQIGYAAIGILEEKISQRGGKTVGKFGLALDEDNQDSLTDDRIKSWVA
FLAV_ANASP KIGLFFGSNTGKTRKVAKSIKKRFDDET.ALNVNRVSAEDFAILGTPTLGBGQPENESWEEFLPKIEGLDFSGKGLGDQVGYPALGELYSFFKDRGAKIVGSWGKFGLALDLDNQSGKTDERVAAWLA
2fcr       TIGIFFGSDTQCTRKVALLSLLGTPTLGDGPQYDSWQEFTNTLSEADLTGKDLNIIGCPTWNTGT...SWDEFLYKLPEVDMKDLGLGDABGYPAIEBIHDCFAKQGAKPVGPSKPGLPLDNQDAPECKE
FLAV_ENTAG ITGIFFGSDTGNTENIAKMIQKQLGKDV.VHDIAKSSKEDLELLGIPTWYYQQ...WDDFPFTLEEIDFNGKGCGDQEDYAALGTIRDIIEPRGATIVGHWDHFGLAIDEDRQPELTAERVEKWVK
dsc        --EEE-----------------HHHHHHHHHH-----------HHE--------------------HHHHHHHHH
mul        -EEEEE-----------HHHHHHHHHHH----------HHH----HHHHHHHHHHH-----HHHHHHHHHH-----HHE
phd        -EEEEEB-----------HHHHHHHHHHH----------HHH----EEEE--BB-------EEEE---HHHHHHHHHHH---HHHHHHHHHHH--
pred       EEEEEE-----------HHHHHHHHHH-----------HHHHH----EEEEE-B-----HHHHH-----HHHHHHHHHHH-----HHHHHHHHH----
zpred      EEEEEB-----------HHHHHHHHHH-----------HHHHHHHHHHHHHEEEEEE----HHHHHHHHHHHH
```

JPRED CONSENSUS PREDICTION:

```
3chy            ADKELKFLIVDDFSTMRRIVRNLLKELGFNNVEEAEDGVDALNKLQAGGYGFVISDWNMPNMDGLELLKTIRADGAMSALPVLMVTAEAKKENIIAAAQAGASGYVVKPFTAATLEEKLNKIFEKLGM
cons HOMOLOGS  -----EEEEE-----------HHHHHHHHHH---EEEHH--------------------E-------HHHHHHHHHH-----------HHHHHHHHHHHHH
cons PRALINE  ----EEEEEEE----------------HHHHHHHHHHH-------E---HHHH-----EEEEEE--------HHHHHHHHHHHHH
cons CLUSTALX -EEEEE-----------HHHHHHHHHHH-----------HH----HHHHHHHHHHHHEE-B-------HHHHHHHHHHHH---HHHHHHHHHHHHHHH
DSSP            |  EEEE  HHHHHHHHH    EEEE  HHHHHHHHH        EEEE     HHHHHHHHH    EEEE   HHHHHHHHHHHHH
```

processed with the user-provided single sequence on top in the alignment. If a sufficient number of methods predict an identical secondary structure for a given alignment position, that structure is taken as the consensus prediction for the position. If insufficient agreement is reached, the PHD prediction is taken. This consensus prediction is somewhat less accurate when the NNSSP method is not invoked or is not completed in the computer time slot allocated to the user. An example of predictions by the Jpred server for the signal transduction protein cheY (PDB code 3chy) is given in Figure 11. The Jpred consensus approach was evaluated at 72.9% correct prediction using a database of 396 domains, which was 1% more accurate than the best individual method (PHD) among those included [192].

Secondary structure prediction methods depend crucially on the quality of the input multiple alignment as well as the sequences represented in it. Heringa demonstrated the importance of multiple alignment using the signal transduction protein cheY (PDB code 3chy) and 13 distant flavodoxin sequences (Fig. 6). The 3chy structure adopts a flavodoxin fold despite very low sequence similarities with genuine flavodoxins. The popular multiple alignment program ClustalX and the method PRALINE [66] were used to construct a multiple alignment for this distance sequence set. Figure 9 shows the ClustalX and PRALINE alignments with corresponding secondary structure predictions made by the Jpred server. The difference in accuracy of the consensus predictions for these two alignments is dramatic and amounts to more than 30% (Fig. 11). The Jpred server was also activated with only the 3chy sequence as input, upon which it constructed a set of putative homologs through a BLAST search and then aligned the resulting sequences using the ClustalX method. For the 3chy sequence, 32 related sequences were thus found and aligned. The accuracy of the consensus secondary structure prediction by Jpred for this alignment was only 3%

FIGURE 11 Secondary structure prediction for chemotaxis protein cheY (PDB code: 3chy). The top alignment block represents the multiple alignment of the 3chy sequence with 13 distant flavodoxin sequences by the method PRALINE. The middle block represents the same sequence set aligned by ClustalX. Under both alignments are given the alignments by five secondary structure prediction methods. The bottom block depicts consensus secondary structures deter-mined by Jpred using five prediction methods, respectively for a set of 32 homologs to 3chy found by the BLAST method and aligned by ClustalX (cons HOMOLOGS), and those for the PRALINE and ClustalX alignments. Vertical bars under consensus predictions indicate correct predictions. The bottom line identifies the standard of truth as obtained from the 3chy tertiary structure by the DSSP program [181]. The secondary structure states assigned by DSSP other than H and E were set to coil, designated by blanks.

higher than that obtained for the PRALINE alignment of the cheY-flavo-doxin set (Fig. 11). However, the second β-strand of the 3chy structure was recognized in the alignment of homologs but missed by the predictions based on both the ClustalX and PRALINE alignments. It is important to note that the flavodoxin sequences are evolutionarily extremely distant from the cheY sequence, so this set of sequences is likely to be difficult for multiple alignment methods and consequently also for secondary structure prediction techniques, which are generally trained on homologous families of less extreme divergence.

3. PREDICTION OF TRANSMEMBRANE SEGMENTS

Membrane proteins (MPs) contain one or more transmembrane (TM) seg-ments. Although, in principle, all possible mutual orientations of individual structural elements are possible in globular proteins, the organization of MP transmembrane segments is severely restricted by the lipid bilayer of the cell membrane. They can therefore be regarded as a distinct topological class. Compared to soluble proteins, few X-ray or NMR data regarding the tertiary structure of TM proteins are yet available [219]. The most frequently observed secondary structure in transmembrane segments is the α-helix, but in addition four families involving transmembrane structures based on β-strands that constitute a β-barrel have recently been identified. Whereas α-helical TM proteins generally interact with the lipid bilayer of the cytoplasmic membrane of all cells, TM proteins comprising antiparallel β-strands have been localized in the outer membrane of bacteria, mitochondria, and chloroplasts. The initial idea that TM segments are either completely of an α-helical nature or consist of β-strands exclusively was challenged by electron microscopic data for the nicotinic acetylcholine receptor [220]. These data, albeit with a low resolution of 9 Å, were interpreted as a central five-helix bundle surrounded by β-strands. A more refined structure at 4.6 Å [221], but the controversial mixed α/β TM structure, was not discussed. Another example of a mixed TM structure was provided by Doyle et al. [222], who described a tetrameric potassium channel structure consisting of inner and outer layers, each consisting of four helices. At the extracellular side of the pore, a short four-stranded β-barrel-like selectivity filter protrudes into the pore.

Fortunately, the location of the transmembrane segments in the pri-mary structure of the MP is relatively easy to predict owing to the rather strong tendency of certain hydrophobic amino acids with special physico-chemical properties to occur in membrane-spanning regions. Predictions have generally been focused on the determination of the membrane sequence segment boundaries and their tentative orientation with respect to the membrane, mostly assuming an α-helical structure.

3.1.   Prediction of α-Helical TM Segments

The following considerations are important for transmembrane sequence prediction. Amino acids in contact with the lipid phase are likely to be hydrophobic. Therefore, any measure of amino acid hydrophobicity derived from physical calculations and/or experimental data can serve as an indicator of the propensity for a residue type to occur in a membrane-spanning segment.

Transmembrane segments are believed to adopt the α-helical conformation in most cases. An α-helix is the most suitable local arrangement because, in the absence of water molecules inside the membrane, all main-chain polypeptide donors and acceptors can mutually satisfy each other through formation of hydrogen bonds. This energetic argument is supported by experimental evidence that polypeptide chains tend to adopt the helical conformation when immersed in a nonpolar medium [223]. Therefore, α-helical propensities of amino acids derived from the analysis of globular proteins can be considered in MP structure prediction.

Experimental data on the boundaries of the transmembrane segments are not very precise, because they are acquired from site-directed mutagenesis, enzymatic cleavage, immunological methods, etc. This contrasts with the standard secondary structure prediction methods for soluble proteins, with which statistical propensities of different amino acids to form one of the major secondary structure elements are derived from much more accurate protein tertiary structural data from X-ray crystallography and NMR spectroscopy.

Although the globular interior of soluble proteins is less apolar than the lipid bilayer, extensive use of these data has been made for MP structure prediction. A particular example is the widespread use of the classical hydrophobicity scale of Kyte and Doolittle [224]. Other techniques also became available early on that were more specifically aimed at searching MP transmembrane regions [225–228]. Generally, hydrophobic scales are used to build a smoothed curve, often called a hydropathic profile, by averaging over a sliding window of given length that is slid along the query sequence, to predict transmembrane regions. Stretches of hydrophobic amino acids likely to reside in the lipid bilayer then appear as peaks with lengths corresponding to those expected for transmembrane segments, typically 16–25 residues. The choice of window length should correspond to the expected length of a transmembrane segment. Given that the average membrane thickness is about 30 Å, approximately 20 residues form a helix reaching from one lipid bilayer surface to another. To determine the boundaries of a membrane-spanning segment, a cutoff value for the hydrophobic peaks is also required. Kyte and Doolitle [224] based their cutoff value on the hydropathic character of just a few available membrane proteins. Later, a much larger learning set was used by Klein et al. [229], who applied discriminant analysis to aid prediction. Rao and Argos [230] suggested a minimum value for the peak hydrophobicity and two more cutoff values for either end of the peak to terminate the helix. It must be emphasized that the relatively simple physical considerations forming the basis of the above prediction methods do not exhaust the whole variety of possible situations. Some commonly used hydrophobicity scales are given in Table 2.

The relative orientation of the helices and the interaction of the corresponding side chains in membrane proteins with more than one trans-membrane helix are important features to be included in structure prediction efforts. The structures of membrane proteins determined to date and also theoretical evidence [231] support the view that α-helices in membranes form compact clusters. TM residues facing the lipid environment conform to the hydrophobic preferences described above, but interface residues between different helices do not necessarily have contact with the membrane and

TABLE 2 Hydrophobicity Scales

Amino acid	Janin, 1979	von Heijne, 1981	Kyte and Doolitle [224]	Sweet and Eisenberg [226]	Engelman et al., 1986	Rao and Argos [230]
Ala	0.3	−4.2	1.8	−0.4	−1.6	1.4
Cys	0.9	−6.3	2.5	0.2	−2	1.3
Asp	−0.6	31.0	−3.5	−1.3	9.2	0.1
Glu	−0.7	24.7	−3.5	−1.2	8.2	0.3
Phe	0.5	−14.2	2.8	1.9	−3.7	1.6
Gly	0.3	0.0	−0.4	−0.7	−1	1.1
His	−0.1	14.3	−3.2	−0.6	3	0.7
Ile	0.7	−10.5	4.5	1.3	−3.1	1.4
Lys	−1.8	17.6	−3.9	−0.7	8.8	0.1
Leu	0.5	−10.1	3.8	1.2	−2.8	1.5
Met	0.4	−11.3	1.9	1.0	−3.4	1.4
Asn	−0.5	12.2	−3.5	−0.9	4.8	0.3
Pro	−0.3	13.9	−1.6	−0.5	0.2	0.5
Gln	−0.7	10.1	−3.5	−0.9	4.1	0.3
Arg	−1.4	47.3	−4.5	−0.6	12.3	0.2
Ser	−0.1	6.3	−0.8	−0.6	−0.6	1
Thr	−0.2	3.8	−0.7	−0.3	−1.2	1.1
Val	0.6	−8.4	4.2	0.9	−2.6	1.4
Trp	0.3	−8.4	−0.9	0.5	−1.9	1
Tyr	−0.4	4.7	−1.3	1.7	0.7	0.8

therefore can display different tendencies. For example, charged residues can occur in TM helices in a coordinated fashion, such that positively charged side groups on one helix will have their negatively charged counterparts on another helix [219]. Charged residues can also constitute a membrane channel. TM α-helices can thus show an amphipathic behavior similar to helices in soluble proteins. In such cases, hydropathic profiles can fail to detect the transmembrane segments. In cases where the number of transmembrane segments is large (more than 20 in some channel proteins), inner helices of the transmembrane helical bundle have been observed to completely avoid contact with the lipid bilayer. As a result, any restrictions on the amino acid content or even length of such helices do not apply in such cases.

Eisenberg et al. [232] introduced a quantitative measure of helix amphipathicity that was named the hydrophobic moment. It is defined as the vector sum of the individual amino acid hydrophobicities radially directed from the helical axis. In general, the hydrophobic moment provides sufficient sensitivity to discriminate between amphipathic α-helices of globular, surface, and membrane proteins. Many methods for amphipathic analysis were developed based on Fourier analysis of the residue hydrophobicities [232–234] and the average hydrophobicity on the helix face [235].

Several prediction methods have emerged that are based on multiple factors, complex decision rules, and large learning sets. For example, von Heijne [236] proposed a combinatorial technique that supplements a standard hydrophobicity analysis with charge bias analysis. Other synthetic methods include the joint use of several selected hydrophobicity scales [237], utilization of optimization techniques with membrane segments as defined by X-ray analysis serving as reference examples [238], and the application of neural networks trained on secondary structural elements of globular proteins for membrane protein structure analysis [239,240].

In the method TMAP [241], information from multiple alignments is used to aid TM prediction. The propensities of amino acids to be positioned in either the central or flanking regions of a transmembrane segment were calculated using more then 7500 individual TM helices as annotated in the Swiss-Prot sequence databank. These propensities were then used to build a prediction algorithm wherein for each segment of a multiple sequence alignment and for each sequence included therein, average values of the central and flanking propensities are calculated over sliding windows. The optimal window lengths were found to be 15 and four residues for central and flanking propensities, respectively. If the peak value for a central transmembrane region exceeds a certain threshold, this region is considered a possible candidate to be membrane-spanning. The algorithm then attempts to expand this region in either sequence direction until a flanking peak is reached or the central propensity average falls below a certain value. Some further restraints are imposed on the possible length of a tentative TM segment. The added sensitivity compared to other sliding window approaches is a result of using multiple alignment information as well as the second propensity for flanking regions.

Rost et al. [240] also used multiple sequence information and trained the PHD method [79] on multiple alignments for 69 protein families with known TM helices. They achieved a prediction accuracy of 95% as assessed through the jackknife test.

The methods for MP structure analysis discussed here can be used to create constraints for homology modeling and tertiary structure prediction. After generating an initial approximation of the topology of a protein's TM regions and their relative orientation based on hydrophobicity analysis and available experimental evidence, the structure can be optimized using energy minimization and molecular dynamics techniques. Early examples of such modeling efforts for molecules of pharmacological interest include G protein–coupled receptors [242–244] and TM channels [245].

3.2. Predicting the Orientation of Transmembrane Helices

An issue closely related to transmembrane segment prediction is the prediction of the intra- and extracellular membrane side. In bacterial membrane proteins, intracellular loops between transmembrane helices were found to contain arginine and lysine residues much more frequently than the extracellular exposed loops [228,246]. Although less pronounced, this tendency has been shown to also occur in eukaryotic membrane proteins [247]. For eukaryotic proteins, it was further observed that the difference in the total charge within stretches of about 15 residues flanking the transmembrane region on both sides of the membrane coincides with the orientation of the protein [248] in that the side most positively charged shows a tendency to reside in the cytosol. This tendency, which was named the "positive inside rule," aids prediction schemes for MP topology. The nonrandom flanking charge distribution may also play an important role in the physical insertion of the protein into the membrane. However, the positive inside rule is applicable only to α-helical TM regions.

3.3. Prediction of β-Strand Transmembrane Regions

The prediction of TM segments constituted by β-strands by the methods described above is not likely to be successful, because they all assume the α-helical conformation. Although relatively underrepresented, four different families of β-barrel-containing membrane proteins are known to date. The earliest discovered family comprised archetypal trimeric proteins of Gram-negative bacteria, called porins, that can nonspecifically mediate passive transport of small hydrophilic molecules (< 6 kDa), whereas larger molecules, such as malto-oligosaccharides, can be transported selectively. Porins play their important cellular roles by forming voltage-dependent and water-filled membrane channels constituted by a β-barrel consisting of 16 β-strands [249], in which adjacent β-strands are linked by hydrogen bonds. Further families comprise transporter proteins from enteric bacteria (e.g., FepA and FhuA), proteins involved in bacterial conjugation (OmpA and OmpX in *E. coli*), and the lytic outer TM domain of staphylococcal α-hemolysin. The β-strands constituting the barrels seen in the known families are even in number and vary from 8 to 22. Most of the outer surface of the β-barrel interacts with the membrane's lipid environment, while the internal part serves as an aqueous pore. Each individual β-strand could thus be expected to be amphipathic with a period of two residues. However, although every second residue facing the lipid bilayer tends to be hydrophobic, the side chains facing the interior of the barrel display no definitive tendency and can therefore distinctly lower the amphipathic signal. This compromised amphiphatic behavior is observed particularly in families other than porins. Moreover, the number of amino acid residues in a β-strand needed to span the membrane is much smaller than that for the helical conformation, typically only about 10, so they can be missed easily by smoothed hydropathic profiles.

Jacoboni et al. [250] devised a method based on neural networks to predict the location of TM β-strands. The accuracy of their method was reported to vary from 69% for single-sequence prediction to 78% for predictions using alignments generated by PSI-BLAST [168], albeit these statistics were based on jackknife tests over a critically small database.

4. PREDICTION OF PROTEIN ANTIGENIC SITES

Antigenic sites (ASs) are locations on the protein molecule that are responsible for specific antibody binding. Their detection is an important step in biochemical characterization of a protein. Theoretical prediction of the sequence positions of ASs exploits mostly their preferred location on the surface of the protein [251]. In contrast to transmembrane prediction methods based on the most hydrophobic sequence spans, the AS prediction techniques are based on the preferred location of antigenic sites on the surface of the protein. The methods therefore search for the most hydrophilic sequence regions associated potentially with solvent-exposed regions. An important early approach to this problem was that by Hopp and Woods [252], who calculated a smoothed curve (similar to the hydropathic profile considered in Sec. 3.1) employing a sliding window approach that used hydrophilicity values given by Levitt [253], that were slightly modified.

As in the case of membrane segment prediction, the averaging window length must be optimized. Empirical tests showed that it corresponded to the expected size of continuous surface epitopes typically five to eight amino acids long. Hopp and Woods [252] found that a window length of six residues produced the best results; therefore that window length was consistently used. Based on evidence from 12 proteins with known antigenic structure, Hopp and Woods could not detect a one-to-one correspondence between the AS and high peaks on the hydrophilicity curve: Not all high peaks appeared to coincide with ASs and, conversely, not all ASs were reflected as high peaks in the Hopp–Woods method (Fig. 12). Although the highest peak on the curve nearly always lay within one of the antigenic sites or close to it in sequence, the method generally was able to predict only one out of many potentially protein antigenic determinants. Following Hopp and Woods [252], Parker et al. [255] built surface profiles using an alternative set of hydrophilicity values derived

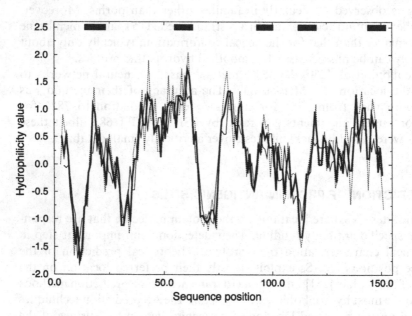

FIGURE 12 Detecting antigenic sites. Plotted are running averages of amino acid hydrophilicity values [252] for the sperm whale myoglobin sequence. Running averages are given using window lengths of 5 (dotted line) and 8 (thin solid line), as well as for optimal six-residue windows (thick solid line). The location of the five known antigenic sites [254] is indicated with black bars. Note that the highest peak corresponds to one of the antigenic regions. The correspondence of other high peaks and antigenic sites is only partial.

from retention times in high-performance liquid chromatography, which resulted in more accurate predictions.

The efficiency of antigen–antibody binding was found to correlate with the mobility of protein sites constituting antigenic determinants [256,257]. Their high flexibility could allow ASs to develop a better fit to the corresponding location on the antibody molecule. This notion of induced fit is supported by experimental evidence [258]. Karplus and Schultz [259] therefore derived amino acid propensities aimed at localizing protein sites with high polypeptide chain flexibility. They used empirically determined crystallographic temperature factors that correspond to mean-square atomic displacement. Jameson and Wolf [260] combined the various signals in a single so-called antigenic index and included the acid flexibility propensities of Karplus and Schultz [259], surface probabilities [251], and residue hydrophilicities [252]. They also devised a weighting scheme for these three signals to optimize their prediction results.

Although the AS prediction methods can give reasonable results, particularly when used jointly [260,261], additional biochemical evidence is required for reliable conclusions. Further, these sliding window methods are not applicable to discontinuous antigenic determinants when the residues that constitute the antibody-binding pocket are not close in sequence but are close in proximity in the tertiary structure of the protein.

5. PREDICTING COILED-COIL STRUCTURES

Another feature closely related to secondary structure prediction is the prediction of coiled-coil structures. If a soluble protein is predicted to contain α-helices, higher order information as well as increased confidence in predictions could be gained from testing the possibility that a pair of helices adopt a superhelical twist that results in a coiled-coil conformation. The usual left-handed coiled-coil interaction involves a repeated motif of seven helical residues (*abcdefg*), where the *a* and *d* positions are normally occupied by hydrophobic residues constituting the hydrophobic core of the helix/helix interface, while the other positions display a high likelihood of containing polar residues. Another feature is that the heptad *e* and *g* positions are often charged and can form salt bridges. The program COILS2 [262,263] exploits this information and compares a query sequence with a database of known parallel two-stranded coiled coils. A similarity score is derived and compared to two score distributions, one for globular proteins (without coiled coils) and one for known coiled-coil structures. The two scores are then converted to a probability for the query sequence to adopt a coiled-coil conformation. Because the program assumes the presence of heptad repeats, probabilities are derived using default window lengths of 14, 21, and 28 amino acids. The program can also use user-defined window lengths for the prediction of extreme coiled-coil lengths. A recently updated scoring matrix, based on data from recent coiled-coil structures and containing amino acid type propensities for various positions in the heptad repeat, shows better recognition of coiled-coil elements. The COILS2 method accurately recognizes left-handed two-stranded coiled coils but loses sensitivity for coiled-coil structures consisting of more than two strands. Also, it is not able to recognize right-handed or buried coiled–coil helices and therefore is not applicable to transmembrane coiled-coil structures known to show coiled-coil conformations basically similar to those of soluble proteins, albeit with dramatically different and more hydrophobic constituent amino acids [219].

REFERENCES

1. Fitch W. An improved method of testing for evolutionary homology. J Mol Biol 1966; 16:9–16.
2. Gibbs AJ, McIntyre GA. The diagram, a method for comparing sequences. Its use with amino acid and nucleotide sequences. Eur J Biochem 1970; 16:1–11.
3. McLachlan AD. Tests for comparing related amino acid sequences: cytochrome c and cytochrome c551. J Mol Biol 1971; 61:409–424.
4. McLachlan AD. Repeating sequences and gene duplications in proteins. J Mol Biol 1972; 72:417–437.
5. McLachlan AD. Analysis of gene duplication repeats in the myosin rod. J Mol Biol 1983; 169:15–30.
6. Argos P. A sensitive procedure to compare amino acid sequences. J Mol Biol 1987; 193:385–396.
7. Staden R. An interactive graphics program for comparing and aligning nucleic acid and amino acid sequences. Nucleic Acids Res 1982; 10:2951–2961.
8. Pustell J, Kafatos FC. A high speed, high capacity homology matrix: zooming through SV40 and polyoma. Nucleic Acids Res 1982; 10:4765–4782.
9. Pustell J, Kafatos FC. A convenient and adaptable package of computer programs for DNA and protein sequence management, analysis and homology determination. Nucleic Acids Res 1984; 12:643–655.
10. Schultz GE, Schirmer RH. Principles of Protein Structure. New York: Springer-Verlag, 1979.
11. Gilbert W. The exon theory of genes. Cold Spring Harbor Symp Quant Biol 1987; 52:901–905.
12. Waterman MS. Sequence alignments. In: Waterman MS, ed. Mathematical Methods for DNA Sequences. Boca Raton, FL: CRC Press, 1989.
13. Needleman SB, Wunsch CD. A general method applicable to the search for similarities in the amino acid sequence of two proteins. J Mol Biol 1970; 48: 443–453.
14. Dayhoff MO, Barker WC, Hunt LT. Establishing homologies in protein sequences. Methods Enzymol 1983; 91:524–545.

15. Sellers P. On the theory and computation of evolutionary distances. SIAM J Appl Math 1974; 26:787–793.
16. Gotoh O. Alignment of three biological sequences with an efficient traceback procedure. J Theor Biol 1986; 121:327–337.
17. Gotoh O. Pattern matching of biological sequences with limited storage. CABIOS 1987; 3:17–20.
18. Myers EW, Miller W. Optimal alignment in linear space. CABIOS 1988; 4: 11–17.
19. Hirschberg DS. A linear space algorithm for computing longest common subsequences. Commun Assoc Comput Mach 1975; 18:341–343.
20. Sankoff D, Kruskal JB. Time Warps, String Edits and Macromolecules: The Theory and Practise of Sequence Comparison. Reading, MA: Addisson-Wesley, 1983.
21. Ukkonen E. On approximate string matching. Proc Int Conf Found. Comput Theor. Lectures Comput Sci 1983; 158:487–493.
22. Ficket JW. Fast optimal alignment. Nucleic Acids Res 1984; 12:175–180.
23. Heringa J, Taylor WR. Three-dimensional domain duplication, swapping and stealing. Curr Opin Struct Biol 1997; 7:416–421.
24. Heringa J. Detection of internal repeats: how common are they? Curr Opin Struct Biol 1998; 8:338–345.
25. Smith TF, Waterman MS. Identification of common molecular subsequences. J Mol Biol 1981; 147:195–197.
26. Waterman MS, Eggert M. A new algorithm for best subsequences alignment with applications to the tRNA-rRNA comparisons. J Mol Biol 1987; 197: 723–728.
27. Huang X, Hardison RC, Miller W. A space-efficient algorithm for local similarities. CABIOS 1990; 6:373–381.
28. Huang X, Miller W. A time-efficient, linear-space local similarity algorithm. Adv Appl Math 1991; 12:337–357.
29. Pearson WR, Lipman DJ. Improved tools for biological sequence comparison. Proc Natl Acad Sci USA 1988; 85:2444–2448.
30. Vingron M, Argos P. Determination of reliable regions in protein sequence alignments. Protein Eng 1990; 3:565–569.
31. Zuker M. Suboptimal sequence alignment in molecular biology. Alignment with error analysis. J Mol Biol 1991; 221:403–420.
32. Mott R. Maximum-likelihood estimation of the statistical distribution of Smith-Waterman local sequence similarity scores. Bull Math Biol 1992; 54: 59–75.
33. Bucher P, Hofmann K. A sequence similarity approach based on a probabilistic interpretation of an alignment scoring system. In: States DJ, Agarwal P, Gaasterland T, Hunter LR, eds. Proceedings of the Fourth International Conference on Intelligent Systems for Molecular Biology (ISMB). AAAI Press, 1996:44–51.
34. Baldi P, Chauvin Y, Hunkapiller T, McClure MA. Hidden Markov models of biological primary sequence information. Proc Natl Acad Sci USA 1994; 91: 1059–1063.

35. Krogh A, Mian IS, Sjölander K, Haussler D. Hidden Markov models in computational biology. J Mol Biol 1994; 235:1501–1531.

36. Notredame C, Holm L, Higgins DG. COFFEE: an objective function for multiple sequence alignments. Bioinformatics 1998; 14:407–422.

37. Thompson JD, Plewniak F, Poch O. A comprehensive comparison of multiple sequence alignment programs. Nucleic Acids Res 1999; 27:2682–2690.

38. Murata M, Richardson JS, Sussman JL. Simultaneous comparison of three protein sequences. Proc Natl Acad Sci USA 1985; 82:3073–3077.

39. Carillo H, Lipman DJ. The multiple sequence alignment problem in biology. SIAM J Appl Math 1988; 48:1073–1082.

40. Lipman DJ, Altschul SF, Kececioglu JD. A tool for multiple sequence alignment. Proc Natl Acad Sci USA 1989; 86:4412–4415.

41. Altschul SF, Carrol RJ, Lipman DJ. Weights for data related by a tree. J Mol Biol 1989; 207:647–653.

42. Stoye J, Moulton V, Dress AWM. DCA: an efficient implementation of the divide-and-conquer approach to simultaneous multiple sequence alignment. Comput Appl Biosci 1997; 13:625–626.

43. Stoye J. Multiple sequence alignment with the divide-and-conquer method. Gene 1998; 211:GC45–GC56.

44. Johnson MS, Doolittle RF. A method for the simultaneous alignment of three or more amino acid sequences. J Mol Evol 1986; 23:267–278.

45. Sobel E, Martinez HM. A multiple sequence alignment program. Nucleic Acids Res 1986; 14:363–374.

46. Waterman MS. Multiple sequence alignment by consensus. Nucleic Acids Res 1986; 14:9095–9102.

47. Waterman MS, Jones R. Consensus methods for DNA and protein sequence alignment. Methods Enzymol 1990; 183:221–237.

48. Vingron M, Argos P. A fast and sensitive multiple sequence alignment program. CABIOS 1989; 5:115–121.

49. Wilbur WJ, Lipman DJ. The context dependent comparison of biological sequences. SIAM J Appl Math 1984; 44:557–567.

50. Hogeweg P, Hesper B. The alignment of sets of sequences and the construction of phyletic trees: an integrated method. J Mol Evol 1984; 20:175–186.

51. Feng DF, Doolittle RF. Progressive sequence alignment as a prerequisite to correct phylogenetic trees. J Mol Evol 1987; 25:351–360.

52. Thompson JD, Gibson TJ, Plewniak F, Jeanmougin F, Higgins DG. The ClustalX windows interface: flexible strategies for multiple sequence alignment aided by quality analysis tools. Nucleic Acids Res 1997; 25:4876–4882.

53. Sneath PH, Sokal RR. Numerical Taxonomy. San Francisco, CA: Freeman, 1973.

54. Blanken RL, Klotz LC, Hinnebusch AG. Computer comparison of new and existing criteria for constructing evolutionary trees from sequence data. J Mol Evol 1982; 19:9–19.

55. Saitou N, Nei M. The neighbour-joining method: a new method for reconstructing phylogenetic trees. Mol Biol Evol 1987; 4:406–425.

56. Barton GJ, Sternberg MJE. A strategy for the rapid multiple alignment of

protein sequences: confidence levels from tertiary structure comparisons. J Mol Biol 1987; 198:327–337.

57. Fitch WM, Margoliash E. Construction of phylogenetic trees. Science 1967; 155:279–284.

58. Corpet F. Multiple sequence alignment with hierarchical clustering. Nucleic Acids Res 1988; 16:10881–10890.

59. Gribskov M, McLachlan AD, Eisenberg D. Profile analysis: detection of distantly related proteins. Proc Natl Acad Sci USA 1987; 84:4355–4358.

60. Higgins DG, Sharp PM. CLUSTAL: a package for performing multiple sequence alignment on a microcomputer. Gene 1988; 73:237–244.

61. Wilbur WJ, Lipman DJ. Rapid similarity searches of nucleic acid and protein data banks. Proc Natl Acad Sci USA 1983; 80:726–730.

62. Higgins DG, Bleasby AJ, Fuchs R. CLUSTAL V: improved software for multiple sequnce alignment. CABIOS 1992; 8:189–191.

63. Thompson JD, Higgins DG, Gibson TJ. CLUSTAL W: improving the sensitivity of progressive multiple sequence alignment through sequence weighting, positions-specific gap penalties and weight matrix choice. Nucleic Acids Res 1994; 22:4673–4680.

64. Genetics Computer Group. Program manual for the GCG Package, Version 8. 575 Science Drive Madison, WI, USA, 1993:53711.

65. Taylor WR. A flexible method to align large numbers of biological sequences. J Mol Evol 1988; 28:161–169.

66. Heringa J. Two strategies for sequence comparison: profile-preprocessed and secondary structure-induced multiple alignment. Comput Chem 1999; 23:341–364.

67. Taylor WR. Identification of protein sequence homology by consensus template alignment. J Mol Biol 1986; 188:233–258.

68. Depereiux E, Feytmans E. MATCH-BOX—a fundamentally new algorithm for the simultaneous alignment of several protein sequences. Comput Appl Biosci 1992; 8:501–509.

69. Boguski MS, Hardison RC, Schwartz S, Miller W. Analysis of conserved domains and sequence motifs on cellular regulatory proteins and locus control regions using new software tools for multiple alignment and visualization. New Biol 1992; 4:247–260.

70. Schuler GD, Altschul SF, Lipman DJ. A workbench for multiple alignment construction and analysis. Proteins 1991; 9:180–190.

71. Lawrence CE, Altschul SF, Boguski MS, Liu JS, Neuwald AF, Wootton JC. Detecting subtle sequence signals: a Gibbs sampling strategy for multiple alignment. Science 1993; 262:208–214.

72. Morgenstern B, Dress A, Werner T. Multiple DNA and protein sequence alignment based on segment-to-segment comparison. Proc Natl Acad Sci USA 1996; 93:12098–12103.

73. Morgenstern B. DIALIGN2: improvement of the segment-to-segment approach to multiple sequence alignment. Bioinformatics 1999; 15:211–218.

74. Brocchieri L, Karlin S. A symmetric-iterated multiple alignment of protein sequences. J Mol Biol 1998; 276:249–264.

75. Gotoh O. Significant improvement in accuracy of multiple protein sequence alignments by iterative refinement as assessed by reference to structural alignments. J Mol Biol 1996; 264:823–838.

76. Heringa J. Local weighting schemes for protein multiple sequence alignments. Comput Chem 2001. In press.

77. Frishman D, Argos P. Incorporation of long-distance interactions in a secondary structure prediction method. Protein Eng 1996; 9:133–142.

78. Frishman D, Argos P. Seventy-five percent accuracy in protein secondary structure prediction. Proteins 1997; 27:329–335.

79. Rost B, Sander C. Prediction of protein secondary structure at better than 70% accuracy. J Mol Biol 1993; 232:584–599.

80. Lüthy R, McLachlan AD, Eisenberg D. Proteins Struct Funct Genet 1991; 10:229.

81. Notredame C, Higgins DG, Heringa J. T-Coffee: a novel method for fast and accurate multiple sequence alignment. J Mol Biol 2000; 302:205–217.

82. Notredame C, Higgins DG. SAGA: sequence alignment by genetic algorithm. Nucleic Acids Res 1996; 24:1515–1524.

83. Eddy SR. Multiple alignment using hidden Markov models. Proc. Int. Conf. on Intelligent Systems in Mol Biol (ISMB) 1995; 3:114–120.

84. Thompson JD, Plewniak F, Poch O. BAliBASE: a benchmark alignment database for the evaluation of multiple sequence alignment programs. Bioinformatics 1999; 15:87–88.

85. Feng D-F, Johnson MS, Doolittle RF. Aligning amino acid sequences: comparison of commonly used methods. J Mol Evol 1985; 25:351–360.

86. Doolittle RF. Similar amino acid sequences: chance or common ancestry. Science 1981; 214:149–159.

87. Dayhoff MO, Schwartz RM, Orcutt BC. A model of evolutionary change in proteins. In: Dayhoff MO, ed. Atlas of Protein Sequence and Structure. Vol. 5. Washington, DC: Natl. Biomed. Res. Found, 1978; (suppl 3):345–352.

88. Jones DT, Taylor WR, Thornton JM. The rapid generation of mutation matrices from protein sequences. CABIOS 1992; 8:275–282.

89. Gonnet GH, Cohen MA, Benner SA. Exhaustive matching of the entire protein sequence database. Science 1992; 256:1443–1445.

90. Henikoff S, Henikoff JG. Amino acid substitution matrices from protein blocks. Proc Natl Acad Sci USA 1992; 89:10915–10919.

91. Hofmann K, Bucher P, Falquet L, Bairoch A. The PROSITE database, its status in 1999. Nucleic Acids Res 1999; 27:215–219.

92. Risler JL, Delormo MO, Delacroix H, Henaut A. Amino acid substitutions in structurally related proteins. A pattern recognition approach. Determination of new and efficient scoring matrix. J Mol Biol 1988; 204:1019–1029.

93. Johnson MS, Overington JP. A structural basis for sequence comparisons. An evaluation of scoring methodologies. J Mol Biol 1993; 233:716–738.

94. Šali A, Blundell TL. Comparative protein modelling by satisfaction of spatial restraints. J Mol Biol 1993; 234:779–815.

95. Henikoff S, Henikoff JG. Performance evaluation of amino acid substitution matrices. Proteins Struct Funct Genet 1993; 17:49–61.

96. Altschul SF, Gish W, Miller W, Meyers EW, Lipman DJ. Basic local alignment search tool. J Mol Biol 1990; 215:403–410.

97. Vogt G, Etzold T, Argos P. An assessment of amino acid exchange matrices in aligning protein sequences: the twilight zone revisited. J Mol Biol 1995; 249:816–831.

98. Pascarella S, Argos P. A data bank merging related protein structures and sequences. Protein Eng 1992; 5:121–137.

99. Mehta PK, Heringa J, Argos P. A simple and fast approach to prediction of protein secondary structure from multiply aligned sequences with accuracy above 70%. Protein Sci 1995; 4:2517–2525.

100. Thompson MJ, Goldstein RA. Constructing amino acid residue substitution classes maximally indicative of local protein structure. Proteins 1996; 25:28–37.

101. Koshi JM, Goldstein RA. Models of natural mutations including site heterogeneity. Proteins 1998; 32:289–295.

102. Rooman MJ, Wodak SJ. Identification of predictive sequence motifs limited by protein structure data base size. Nature 1988; 335:45–49.

103. Presnell SR, Cohen BI, Cohen FE. A segment-based approach to protein secondary structure prediction. Biochemistry 1992; 31:983–993.

104. Staden R. Methods to define and locate patterns of motifs in sequences. Comput Appl Biosci 1988; 4:53–60.

105. Day WHE, McMorris FR. Critical comparison of consensus methods for molecular sequences. Nucleic Acids Res 1992; 20:1093–1099.

106. Taylor WR. The classification of amino acid conservation. J Theor Biol 1986; 119:205–218.

107. Patthy L. Detecting homology of distantly related proteins with consensus sequences. J Mol Biol 1987; 198:567–577.

108. Smith RF, Smith TF. Automatic generation of primary sequence patterns from sets of related sequences. Proc Natl Acad Sci USA 1990; 87:118–122.

109. Vihinen M. An algorithm for simultaneous comparison of several sequences. Comput Appl Biosci 1988; 4:89–92.

110. Vingron M, Argos P. Motif recognition and alignment for many sequences by comparison of dot-matrices. J Mol Biol 1991; 218:33–43.

111. Chappey C, Hazout S. A method for delineating structurally homogeneous regions in protein sequences. Comput Appl Biosci 1992; 8:255–260.

112. Dodd IB, Egan JB. Improved detection of helix-turn-helix DNA-binding motifs in protein sequences. Nucleic Acids Res 1990; 18:5019–5026.

113. Attwood TK, Findlay JBC. Design of a discriminating fingerprint for G-protein-coupled receptors. Protein Eng 1993; 6:167–176.

114. Bairoch A. The PROSITE dictionary of sites and patterns in proteins, its current status. Nucleic Acids Res 1993; 21:3097–3103.

115. Bucher P, Karplus K, Moeri N, Hofmann K. A flexible motif search technique based on generalized profiles. Comput Chem 1996; 20:3–24.

116. Henikoff S, Henikoff JG. Automated assembly of protein blocks for database searching. Nucleic Acids Res 1991; 19:6565–6572.

117. Smith HO, Annau TM, Chandrasegaran S. Finding sequence motifs in

groups of functionally related proteins. Proc Natl Acad Sci USA 1990; 87: 826–830.

118. Wallace JC, Henikoff S. PATMAT: a searching and extraction program for sequence, pattern and block queries and databases. Comput Appl Biosci 1992; 8:249–254.

119. Henikoff JG, Henikoff S. New features of the Blocks database server. Nucleic Acids Res 1999; 27:226–228.

120. Attwood TK, Beck ME. PRINTS—a protein motif fingerprint database. Protein Eng 1994; 7:841–848.

121. Bateman A, Birney E, Durbin R, Eddy SR, Finn RD, Sonnhammer ELL. Pfam 3.1: 1313 multiple alignments and profile HMMs match the majority of proteins. Nucleic Acids Res 1999; 27:260–262.

122. Corpet F, Gouzy J, Kahn D. Recent improvements of the ProDom database of protein domain families. Nucleic Acids Res 1999; 27:263–267.

123. Gracy J, Argos P. Automated protein sequence database classification. II. Delineation of domain boundaries from sequence similarities. Bioinformatics 1998; 14:174–187.

124. Ogiwara A, Uchiyama I, Seto Y, Kanehisa M. Construction of a dictionary of sequence motifs that characterise groups of related proteins. Protein Eng 1992; 5:479–488.

125. Sheridan RP, Venkataraghavan R. A systematic search for protein signature sequences. Proteins 1992; 14:16–28.

126. Sonnhammer ELL, Kahn D. Modular arrangement of proteins as inferred from analysis of homology. Protein Sci 1994; 3:482–492.

127. Altschul SF, Lipman DJ. Protein database searches for multiple alignments. Proc Natl Acad Sci USA 1990; 87:5509–5513.

128. Gouzy J, Eugène P, Greene EA, Kahn D, Corpet F. XDOM, a graphical tool to analyse domain arrangements in protein families. CABIOS 1997; 13:601–608.

129. Park J, Teichmann SA. DIVCLUS: an automatic method in the GEANFAM-MER package that finds homologous domains in single- and multi-domain proteins. Bioinformatics 1998; 14:144–150.

130. Katz WT, Snell JW, Merickel MB. Artificial neural networks. Methods Enzymol 1992; 210:610–636.

131. Hirst JD, Sternberg MJE. Prediction of structural functional features of protein and nucleic acid sequences by artificial neural networks. Biochemistry 1992; 31:7211–7218.

132. Frishman DI, Argos P. Recognition of distantly related protein sequences using conserved motifs and neural networks. J Mol Biol 1992; 228:951–962.

133. Bailey TL, Elkan C. Fitting a mixture model by expectation maximization to discover motifs in biopolymers. Proceedings of the Second International Conference on Intelligent Systems for Molecular Biology. AAAI Press, 1994:28–36.

134. Bailey TL, Elkan C. The value of prior knowledge in discovering motifs with MEME. Proceedings of the Third International Conference on Intelligent Systems for Molecular Biology. AAAI Press, 1995:21–29.

135. Cockwell KY, Giles IG. Software tools for motif and pattern scanning: program descriptions including a universal sequence reading algorithm. Comput Appl Biosci 1989; 5:227–232.
136. Sibbald PR, Argos P. Scrutineer: a computer program that flexibly seeks and describes motifs and profiles in protein sequence databases. Comput Appl Biosci 1990; 6:279–288.
137. Sternberg MJE. Library of common protein motifs. Nature 1991; 349:111.
138. Staden R. Screening protein and nucleic acid sequences against libraries of patterns. DNA Sequence 1991; 1:369–374.
139. Fuchs R. MacPattern: protein pattern searching on the Apple MacIntosh. Comput Appl Biosci 1991; 7:105–106.
140. Venezia D, O'Hara PJ. Rapid motif compliance scoring with match weight sets. Comput Appl Biosci 1993; 9:65–69.
141. Mehldau G, Myers G. A system for pattern matching applications of bio-sequences. Comput Appl Biosci 1993; 9:299–314.
142. Guigo R, Johansson A, Smith TF. Automatic evaluation of protein sequence functional patterns. Comput Appl Biosci 1991; 7:309–315.
143. Rhode K, Bork P. A fast sensitive pattern-matching approach for protein sequences. Comput Appl Biosci 1993; 9:183–189.
144. van Belkum A, Scherer S, van Alphen L, Verbrugh H. Short sequence DNA repeats in prokaryotic genomes. Microbiol Mol Biol 1998; 28:905–916.
145. Huang C, Lin Y, Yang Y, Huang S, Chen C. The telomeres of *Streptomyces* chromosomes contain conserved palindromic sequences with potential to form complex secondary structures. Mol Microbiol 1998; 28:905–916.
146. Kurtz S, Schleiermacher C. REPuter: fast computation of maximal repeats in complete genomes. Bioinformatics 1999; 15:426–427.
147. Heringa J. The evolution and recognition of protein sequence repeats. Comput Chem 1994; 18:233–243.
148. McLachlan AD. Gene duplications in the structural evolution of chymo-trypsin. J Mol Biol 1979; 128:49–79.
149. McLachlan AD, Stewart M. The 14-fold periodicity in α-tropomyosin and the interaction with actin. J Mol Biol 1976; 103:271–298.
150. McLachlan AD. Analysis of periodic patterns in amino acid sequences: collagen. Biopolymers 1977; 16:1271–1297.
151. Kolaskar AS, Kulkarni-Kale U. Sequence alignment approach to pick up conformationally similar protein fragments. J Mol Biol 1992; 223:1053–1061.
152. Boswell DR, McLachlan AD. Sequence comparison by exponentially-damped alignment. Nucleic Acids Res 1984; 12:457–464.
153. Heringa J, Argos P. A method to recognize distant repeats in protein sequences. Proteins Struct Funct Genet 1993; 17:391–411.
154. George RA, Heringa J. The REPRO server: finding protein internal sequence repeats through the web. Trends Biochem Sci 2000; 25:515–517.
155. Pellegrini M, Marcotte EM, Yeates TO. A fast algorithm for genome-wide analysis of proteins with repeated sequences. Proteins Struct Funct Genet 1999; 35:440–446.

156. Marcotte EM, Pellegrini M, Yeates TO, Eisenberg D. A census of protein repeats. J Mol Biol 1998; 293:151–160.
157. Heger A, Holm L. Rapid automatic detection of repeats in protein sequences. Proteins Struct Funct Genet 2000; 41:224–237.
158. Barton GJ, Sternberg MJE. Flexible protein sequence patterns: a sensitive method to detect weak structural similarities. J Mol Biol 1990; 212:389–402.
159. Sibbald PR, Argos P. Weighting aligned protein or nucleic acid sequences to correct for unequal representation. J Mol Biol 1990; 216:813–818.
160. Lüthy R, Xenarios I, Bucher P. Improving the sensitivity of the sequence profile method. Protein Sci 1994; 3:139–146.
161. Sunyaev SR, Rodchenkov IV, Eisenhaber F, Kuznetsov EN. Analysis of the position dependent amino acid probabilities and its application to the search for remote homologues. Proc 2nd Annu Int Conf Computers in Molecular Biology (RECOMB98), 1998:258–264.
162. Sunyaev SR, Eisenhaber F, Rodchenkov IV, Eisenhaber B, Tumanyan VG, Kuznetsov EN. PSIC: profile extraction from sequence alignments with position-specific counts of independent observations. Protein Eng 1999; 12: 387–394.
163. Tatusov RL, Altschul SF, Koonin EV. Proc Natl Acad Sci USA 1994; 91: 12091–12095.
164. Bruno WJ. Modeling residue usage in aligned protein sequences via maximum likelihood. Mol Biol Evol 1996; 13:1368–1374.
165. Henikoff JG, Henikoff S. Using substitution probabilities to improve position-specific scoring matrices. CABIOS 1996; 12:135–143.
166. Sjölander K, Karplus K, Brown M, Hughly R, Krogh A, Mian I, Haussler D. Dirichlet mixtures: a method for improved detection of weak but significant protein sequence homology. CABIOS 1996; 12:327–345.
167. Henikoff S, Henikoff JG. Embedding strategies for effective use of information from multiple sequence alignments. Protein Sci 1997; 6:698–705.
168. Altschul SF, Madden TL, Schäffer AA, Zhang J, Zhang Z, Miller W, Lipman DJ. Gapped BLAST and PSI-BLAST: a new generation of protein database search programs. Nucleic Acids Res 1997; 25:3389–3402.
169. Durbin R, Eddy S, Krogh A, Mitchison G. Biological Sequence Analysis. Probabilistic Models of Proteins and Nucleic Acids. London, UK: Cambridge Univ Press, 1998:356.
170. Eddy SR. Hidden Markov models. Curr Opin Struct Biol 1996; 6:361–365.
171. Karplus K, Barrett C, Hughey R. Hidden Markov models for detecting remote protein homologies. Bioinformatics 1998; 14:846.
172. Chao K-M, Pearson WR, Miller W. Aligning two sequences within a specified diagonal band. CABIOS 1992; 8:481–487.
172a. Eddy SR. Profile hidden Markov models. Bioinformatics 1998; 14:755–763.
173. Karlin S, Altschul SF. Methods for assessing the statistical significance of molecular sequence features by using general scoring schemes. Proc Natl Acad Sci USA 1990; 87:2264–2268.
174. Altschul SF, Koonin EV. Iterated profile searches with PSI-BLAST—a tool for discovery in protein databases. TIBS 1997; 23:444–447.

175. Jones DT. Protein secondary structure prediction based on position specific scoring matrices. J Mol Biol 1999; 292:195–202.
176. Collins JF, Coulson AFW. Significance of protein sequence similarities. Methods Enzymol 1990; 183:474–486.
176a. Sturrock SS, Collins JF. MPsrch version 1.3. Biocomputing Research Unit. Edinburgh, Scotland: University of Edinburgh, 1993.
177. Barton GJ. An efficient algorithm to locate all locally optimal alignments between two sequences allowing for gaps. CABIOS 1993; 9:729–734.
178. Shi J, Blundell TL, Mizuguchi K. FUGUE: sequence-structure homology recognition using environment-specific substitution tables and structure-dependent gap penalties. J Mol Biol 2001; 310:243–257.
179. Bernstein FC, Koetzle TF, Williams GJ, Meyer EF, Brice MD, Rodgers JR, Kennard O, Shimanouchi T, Tasumi M. The protein data bank: a computer-based archival file for macromolecular structures. J Mol Biol 1977; 112:535–542.
180. Schiffer M, Edmundson AB. Use of helical wheels to represent the structures of proteins and to identify segments with helical potential. Biophys J 1967; 7:121–135.
181. Kabsch W, Sander C. Dictionary of protein secondary structure: pattern recognition of hydrogen-bonded and geometrical features. Biopolymers 1983; 22:2577–2637.
182. Sklenar H, Etchebest C, Lavery R. Describing protein structure: a general algorithm yielding complete helicoidal parameters and a unique overall axis. Proteins 1989; 6:46–60.
183. Woodcock S, Mornon J-P, Henrissat B. Detection of secondary structure elements in proteins by hydrophobic cluster analysis. Protein Eng 1992; 5:629–635.
184. Colloc'h N, Etchebest C, Thoreau E, Henrissat B, Mornon J-P. Comparison of three algorithms for the assignment of secondary structure in proteins: the advantage of a consensus assignment. Protein Eng 1993; 6:377–382.
185. Russell RB, Barton GJ. The limits of protein secondary structure prediction accuracy from multiple sequence alignment. J Mol Biol 1993; 234:951–957.
186. Taylor WR. An algorithm to compare secondary structure predictions. J Mol Biol 1984; 173:512–521.
187. Cohen FE, Abarbanel RM, Kuntz ID, Fletterick RJ. Turn prediction in proteins using a pattern-matching approach. Biochemistry 1986; 25:266–275.
188. Cohen FE, Kuntz ID. Tertiary structure prediction. In: Fasman GD, ed. Prediction of Protein Structure and the Principles of Protein Conformation. New York: Plenum, 1989:647–706.
189. Sternberg MJE. Secondary structure prediction. Curr Opin Struct Biol 1992; 2:237–241.
190. Benner SA, Cohen MA, Gerloff D. Predicted secondary structure for the Src homology 3 domain. J Mol Biol 1993; 229:295–305.
191. Rost B, Sander C, Schneider R. Redefining the goals of secondary structure prediction. J Mol Biol 1994; 235:13–26.
192. Cuff JA, Barton GJ. Evaluation and improvement for multiple sequence methods for protein secondary structure prediction. Proteins Struct Funct Genet 1999; 34:508–519.

193. Frishman D, Argos P. Knowledge-based secondary structure assignment. Proteins Struct Funct Genet 1995; 25:633.

194. King SM, Johnson WC. Assigning secondary structure from protein coordinate data. Proteins Struct Funct Genet 1999; 35:313–320.

195. Taylor WR. Defining linear segments in protein structure. J Mol Biol 2001; 310:1135–1150.

196. Szent-Györgyi AG, Cohen C. Role of proline in polypeptide chain configuration of proteins. Science 1957; 126:697.

197. Periti PF, Quagliarotti G, Liquori AM. Recognition of α-helical segments in proteins of known primary structure. J Mol Biol 1967; 24:313–322.

198. Nagano K. Logical analysis of the mechanism of protein folding. J Mol Biol 1973; 75:401–420.

199. Lim VI. Structural principles of the globular organization of protein chains. A stereochemical theory of globular protein secondary structure. J Mol Biol 1974; 88:857–872.

200. Chou PY, Fasman GD. Prediction of protein conformation. Biochemistry 1974; 13:211–215.

201. Garnier J, Osguthorpe DJ, Robson B. Analysis of the accuracy and implications of simple methods for predicting the secondary structure of globular proteins. J Mol Biol 1978; 120:97–120.

202. Gibrat J-F, Garnier J, Robson B. Further developments of protein secondary structure prediction using information theory. New parameters and consideration of residue pairs. J Mol Biol 1987; 198:425–443.

203. Garnier JG, Gibrat J-F, Robson B. GOR method for predicting protein secondary structure from amino acid sequence. Methods Enzymol 1996; 266:540–553.

204. Schultz GA. A critical evaluation of methods for prediction of protein secondary structures. Annu Rev Biophys Chem 1988; 17:1–21.

205. Zvelebil MJ, Barton GJ, Taylor WR, Sternberg MJE. Prediction of protein secondary structure and active sites using the alignment of homologous sequences. J Mol Biol 1987; 195:957.

206. Levin JM, Pascarella S, Argos P, Garnier J. Quantification of secondary structure prediction improvement using multiple alignments. Protein Eng 1993; 6:849–854.

207. Cohen FE, Abarbanel RM, Kuntz ID, Fletterick RJ. Secondary structure assignment for α/β proteins by a combinatorial approach. Biochemistry 1983; 25:4894–4904.

208. Taylor WR, Thornton JM. Prediction of super-secondary structures in proteins. Nature 1983; 354:105–106.

209. Rooman MJ, Wodak S, Thornton JM. Amino acid templates derived from recurrent turn motifs in proteins: critical evaluation of their predictive power. Protein Eng 1989; 23–27.

210. Chandonia J-M, Karplus M. New methods for accurate prediction of protein secondary structure. Proteins Struct Funct Genet 1999; 35:293–306.

211. Yi T-M, Lander ES. Protein secondary structure prediction using nearest-neighbor methods. J Mol Biol 1993; 232:1117–1129.

212. Salamov AA, Solovyev VV. Prediction of protein secondary structure by combining nearest-neighbor algorithms and multiple sequence alignments. J Mol Biol 1995; 247:11–15.

213. King RD, Sternberg MJE. Identification and application of the concepts important for accurate and reliable protein secondary structure prediction. Protein Sci 1996; 5:2298.

214. Muggleton S, King R, Sternberg MJE. Protein secondary structure prediction using logic-based machine learning. Protein Eng 1992; 5:647–657.

215. Minsky M, Papert S. Perceptrons. Cambridge, MA: MIT Press, 1988.

216. Sander C, Schneider R. Database of homology derived protein structures and the structural meaning of sequence alignment. Proteins 1991; 9:56–68.

217. Bowie JU, Lüthy R, Eisenberg D. A method to identify protein sequences that fold into a known three-dimensional structure. Science 1991; 253:164–170.

218. Hutchinson EG, Thornton JM. A revised set of potentials for beta-turn formation in proteins. Protein Sci 1994; 3:2207.

219. Langosch D, Heringa J. Interaction of transmembrane helices by a knobs-into-holes packing characteristic of soluble coiled coils. Proteins Struct Funct Genet 1998; 31:150–159.

220. Unwin N. Nicotinic acetylcholine receptor at 9 Å resolution. J Mol Biol 1993; 229:1101–1124.

221. Miyazawa A, Fuyiyoshi Y, Stowell M, Unwin N. Nicotinic acetylcholine receptor at 4.6 Å resolution: transverse tunnels in the channel wall. J Mol Biol 1999; 288:765–786.

222. Doyle DA, Cabral JM, Pfuetzner RA, Kuo A, Gulbis JM, Cohen SL, Chait BT, MacKinnon R. The structure of the potassium channels: molecular basis of K^+ conduction and selectivity. Science 1998; 280:69–77.

223. Singer SJ. The properties of proteins in nonaqueous solvents. Adv Protein Chem 1962; 17:1–68.

224. Kyte J, Doolittle RF. A simple method for displaying the hydropathic character of a protein. J Mol Biol 1982; 157:105–132.

225. Argos P, Rao MJK, Hargrave PA. Structural prediction of membrane-bound proteins. Eur J Biochem 1982; 128:565–575.

226. Sweet RM, Eisenberg D. Correlation of sequence hydrophobicities measures similarity in three-dimensional protein structure. J Mol Biol 1983; 171:479–488.

227. Cornette JL, Cease KB, Margalit H, Spouge JL, Berzofsky JA, DeLisi C. Hydrophobicity scales and computational techniques for detecting amphipathic structures in proteins. J Mol Biol 1987; 195:659–685.

228. von Heijne G. The distribution of positively charged residues in bacterial inner membrane proteins correlates with the trans-membrane topology. EMBO J 1986; 5:3021–3027.

229. Klein P, Kanehisa MI, DeLisi C. The detection and classification of membrane-spanning proteins. Biochim Biophys Acta 1985; 815:468–476.

230. Rao MJK, Argos P. A conformational preference parameter to predict helices in integral membrane proteins. Biochim Biophys Acta 1986; 869:197–214.

231. Wang J, Pullman A. Do helices in membranes prefer to form bundles or stay dispersed in the lipid phase. Biochim Biophys Acta 1991; 1070:493–496.

232. Eisenberg D, Weiss RM, Terwilliger TC. The helical hydrophobic moment: a measure of the amphiphilicity of a helix. Nature 1982; 299:371–374.

233. Eisenberg D, Weiss RM, Terwilliger TC. The hydrophobic moment detects periodicity in protein hydrophobicity. Proc Natl Acad Sci USA 1984; 81:140–144.

234. Finer-Moore J, Stroud RM. Amphipathic analysis and possible formation of the ion channel in an acetylcholine receptor. Proc Natl Acad Sci USA 1984; 81:155–159.

235. Vogel H, Wright JK, Jähnig F. The structure of the lactose permease derived from Raman spectroscopy and prediction methods. EMBO J 1985; 4:3625–3631.

236. von Heijne G. Membrane protein structure prediction. Hydrophobicity analysis and the positive-inside rule. J Mol Biol 1992; 225:487–494.

237. Esposti MD, Crimi M, Venturoli G. A critical evaluation of the hydropathy profile of membrane proteins. Eur J Biochem 1990; 190:207–219.

238. Edelman J. Quadratic minimization of predictors for protein secondary structure. Application to transmembrane α-helices. J Mol Biol 1993; 232:165–191.

239. Fariselli P, Compiani M, Casadio R. Predicting secondary structure of membrane proteins with neural networks. Eur Biophys J 1993; 22:41–51.

240. Rost B, Casadio R, Fariselli P, Sander C. Transmembrane helices predicted at 95% accuracy. Protein Sci 1995; 4:521.

241. Persson B, Argos P. Prediction of transmembrane segments in proteins utilizing multiple sequence alignments. J Mol Biol 1994; 237:182–192.

242. Cronet P, Sander C, Vriend G. Modeling of transmembrane seven helix bundles. Protein Eng 1993; 6:59–64.

243. Kontoyianni M, Lybrand TP. Three-dimensional models for integral membrane proteins: possibilities and pitfalls. Perspect Drug Discov Des 1993; 1:291–300.

244. Taylor WR, Jones DT, Green NM. A method for α-helical integral membrane protein fold prediction. Proteins 1994; 18:281–294.

245. Durell SR, Guy HR. Atomic scale structure and functional models of voltage-gated potassium channels. Biophys J 1992; 62:238–250.

246. Boyd D, Beckwith J. Positively charged amino acid residues can act as topogenic determinants in membrane proteins. Proc Natl Acad Sci USA 1989; 86:9446–9450.

247. Sipos L, von Heijne G. Predicting the topology of eukaryotic proteins. Eur J Biochem 1993; 213:1333–1340.

248. Hartmann E, Rapoport TA, Lodish HF. Predicting the orientation of eukaryotic membrane-spanning proteins. Proc Natl Acad Sci USA 1989; 86:5786–5790.

249. Schirmer T, Rosenbusch JP. Prokaryotic and eukaryotic porins. Curr Opin Struct Biol 1991; 1:539–545.

250. Jacoboni I, Martelli PL, Fariselli P, De Pinto V, Casadio R. Prediction of the transmembrane regions of β-barrel membrane proteins with a neural network-based predictor. Protein Sci 2001; 10:779–787.

251. Janin J, Wodak S, Levitt M, Maigret M. Conformation of amino acid side-chains in proteins. J Mol Biol 1978; 125:357–386.

252. Hopp TP, Woods KR. Prediction of protein antigenic determinants from amino acid sequences. Proc Natl Acad Sci USA 1981; 78:3824–3828.

253. Levitt M. A simplified representation of protein conformations for rapid simulation of protein folding. J Mol Biol 1976; 104:59–107.

254. Atassi MZ. Antigenic structures of proteins. Eur J Biochem 1984; 145:1–20.

255. Parker JMR, Guo D, Hodges RS. New hydrophilicity scale derived from high-perfomance liquid chromatography peptide retention data: correlation of predicted surface residues with antigenicity and X-ray-derived accessible sites. Biochemistry 1986; 25:5425–5432.

256. Tainer JA, Getzoff ED, Alexander H, Houghten RA, Olson AJ, Lerner RA, Hendrickson WA. The reactivity of anti-peptide antibodies is a function of the atomic mobility of sites in a protein. Nature 1984; 312:127–134.

257. Westhof E, Altschuh D, Moras D, Bloomer D, Mondragon A, Klug A, Van Regenmortel MHV. Correlation between segmental mobility and the location of antigenic determinants in proteins. Nature 1984; 311:123–126.

258. Rini JM, Schulze-Gahmen U, Wilson IA. Structural evidence for induced fit as a mechanism for antibody-antigen recognition. Science 1992; 255:959–965.

259. Karplus PA, Schultz GE. Prediction of chain flexibility in proteins. Naturwissenschaften 1985; 72:212–213.

260. Jameson BA, Wolf H. The antigenic index: a novel algorithm for predicting antigenic determinants. Comput Appl Biosci 1988; 4:181–186.

261. Maksyitov AZ, Zagrebelnaya ES. ADEPT: a computer program for prediction of protein antigenic determinants. Comput Appl Biosci 1993; 9:291–297.

262. Lupas A, van Dyke M, Stock J. Predicting coiled-coils from protein sequences. Science 1991; 252:1162.

263. Lupas A. Prediction and analysis of coiled-coil structures. Methods Enzymol 1996; 266:513–525.

264. Heringa J, Frishman D, Argos P. Computational methods relating proteins sequence and structure. In: Allen G, ed. Proteins: A Comprehensive Treatise. Vol. I. Greenwich, CT: JAI Press, 1997:165–268.

265. Lipman DJ, Pearson WR. Rapid and sensitive protein similarity searches. Science 1985; 227:1435–1441.

5

Discrete Models of Biopolymers

Peter Schuster
University of Vienna, Vienna, Austria

Peter F. Stadler
University of Leipzig, Leipzig, Germany

1. INTRODUCTION

Crystallography has already revealed a great number of biopolymer structures at full atomic resolution, and the productivity of structural biologists is currently increasing at a breathtaking pace. The enormous amounts of data collected in structural databanks contain a true wealth of information. They are readily used in discussions of catalytic mechanisms of enzymes and ribozymes and provide the basis for models of molecular recognition. Many other applications of structural data in biochemistry and molecular biology, however, require fewer details and thus call for notions of coarse-grained structure. Too many data obscure common structural features in related biopolymers and impede comparisons that are of fundamental importance, for example, in molecular evolution. Discretized structure models are particularly interesting because they not only meet the need for straightforward recognition of basic features but also by their nature they can be enumerated and accessed by combinatorial and other rigorous mathematical techniques.

In this chapter we present models of discrete protein and RNA structures and review a few prominent results derived from them. In Sec. 2 we introduce three classes of discretized structures: (1) lattice models that retain only information on spatially coarse-grained structures, (2) contact graphs that reduce spatial information to local nearest-neighbor interactions, and (3) hypergraph models, which are a multidimensional extension of class 2. Answers to counting problems can often be given by combinatorics. Examples are presented in Sec. 3: RNA secondary structure graphs and self-avoiding walks as models for protein structures. Random graph theory is used in Sec. 4 to model the mapping of sequences into structures. The random graph model is then applied to RNA secondary structures in Sec. 5. The last section provides a brief conclusion and an outlook to further developments.

2. DISCRETIZED STRUCTURE MODELS

The *fine-grained* description of a molecular structure is simply the list of three-dimensional coordinates of each individual atom. This level of detail, however, is not suitable for all purposes. Indeed, coarse-grained representations such as *ribbon diagrams* are oftentimes used to interpret and compare protein folds. Ribbon diagrams are obtained by retaining only the coordinates of the backbone atoms, which are still represented by three-dimensional vectors. In this section we shall be concerned with an alternative approach, *discretized structure models*.

We may distinguish two major classes: combinatorial models that encode only local geometric information and models that explicitly retain information about the global three-dimensional embedding of the structure. Contact graphs and their hypergraph generalizations fall into the first class, whereas lattice models (mostly of proteins) belong to the second class. We restrict ourselves to the simplest cases, in which each monomer is represented by a single point or letter.

2.1. Lattice Proteins

Lattice models [1–12] provide a coarse-grained view of protein structure. The structure is represented by a *self-avoiding walk* (SAW), i.e., a path on a lattice that does not visit the same site more than once [13]. SAWs play a major role in polymer physics, where the main interest centers on equilibrium properties such as the number of configurations or the end-to-end distance of a polymer consisting of a fixed number n of monomers [14,15].

2.2. Contact Graphs

The three-dimensional structure of a linear biopolymer such as RNA, DNA, or a protein can be approximated by its *contact structure*, i.e., by the list of all pairs of monomers that are spatial neighbors. Contact structures of poly-peptides were introduced by Dill and coworkers in the context of lattice models of protein folding [16,17]. The secondary structures of single-stranded RNA and RNA form a special class of contact structures.

We assume that the monomers, amino acids and nucleotides alike, are numbered from 1 to n along the backbone. For simplicity we shall write $[n] = \{1,\ldots, n\}$. The adjacency matrix of the backbone $\mathbf{B}$ has the entries $\mathbf{B}_{i,i+1} = \mathbf{B}_{i+1,i} = 1$, $i \in [n-1]$. In a more general context, polymers with cyclic or branched backbones could be considered, see, e.g., Ref. 12.

A contact structure is faithfully represented by the *contact matrix* $\mathbf{C}$ with the entries $\mathbf{C}_{ij} = 1$ if the monomers i and j are spatial neighbors without being adjacent along the backbone, and $\mathbf{C}_{ij} = 0$ otherwise. Hence $\mathbf{C}_{ij} = 0$ if $|i - j| \le 1$. Note that both $\mathbf{B}$ and $\mathbf{C}$ are symmetrical matrices. We define the *(contact) diagram* $([n], \Omega)$ to consist of n vertices labeled 1 to n and a set Ω of *arcs* that connect nonconsecutive vertices. The diagram is simply a graphical represen-tation of the contact matrix. As an example we show the conventional ribbon diagram of the protein ubiquitin together with its discretized structure represented by a contact matrix and contact graph in Figure 1. A closely related class of diagrams that also allow arcs between consecutive vertices are the *linked diagrams* introduced by Touchard [18]. These are studied in some detail in Refs. 19–22.

The *contact graph* has the adjacency matrix $\mathbf{A} = \mathbf{B} + \mathbf{C}$. The familiar drawings of RNA secondary structures are a much used example of biomo-lecular contact graphs. The classical definition of a secondary structure [23] requires that each base pair with at most one other nucleotide. Thus nucleic acid secondary structures are special types of 1-diagrams. The second defining condition is that arcs do not cross. In terms of the contact matrix this means that if $\mathbf{C}_{ij} = \mathbf{C}_{kl} = 1$ and $i < k < j$, then $i < l < j$. Secondary structure (contact) graphs are outerplanar, i.e., they can be drawn in such a way that the backbone forms a circle and all base pairs are represented by chords that must not cross each other; see the example of phenylalanyl-tRNA in Figure 2.

An increasing number of experimental findings, as well as results from comparative sequence analysis, suggest that pseudoknots are important structural elements in many RNA molecules [24]. Notably, functional RNAs such as RNAseP RNA [25] and ribosomal RNA [26] contain pseudoknots. Almost all known pseudoknotted structures, with the notable exception of the *E. coli* αmRNA [27], belong to the class of *bisecondary structures* [28] that generalizes the notion of secondary structures to include pseudoknots without allowing overly involved knotted structures or nested pseudoknots. More precisely, a bisecondary structure can be understood as a superposition of two disjoint secondary structures. Their contact graphs are still planar, but now the chords may be drawn on both the inside and outside of the circle that represents the backbone.

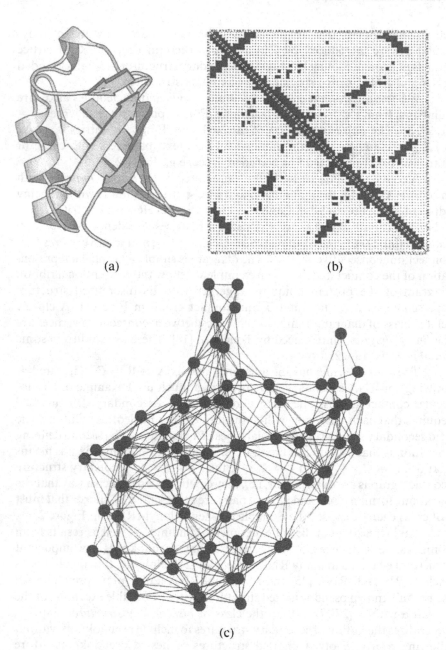

(a)

(b)

(c)

FIGURE 1 The structure of the ubiquitin molecule, pdb entry 1ubq. (a) Conventional ribbon diagram; (b) contact matrix; (c) contact graph.

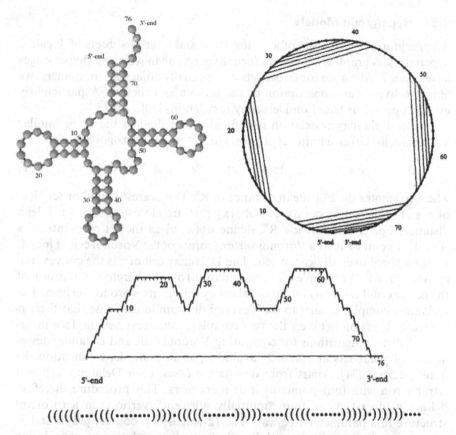

$$(((((((\cdots((((\cdots\cdots))))) \cdot (((((\cdots\cdots)))))\cdots((((\cdots\cdots)))))\cdot)))))\cdots$$

FIGURE 2 A few representations of RNA secondary structures. As an example we show the structure of phenylalanyl-transfer RNA (tRNA^{phe}). The conventional graph representation (upper left) is equivalent to the cyclic representation (upper right), the mountain representation (middle), and the parentheses representation (bottom). The contact matrix of tRNA^{phe} is shown in Figure 4. Each representation has a specific advantage. The conventional graph representation has been used successfully by biochemists in the interpretation of RNA reactivity, the cyclic representation allows the detection of pseudoknots as intersecting chords, the mountain representation is particularly useful for the detection of folding patterns in long RNA stretches, and a distance between structures can be easily defined in the parentheses representation as the Hamming distance between strings. We remark that graph and cyclic representations are two-dimensional and thus allow us to describe and detect pseudoknots, whereas mountain and parentheses representations are one-dimensional and become ambiguous in the case of pseudoknots.

2.3. Hypergraph Models

A hypergraph [29] consists of a vertex set V and a set of subsets of V called hyperedges. A graph is hence a uniform hypergraph in which all (hyper)edges have order 2. Allowing for larger sets of "mutually adjacent" monomers, we obtain a hypergraph description of the molecular structure. A particularly useful approach is based on Delaunay tesselations [30].

The Delaunay tesselation is defined as the dual of the more familar Voronoi cells: Given a finite set of points in $A \subseteq R^n$, the Voronoi cell of $x \in A$ is

$$D(x) = \{y \in R^n \mid d(x,y) \leq d(x',y) \qquad \forall x' \in A \setminus \{x\}\} \qquad (1)$$

where d denotes the Euclidean distance in R^n. The nearest-neighbor set $N(x)$ of $x \in A$ is the set of points $x' \in A \setminus \{x\}$ that are closest to x in Euclidean distance. For each point $u \in R^n$, define $nb(A, u)$ as the set of points $x' \in A \setminus \{u\}$. A point $v \in R^n$ is a Voronoi vertex (corner of the Voronoi cell) if $|nb(A, v)|$ is maximal over all nearest sets. The Delaunay cell of v is the convex hull $conv(nb(S, v))$. The complex (or triangulation) of A is therefore a partition of the convex hull $conv(A)$ into the Delaunay cells of its Voronoi vertices. The Delaunay complex is dual to the Voronoi diagram in the sense that there is a natural bijection between the two complexes that reverses the face inclusions. Efficient algorithms for computing Voronoi cells and Delaunay tesselations of point sets are publicly available; as an example, we mention the qhull package [31]. Apart from degenerate cases, each Delaunay cell is a tetrahedron with four points of A at its corners. This procedure therefore defines 4-edges (sets of four "mutually adjacent" vertices) in a (protein) structure in a parameter-free way. The (2-)edges of a contact graph and 3-edges can, of course, be derived directly from the tesselation by considering subsets.

Delaunay tesselations of protein structures have been used as the basic building block for designing knowledge-based potentials for protein threading and inverse folding [30,32–34]. The secondary structure model of nucleic acids could be extended to hypergraphs to include, e.g., base triplets, guanine quartetts, or adenine platforms [35].

3. COMBINATORIAL CONSIDERATIONS

3.1. Secondary Structure Graphs

3.1.1. Enumeration

A secondary structure on $n + 1$ digits can be obtained from a structure on n digits either by adding a free end at the right-hand end or by inserting a base pair $1 \equiv (k + 2)$. In the second case the substructure enclosed by this pair is an arbitrary structure on k digits, and the remaining part of length $n - k - 1$ is also an arbitrary valid secondary structure. Therefore, we obtain the following recursion formula for the number S_n of secondary structures:

$$S_{n+1} = S_n + \sum_{k=m}^{n-1} S_k S_{n-k-1}, \qquad n \geq m+1$$

(2)

$$S_0 = S_1 = \cdots = S_{m+1} = 1$$

Equation (2) was first derived by Waterman [23]; m denotes the minimum number of unpaired digits in a hairpin loop. Note that our definition of S_n differs from Waterman's for $n < m$; he used $S_n = 0$.

The above recursion can be used to develop an algorithm for generating random secondary structures with a uniform distribution

$$Prob\{S\} = 1/S_n$$

(3)

in the *shape space* of all secondary structures over a given chain length (see Ref. 36). Related recursions can be obtained for restricted classes of structures (see Table 1 and Ref. 37).

TABLE 1   Recursions for Restricted Structures

Structures with b components:

$$J_{n+1}(b) = J_n(b) + \sum_{k=m}^{n-1} S_k J_{n-k-1}(b-1), \quad b > 0, n \geq m+1$$

$$J_n(b) = 0, \quad b > 0, \quad n \leq m+1; \quad J_n(0) = 1, \quad n \geq 0$$

Structures with b base pairs (bonds):

$$H_{n+1}(b) = H_n(b) + \sum_{k=m}^{n-1} \sum_{\ell=0}^{b-1} H_k(\ell) H_{n-k-1}(b-\ell-1), \quad b > 0, \quad n \geq m+1$$

$$H_n(b) = 0, \quad b > 0, \quad n \leq m+1; \quad H_n(0) = 1, \quad n \geq 0$$

Structures with b stacks:

$$N_{n+1}(b) = N_n(b) + \sum_{k=m}^{n-1} \sum_{\ell=0}^{b} Z_{k+2}(\ell) N_{n-k-1}(b-\ell), \quad b > 0, \quad n \geq m+1$$

$$N_n(0) = 1; \quad N_n(b) = 0; \quad b > 0, \quad n \leq m+1$$

where $Z_n(b)$, the number of structures with b stacks given that the 3' and 5' ends are paired, satisfies

$$Z_n(b) = Z_{n-2}(b) + N_{n-2}(b-1) - Z_{n-2}(b-1),$$

$$Z_0(b) = Z_1(b) = 0$$

Structures with exactly b hairpins:

$$A_{n+1}(b) = A_n(b) + \sum_{k=m}^{n-1} \left[\sum_{\ell=1}^{b} A_k(\ell) A_{n-k-1}(b-\ell) + A_{n-k-1}(b-1) \right],$$

$$n \geq m+1$$

$$A_n(b) = \delta_{0,b}, \quad n \leq m+1$$

The recursion for the number of structures with b base pairs, $H_n(b)$, was also considered in Ref. 38. More recently, Schmitt and Waterman [39] obtained the closed expression

$$H_n(b) = \frac{1}{b}\binom{n-b}{b+1}\binom{n-b-1}{b-1}$$

for the special case $m = 1$. Recursions for some other types of structures, including the number $\psi_n^{m,l}$ of structures in which all stacks have predefined minimum length l, can be found in Ref. 37.

Most of the published work on the asymptotic behavior of RNA-related counting series [23,39–44] makes use of a proposition by Bender [45, Theorem 5], which was later found to be true only under more restrictive conditions than the published ones. If follows from the counterexamples discussed in Refs. 46 and 47 that Bender's result cannot be applied directly to the RNA problem. Starting from a simplified version of Darboux's theorem [48] (see also Ref. 49, p. 205), it is shown in Ref. 37 that the published expressions for the RNA counting series (see, e.g., Ref. 23) are nevertheless correct.

The series S_n was extensively studied in Ref. 23. The asymptotics of the more general series $\psi_n^{m,l}$ are determined [37, Theorem 4.8] as

$$\Psi_n^{m,l} \sim \frac{-g(\alpha)}{2\sqrt{\pi}} n^{-3/2}\left(\frac{1}{\alpha}\right)^n \tag{4}$$

where α is the smallest positive solution of

$$p(x) = [(1-x)(1-x^2+x^{2l}) + x^{2l}t_m(x)]^2 - 4x^{2l}(1-x^2+x^{2l})$$
$$= 0 \tag{5}$$

that satisfies

$$g(\alpha) = \frac{-1}{x^{2l}}\left[-\frac{1}{\alpha}\left.\frac{dp(x)}{dx}\right|_\alpha\right]^{1/2} \neq 0 \tag{6}$$

With $l = 1$, the recursions tabulated in Table 1 give rise to the asymptotic expressions

$J_n b$

$$S_n \sim \frac{\alpha^2}{(1-\alpha)^3} b \left(\frac{1-2\alpha}{1-\alpha}\right)^{b-1} \tag{7a}$$

$$H_n(b) \sim \frac{1}{(b+1)!b!} n^{2b} \tag{7b}$$

$$N_n(b) \sim \frac{C_b}{2^b(3b)!} n^{3b} \tag{7c}$$

$$A_n(b) \sim \frac{4}{2^{(3+m)b} b! (b-1)!} n^{2(b-1)} 2^n \tag{7d}$$

Here C_k denotes the Catalan numbers.

Numerical values of $1/\alpha$, which determines the growth of S_n and $\psi_n^{m,l}$ with sequence length n, are tabulated in Table 2. For comparison, we also list numerical estimates for bisecondary structures [28].

3.1.2. Energy Functions

The standard energy model for RNA and DNA secondary structures relies on the decomposition of the structure into "loops" (see Fig. 3). As shown in Ref. 50, these "loops" coincide with the unique minimal cycle basis. The most direct approach to the loop decomposition of a secondary structure uses the following partial order on the set of bonds (base pairs): A base pair k,l is *interior* to the base pair i, j if $i < k < l < j$. It is *immediately interior* if there is no base pair p, q such that $i < p < k < l < q < j$. For each base pair i, j, the corresponding loop is defined as consisting of i, j itself, the base pairs immediately interior to i, j, and all unpaired regions connecting these base pairs.

The energy of an RNA secondary structure is assumed to be the sum of the energy contributions of all loops. The most recent compilation of RNA energy parameters is that of Mathews et al. [51]. Current folding programs mostly rely on the parameter set discussed in Ref. 52, which extends earlier studies [53–55] by the systematic treatment of coaxial stacking. Parameters for DNA folding can be found in Refs. 56 and 57.

3.1.3. The RNA Folding Problem

The additive form of the energy model set the stage for an efficient solution of the minimum energy folding problem by means of a dynamic programming scheme similar to sequence alignment. This similarity was first realized and

TABLE 2 Numerical Values of $1/\alpha$[a]

	Secondary			Bisecondary		
m	$l = 1$	$l = 2$	$l = 3$	$l = 1$	$l = 2$	$l = 3$
1	2.618	1.986	1.716	4.42	2.49	2.00
2	2.414	1.899	1.680	4.03	2.43	1.94
3	2.289	**1.849**	1.652	3.81	**2.35**	1.89
5	2.147	1.783	1.612	3.44	2.22	1.74

[a] The values for the biophysically most relevant case, $l = 2$ and $m = 3$, are in **boldface** type.

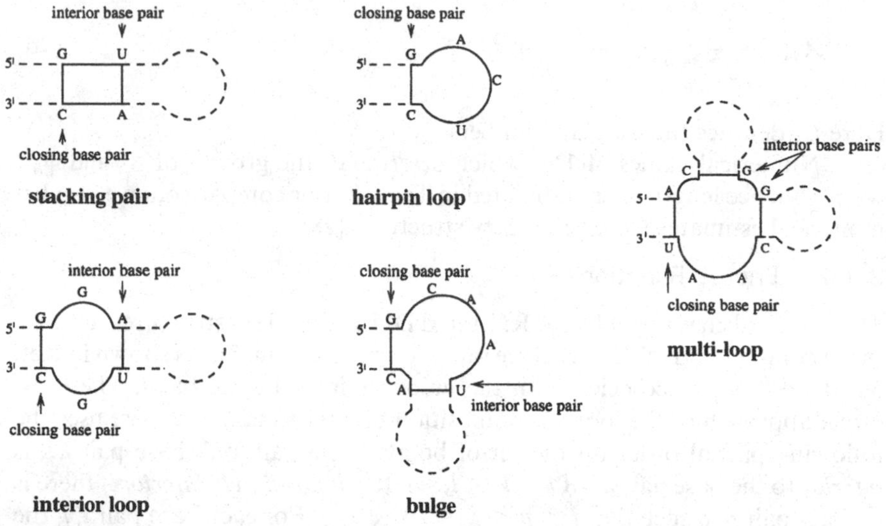

interior base pair

stacking pair

closing base pair

hairpin loop

interior base pairs

closing base pair

multi-loop

interior base pair

closing base pair

interior loop

closing base pair

interior base pair

bulge

FIGURE 3   RNA secondary structure elements. Any secondary structure can be uniquely decomposed into these types of loops.

exploited by Waterman [23] (see also Ref. 42); the first dynamic programming solution was proposed by Nussinor and Jacobson [58], originally for the "maximum matching" problem of finding the structure with the maximum number of base pairs [59]. Zuker and coworkers [60,61] formulated the algorithm for the minimum energy problem using the now standard energy model.

Since then several variations have been developed: Zuker [62] devised a modified algorithm that generates a subset of suboptimal structures within a prescribed increment of the minimum energy (see also Ref. 63). The algorithm will find any structure ψ that is optimal in the sense that there is no other structure ψ' with lower energy containing all base pairs that are present in ψ.

McCaskill [64] noted that the partition function over all secondary structures,

$$Q = \sum_{\Psi} \exp\left[\frac{-\Delta G(\Psi)}{kT}\right] \tag{8}$$

can be calculated by dynamic programming as well. In addition his algorithm can calculate the frequency with which each base pair occurs in the Boltzmann weighted ensemble of all possible structures, which can be conveniently represented in a "dot plot" (see Fig. 4). A related approach can be used to compute the complete density of states of an RNA sequence at predefined

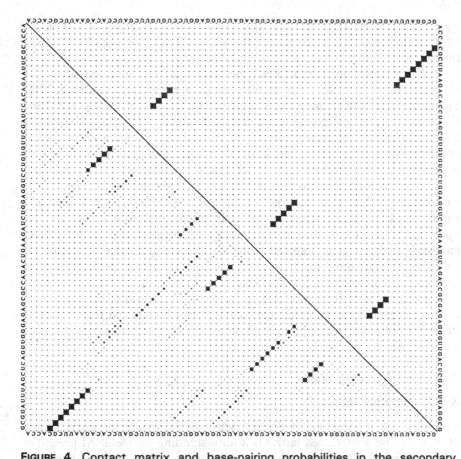

FIGURE 4 Contact matrix and base-pairing probabilities in the secondary structure of phenylalanyl-tRNA. All nonzero entries are indicated as black squares. In the lower (left) triangle we show the contact matrix. The matrix elements are 0 or 1 corresponding to empty or full squares, respectively. The upper (right) triangle contains the partition function. Here the size of the square is representative of the base-pairing probability.

energy resolution [65,66]. Another method for calculating the density of states, based on enumeration of structures, was proposed earlier [67]. However, this algorithm is restricted to subsets of structures containing no helices shorter than three and uses a simplified energy model.

Most recently, a program was designed by the Vienna group that can generate all secondary structures within some interval of the minimum energy based on dynamic programming and multiple backtracking [68,69]. In practice, suboptimal folding can handle millions of structures, corresponding, e.g., to an energy range of, say, 12kcal/mol at a chain length of 100 bases. Most of these algorithms are part of the Vienna RNA Package [70], which is freely available from http://www.tbi.univie.ac.at/.

The assumptions that an RNA molecule folds into its thermodynamic ground state may well be wrong even for moderately long sequences [71]. Simulations of the folding process itself can be used to avoid this problem. Consequently, several groups have designed kinetic folding algorithms for RNA secondary structures, mostly in an attempt to get more accurate predictions or to include pseudoknots (see, e.g., Refs. 72–76). Only a few papers have attempted to reconstruct folding pathways [77–79]. A more recent approach resolves the folding process to three elementary steps: base pair formation, base pair cleavage, and base pair shift [80,81]. RNA folding is simulated as a stochastic process starting from an initial state (commonly the open chain) to the minimum free energy conformation or a long-lived metastable state that is assumed to be an absorbing barrier. Sampling of sufficiently large numbers of folding trajectories yields probabilities of formation for different conformations.

In the case of functional RNAs, and provided a sufficient number of related sequences are available, the structure can be inferred from covariations. This phylogenetic approach is beyond the scope of this review, but see, e.g., Ref. 82.

3.2. Self-Avoiding Walks

3.2.1. Enumeration

Counting the number c_N of distinct self-avoiding walks of given length $N = n - 1$ on a prescibed lattice is a long-standing problem. At present a complete solution to this problem is unknown. It is easy to show, however, that for each lattice A there is a constant

$$\mu = \lim_{N \to \infty} c_N^{1/N} < z - 1 \tag{9}$$

where z is the connectivity of the lattice. The exact values of μ, however, are unknown even for the simplest lattices. Tight analytical bounds on μ have been obtained for a variety of lattices (see Ref. 13). It is commonly believed that the asymtotic behavior of c_N depends only on the spatial dimension d of the lattice:

$$c_N \sim \begin{cases} B\, N^{\gamma-1}\mu^N & for \quad d = 2, 3 \\[2mm] B\mu^N \,(\log N)^{1/4} & for \quad d = 4 \\[2mm] B\mu^N & for \quad d \geq 5 \end{cases} \tag{10}$$

The exponent γ probably depends only on the dimension of the lattice. The logarithmic correction for $n = 4$ was predicted by a renormalization group analysis (see, e.g., Ref. 83). Estimates for the parameters μ, γ, and B are compiled in Table 3.

The effective number μ of conformational isomes per amino acid in a protein structure has been estimated by various authors. For instance, Dill [96] reports $\mu = 3.8$, whereas $\mu \approx 10$ is obtained for the free chain by Flory [97].

3.2.2. Energy Functions

In contrast to the rather elaborate standard energy model for nucleic acids, most lattice protein models use simple contact potentials of the form

$$E(x) = \sum_{i<j} E(x_i, x_j) C_{ij} \tag{11}$$

which depends only on the amino acids x_i, x_j that form a contact (i, j). Most studies distinguish between only two clases of amino acids,

$$\mathbf{H} = \{A, C, I, L, M, F, W, Y, V\}$$
$$\mathbf{P} = \{R, N, D, E, Q, G, H, K, P, S, T\} \tag{12}$$

H (hydrophobic) and **P** (polar), with $E(\mathbf{H}, \mathbf{H}) = -1$ and $E(.,.) = 0$ otherwise; see e.g., Ref. 98. Alternative potentials for two-letter alphabets are studied systematically in Ref. 9.

TABLE 3 Combinatorial Parameters of SAWs in Two and Three Dimensions[a]

d	z	Lattice	μ	γ	B	Ref.
2	3	HEX	1.8477	0.345	1.28	84,85
2	4	SQ	2.6382	0.34275	1.93	84–88
2	6	TRI	4.1507	0.343	1.69	84–86,89
2	8	KM	6.62	~0.15	~1.15	
3	4	TET	2.621	1.164	1.48	85,88,90
3	6	SC	4.6839	1.161	1.39	85,91–93
3	8	BCC	6.5291	1.163	1.25	83,85,92,94
3	12	FCC	10.0364	1.162	1.26	85,91
3	24	TDKM	22.66	1.162	~1.14	95

[a] Lattices in the plane: Hexagonal (honey comb) HEX, square SQ, triangular TRI, and Knight's move KM, Lattices in three dimensions: Diamond (tetrahedral) TET, simple cubic SC, body-centered cubic BCC, face-centered cubic FCC, and a three-dimensional generalization of the Knight's move lattice TDKM.

These models allow the study of hydrophobic collapse. Furthermore, they admit an intrinsic distinction between folding and nonfolding sequences (a sequence folds into a native structure if the lowest energy structure is unique); it is not clear how well this approach will generalize to more complex potential functions and larger alphabets, which would lead to nondegenerate ground states for most sequences [99].

As an example of a more sophisticated contact potential, we mention Crippen's [4] ansatz

$$
E(x_i, x_j) = \begin{cases} \begin{array}{ccc} & -0.008 & \text{if } |i-j| = 3 \\ & 0.004 & \text{if } |i-j| = 4 \\ & 0.021 & \text{if } |i-j| = 5,6,7 \end{array} \\ \begin{pmatrix} -0.012 & -0.074 & -0.054 & 0.123 \\ -0.074 & 0.123 & -0.317 & 0.156 \\ -0.054 & -0.317 & -0.263 & -0.010 \\ 0.123 & 0.156 & -0.010 & -0.004 \end{pmatrix} & \text{if } |i-j| \geq 8 \end{cases}
$$

$$(13)$$

where the matrix entries correspond to the four amino acid classes

$$
\begin{aligned}
1 &= \{G, Y, H, S, R, N, E\} \\
2 &= \{A, V\} \\
3 &= \{L, I, C, M, F\} \\
4 &= \{P, W, T, K, D, Q\}
\end{aligned}
$$

$$(14)$$

The parameters of such potential functions are extracted from databases of known protein structures as log likelihood estimates or by means of the *inverse Boltzmann law* as described, e.g., in Refs. 100–103.

3.2.3. The Lattice Protein Folding Problem

The *lattice folding problem* consists of finding, for a prescribed amino acid sequence, a self-avoiding walk on a given lattice that minimizes energy. This combinatorial problem is NP-hard [104–106] even for simple quadratic and cubic lattices and very simple energy functions, including the **HP** model.

For short sequences and lattices with small effective connectivities μ, all possible conformations can be evaluated. In the case of moderate sequences, sometimes strongly constrained subsets of sequences, such as 27-mers that fill a $3 \times 3 \times 3$ cube, are considered (see e.g., Ref. 107). Heuristic algorithms such as CHCC [108] try to construct good approximations of the ground state using "compactness" as an additional criterion. Simple chain growth algorithms seem to yield fairly good results on average. A series of fast algorithms with exact performance bounds have been devised by Istrail and coworkers [12,109]. These produce solutions within a constant factor $c < 1$ of the maximal number of contacts.

4. RANDOM GRAPH MODELS OF SEQUENCE-STRUCTURE MAPS

4.1. The Random Graph Model

The numbers listed in Tables 2 and 3, together with the observation that the effective value of z for proteins appears to be somewhere in the range of $z = 3$–12, imply that sequence–structure maps are many-to-one, i.e., $f^{-1}(s)$ is a large set, at least for the more common structures.

This observation poses the question of how $f^{-1}(s)$ is embedded in the space of biopolymer sequences, i.e., what can we say in general about the set of sequences folding into s? In the absence of further information, we assume that $f^{-1}(s)$ is uniformly distributed in sequence space. In other words, we assume that the preimage of a structure s can be regarded as a suitable random subgraph Γ of the underlying sequence space. Here we restrict our attention to "host graphs" that are sequence spaces (Hamming graphs) Q_a^n an with a fixed alphabet of size a and fixed sequence length n.

Typically, random graph models assume a fixed vertex set V into which edges are introduced [110]. The appropriate model for preimages in sequence–structure maps, however, are the subgraphs Γ_x induced by randomly selected vertex sets X in the underlying sequence space [111,112].

Definition 1. *Let $\mathcal{B}$ (Q_a^n) be the set of all induced subgraphs of Q_a^n, and let $0 \le \lambda \le 1$ be a constant. Then we set for $\Gamma \in \mathcal{B}$ (Q_a^n)*

$$\mu_\lambda\{\Gamma\} = \lambda^{a^n}(1 - \lambda)^{a^n - |\Gamma|} \tag{15}$$

where $|\Gamma|$ is the size, i.e., the number of vertices, of the subgraph Γ. The random subgraph model is the probability space $\Omega_{n,\lambda} = (\mathcal{B}(Q_a^n), \mu_\lambda)$ of subgraphs of Q_a^n an with the measure μ_λ. We shall write Γ_n, for a random graph drawn from $\Omega_{n,\lambda}$.

The parameter λ can be interpreted as a fraction of neutral neighbors; i.e., $(n - 1)a\lambda$ is the expected vertex degree of the random induced subgraph Γ.

Let Q be a property of Γ_n. We say that Γ has property Q *asymptotically almost surely* (a.a.s.) if

$$\lim_{n \to \infty} \mu\{\Gamma_n \text{ has property Q}\} = 1 \tag{16}$$

4.2. Predictions

A subgraph Γ' is *dense* in Γ if each vertex of Γ is a vertex of Γ' or if it has at least an adjacent vertex in Γ'. A (sub)graph Γ' is *connected* if there is a path (of edges in Γ') connecting any two vertices of Γ'.

The parameter

$$\lambda^* = 1 - \left(\frac{1}{a}\right)^{a-1} \tag{17}$$

plays a crucial role in the random subgraph model.

Theorem 1. *If λ then Γ_n is connected and dense in Q_a^n a.a.s. If $\lambda > \lambda^*$, then Γ_n is neither connected nor dense in Q_a^n a.a.s.*

Proof. The proof of this theorem is quite lengthy and technical [112]. Hence we give only a brief sketch here. In order to deal with denseness, one considers the random variable $Z(\Gamma_n)$, counting the vertices of Q_a^n that are neither in Γ_n nor have an adjacent vertex in Γ_n. Using the "sieve formula" [110, p. 17] it is possible to derive the limit distribution of $Z(\Gamma_n)$ through its factorial moments. One then finds that $\lim_{n \to \infty} \mathbb{E}\left[Z(\Gamma_n)\right]$ is either 0 or ∞ depending on whether λ is larger or smaller than the threshold value λ^*.

The proof of the connectedness part proceeds via an analysis of the sizes of the connected components. In the first step one shows that for $\lambda > \lambda^*$ there are aymptotically almost sure (a.a.s.) no very small components, whereas below the threshold there are many of them. Furthermore, assymptotically almost all (a.a.all) vertices of Γ_n have large degrees above the threshold. The next step is to show that in this case a.a.all vertices of Q_a^n have many adjacent vertices in Γ_n. Then one shows that a.a.s. every pair of vertices in Γ_n with a finite distance k in Q_a^n is connected by a finite path in Γ_n. Finally, one shows that there are large enough subsets of vertices with mutually finite distances that can be connected by such paths.

A related result in the special case of the Boolean hypercube with a different random graph model based on independently drawing edges instead of vertices with probability p can be found in Ref. 110.

A connected component Γ' of graph Γ is a giant component if $|\Gamma'| > c|\Gamma|$ for some fixed constant $c > 0$. It is shown in Ref. 112 that Γ_n a.a.s. has a giant component for whenever $\lambda > 0$ is a constant. For Boolean hypercubes Ajtai et al. [113] proved in the edge-drawing model that there is a component with size $g2^n$, $g > 0$, provided p $= c/n$ and $c > 1$.

The component structure of Γ_n is discussed in more detail in Ref. 111.

Theorem 2. *There is a c > 0 such that for $\lambda_n = c(\ln n)/n$, the largest component X_1 of $\Gamma_n \subset Q_a^n$, for all $\varepsilon > 0$, satisfies a.a.s.*

$$|X| \geq (1 - \varepsilon)|\Gamma_n| \qquad (18)$$

The size of the second largest component X_2 is bounded by $|X_2| \leq Cn/\ln n$, where $C > 0$ is a constant depending only on a and c.

Application of these ideas to biological speciation is discussed in Refs. 114 and 115.

4.3. Neutral Paths

Neutral walks were used to gain information about the structure of the (connected components of) neutral networks in a series of computer experiments on RNA folding landscapes [116–118]. In each step we attempt to find a neutral neighbor such that the distance from the starting point increases. Therefore neutral walks on Q_a^n terminate at the latest after n steps.

The probability that a neutral walk with d steps cannot be elongated any further equals $(1 - \lambda)^{\alpha(d)}$, where $\alpha(d) = (a - 1)(n - d)$ denotes the number of "forward steps" increasing the distance from the starting point. The probability that a neutral walk of a Hamming graph terminates after exactly d steps is therefore [119]

$$\text{Prob}[\mathcal{L} = d] = (1 - \lambda)^{a(d)} \prod_{d'=1}^{d} \left[1 - (1 - \lambda)\alpha(d' - 1) \right] \qquad (19)$$

From Eq. (19) one can infer that there are long neutral paths with typical length n if $\lambda n/\ln n \to \infty$, and the walks are typically short ($\mathcal{L}/n \to 0$) for $\lambda < (\ln n)/n$. In the intermediate regime, $\lambda \sim C(\ln n)/n$ with $C > 1$, the typical neutral path length is proportional to n.

5. RNA SECONDARY STRUCTURES AND THE RANDOM GRAPH MODEL

Mappings of RNA sequence space onto shape space, $Q_a^n \to S^n$, were studied by the approaches summarized in Table 4. The random graph approach introduced in Sec. 4.1 yields information on the generic properties of sequence–structure mappings. Here we are more concerned with the specific features of RNA mappings, in particular with the consequences of the base-pairing logic.

5.1. The Product Space Model

As a consequence of the base-pairing logic, not every sequence is *compatible* with every structure. Although an arbitrary nucleotide may be located at each unpaired position of a structure ϕ, base-pairing positions are constrained to AU, UA, GC, CG, GU, or UA. In the following we shall write $C(\phi)$ for the set

TABLE 4   Various Strategies Applied to Study Sequence-Structure Maps of RNA

	Method	Advantage	Disadvantage	Ref.
Mathematical model	Random graph theory	Analytical expressions	Limited validity of model assumptions	112
Exhaustive folding and enumeration	Folding algorithm and handling of large samples (>109 objects)	Exact results	Limited to short chains: GC, $\ell \leq 32$ AUGC, $\ell \leq 16$	117,118
Statistical evaluation	Inverse folding or random walks in sequence space	Applicability to longer sequences	Limited accuracy due to statistics	116,120
Simulation of evolutionary dynamics	Chemical kinetics of replication and mutation	Evolutionary relevance	Restriction to small parts of sequence space	121–124

of all sequences that are compatible with ϕ. Clearly, only sequences that are compatible with ϕ can actually fold into this structure; thus, $f^{-1}(\phi) \subseteq C(\phi)$.

The distinction between paired and unpaired positions in a structure suggests a factorization of RNA sequence space into a space of unpaired bases and a space of base pairs, $\mathcal{Q}_\phi^n = \mathcal{Q}_{a_u}^{n_u} \times \mathcal{Q}_{a_n}^{n_p}$, with n_u and n_p being the numbers of unpaired bases and base pairs, respectively, in the secondary structure ϕ; hence $n = n_u + 2n_p$. For natural RNA molecules we have $a_u = 4$ and $a_p = 6$, because six base pairs are allowed in stacks. The vertex set of $\mathcal{Q}_\phi^n$ is $C(\phi)$. Two compatible sequences are neighbors of each other if they differ by either a point mutation in the unpaired part or by the exchange of one type of possible base pair for another one. Note that two sequences can be neighbors in $\mathcal{Q}_\phi^n$ if their Hamming distance in $\mathcal{Q}_4^n$ is 2; assume, for instance, that a GC pair is replaced by a UA pair.

The random graph model described in Sec. 4.1 can be customized to fit the situation in RNA more closely by taking the factorization $\mathcal{Q}_\phi^n = \mathcal{Q}_{a_u}^{n_u} \times \mathcal{Q}_{a_n}^{n_p}$ into account. Instead of a random subgraph of $\mathcal{Q}_\alpha^n$ we model the neutral network $f^{-1}(\phi)$ by a random induced subgraph $\Gamma[\phi] \subset \mathcal{Q}_\phi^n$. Two slightly different probability measures for $\Gamma[\phi]$ are considered in Ref. 112 with essentially the same qualitative results: One may conclude that if the restriction of the random graph $\Gamma[\phi]$ to both factors $\mathcal{Q}_{a_u}^{n_u}$ and $\mathcal{Q}_{a_n}^{n_p}$ is dense and connected, then $\Gamma[\phi]$ itself is dense and connected. Hence the discussion in Sec. 4.1 remains valid; one just has to take into account that we have different threshold values for the paired and unpaired factors.

5.2. Shape Space Covering

The random graph model can also be used to address the mutual location of the neutral networks of two different structures ϕ and ψ. The basic fact in this context is the so-called *intersection theorem* (Theorem 5 of Ref. 112):

Theorem 3. *Let ϕ and ψ be two secondary structures with the same length. Then $C(\phi) \cap C(\psi) \neq \emptyset$.*

The random graph approach then provides the following result (Theorem 8 of Ref. 112):

Theorem 4. *Let ϕ, ψ be two secondary structures with the same length, and suppose the neutral networks $\Gamma[\phi]$ and $\Gamma[\psi]$ are dense and connected almost surely. Then*

1. *The minimum distance of $\Gamma[\phi]$ and $\Gamma[\psi]$ in $\mathcal{Q}_4^{n_u} \times \mathcal{Q}_6^{n_p}$ is a.a.s. at most 2.*
2. *The expected Hamming distance from a randomly chosen sequence to the neutral network is a.a.s. at most*

$$\mathbb{E}[r] < (1 - 6/16)n_p + o(1) \tag{20}$$

This predicts that the neutral networks of any two secondary structures come very close together at least somewhere in sequence space. As a consequence, any two common secondary structures should be *accessible* from each other. We shall return to this topic in Sec. 5.3.5. Furthermore, Eq. (20) predicts that we can find sequences that fold into almost all common secondary structures within a ball of radius $\mathbb{E}[r]$ centered at any given point in sequence space. This phenomenon was termed *shape space covering* in Ref. 116 and was studied in detail in Ref. 118.

5.3. Comparison of Random Graph Models with Data from RNA

5.3.1. Exhaustive Enumeration

One of the few examples that allow direct testing of the prediction of random graph models is the mapping of RNA sequences into secodary structures. The most straightforward strategy is exhaustive folding of complete sequence spaces ($\mathcal{Q}_a^n$) and enumeration of results (Table 4). Because of the exponential increase in the number of sequences with chain length n and the limitation of efficient retrieval of data at sample sizes of a few 10^9 objects, this strategy is limited to rather small molecules. This implies restriction to chain lengths $n \leq 16$ for AUGC and $n \leq 32$ for AU or GC sequences. Table 5 contains a comparison of selected data on the numbers of minimum free energy RNA structures from exhaustive folding with the numbers $\psi_n^{3,2}$ of all secondary structure graphs with minimum stack length $l = 2$ and minimum length $m = 3$

TABLE 5   Comparison of Exhaustively Folded Sequence Spaces[a]

Chain length n	Number of sequences		Number of structures			
	2^n	4^n	$\psi_n^{3,2}$	AUGC	GC	AU
7	128	4.29×10^9	2		2	1
10	1,024	1.05×10^6	14		11	1
12	4,096	1.68×10^7	37		31 (29)	1
15	3.28×10^4	1.07×10^9	174		116	2
16	6.55×10^4	4.29×10^9	304	274 (223)	195 (186)	4
17	1.31×10^5	1.73×10^{10}	530		340	8
20	1.05×10^6	1.10×10^{12}	2,741		1,601	35
25	3.36×10^7	1.13×10^{15}	44,695		18,590	164
30	1.07×10^9	1.15×10^{18}	760,983		218,820	1,064

[a] Values given in parentheses are the counted numbers of actually occurring minimum free energy structures without isolated base pairs that are directly comparable to the numbers $\psi_n^{3,2}$.
Source: Refs. 117,118,125,126.

of the unpaired stretch in a hairpin loop. These were chosen according to empirical experience: Very small hairpin loops, $m < 3$, and isolated base pairs, $l = 1$, are highly unstable and occur only in exceptional cases such as short sequences and sequences with an extremely biased base composition. The examples shown in Table 5 contain only two minimum free energy structures with isolated base pairs formed by GC sequences of chain length $n = 12$, nine structures for GC sequences of chain length $n = 16$, and 51 structures for AUGC sequences of chain length $n = 16$.

Depending on the base-pairing alphabet, only a certain fraction of all structures will actually appear as most stable conformation. We see also that AUGC sequences sustain substantially more minimum free energy structures than GC sequences. The number of structures formed by AU sequences is rather small as a result of the relative weakness of AU base pairing and base pair stacking (in comparison to GC). We shall compare in more detail two cases with the prediction from random graph theory: (1) all sequences of chain length $n = 16$ and (2) GC sequences of chain length $n = 30$. For longer sequences we have to rely on statistical method to obtain direct information.

5.3.2.   Sequences of Chain Length $n = 16$

Structures, α_k, in Tables 6–8 are ranked according to their probability of formation from random sequences. These probabilities are simply derived by dividing the size of the preimages in sequence space by the total number of sequences, $p(\alpha_k) = |f^{-1}(\alpha_k)|/a^n$. Neutral networks in sequence space, corresponding, to the structure α_k, are characterized by their sequence of

TABLE 6 Shapes Frequently Formed by GC Sequences of Chain Length $n = 16$ as Minimal Free Energy Structure

Rank	Structure	Number of sequences	Number of components	Sequence of components
1	((((•••))))••••••	2568	1	2568
2	••••••((((•••))))	2541	1	2541
3	••••((((•••••))))	1895	1	1895
4	(((((•••)))))•••	1881	1	1881
5	•••(((((•••)))))	1880	1	1880
6	((((•••••))))••••	1803	1	1803
7	••(((((•••••)))))	1759	1	1759
8	(((((•••••)))))••	1738	1	1738
9	••••••••••••••••	1427	13	1016 358 16 11 10 4 3 2 2 2 1 1 1
10	••((((•••))))•••	1316	2	695 621
11	•((((•••))))•••••	1316	2	732 582
12	•••••••(((•••)))	1314	10	1292 5 4 3 3 2 2 1 1 1
13	(((•••)))••••••••	1310	9	1293 4 4 2 2 2 1 1 1
14	••••((((•••))))•	1293	2	691 602
15	•••((((•••))))••	1290	2	647 643
16	•••((((•••••))))	1231	2	658 573
17	((((•••••))))•••	1205	2	664 541
18	••((((•••••••))))	1099	2	603 496
19	((((•••••••))))••	1075	2	560 515
20	•((((•••••••))))	1064	2	574 490
⋮	⋮	⋮	⋮	⋮
39	••((((•••••))))•	659	4	181 171 157 150
40	•((((•••••))))••	647	4	174 166 160 147

components, which are listings of component sizes. What we expect to observe are either connected networks above the connectivity threshold or networks consisting of several components with one largest *giant* component. We have to recall, however, that the connectivity phenomenon discussed in Sec. 4.1 is an asymptotic property and finite size effects may easily override it in the case of short sequences. The most dramatic example is the sequence space $\mathcal{Q}_{AU}^{16}$: 96.8% of the sequences do not form a stable secondary structure at all. For GC sequences the open chain amounts to only 2.2%, and in $\mathcal{Q}_{AUGC}^{16}$ we have 63.1% sequences with a nontrivial minimum free energy structure.

The first eight most frequent shapes formed by sequences from $\mathcal{Q}_{AU}^{16}$ have a single connected component. The neutral network of the open chain

TABLE 7 All Shapes Formed by AU Sequences of Chain Length $n = 16$ as Minimal Free Energy Structures

Rank	Structure	Number of sequences	Number of components
1	•••••••••••••••••	63 488	1
2	•(((((((((•••)))))))))	1020	1
3	(((((•••)))))•	1012	1
4	((((((••••))))))	16	1

TABLE 8 Shapes Frequently Formed by AUGC Sequences of Chain Length $n = 16$ as Minimal Free Energy Structures

Rank	Structure	Number of sequences	Number and sequence of components
1	•••••••••••••••••	2,709,560,048	1
2	(((•••)))••••••••	52,505,831	1
3	••••••••(((•••)))	52,376,319	1
4	•••••((((•••))))	44,544,114	1
5	((((•••))))••••••	44,273,764	1
6	••(((•••)))••••••	33,131,192	1
7	•••••(((•••)))••	32,883,686	1
8	•(((•••)))•••••••	32,878,614	1
9	••••••••(((•••)))	32,800,711	1
10	•••(((•••)))•••••	31,738,681	1
11	••••(((•••)))•••	31,720,954	1
12	••((((•••))))•••	27,886,795	1
13	•((((•••))))••••	27,835,512	1
14	••••((((•••))))•	27,791,612	1
15	•••((((•••))))••	27,778,147	1
⋮	⋮	⋮	⋮
93	••••••••((••••))	2,329,003	2 (2,034,559; 294,444)
⋮	⋮	⋮	⋮
97	((••••))••••••••	2,254,841	2 (1,906,756; 348,085)
⋮	⋮	⋮	⋮
174	•(((•(••••)•)))•	87,295	3 (76,755; 10,222; 318)

structure (rank 9), however, is partitioned into 13 components with a largest one containing 71.2% of the sequences. Unexpected partitions of neutral networks are found with two structures (ranks 10 and 11): They consist of two components of almost equal size. Further down in the probabilities of structures we observe many examples of this kind (ranks 14–20), and eventually structures appear whose neutral networks are split evenly into four equal-sized components (ranks 39 and 40). These clear deviations from the generic properties predicted by random graph theory found a straight-forward biophysical explanation [118]. All structures containing a stack that cannot be elongated (class I in Fig. 5) behave perfectly normally in the sense tha they form generic networks. The distribution of sequences belonging to such a network closely resembles the symmetrical binomial distribution, which is also the distribution of random sequences. Structures of class II, however, can form an additional base pair on one side of the stack, and, in general, they will do so when complementary bases are in the opposing positions. This is most likely the case when the overall base composition is 50% G and 50% C, and hence class II structures are less likely to be formed by sequences of equal percentages of G and C. The highest probability to form class II structures is thus expected to lie at a certain distance displaced from the middle of sequence space. Indeed, the two components of the class II structures have maxima of the distribution functions at excess G or excess C [(50 + δ)% G or (50 − δ)% G, respectively]. The distribution of each component is close to binomial, with equal offset from the center of sequences space (50% G/50% C). By the same token, structures of class III have two independent possibilities of stack elongation at each end, and thus the probability of their formation is largest if the sequence are displaced from

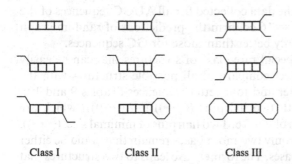

Class I   Class II   Class III

FIGURE 5   Three classes of RNA stacks. Stacks are classified with respect to their compatibility with stack elongation on the two ends. Class I stacks cannot be elongated, class II stacks are compatible with elongation at one end, whereas class III stacks can add base pairs at both ends of the stacks.

the uniform distribution by δ and ε (for the left-and right-hand ends, respectively). Without further information we assume $\delta = \varepsilon$. Independent superposition then yields four components with maximal probability densities at the following G/C ratios: $(50 + 2\delta)/(50 - 2\delta)$, 50/50, 50/50, and $(50 - 2\delta)/(50 + 2\delta)$. These are precisely the positions of the peaks observed with four-component networks. The structural details of neutral networks, we may conclude, are well described by the random graph model unless special structural features lead to systematic biases that can be interpreted straightforwardly.

Minimum free energy structures over the sequence space $\mathcal{Q}_{AU}^{16}$ are little more than an exercise in finding the most stable hairpin loops with the largest possible number of base pairs. As said above, the shape space is dominated by the open chain that expresses the overwhelming influence of finite size. Stable structures are the three hairpin loops with six base pairs, the two triloops (ranks 2 and 3), and the tetraloop (rank 4). The other structures with less than six base pairs are apparently unstable.

The essential difference between $\mathcal{Q}_{GC}^{16}$ or $\mathcal{Q}_{AU}^{16}$ and $\mathcal{Q}_{AUGC}^{16}$ lies in the cardinality, 65,536 versus 4.29×10^9 sequences. This has to be compared with a rather small difference in the numbers of structures, 195 versus 274, and leads to average numbers of 336 and 15.7×10^6 sequences per structure, respectively. Distances in sequence space, however, are the same in $\mathcal{Q}_{GC}^{16}$ and $\mathcal{Q}_{AUGC}^{16}$, and thus we suspect substantial differences in the sequence of components. Indeed, most neutral networks in $\mathcal{Q}_{AUGC}^{16}$ belonging to frequent structures are connected. The rank of the first network with two components is 93, and the two components have a size ratio of about 7. Smaller networks have numbers of components up to five, but nowhere did we find a situation of two or four equal-sized components as in the $\mathcal{Q}_{GC}^{16}$ case. A straightforward interpretation is based on the much higher cardinality of neutral networks in the AUGC case, which leads to merging of components compared to networks in $\mathcal{Q}_{GC}^{16}$. In summary, the data collected for all AUGC sequences of the small chain length of only $n = 16$ confirm the predictions of random graph theory rather well and certainly better than those for GC sequences.

Finally, we choose a special rare class of structures that can be easily counted and thus allows direct compare of all possible structures with the results derived from two-letter and four-letter sequences (Tables 9 and 10). These are the structures with two hairpins $(((((\bullet\bullet\bullet))))) \bullet\bullet (((((\bullet\bullet\bullet)))))$ which are hard to form at a chain length of $n = 16$. Two hairpins of minimal size, $((\bullet \ \bullet \ \bullet))$, require 2×7 bases, and thus only two more bases remain that could be either a base pair or two unpaired bases. The former case leads to two structures that are recognized as the most common structures of this class on both sequence spaces $\mathcal{Q}_{GC}^{16}$ (ranks 78 and 80) and $\mathcal{Q}_{AUGC}^{16}$ (ranks 144 and 145). All other 15 two-hairpin structures are readily derived from the shorthand diagram by inserting the two unpaired bases at all possible positions. It is worth noticing

TABLE 9 Structures of GC Sequences of Chain Length $n = 16$ with Two Hairpins

Rank	Structure	Number of sequences	Number of components	Sequence of components
78	(((•••)))((•••))	135	4	132; 1; 1; 1
80	((•••))(((•••)))	123	3	120; 2; 1
164	((•••))••((•••))	12	4	4; 3; 3; 2
178	••((•••))((•••))	4	2	3; 1
179	•((•••))•((•••))	4	3	2; 1; 1
184	((•••))•((•••))•	3	2	2; 1
195	((••••••))((•••))	2	1	2

that all of them are formed by the four-letter sequences, whereas only five of them appear on Q_{GC}^{16}. Interestingly, the stuctures formed by GC sequences are in the same sequence (with only one exception) as are the most common structures of AUGC sequences.

5.3.3. GC Sequences of Chain Length $n \leq 30$

Data derived from folding all GC sequences into secondary structures have been reported in detail [117,118]. We shall consider here mainly the chain length dependence of the most prominent features of sequence–structure mappings in order to be able to predict the behavior in the limit of long chains and to eliminate thereby the finite size effects. First the fraction of sequences forming no structure—i.e., the cardinality of the preimage of the open chain—decrease exponentially with increasing chain length n. It already contains less than 0.01% on Q_{GC}^{16}. Second, careful inspection of the fraction of sequences forming common structures allows us to extrapolate to long chains and leads to the following conjecture. In the limit of long chains, almost all sequences fold into common structures, which constitute only a minute fraction of all structures, or, in other words, the fraction of sequences folding into common structures approaches 1 in the lim $n \to \infty$, whereas at the same time the fraction of structures fulfilling the condition of being common goes to zero. The results derived from exhaustive folding of binary sequences (GC and AU) with $n \leq 30$ still show tremendous finite size effects, but the general trends are already clear at the long chain ends of the diagrams in Ref. 117.

5.3.4. Statistical Evaluation of Sequence Spaces with Chain Lengths $n > 30$

Exhaustive techniques become infeasible when the total number of sequences exceeds $\sim 10^{10}$ and one has to resort to sampling techniques [116]. Neutral

TABLE 10 Structures of AUGC Sequences of Chain Length $n = 16$ with Two Hairpins[a]

Rank	Structure	Number of sequences	Number and Sequence of components
144 (78)	(((•••)))((•••))	257 506	1
145 (80)	((•••))(((•••)))	254 456	1
188 (164)	((•••))••((•••))	57 398	1
196 (179)	•((•••))•((•••))	32 528	1
197 (178)	••((•••))((•••))	31 533	1
198 (184)	((•••))•((•••))•	31 429	1
199	((•••))((•••))••	30 367	1
223 (195)	((•••••))((•••))	15 048	1
224	((•••))((••••••))	14 625	2
			(13968; 657)
225	•((•••))((•••))•	14 497	1
229	((•••))•((••••))	11 518	2
			(10226; 1292)
233	((••••))•((•••))	10 762	2
			(8880; 1846)
236	((•••))((••••))•	7318	2
			(6590; 728)
238	((••••))((•••))•	6855	2
			(5822; 1063)
239	•((•••))((••••))	6739	4
			(6329; 217; 183; 10)
241	•((••••))((•••))	6466	2
			(5423; 1043)
270	((••••))((•••••))	1837	5
			(1344; 245; 204; 41; 3)

[a] For structures that are also formed by GC sequences, the rank is given in parentheses.

paths (Sec. 4.3), for instance, can be used to detect neutral networks. The covering radius (Sec. 5.2) can be estimated by measuring the minimum distance that is necessary to find a given structure from a chosen starting sequence and averaging over the starting sequences and target structures weighted by their preimage sizes. This provides an upper bound for the mean covering radius. Extensive computer simulations reported in Refs. 70,116,120, and 127–129 provide strong evidence for the existence of sequence space percolating neutral networks and shape space covering.

5.3.5. Shape Space Topology

The topological (and possibly metric) properties of phenotype spaces are still largely uncharted territory. In fact, the description of the genotype–phenotype maps of RNA so far has made no reference to the structure of shape space itself beyond a definition of equality of structures (Fig. 6).

To understand the sequence of phenotypic changes along an evolutionary trajectory, however, it is necessary to know which phenotypes are *accessible* from which genotypes. Accessibility can then be used to define a relation of "nearness" among phenotypes, independently of their geometrical, biophysical, or biological similarities [123,124]. In the simplest case, we might say that ψ is *accessible* from ϕ if it is possible to jump from $f^{-1}(\phi)$ to $f^{-1}(\psi)$ by means of a point mutation. Shape space covering (Sec. 5.2) suggests that each structure should be accessible from any other structure. However, sequence space is so large that not all possible sequences are ever realized in the course of a simulation run (or during the history of evolution). Fontana and Schuster [124] argue that a more restrictive condition for accessibility is more suitable—for instance, a minimum number of sequences in $f^{-1}(\psi)$ that are neighbors of sequences folding into ϕ.

The evolutionary trajectories observed in computer simulations can be regarded as a sequence $(x_0, x_1, \ldots)$ of those phenotypes on whose neutral networks the population is concentrated during subsequent diffusion phases. The question hence becomes whether there is a meaningful way of distinguishing between continuous (smooth, expectable) and discontinuous (surprising) evolutionary transitions. From a more abstract point of view, *continuity* is a topological property of a map from one topological space into another one. Having defined the topology by specifying a suitable notion of accessibility, it

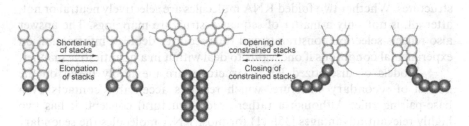

FIGURE 6 Structural changes corresponding to continuous evolutionary transitions. Shortening and elongation of stacks as well as opening of constrained stacks in general lead to easily accessible structures. Closing a constrained stack, on the other hand, leads to inaccessible structures and hence corresponds to discontinuous transitions.

becomes a matter of observation or computer simulation to find out whether "real" evolutionary trajectories are in fact continuous. We find that most evolutionary transitions are, indeed, continuous most of the time. Rare discontinuous transitions are often associated with major structural transitions [123,124,130]. We note, finally, that the topological notion of continuity might sometimes be too restrictive. Weaker mathematical structures such as filter spaces or convergence spaces, as introduced, for instance, in Refs. 131–134, appear to be promising starting points for a generalization of this approach.

6. CONCLUSIONS AND OUTLOOK

The lack of complementarity rules in discrete protein models makes the folding problem much harder than in RNA and less straightforwardly accessible to combinatorics. Some results, such as the relatively small extensions and clustering of the neutral networks that have been observed in some lattice models [11], are not very compatible with the simulations based on knowledge-based potentials [135,136], suggesting that proteins and RNA behave in essentially the same way. This discrepancy might be explained by the short chains, $n < 30$, and the two-letter **HP** alphabet used in the lattice models. Whereas native-like proteins can be designed from reduced alphabets, experiments [137] as well as computer simulations [135] suggest that two letters are not sufficient.

The notion of neutral networks in RNA sequence space requires modification when suboptimal conformations or folding behavior of molecules are taken into account as an additional constraint. The degree of neutrality will certainly be smaller than in the case of the minimum free energy structures. Whether two folded RNA molecules are selectively neutral or not, after all, is not only a matter of sequence–structure mappings. The answer also reflects selection constraints and thus requires detailed information on experimental conditions if one wants to deal with it in a quantitative manner.

Models of discretized RNA structures are inevitably based on the notion of secondary structure, which restricts acceptable contacts by a base-pairing rule. Although a rather crude structural concept, it has two highly relevant advantages [35]: (1) for most RNA molecules the secondary structure is a folding intermediate that becomes three-dimensional by the formation of tertiary contacts; (2) the majority of tertiary contacts can be classified by a few simple principles such as pseudoknots, terminal (non–Watson–Crick) base pairs, base triplets, base quartets, and coaxial stacks. Making use of algorithms that are not restricted by the conventional secondary structure concept, such as the kinetic folding algorithm [81], these tertiary interactions can be incorporated into structures. Thereby one would still stay within the realm of discreteness and at the same time approach a more realistic concept of RNA structures.

REFERENCES

1. Lau KF, Dill KA. Theory for protein mutability and biogenesis. Proc Natl Acad Sci USA 1990; 87:638–642.
2. Chan HS, Dill KA. Origins of structure in globular proteins. Proc Natl Acad Sci USA 1990; 87:6388–6392.
3. Chan HS, Dill KA. Sequence space soup. J Chem Phys 1991; 95:3775–3787.
4. Crippen GM. Prediction of protein folding from amino acid sequences of discrete conformation spaces. Biochemistry 1991; 30:4232–4237.
5. Lipman DJ, Wilbur WJ. Modelling neutral and selective evolution of protein folding. Proc Roy Soc Lond B 1991; 245:7–11.
6. Camacho CJ, Thirumalai D. Minimum energy compact structures of random sequences of heteropolymers. Phys Lett 1993; 71:2505–2508.
7. Šali A, Shakhnovich E, Karplus M. Kinetics of protein folding. A lattice model study on the requirements for folding of native states. J Mol Biol 1994; 253:1614–1636.
8. Dill KA, Bromberg S, Yue K, Fiebig KM, Yeo DP, Thomas PD, Chan HS. Principles of protein folding: a perspective from simple exact models. Protein Sci 1995; 4:561–602.
9. Chan HS, Dill KA. Comparing folding codes for proteins and polymers. Proteins 1996; 24:335–344.
10. Li H, Helling R, Tang C, Wingreen N. Emergence of preferred structures in a simple model of protein folding. Science 1996; 273:666–669.
11. Renner A, Bornberg-Bauer E. Exploring the fitness landscapes of lattice proteins. In: Altman R, Dunker AK, Hunter L, Klein TE, eds. Pacific Symposium on Biocomputing '97. Singapore: World Scientific, 1997: 361–372.
12. Hart WE, Istrail S. Lattice and off-lattice side chain models of protein folding: linear time structure prediction better than 86% of optimal. J Comput Biol 1997; 4:241–259.
13. Madras N, Sokal G. The Self-Avoiding Walk. Boston: Birkhäuser, 1993.
14. Flory PJ. Principles of Polymer Chemistry. Ithaca, NY: Cornell Univ. Press, 1971.
15. Whittington SG. Statistical mechanics of polymer solutions and polymer adsorption. Adv Chem Phys 1982; 51:1–48.
16. Chan HS, Dill KA. Interchain loops in polymers: effects of excluded volume. J Chem Phys 1988; 90:492–508.
17. Chen S-J, Dill KA. Statistical thermodynamics of double-stranded polymer molecules. J Chem Phys 1995; 103:5802–5808.
18. Touchard J. Sur une problème de configurations et sur les fractions continues. Can J Math 1952; 4:2–25.
19. Hsieh WN. Proportions of irreducible diagrams. Studies Appl Math 1973; 52:277–283.

20. Kleitman D. Proportions of irreducible diagrams. Studies Appl Math 1970; 49:297–299.

21. Stein PR. On a class of linked diagrams, I. Enumeration. J Comb Theory A 1978; 24:357–366.

22. Stein PR, Everett CJ. On a class of linked diagrams II. Asymptotics Disc Math 1978; 22:309–318.

23. Waterman MS. Secondary structure of single-stranded nucleic acids. Adv Math Suppl Studies 1978; 1:167–212.

24. Westhof E, Jaeger L. RNA pseudoknots. Cur Opin Struct Biol 1992; 2:327–333.

25. Loria A, Pan T. Domain structure of the ribozyme from eubacterial ribonuclease. P RNA 1996; 2:551–563.

26. Konings DAM, Gutell RR. A comparison of thermodynamic foldings with comparatively derived structures of 16S and 16S-like rRNAs. RNA 1995; 1:559–574.

27. Tang CK, Draper DE. Evidence for allosteric coupling between the ribosome and repressor binding sites of a translationally regulated mRNA. Biochemistry 1990; 29:4434–4439.

28. Stadler PF, Haslinger C. RNA structures with pseudo-knots: graph-theoretical and combinatorial properties. Bull Math Biol 1999; 61:437–467.

29. Berge C. Hypergraphs. Amsterdam: Elsevier, 1989.

30. Singh RK, Tropsha A, Vaisman II. Delaunay tessellation of proteins: four body nearest neighbor propensities of amino acid residues. J Comput Biol 1996; 3:213–221.

31. Barber CB, Dobkin DP, Huhdanpaa H. The quickhull algorithm for convex hulls. ACM Trans Math Software 1996; 22:469–483.

32. Zheng W, Cho SJ, Vaisman II, Tropsha A. A new approach to protein fold recognition based on Delaunay tessellation of protein structure. In: Altman R, Dunker AK, Hunter L, Klein TE, eds. Pacific Symposium on Biocomputing '97. Singapore: World Scientific, 1997:487–496.

33. Munson PJ, Singh RK. Statistical significance of hierarchical multi-body potentials based on Delaunay tessellation and their application in sequence-structure alignment. Protein Sci 1997; 6:1467–1481.

34. Weberndorfer G, Hofacker IL, Stadler PF. An efficient potential for protein sequence design. In: Bielefeld D, ed. Computer Science in Biology. Proceedings of the GCB'99. Hannover: Univ. Bielefeld, 1999:107–112.

35. Batey RT, Rambo RP, Doudna JA. Tertiary motifs in structure and folding of RNA. Angew Chem Int Ed 1999; 38:2326–2343.

36. Tacker M, Stadler PF, Bornberg-Bauer EG, Hofacker IL, Schuster P. Algorithm independent properties of RNA structure prediction. Eur Biophys J 1996; 25:115–130.

37. Hofacker IL, Schuster P, Stadler PF. Combinatorics of RNA secondary structures. Discr Appl Math 1998; 88:207–237.

38. Howell JA, Smith TF, Waterman MS. Computation of generating functions for biological molecules. SIAM J Appl Math 1980; 39:119–133.

39. Schmitt WR, Waterman MS. Linear trees and RNA secondary structure. Discr Appl Math 1994; 12:412–427.

40. Penner RC, Waterman MS. Spaces of RNA secondary structures. Adv Math 1993; 101:31–49.
41. Stein PR, Waterman MS. On some new sequences generalizing the Catalan and Motzkin numbers. Disc Math 1978; 26:261–272.
42. Waterman MS, Smith TF. Combinatorics of RNA hairpins and cloverleaves. Studies Appl Math 1978; 60:91–96.
43. Waterman MS, Smith TF. RNA secondary structure: a complete mathematical analysis. Math Biosci 1978; 42:257–266.
44. Waterman MS. Introduction to Computational Biology: Maps, Sequences, and Genomes. London: Chapman & Hall, 1995.
45. Bender EA. Asymptotic methods in enumeration. SIAM Rev 1974; 16:485–515.
46. Canfield ER. Remarks on an asymptotic method in combinatoric. J Comb Theory A 1984; 37:348–352.
47. Meir A, Moon JW. On an asymptotic method in enumeration. J Comb Theory A 1989; 51:77–89.
48. Darboux G. Mémoir sur l'approximation des fonctions de très grande nombres, et sur une classe étendu de développements en série. J Math Pures Appl 1878; 4:5–56.
49. Szegö G, Orthogonal Polynomials, volume XXIII of Am Math Soc Coll Publ Am Math Soc, New York, 1959.
50. Leydold J, Stadler PF, Minimal cycle basis of outerplanar graphs. Elec. J Comb., 5. 1998. R16. See http://www.combinatorics.org and Santa Fe Institute Preprint 98-01-011.
51. Mathews DH, Sabina J, Zuker M, Turner DH. Expanded sequence dependence of thermodynamic parameter improves prediction of RNA secondary structure. J Mol Biol 1999; 288:911–940.
52. Walter AE, Turner DH, Kim J, Lyttle MH, Müller P, Mathews DH, Zuker M. Co-axial stacking of helixes enhances binding of oligoribonucleotides and improves predictions of RNA folding. Proc Natl Acad Sci USA 1994; 91: 9218–9222.
53. Jaeger JA, Turner DH, Zuker M. Improved predictions of secondary structures for RNA. Proc Natl Acad Sci USA 1989; 86:7706–7710.
54. He L, Kierzek R, SantaLucia J, Walter AE, Turner DH. Nearest-neighbor parameters for GU mismatches. Biochemistry 1991; 30:11124–11132.
55. Peritz AE, Kierzek R, Sugimoto N, Turner DH. Thermodynamic study of internal loops in oligoribonucleotides: symmetric loops are more stable than asymmetric loops. Biochemistry 1991; 30:6428–6436.
56. SantaLucia J Jr, Allawi HT, Seneviratne PA. Improved nearest-neighbor parameters for predicting DNA duplex stability. Biochemistry 1996; 35:3555–3562.
57. SantaLucia J Jr. A unified view of polymer, dumbbell, and oligonucleotide DNA nearest-neighbor thermodynamics. Proc Natl Acad Sci USA 1998; 95:1460–1465.
58. Nussinov R, Jacobson AB. Fast algorithm for predicting the secondary structure of single-stranded RNA. Proc Natl Acad Sci USA 1980; 77 (11):6309–6313.

59. Nussinov R, Piecznik G, Griggs JR, Kleitman DJ. Algorithms for loop matching. SIAM J Appl Math 1978; 35:68–82.
60. Zuker M, Stiegler P. Optimal computer folding of larger RNA sequences using thermodynamics and auxiliary information. Nucleic Acids Res 1981; 9:133–148.
61. Zuker M, Sankoff D. RNA secondary structures and their prediction. Bull Math Biol 1984; 46(4):591–621.
62. Zuker M. On finding all suboptimal foldings of an RNA molecule. Science 1989; 244:48–52.
63. Schmitz M, Steger G. Base-pair probability profiles of RNA secondary structures. Comput Appl Biosci 1992; 8:389–399.
64. McCaskill JS. The equilibrium partition function and base pair binding probabilities for RNA secondary structure. Biopolymers 1990; 29:1105–1119.
65. Cupal J, Hofacker IL, Stadler PF. Dynamic programming algorithm for the density of states of RNA secondary structures. In: Hofstädt R, Lengauer T, Löffler M, Schomburg D, eds. Computer Science and Biology 96 (Proceedings of the German Conference on Bioinformatics). Leipzig, Germany: Universität Leipzig, 1996:184–186.
66. Cupal J. The density of states of RNA secondary structures. Master's Thesis, Univ Vienna, 1997.
67. Higgs PG. RNA secondary structure: a comparison of real and random sequences. J Phys I (France) 1995; 3:43–59.
68. Wuchty S, Fontana W, Hofacker IL, Schuster P. Complete suboptimal folding of RNA and the stability of secondary structures. Biopolymers 1999; 49:145–165.
69. Wuchty S. Suboptimal secondary structures of RNA. Master's Thesis, Univ Vienna, 1998.
70. Hofacker IL, Fontana W, Stadler PF, Bonhoeffer S, Tacker M, Schuster P. Fast folding and comparison of RNA secondary structures. Monatsh Chem 1994; 125:167–188.
71. Morgan SR, Higgs PG. Evidence for kinetic effects in the folding of large RNA molecules. J Chem Phys 1996; 105:7152–7157.
72. Martinez HM. An RNA folding rule. Nucleic Acid Res 1984; 12:323–335.
73. Mironov AA, Dyakonova LP, Kister AE. A kinetic approach to the prediction of RNA secondary structures. J Biomol Struct Dynam 1985; 2:953.
74. Abrahams JP, van den Berg M, van Batenburg E, Pleij C. Prediction of RNA secondary structure, including pseudoknotting, by computer simulation. Nucleic Acids Res 1990; 18:3035–3044.
75. Gultyaev AP. The computer simulation of RNA folding involving pseudoknot formation. Nucleic Acids Res 1991; 19:2489–2493.
76. Tacker M, Fontana W, Stadler PF, Schuster P. Statistics of RNA melting kinetics. Eur Biophys J 1994; 23:29–38.
77. Higgs PG. Thermodynamic properties of transfer RNA: a computational study. J Chem Soc Faraday Trans 1995; 91:2531–2540.
78. Gultyaev AP, van Batenburg, Pleij CWA. The computer simulation of RNA folding pathways using a genetic algorithm. J Mol Biol 1995; 250:37–51.
79. Suvernev AA, Frantsuzov PA. Statistical description of nucleic acid secondary structure folding. J Biomol Struct Dynam 1995; 13:135–144.

80. Flamm C. Kinetic folding of RNA. PhD Thesis, Universität Wien, 1998.

81. Flamm C, Fontana W, Hofacker IL, Schuster P. Elementary step dynamics of RNA folding. RNA 2000; 6:325–338.

82. Gutell RR. Evolutionary characteristics of RNA: inferring higher-order structure from patterns of sequence variation. Curr Opin Struct Biol 1993; 3:313–322.

83. Chaves C, Riera R. Correction to scaling for the self-avoiding walk in d = 2: result based on a cell renormalization group. Phys Rev B 1993; 48:16084–16087.

84. Beretti A, Sokal A. New Monte Carlo method for the self-avoiding walk. J Stat Phys 1985; 40:483–531.

85. Watts MG. Application of the method of pad e approximants to the excluded volume problem. J Phys A 1975; 8:61–66.

86. Guttmann AJ. The high-temperature susceptibility and spin-spin correlation function of the three-dimensional sing model. J Phys A 1987; 20:1855.

87. Enting IG, Guttmann AJ. Self-avoiding polygons on the square, 1 and manhatten lattice. J Phys A 1985; 18:1007.

88. Ishinabe T, Chikahisa Y. Exact enumeration of self-avoiding lattice walks with different nearest-neighbor contacts. J Chem Phys 1986; 85(2):1009–1017.

89. Guttmann AJ, Osborn TR, Sokal AD. Connective constant of the self-avoiding walk on the triangular lattice. J Phys A: Math Gen 19: 2591–2598.

90. Kremer K, Baumgärtner A, Binder K. Collapse transition and crossover scaling for self-avoiding walks on the diamond lattice. J Phys A: Math Gen 1981; 15:2879–2897.

91. Guttmann AJ. On the critical behaviour of self-avoiding walks. J Phys A July 1986; 20:1839–1854.

92. Rapaport DC. On three-dimensional self-avoiding walks. J Phys A 1985; 18:113–126.

93. JD et al. Scaling exponents of the self-avoiding-walk problem in three dimensions. J. Dayantis and J. Palierne Phys Rev B 1994; 49:3217–3225.

94. Guttmann A, Ninham B, Thompson C. Determination of critical behaviour in lattice statistics from series expansions. I Phys Rev 1968; 172:554–558.

95. Kolinski A, Milik M, Skolnick J. Static and dynamic properties of a new lattice model of polypeptide chains. J Chem Phys 1991; 94(5):3978–3985.

96. Dill KA. Theory for the folding and stability of globular proteins. Biochemistry 1985; 24:1501–1509.

97. Flory PJ. Statistical Mechanics of Chain Molecules. New York: Wiley, 1969.

98. Sun S, Brem R, Chan HS, Dill KA. Designing amino-acid sequences to fold with good hydrophobic cores. Protein Eng 1996; 8:1205–1213.

99. Buchler NEG, Goldstein RA. The effect of alphabet size and foldability requirements on protein structure designability. Proteins 1999; 34:113–124.

100. Miyazawa S, Jernigan RL. Estimation of effective interresidue contact energies from protein crystal structures: quasi-chemical approximation. Macromolecules 1985; 18:534–552.

101. Sippl MJ. Calculation of conformational ensembles from potentials of mean force—an approach to the knowledge-based prediction of local structures in globular proteins. J Mol Biol 1990; 213:859–883.

102. Casari G, Sippl MJ. Structure-derived hydrophobic potentials—hydrophobic potentials derived from X-ray structures of globular proteins are able to identify native folds. J Mol Biol 1992; 224:725–732.

103. Sippl MJ. Recognition of errors in three-dimensional structures of proteins. Proteins 1993; 17:355–362.

104. Ngo JT, Marks J. Computational complexity of a problem in molecular structure prediction. Protein Eng 1992; 5:313–321.

105. Unger R, Moult J. Finding the lowest free energy conformation of a protein is an NP-hard problem: proof and implications. Bull Math Biol 1993; 55:1183–1198.

106. Hart WE, Istrail S. Robust proofs of NP-hardness for protein folding: general lattices and energy potentials. J Comput Biol 1997; 4:1–22.

107. Li H, Helling R, Tang C, Wingreen N. Emergence of preferred structures in a simple model of protein folding. Science 1996; 273:666–669.

108. Yue K, Dill KA. Sequence structure relationships in proteins and copolymers. Phys Rev E 1993; 48:2267–2278.

109. Hart WE, Istrail S. Fast protein folding in the hydrophilic-hydrophobic model within three-eights optimal. Extended Abstract published in the Proceedings of the 27th Annual ACM Symposium on Theory of Computation, May 1995, 1994.

110. Bollobás B. Random Graphs. London: Academic Press, 1985.

111. Reidys CM. Random induced subgraphs of generalized n-cubes. Adv Appl Math 1997; 19:360–377.

112. Reidys CM, Stadler PF, Schuster P. Generic properties of combinatory maps: neutral networks of RNA secondary structures. Bull Math Biol 1997; 59:339–397.

113. Ajtai M, Komlós J, Szemerédi E. Largest random component of a k-cube. Combinatorica 1982; 2:1–7.

114. Gavrilets S, Gravner J. Percolation on the fitness hypercube and the evolution of reproductive isolation. J Theor Biol 1997; 184:51–64.

115. Gavrilets S, Li H, Vose MD. Rapid parapatric speciation on holey adaptive landscapes. Proc Roy Soc Lond B 1998; 265:1483–1489.

116. Schuster P, Fontana W, Stadler PF, Hofacker IL. From sequences to shapes and back: a case study in RNA secondary structures. Proc Roy Soc Lond B 1994; 255:279–284.

117. Grüner W, Giegerich R, Strothmann D, Reidys C, Weber J, Hofacker IL, Stadler PF, Schuster P. Analysis of RNA sequence structure maps by exhaustive enumeration. I. Neutral networks. Monatsh Chem 1996; 127:355–374.

118. Grüner W, Giegerich R, Strothmann D, Reidys C, Weber J, Hofacker IL, Stadler PF, Schuster P. Analysis of RNA sequence structure maps by exhaustive enumeration. II. Structures of neutral networks and shape space covering. Monatsh Chem 1996; 127:375–389.

119. Reidys CM, Stadler PF. Neutrality in fitness landscapes. Appl Math & Comput 2001; 117:321–350.

120. Fontana W, Stadler PF, Bornberg-Bauer EG, Griesmacher T, Hofacker IL, Tacker M, Tarazona P, Weinberger ED, Schuster P. RNA folding and combinatory landscapes. Phys Rev E 1993; 47:2083–2099.

121. Huynen MA, Stadler PF, Fontana W. Smoothness within ruggedness: the role of neutrality in adaptation. Proc Natl Acad Sci USA 1996; 93:397–401.

122. van Nimwegen E, Crutchfield JP, Mitchell M. Finite populations induce metastability in evolutionary search. Phys Lett A 1997; 229:144–150.

123. Fontana W, Schuster P. Shaping space. The possible and the attainable in RNA genotype-phenotype mapping. J Theor Biol 1998; 194:491–515.

124. Fontana W, Schuster P. Continuity in evolution. On the nature of transitions. Science 1998; 280:1451–1455.

125. Schuster P. How to search for RNA secondary structures. Theoretical concepts in evolutionary biotechnology. J Biotechnol 1995; 41:239–257.

126. Göbel U. Neutral Networks of Minimum Free Energy RNA Secondary Structures. PhD thesis, Universität Wien, 2000. http://www.tbi.univie.ac.at/papers/PhD_theses.html.

127. Fontana W, Griesmacher T, Schnabl W, Stadler PF, Schuster P. Statistics of landscapes based on free energies, replication and degradation rate constants of RNA secondary structures. Monatsh Chem 1991; 122:795–819.

128. Fontana W, Konings DA, Stadler PF, Schuster P. Statistics of RNA secondary structures. Biopolymers 1993; 33:1389–1404.

129. Göbel U, Forst CV, Schuster P. Structural constraints and neutrality in RNA. In: Hofestädt R, Lengauer T, Löffler M, Schomburg D, eds. Bioinformatics. German Conference on Bioinformatics, GCB'96, volume 1278 of Lecture Notes Comput Sci. Berlin: Springer-Verlag, 1997:156–165.

130. Cupal J, Kopp S, Stadler PF. RNA shape space topology. Alife 2000; 6:3–23. In press. SFI Preprint 99-03-022.

131. Hausdorff F. Gestufte Räume. Fund Math 1935; 25:486–502.

132. Choquet G. Convergences. Ann Univ Grenoble 1947; 23:55–112.

133. Kent DC. Convergence functions and their related topologies. Fund Math 1964; 54:125–133.

134. Katetov M. On continuity structures and spaces of mappings. Commun Math Univ Carolinae 1965; 6:257–278.

135. Babajide A, Hofacker IL, Sippl MJ, Stadler PF. Neutral networks in protein space: a computational study based on knowledge-based potentials of mean force. Folding Des 1997; 2:261–269.

136. Babajide A, Farber R, Hofacker IL, Inman J, Lapedes AS, Stadler PF. Exploring protein sequence space using knowledge based potentials. J Comput Biol 1999. J Theor Biol 2001; 212:35–46. Santa Fe Preprint 98-11-103.

137. Plaxco K, Riddle D, Grantcharova V, Baker D. Simplified proteins: minimalist solutions to the "protein folding problem". Curr Opin Struct Biol 1998; 8:80–85.

6

Protein Structure Folding and Prediction

William R. Taylor
National Institute for Medical Research, London, England

1. PROTEIN STRUCTURE

1.1. The Shapes and Sizes of Proteins

Proteins are linear heteropolymers that incorporate a sequence of the 20 naturally occurring amino acids. They vary greatly in length, from just a few amino acids to many thousands of residues.* The larger extreme of the range is usually confined to long (extended) fibrous proteins in which, typically, a short unit is repeated many times. This unit can be from as small as three residues (as in collagen) to about 100 (as in the muscle protein titan). The proteins that are of primary interest in this chapter, however, lie in the middle of the range, with lengths of several tens of residues to several hundred. They tend to have more compact dimensions and are accordingly referred to as globular.

This class includes the majority of proteins that perform the metabolic and other active functions of the cell and also exhibit the greatest variety of

---

*Amino acids are referred to as residues when they are linked in a polypeptide chain. This unexpected term derives from a description of the material left at the bottom of the test tube in the early chemical analysis of the protein amino acid sequence.

structure—hence their importance from both biological and physicochemical viewpoints. A further class of proteins are intimately associated with the phospholipid (bilayer) membranes and are of great biological importance in cell–cell communication and environmental sensing because of their ability to bridge the external and internal environments. Their structures appear to be more limited (relative to globular proteins), either as a result of the constraints imposed by the planar membrane or perhaps because we know the structure of only a few members of this class.

(See Brändén and Tooze [1] for a basic introduction to protein structure and Kyte [2] for a more detailed approach.)

1.2. Amino Acid Properties

Before considering the structure of proteins in more detail, it is worthwhile to review the individual properties of their constituent amino acids, because it is the physicochemical properties of these that dictate how they will interact to form the protein structure. (See Taylor [3,4] for a fuller analysis of this topic.)

The most fundamental distinction among the amino acids is whether their side chains incorporate "active" (or polar) atoms such as N, S, or O or whether they are composed of "inert" hydrocarbon. The former class of atoms gives rise to electronic bond polarization, allowing the possibility of forming hydrogen bonds and participating in (usually catalyzing) chemical reactions. By contrast, the hydrocarbon residues tend to form the core of the protein, where their mutual repulsion of water (hydrophobic effect) allows them to stabilize the overall globular structure by packing tightly together in a dehydrated mass (Fig. 1).

When related protein sequences are aligned, the degree to which these properties are conserved often provides a good guide to the role of the position in the structure. Specifically, conserved polar residues are likely to be involved in a specific function (ligand binding or catalysis), whereas conserved hydrophobic residues are more likely to perform a structural role in the core.

Two residues, glycine (Gly) and proline (Pro), have unique stereochemical properties and are often exceptions to this general trend. Glycine, which has no side chain, can pack in the core, but it is also flexible (greater conformational freedom around its main-chain torsion angles), making it favored in tight turns on the surface of the protein. Proline, although hydrophobic, is also often found on the surface because it has one of its torsional freedoms fixed in a turn conformation, again making it favored in surface turns.

1.3. Secondary Structure

The basic organizing principle of protein structure places the hydrophobic residues in the core and surrounds them with a shell of hydrophilic (polar)

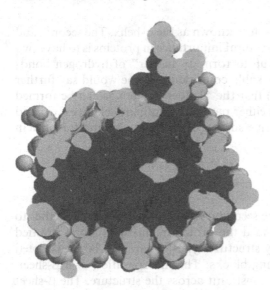

FIGURE 1 The hydrophobic core. A section (slab) has been taken through the core of a small protein (PDB code: 3chy) and displayed (using RASMOL) to show the van der Waals surface of all the (non-hydrogen) atoms. These are shown (using the defaut RASMOL selections) gray for polar amino acids and black for hydrophobic amino acids. The black hydrophobic core can be clearly seen, but (as with all "rules" concerning protein structure) there are some exceptions, and a (gray) hydrophilic residue can be seen in the core and a (black) hydrophobic residue on the surface. The former probably is hydrogen bonded to another hydrophilic side chain or to main-chain polar groups.

residues that provide an interface to the solvent. One complication of this simple scheme, however, is that all residues also have polar atoms in their main chain, which, like the hydrophilic side chains, cannot simply be buried in the core.

A solution to this problem can be found by forming a hydrogen bond between unlike charges (using carbonyl and amide groups from different parts of the main chain: $> = O-H-N<$). When mutually satisfied in this way, the bonded pair can then be buried away from solvent. One might imagine that this solution could be achieved in an ad hoc manner (simply matching up whatever pairs came nearby), but whether as a consequence of the complexity of connecting such a network, or whether guided by divine symmetry, the hydrogen-bonded networks found in proteins are remarkably ordered.

Hydrogen-bonded pairings are dominated by the shortest (unstrained) local connection along the chain, bonding the carbonyl group of residue i to the amide group of residue $i + 4$, giving (when repeated along the chain) a helical structure of period 3.6 residues known as the α-helix. The second, and almost only, other solution of structural importance in proteins is to have two remote parts of the chain line up to form a "ladder" of hydrogen bonds between them. It is also a remarkable coincidence (some would say further evidence of divine intervention) that the "ladder" of bonds can be formed when the juxtaposed chains run either parallel or antiparallel. (See Reid and Franchini [5] for a comprehensive survey of these constraints for protein design.)

1.4. Packed Layers

The units of globular protein are secondary structures that pack together to form a hydrophobic core. Provided the protein main-chain atoms are tied up in one of the two secondary structure types, a core can be constructed using any mix of α or β building blocks. The incorporation of a β-sheet, however, imposes a long-range constraint across the structure. The β-sheet has free hydrogen bonds on its two edges that consequently prevent the sheet from terminating in the core. This divides the core into two and, if considered more generally, imposes a layered structure onto the further arrangement of secondary structures in the protein. (See Figure 2 for examples.) (See Refs. 6 and 7 for further consideration of protein structure along these lines.)

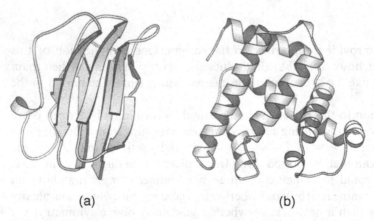

(a)   (b)

FIGURE 2 Protein structures with one secondary structure type. (a) An all-β protein (immunoglobulin) with two packed β-sheets. (b) An all-α protein (globin) showing packed α-helices.

Seldom are more than four layers seen in proteins, and because these can be composed of only one of two secondary structures (i.e., no mixed layers), the possibilities are few enough to enumerate.

Two layers: BB; AB; AA
Three layers: BBB; ABB, BAB; AAB, ABA; AAA
Four layers: BBBB; ABBB, BABB; AABB, BAAB, ABAB, ABBA; AAAB, AABA; AAAA

(These combinations allow for reversals, because proteins do not distinguish top from bottom.)

This gives 19 possible combinations, but this is something of an overestimate, because adjacent layers of α-helices are not always distinct. (The helices lack the strict registration imposed by the hydrogen bonding through the β-sheet.) Among these, not all possibilities are equally favored in nature; among the three-layer options the ABA combination is very widespread, whereas in the four-layer structures the corresponding ABBA structure is also encountered frequently. (For a recent survey of the current state of known structures see Thornton et al. [8].)

1.5. Barrel Structures and β-Helices

Other solutions can be found to tie up the "loose" hydrogen bonds on the edge of a β-sheet. One commonly encountered solution is to twist the sheet so that the two edges meet and can hydrogen bond to each other, forming a closed barrel-like network of hydrogen bonds. This cannot easily be accomplished with less than six strands. If only β-structure is used, then the barrel must incorporate antiparallel pairings (for example, in the trypsin family of structures, or completely antiparallel as in the calpactin family). However, with a combination of secondary structure, α-helices can link one (open) end of the barrel to the other and allow the formation of a predominantly or pure parallel sheet. This arrangement is seen in the eightfold β-α-barrel $(\beta\alpha)_8$, which was seen originally in the enzyme triosephosphate isomerase (TIM*) and is often referred to as the TIM barrel (Fig. 3).

A barrel can also be formed with the β-strands running in the orthogonal direction (leaving free hydrogen bonds on the open ends of the barrel). This structure, however, completely dictates the course of the protein chain (as a simple helix), giving little scope for evolutionary exploitation of the fold of different functions. (See Ref. 9 for some examples.)

---

*This slightly contrived acronym was derived from the name of the son of one of the authors of the structure!

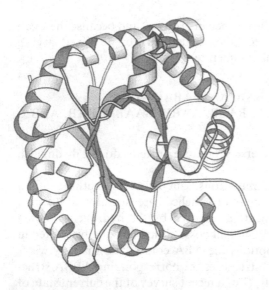

FIGURE 3 Eight-fold alternating β/α barrel protein. The protein chain spirals (as a toroid) while alternating between β and α secondary structure types, giving rise to a closed ring or barrel β-sheet in the center surrounded by a larger ring of α-helices on the outside. The structure, first seen in the enzyme triosephosphate isomerase (after which it is often named the TIM barrel) has been seen many times in unrelated proteins.

1.6. Protein Topology

The preceding analysis of proteins as layers of secondary structure (forming a hydrophobic core) neglected the path of the chain through the various layers (frameworks or architectures) described above. Unfortunately for structure prediction, the course of the chain through these frameworks is largely unrestricted. Two constraints, however, are well observed. The strongest is that two loops cannot cross on the same face between layers. The source of this constraint is a simple consequence of the polypeptide chain: If two loops cross, one will be shielded from solvent by the other, which will be energetically unfavorable unless the buried loop can satisfy its main-chain hydrogen bonds. Having done this, however, the loop is now a secondary structure, so the rule that loops do not cross is preserved [10].

The second strong constraint derives from the chiral nature of the central (α) carbon in each residue. This favors a particular (right) handedness for the α-helix and a corresponding twist to the β-sheet. Together, this higher propagation of the chirality results in a strong preference for connections between strands in the same sheet to be right-handed (even if there is no α-helix involved). The few exceptions to this rule are seen when the chain meanders to a remote part of the structure (another domain) and the context of the local constraint is lost [11] (Fig. 4).

1.7. Domain Structure

Large hydrophobic cores are not found in globular proteins, possibly because of limitations in the folding kinetics and stability. Single compact units of more than 500 residues are rare, with the typical size lying around half this size (200–300 residues). Large proteins are then organized into units of this size referred to as domains [12] (Fig. 5).

The definition of a domain is problematic. One suggestion is that if the chain were to be cut, then the two parts would remain stable (with each having its own hydrophobic core). With well-segregated domains (like beads on a string) this is undoubtedly true, but with more closely interacting domains (in particular, those in which the chain crosses between the domains more than once) such an experiment cannot be carried out without exposing surfaces that are not optimally evolved for solvation.

Various working definitions of domains have been derived, but in the more difficult examples these seldom agree. The problem with all of these methods is that they lack a sense of aesthetics, and either fortunately or unfortunately (depending on your artistic vs. scientific leaning) this is clearly one of the faculties that we (as humans) employ when we decide on the division of proteins into domains. (See Taylor [13] for a recent effort and also references to past approaches.)

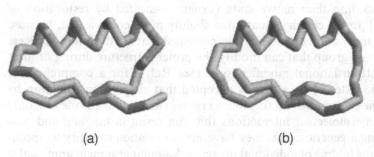

(a) (b)

FIGURE 4 Handedness in secondary structure connections. An α-helix linking two β-strands (hydrogen bonded in a sheet) is shown as a backbone (α-carbon) trace in (a) the common right-handed configuration and (b) the rare left-handed connection. The different chiralities can be appreciated if the whole chain is viewed as a superhelix: In the right-hand form, clockwise rotation would drive it into the page (like a screw or corkscrew), whereas the same rotation would extract the left-hand form.

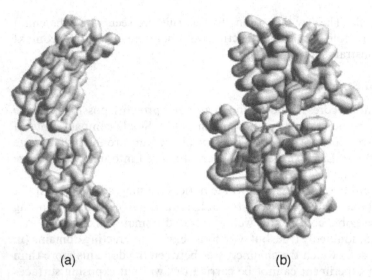

(a)   (b)

FIGURE 5 Simple and complex domain connections. (a) Two immunoglobulin domains linked by a single connection. (b) Two more closely packed domains (arabinose-binding protein) between which the chain passes three times. (The linkers have been drawn thinner for clarity.)

2. PROTEIN FOLDING

2.1. Folding In Vitro and In Vivo

Proteins fold spontaneously, and many can be unfolded (denatured) and refolded back into their native state (usually indicated by restoration of activity) [14]. In the cell the situation is slightly more complicated, because there are a variety of other proteins (chaperonins) that aid the folding process and yet another group that can modify the protein structure during or after folding (posttranslational modification). (See Ref. 2 for a comprehensive survey of the latter.) It is generally accepted that chaperonins function by modifying the environment of the folding protein and that their role is mainly to prevent intermolecular interactions that can result in tangling and misfolding. Being a generic agent, they have no information capacity to specifically direct the folding of individual proteins. A similar argument applies also to the posttranslational modifications. This implies that the complete information encoding the final three-dimensional form of the protein lies in its sequence of amino acid residues.

There has been much debate over the years on the mechanisms of protein folding. (See Ref. 15 for a review of some aspects of protein folding from a computational viewpoint.) One model is that the hydrophobic core nucleates and the secondary structures "grow" out of it, simultaneously expanding the core (Fig. 6). Alternatively, it can be argued that the secondary structures form and pack (condense) together to form a core, allowing some limited diffusion and rearrangement along the way until the correct fold is encountered (Fig. 7). Almost all possible intermediate variants along these lines can be imagined (and have probably been proposed). An attractive simplification containing elements of both models (but lying closer to the condensation model) is the model of Dill and coworkers; in which local hydrophobic residues "zip" together, forming ever more compact structures [16].

Only with the application of NMR have we been properly able to "see" snapshots of the folding process. This has revealed partly formed, packed secondary structures (a bit of both of the above models). (See Ref. 17 for a review of some recent papers reporting these results.)

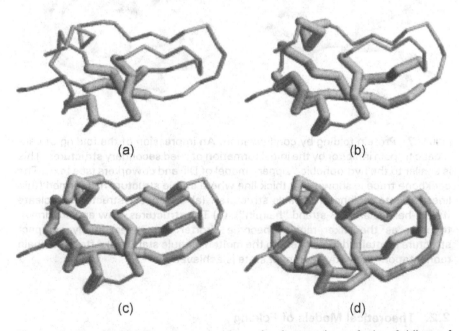

(a)   (b)

(c)   (d)

FIGURE 6 Protein folding by accretion. An impression of the folding of pancreatic trypsin inhibitor by the accretion of stable structure (thick lines) around a nucleating core. (Thin lines indicate regions of variable structure) (a) A hydrophobic core is formed from a three-strand β-sheet and a hydrophobic residue near the carboxyl terminus. (b) The sheet grows, and α-helices grow on its surface. (c) The molten globule state, which, as in Figure 6, compacts to give the native structure (d).

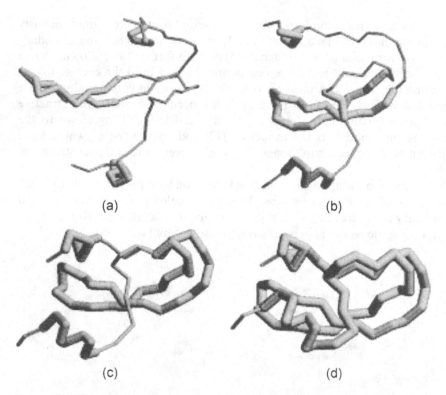

(a)   (b)

(c)   (d)

Figure 7 Protein folding by condensation. An impression of the folding of pancreatic trypsin inhibitor by the initial formation of seed secondary structures. This is similar to the hydrophobic "zipper" model of Dill and coworkers (see text). The backbone trace is shown as a thick line where stable structure has formed (thin lines indicate regions of variable structure). (a) Secondary structures nucleate (two α-helices and a β-strand "hairpin"). (b) The structures grow and are drawn together as the linker regions become shorter. (c) A reasonably compact structure is attained (equivalent to the molten globule state). (d) After side-chain reorientation, the native compact state is achieved.

2.2. Theoretical Models of Folding

From kinetic studies it appears that the protein can attain a globular state very quickly but must then wait a relatively long time for the exact packing interactions that are seen in the native structure to be attained. This intermediate state has been termed the molten globule, and it seems to be essentially native-like in terms of the path of its chain. Although some time may be required to search through all the possible side-chain packings (even in the constrained molten globule state), a factor contributing to the perseverance of this state may be the propensity of prolines to flip the orientation of their peptide bond. Clearly, the native structure cannot be attained until all prolines have been "frozen" into their native conformation (normally trans).

Although the molten globule allows side-chain packing rearrangement, it is likely that at this stage the overall fold of the protein has become fixed (for example, it is hard to imagine rearrangement of a β-sheet strand order). The question still remains as to how the protein gets to this stage. If it follows a "blind" conformational search through all the possible conformations allowed under the flexibility of the polypeptide chain, then, as Leventhal noted, folding could not happen. (See Honig [15] for a review of this topic.) This implies that the chain must follow a directed folding (or kinetic) pathway. A simple model for investigating this is the "hydrophic zipper" of Dill in which local hydrophobic interactions zip together into larger and larger assemblies. Elaboration of this approach, viewed as a two-dimensional energy landscape, leads to the impression of a folding funnel (or perhaps a more complex network of gullies) that directs the conformational evolution of the chain toward the native structure represented by a deep hole. By contrast, the equivalent energy landscape for the undirected conformational search has been likened to the search for the hole over an almost flat golf green [18,19].

As with factors that stabilize the native protein structure, the factors that determine the direction of its folding pathway must also be under selection pressure during evolution. From theoretical studies, it has been argued that the sequences that give rise to fast folding should be favored (i.e., those that create the deepest, most unique funnel to the native structure). As with most aspects of protein folding, the equation is never simple. It might seem that this ideal could be attained simply by stabilizing the native structure, but for every mutation that stabilizes the native state it is necessary to check that it does not also stabilize any competing state. This is virtually impossible to do with any realistic model of a protein, but it can be explicitly tested using very simple models of protein structure in which every conformational possibility can be enumerated. The favorite model is a 3 × 3 × 3 box filled with a chain consisting only of hydrophobic and hydrophilic residues [20,21]. Running many simulations of folding with this simple model revealed that the fastest folders are those that have the biggest energy gap between the native structure and its nearest rival folds.

2.3. Folding In Silico

The ability of the protein chains to self-organize and attain their own tertiary structure (even if they have "a bit of help from their friends") has given rise to the hope that the native structure might be calculated from first principles (ab initio) using only the laws of physics and chemistry. Despite many attempts, including the inclusion of realistic amounts of solvent, this goal has never been attained for any structure of a reasonable size. One argument advanced for the failure of this approach is that the formation of a hydrophobic core is an entropic effect and the enthalpic components roughly balance. Taking a simple tally, for every hydrogen bond found in secondary structure, two protein–water bonds must have been broken and one water–water bond formed. In short, $(P–W)_2 \rightarrow P–P + W–W$, and taking the bonds as of equivalent energy, the enthalpic sum is zero. In addition, the entropic components are not obviously in favor of folding either. As the hydrophobic core condenses, partially restricted (ordered) water is released to bulk solvent (disordered), thus increasing the entropy; however, as the chain compacts, its degrees of freedom are constrained and the side chains become fixed; both effects that decrease entropy. The final energy of the folded protein is therefore a fine balance between the difference of large enthalpic and entropic components, which is why it has proved so hard to calculate with any accuracy. (See Ref. 15 for various pointers into this large body of literature.)

Suggestions to overcome these problems have included a more accurate energetic model (perhaps involving quantum-mechanical approximations) on the enthalpic side; for the entropic component, simulating the protein in sufficient bulk solvent for a long time period has been considered necessary. A long period in folding terms is the time it takes proteins to fold in vitro or in vivo (typically from microseconds to milliseconds). For many years this seemed an unobtainable goal (simulations encompassed only the equivalent of picoseconds); however, with the ever-increasing power of computers, this was recently attained (abeit, only for a small protein). Despite this achievement, the native state of the protein was only partially attained.

3. PROTEIN STRUCTURE PREDICTION

The effective failure of the ab initio approach to protein folding (discussed in the previous section) does not mean that it is impossible to gain some idea of a protein structure from its sequence. The approach that has yielded the most practical advantage simply looks at the structures we know, together with their sequences, and attempts to derive rules, or correlations, between the two. This approach, commonly referred to as empirical, is intellectually less satisfying than the ab initio approach because it does not require understanding of the physical processes tht gave rise to the correlation. However, because many of these processes were until recently unobservable (such as the details of the folding pathway), there is some advantage in using an approach that does not rely on this knowledge.

The empirical approach can be divided into two reasonably distinct branches. One is based on the correlation of (sequentially) local amino acid composition with local structure (typically secondary structure type) and makes little claim to predict the overall fold of the protein. The other is based on the alignment of sequences; when one of these sequences has a known structure, a prediction of the overall fold can be made. Again, all intermediate possibilities can be found. For example, only local alignments might be found corresponding to a fragment of structure, which, in the end, might be only a piece of secondary structure, at which point the two approaches converge.

3.1. Structure Preference Statistics

As noted above, amino acids have a "preference" to be either in or out of the hydrophobic core. For less obvious physical reasons, a preference can also be found for some acids to adopt a particular type of secondary structure. These preferences have been variously encoded over the years with a view to predicting secondary structure type, given the protein sequence. From some poor beginnings, these methods are now reasonably accurate and attain their best performance when multiple, aligned sequences are used [22]. One of the state-of-the-art approaches uses a complicated neural net method and achieves an average accuracy approaching 80% [23,24] However, as Zvelebil et al. [22] point out, this good average is also accompanied by a large standard deviation.

A reasonable prediction of exposure to solvent can also be made by using similar methods that can in turn be used to infer the presence of a particular secondary structure based on patterns of exposure along the sequence. The α-helix is often half-buried in the core, leading to a sequential pattern of exposure with roughly the same period as the α-helix (3.6 residues/ turn). Similarly, many β-sheets have one buried face, leading to an alternating pattern of exposure.

Unfortunately, none of these methods that are based on the properties of relatively local parts of the sequence can place any constraints on the overall fold of the chain. The best that can be expected is perhaps that a predicted turn will be short enough to constrain two secondary structures to pack together. The sequential arrangements of secondary structures along the chain nevertheless gives some idea of the class or structure that can be expected, and, as was seen in Sec. 1.4, this knowledge might confine the protein to one of the few different architectures. Again as we have seen, knowing the architecture (or framework) places few constraints on the possible folds, but with small proteins some useful predictions can be made [25,26]. (See also Taylor and Aszódi [27] and other chapters in the same volume for further reviews.)

Both secondary structure prediction and methods that try to pack the secondary structures together, while perhaps not producing a unique solution, greatly constrain the number of possible solutions. If some additional constraints can be found (for example, common ligand binding residues or disulfide bonding partners), then the number of folds that are consistent with these can be much smaller, possibly only one [28]. This was investigated using a general tertiary structure prediction method based on distance geometry, and it was found that two such constraints per secondary structure were enough to specify the fold. Again, unfortunately, such data are not generally available (except from NMR experiments), and where they can be found they are generally not ideally distributed over the structure [29].

A step toward deriving general constraints from the sequence was made with the investigation of amino acid pairwise interaction preferences (such as those derived by Sippl [30], which are commonly used in the threading methods described below). Although these can restrict the pairing of residues in the structure, they cannot distinguish, say, one pair of valines as any more likely to be found together than any other, which, in a prediction context, is often little better than saying that there is a core of hydrophobic residues. There was some optimism that these vital specific pairwise interactions might be generally derived from the analysis of residue covariation within multiple sequence alignments. However, more recent analyses have cast doubt upon whether this approach can provide data of sufficient specificity [31,32]. Unlike RNA (where residue correlation provides good information), it seems likely that proteins are too "soft" and that if a residue mutates there is no strong pressure for any specific neighbor to compensate for it. Instead, the whole structure might simply shift a little, with perhaps any added strain being compensated for later at a residue position that is not in direct contact with the original mutation site.

3.2. Alignment-Based Methods

3.2.1. Sequence Alignment and 3D–1D Matching

The surest way to predict a protein structure is to find a protein of known structure that shares some sequence similarity with the sequence of the protein one wishes to investigate. Because structure is more strongly preserved under evolution than the sequences that determine it, any reasonable degree of sequence similarity can lead to a good prediction of structure. The key in this approach is to decide what is reasonable.

At one end of the similarity scale (close to 100%), clearly what constitutes a reasonable alignment is obvious. As the sequences diverge, in particular when their difference is greater than 50% in residue identity (counted over all aligned positions), the intuitive guide of "obvious" begins to break down. Indeed, "reasonable" alignments have been made between sequences with over 50% identity that are known from structure to have different folds.

An approach that alleviates much of this difficulty is to align more than two sequences together. A simple strategy for this is to start on firm ground with the most obvious pair and work outward in levels of increasing difficulty. One can also start at more than one point and later align the subalignments. This latter is the approach adopted by most of the practical multiple sequence alignment methods [33,34]. (See also Heringa, Chapter 4 in this volume.) With such data it becomes more obvious which positions are conserved and in what way, allowing like positions to be matched up with greater certainty.

Into this equation, precalculated secondary structure and exposure can also be introduced, and, depending on the degree to which the sequences are similar, these predicted components can be the main determining factors in the alignment [35,36]. This approach of using predicted structure can also be used with sequences on the one hand and a structure on the other. Indeed, the approach is quite general, and any mix of sequences and structure can be used.

3.2.2. Molecular Modeling and Threading

Aligning a sequence to a structure also has independent roots in molecular modeling (sometimes call modeling by homology). This involves "mutating" the side chains of a proteins structure by using the corresponding sequence position of another. This procedure requires a starting alignment, but it can be modified under the influence of the physical constraints imposed by the framework structure [37,38]. This "by-hand" approach has become increasingly automated with methods that combine the three-dimensional fitting aspect simultaneously with the alignment. A major application of this technique is in scanning the known folds to see if a novel sequence has a significant fit (Fig. 8).

This is a computationally difficult procedure, and various heuristic algorithms have been devised. Most simply, one can pick a new residue (fix a point in the alignment) with reference only to the existing structural positions, do this for each position, and then repeat. This is known as the "frozen" approximation. Alternatively, more sophisticated algorithms can be used that approximate a more simultaneous solution of all positions [39]. Where there is any clear sequence similarity, the choice of algorithm makes little difference; and where there is no clear sequence similarity, the rough models of residue interaction, combined with the limitation of considering only pairwise interactions, obscure any performance advantage of the different matching algorithms.

The term "threading" has generally come to be applied to all methods that match a sequence to a structure, whether they have their roots in

(a) Sequence   **W- I- L- L- I- E- T- A- Y- L-...**

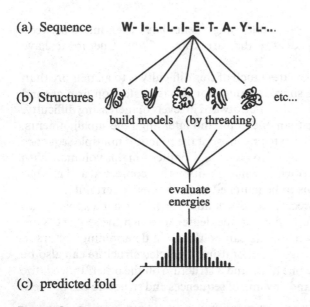

(b) Structures

build models   (by threading)

evaluate
energies

(c) predicted fold

FIGURE 8 Outline of threading used for fold recognition The sequence of unknown structure (a) is compared (fitted or threaded) onto each of a library of representative known structures (b). Each of the resulting models has an associated score or energy that is plotted as a frequency histogram. Good matches should have significantly low energy, and the best can be taken as the predicted fold (c).

sequence alignment or molecular modeling. However, the term was originally used to describe the direct matching of a sequence to a three-dimensional (3D) structure [39], and I consider the term "3D–1D matching" to be a better description for the alignment of derived (predicted) properties. Rather than consider detailed residue side-chain interactions, these methods employed normalized preferences for residues to pack, sometimes called mean force potentials, that were calculated from extensive analysis of the structure databank [30]. Other approaches have attempted to restrict the interactions to residue contacts, but there is no clear advantage either way. The current trend with these methods is toward a more comprehensive inclusion of all available data, including multiple sequences, and predicted secondary structure along with any remnant of sequence similarity [40,41].

ACKNOWLEDGMENTS

I thank David Jones for his help with some of the figures.

REFERENCES

1. Brändén C-I, Tooze J. Introduction to Protein Structure. New York: Garland, 1991.
2. Kyte J. Structure in Protein Chemistry. New York: Garland, 1995.
3. Taylor WR. The classification of amino acid conservation. J Theor Biol 1986; 119:205–218.
4. Taylor WR. The properties of amino acids in sequences. In: Bishop MJ, ed. Nucleic Acid and Protein Databases: A Practical Approach. 2d ed. San Diego: Academic Press, 1999:81–103. Chapter 5.
5. Reid ER, Franchini PL. Practical principles of protein design. In: Reid RE, ed. Peptide and Protein Drug Analysis. New York: Marcel Dekker, 1999:1–80.
6. Chothia C, Finkelstein AV. The classification and origins of protein folding patterns. Annu Rev Biochem 1990; 59:1007–1039.
7. Finkelstein AV, Ptitsyn OB. Why do globular proteins fit the limited set of folding patterns? Prog Biophys Mol Biol 1987; 50:171–190.
8. Thornton JM, Orengo CA, Todd AE, Pearl FMG. Protein folds, functions and evolution. J Mol Biol 1999; 293:333–342.
9. Chothia C, Murzin AG. New folds for all-β proteins. Structure 1993; 1:217–222.
10. Ptitsyn OB, Finkelstein AV. Similarities of protein topologies: evolutionary divergence, functional convergence or principles of folding? Quart Rev Biophys 1980; 13(3):339–386.
11. Sternberg MJE, Thornton JM. On the conformation of proteins: the handedness of the connection between parallel β-strands. J Mol Biol 1977; 110:269–283.
12. Janin J, Chothia C. Domains in proteins: definitions, location and structural principles. Methods Enzymol 1985; 115:420–440.
13. Taylor WR. Protein structure domain identification. Protein Eng 1999; 12:203–216.
14. Anfinsen CB. Principles that govern the folding of protein chains. Science 1973; 181:223–230.
15. Honig B. Protein folding: from Leventhal paradox to structure prediction. J Mol Biol 1999; 293:283–293.
16. Dill KA. Folding proteins—finding a needle in a haystack. Curr Opinion Struct Biol 1993; 3:99–103.
17. Goldberg DP. Finding the right fold. Nature Struct Biol 1999; 6:987–989.
18. Dill KA, Chan HS. From Leventhal to pathways to funnels. Nature Struct Biol 1997; 4:10–19.
19. Dill KA, Fiebig KM, Chan HS. Cooperativity in protein folding kinetics. Proc Natl Acad Sci USA 1993; 90:1942–1946.
20. Godzik A, Kolinski A, Skolnick J. Lattice representations of globular-proteins—how good are they. J Comput Chem 1993; 14:1194–1202.
21. Hinds DA, Levitt M. A lattice model for protein-structure prediction at low resolution. Proc Nat Acad Sci USA 1992; 89:2536–2540.
22. Zvelebil MJ, Barton GJ, Taylor WR, Sternberg MJE. Prediction of protein secondary structure and active sites using the alignment of homologous sequences. J Mol Biol 1987; 195:957–961.

23. Rost B, Sander C. Prediction of protein secondary structure at better than 70-percent accuracy. J Mol Biol 1993; 232:584–599.

24. Mehta PK, Heringa J, Argos P. A simple and fast approach to prediction of protein secondary structure from multiply aligned sequences with accuracy above 70%. Protein Sci 1995; 4:2517–2525.

25. Taylor WR. Towards protein tertiary fold prediction using distance and motif constraints. Protein Eng 1991; 4:853–870.

26. Dandekar T, Argos P. Folding the main-chain of small proteins with the genetic algorithm. J Mol Biol 1994; 5:637–645.

27. Taylor WR, Aszódi A. Building protein folds using distance geometry: towards a general modeling and prediction method. In: Merz KM, LeGrand SM, eds. The Protein Folding Problem and Tertiary Structure Prediction. Boston: Birkhäuser Springer-Verlag, 1994:165–192. Chap 6.

28. Munro REJ, Taylor WR. Structure prediction and molecular modelling. In: Reid RE, ed. Peptide and Protein Drug Analysis. New York: Marcel Dekker, 1999:115–132.

29. Aszódi A, Gradwell MJ, Taylor WR. Global fold determination from a small number of distance restraints. J Mol Biol 1995; 251:308–326.

30. Sippl MJ. Calculation of conformational ensembles from potentials of mean force: an approach to the knowledge-based prediction of local structures in globular proteins. J Mol Biol 1990; 213:859–883.

31. Pollock DD, Taylor WR. Effectiveness of correlation analysis in identifying protein residues undergoing correlated evolution. Protein Eng 1997; 10:647–657.

32. Pollock DD, Taylor WR, Goldman N. Coevolving protein residues: maximum likelihood identification and relationships to structure. J Mol Biol 1999; 287:187–198.

33. Taylor WR. A flexible method to align large numbers of biological sequences. J Mol Evol 1988; 28:161–169.

34. Higgins DG, Sharp PM. Clustal: a package for performing multiple sequence alignment on a microcomputer. Gene 1988; 73:237–244.

35. Nishikawa K, Matsuo Y. Development of pseudoenergy potentials for assessing protein 3D–1D compatibility and detecting weak homologies. Protein Eng 1993; 6:811–820.

36. Rice DW, Eisenberg D. A 3D–1D substitution matrix for protein fold recognition that includes predicted secondary structure of the sequence. J Mol Biol 1997; 267:1026–1038.

37. Šali A, Overington JP, Johnson MS, Blundell TL. From comparisons of protein sequences and structures to protein modeling and design. TIBS 1990; 15:235–240.

38. Pearl LH, Taylor WR. A structural model for the retroviral proteases. Nature 1987; 329:351–354.

39. Jones DT, Taylor WR, Thornton JM. A new approach to protein fold recognition. Nature 1992; 358:86–89.

40. Taylor WR. Multiple sequence threading: an analysis of alignment quality and stability. J Mol Biol 1997; 269:902–943.

41. Jones DT. GenTHREADER: an efficient and reliable protein fold recognition method for genomic sequences. J Mol Biol 1999; 287:797–815.

7

DNA–Protein Interactions: Target Prediction

Akinori Sarai
RIKEN Institute, Tsukuba Life Science Center, Tsukuba, Japan

Hidetoshi Kono
University of Pennsylvania, Philadelphia, Pennsylvania, U.S.A.

1. INTRODUCTION

Regulation of gene expression in higher organisms is achieved by a complex network of transcription factors and their target genes. Large amounts of sequence information for genes and transcription factors from genome analyses are presenting a great challenge in the field of bioinformatics. In spite of the tremendous amount of sequence and structural data for transcription factors, the mechanisms of target recognition by transcription factors are not well understood. Sequence similarity searches are the most commonly used methods for extracting functional information from sequence data. However, we are only beginning to discern meanings encoded in nucleic acid and protein sequences. Structural data contain valuable functional information as well. Inspection of structural data of protein–DNA complexes reveals that there are no simple rules for the interactions between amino acids and base pairs, i.e., the interactions are more redundant and flexible than expected. Sometimes conformation of DNA plays an important role in protein binding. Because transcription factors usually bind to multiple target sequences and regulate multiple genes, cooperativity with other factors should play important roles in target recognition. Because of the contributions from these multiple factors to protein–DNA recognition, target prediction is a rather complicated problem. To tackle such a problem, we need to use as much information as possible. Here, we describe several methodologies of target prediction.

The methods for predicting target sites can be classified into four methods according to the type of information used:

1. The sequence-based method uses sequence information obtained from known binding sequences. From the alignment of collected sequences, a consensus sequence pattern or profile is constructed and used to scan the database for finding potential target binding sites.
2. The ΔG-based method is based on experimental measurements of binding between protein and DNA. The binding affinity data for systematic single-base mutations to a consensus binding site are used for target prediction.
3. The structure-based method is based on the statistical analysis of the structural database of the protein–DNA complex. Empirical potential functions for specific interactions between bases and amino acids are derived and used to evaluate the fitness of sequences to the complex structures of particular transcription factors by a combinatorial threading procedure similar to protein structure prediction.
4. The ab initio method does not rely on experimental data but uses computer simulations to derive contact potentials between bases and amino acids, serving to complement the structure-based method.

The first and second methods do not use structural information, whereas the last two methods are based on the structure of the DNA–protein complex or base–amino acid interactions. The first method is currently the most commonly used of the four. Thus, we will not go into details of this method (see Ref. 1 for review) and will describe the other methods in more detail.

Currently, complete genome sequences have been obtained for many different organisms. Now the main focus of interest is to extract functional information from these genomes. Finding target genes for transcription factors at the genome level will lay a basis for the analysis of the gene regulatory network. In the final section of the chapter we describe a strategy for the application of target prediction to functional genomics.

2. SEQUENCE-BASED METHOD

The sequence-based method relies on sequence information obtained from known binding sites of transcription factors. Multiple sequence alignment of those binding sites usually reveals that bases at some positions are more conservative than others. Assuming that the conserved sites are more important for specificity, we can derive a so-called consensus sequence pattern (e.g., IUPAC ambiguity strings) and more quantitatively a position-specific weight matrix based on the frequency of the occurrence of bases at each position [2–4]. The simplest way to find potential binding sites is to use pattern matching with the consensus sequence pattern [5]. However, this digital detection method may miss many true binding sites, because pattern matching is too sensitive to base changes. On the other hand, the weight matrix calculates the probability for a protein binding to a given sequence; thus it is more tolerant of base changes. More rigorous treatment by statistical mechanics of protein–DNA binding enables one to calculate relative binding affinity for any sequence [6]. Several hundred weight matrices have already been constructed for different transcription factors and compiled in databases such as TRANSFAC [7]. There are also several tools such as SignalScan [5,8], MATRIX SEARCH [3], MatInspector [4], and ConInspector [9] to find binding sites from a sequence database. These tools are widely used for target prediction (see Ref. 1 for review).

The sequence-based method is quite straightforward, but its validity strongly depends on the quality of the sequence information used. Some binding sites were identified as targets for transcription factors from in vivo experiments, whereas others were derived from in vitro screening procedures for particular transcription factors. Although this method was shown to be successful for some simple systems such as bacterial promoters [10], its accuracy for general use needs to be tested. The weight matrices for eukaryotic transcription factors are typically defined in a short range of sequences, indicating that the intrinsic specificity is rather low. This may be partly because the synergistic action of multiple transcription factors on the same promoter may be the strategy for the complex regulation of gene expression. Thus, screening of binding sites of a single transcription factor often produces many false positives and false negatives. To minimize those false positives and false negatives, it is critical to optimize the selection criteria by adjusting threshold parameters. There have been some attempts to automate the adjustment [4,11,12]. However, because the binding of a transcription factor to a given site is likely to be context-dependent, completely automated adjustment will be very difficult.

There may be several ways to improve the method. There have been some attempts to incorporate physical properties of DNA such as conformational properties and stability into the prediction scheme [13,14]. These additional pieces of information may enhance the accuracy by incorporating some cooperative effects among different sequence positions. Furthermore, longer range cooperativity through DNA bending, mediating molecules, etc. may be caused by the synergistic action of multiple transcription factors on the same promoter (see later sections for more discussion). Therefore, considerations of other transcription factors and the context of neighboring sequences will increase the accuracy of binding site prediction. Some tools have incorporated such information [15].

Roulet et al. [16] evaluated currently available prediction tools for a particular transcription factor by comparing their results with experimental binding affinities and found that the values predicted by various methods are internally correlated with one another rather than with the experimental values. The sequence-based method relying only on sequence information may suffer from the same problem as other sequence-based predictions of function and structure, which lack information on physical processes. Thus, it is desirable to incorporate other kinds of information into the prediction scheme.

3. ΔG-BASED METHOD

The sequence-based method is based on the assumption that the conserved sequences are important for specificity in DNA recognition by proteins. This assumption will be valid if the sequence selection during evolution has reached an equilibrium under the pressure of protein binding. In reality, however, conserved bases among a set of sequences do not necessarily reflect specificity. If multiple proteins bind to the same site or different binding sequences overlap, then the situation becomes more complicated. As noted above, the accuracy of the sequence-based method will depend on the quality of sequence data. On the other hand, the binding affinity data for systematic single-base mutations at a consensus binding site can be used to derive matrices similar to the weight matrices in the sequence-based method. Those matrices are based on physical interactions under equilibrium conditions, and their accuracy depends only on experimental errors. Thus, they will be more reliable than the sequence-based weight matrices.

3.1. Binding Free Energy Change and Specificity

The binding free energy change due to single-base mutations is calculated with the equation

$$\Delta\Delta G = RT \ln\left(\frac{K(\text{wild})}{K(\text{mutant})}\right)$$

where R is the gas constant, T is temperature, and K(wild) and K(mutant) are binding constants for wild-type and mutant DNA, respectively. If this quantity is obtained for the complete single-base mutations within the binding region, it will define a kind of matrix in terms of base position vs. base type. Thus, it can be used for the prediction of potential binding sequences for the protein. These $\Delta\Delta G$ data are available for only a limited number of cases such as phage Cro [17] and λ repressor [18], c-Myb transcription factor [19], and ethylene-responsive element binding proteins [20]. The original purpose of these analyses was to examine the specificity of the interactions involved in a DNA–protein complex, because the $\Delta\Delta G$ values reflect the extent of specificity.

Compared with the structures of the DNA–protein complex, the relationship between structure and specificity can be examined in a quantitative manner. Figure 1 shows an interesting example of the comparison between specificity and structure for Cro and λ repressor, which serve as a genetic switch in the phage life cycle. Despite the fact that these regulatory proteins have the same helix–turn–helix motif [21,22], bind as a dimer, and recognize the same six homologous operator sequences (OR1-3, OL1-3), the specificity represented by the pattern of $\Delta\Delta G$ values is quite distinct; i.e., in the case of Cro, most specific interactions come from the recognition helix of the helix–turn–helix motif, whereas λ repressor uses N-terminal arms for the specific interactions rather than the helix–turn–helix region and recognizes the sequences in an asymmetrical manner. These differences enable the two proteins to distinguish sequences of the six operators.

3.2. Application to the Target Prediction

The spin-off of this binding analysis is its application to target prediction. In order for this method to be applicable to the prediction, however, the $\Delta\Delta G$ values have to be independent of one another, i.e., they must be additive. The additivity of $\Delta\Delta G$ values has been tested for Cro and λ repressor by multiple-mutation experiments. These analyses have shown that summations of $\Delta\Delta G$ values agreed, with a few exceptions, with experimental measurements quite well. The experimental $\Delta\Delta G$ values seem to become constant when they exceed a certain limit (in the case of Cro this corresponds to about 5 kcal/mol). This indicates that the binding mode is changed from specific to nonspecific, where the binding no longer depends on sequence. Cro and λ repressor both recognize OR1-3 and OL1-3 and bind with different affinities. The sum of $\Delta\Delta G$ values agrees with the binding orders of Cro and λ repressor to the operators almost perfectly (see Fig. 2). These results suggest that the binding affinity of these proteins to any sequences can be predicted very accurately by adding the $\Delta\Delta G$ values. The calculation of binding affinity of Cro for every

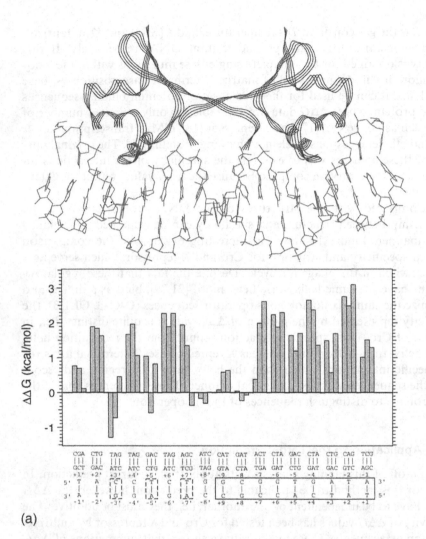

(a)

FIGURE 1 Binding free energy changes ($\Delta\Delta G$) due to systematic single base substitutions for Cro (a) and λ repressor (b). The sequence shown is OR1 with a boxed consensus sequence. The deviations from the consensus are shown by broken-line boxes in (b). The structures of these proteins complexed with DNA (6CRO and 1LMB for Cro and λ repressor, respectively) are aligned with the $\Delta\Delta G$ profiles. ($\Delta\Delta G$ values were taken from Refs. 17 and 18.)

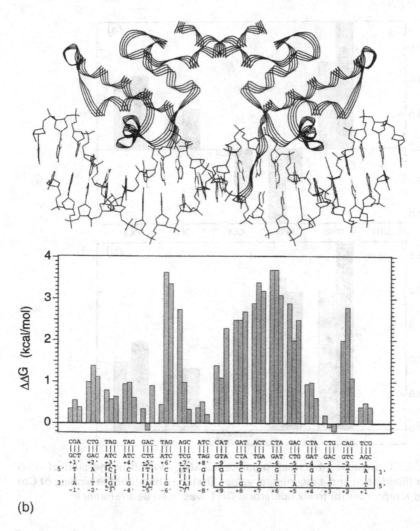

(b)

FIGURE 1 Continued.

base frame in the entire λ phage genome (48 kb) has shown that OR3 is the strongest binding site in the genome and that the OR3 binding site is far from the tail of the distribution in the histogram, as shown in Figure 3. These results demonstrate the accuracy and sensitivity of the method.

The ΔG-based method has also been applied to c-Myb transcription factor [23], which is an oncogene product. The binding experiment and structural analysis for this protein have shown that the specific interactions

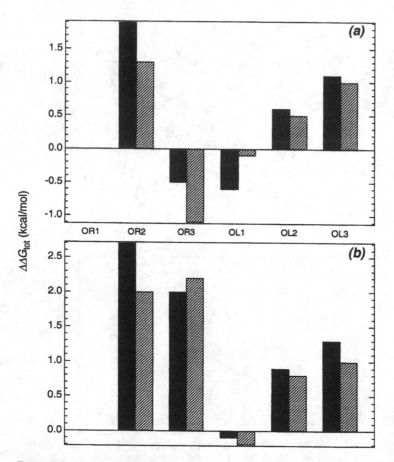

Figure 2 Comparison between predicted $\Delta\Delta G_{tot}$, which is the sum of $\Delta\Delta G$ values (filled bars) and experimental values (hatched bars) for the binding of Cro (a) and λ repressor (b) to six operators. OR1 was used as a reference.

are rather localized in a narrow range (7 bp from $+2'$ to $+8'$ in Fig. 4), compared to the prokaryotic repressors described above. The specific binding region of ethylene-responsive element binding protein in plants is also narrow (about 6 bp). This low intrinsic specificity may be due to the required synergistic action of multiple transcription factors on the same promoter for the complex regulation of gene expression. Therefore, finding potential target sites for transcription factors will be more complicated than in prokaryotic systems. The c-Myb protein has been known to activate or repress the transcription of several potential target genes. We have tested the method for the known binding sites in those promoters [23]. The predicted binding sites agree

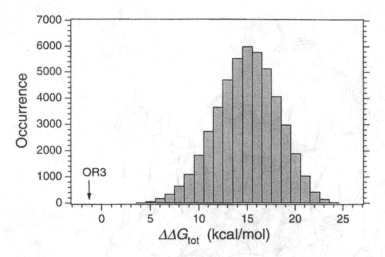

FIGURE 3 Histogram of $\Delta\Delta G_{tot}$ for the binding of Cro to the entire genome of λ phage. The position of $\Delta\Delta G_{tot}$ for OR3 is shown by an arrow.

with many putative binding sites of known target promoters. However, there are some binding sites not predicted by the analysis. When those sequences are aligned, bases at a particular position deviate from the consensus sequence derived from the binding analyses. In light of the structure of the Myb-DNA complex, these results indicate that different DNA-binding modes may be used by c-Myb to recognize different classes of binding sites. Nevertheless, this method has enabled us to screen the sequence database for potential Myb-binding sites and find sequences of several promoters that had not been identified experimentally but could be targets for c-Myb.

The Myb case provides an opportunity to compare the sequence-based and ΔG-based methods, because, in addition to the binding data, sequence information exists on the target sites that was derived from experiments. The results of the two methods may be compared in terms of the probability of base occurrence at each position. In the case of the ΔG-based method, the probability can be calculated from the $\Delta\Delta G$ values by the Boltzmann relation. The probabilities calculated from the alignment of 60 sequences derived from random-oligo screening experiments correlated well with those from the $\Delta\Delta G$ data with a correlation coefficient of 0.94. When the number of aligned sequences is reduced, the correlation decreases monotonously. In order for the correlation to be higher than 0.9, the number of aligned sequences must be greater than 30. This result indicates that a sufficient number of sequence entries are required to derive high quality weight matrices for the accurate target prediction by the sequence-based method.

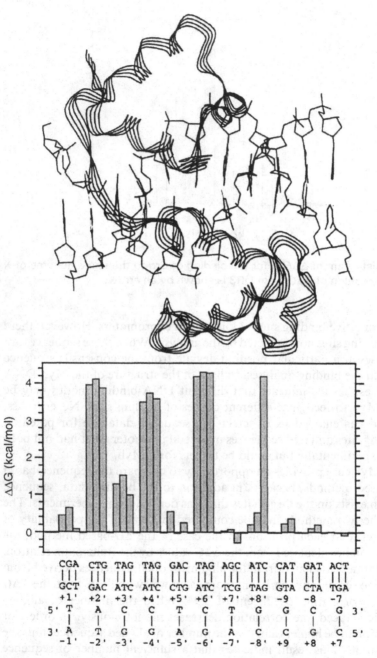

FIGURE 4 The $\Delta\Delta G$ profile for c-Myb. The structure of c-Myb complexed with DNA (1MSE) is aligned with the $\Delta\Delta G$ profiles. $\Delta\Delta G$ values were taken from Ref. 19.

3.3. Limitations and Prospects

The cooperative binding of Myb points out some limitations of the ΔG-based method. The $\Delta\Delta G$ values may not hold strict additivity for some proteins that exhibit cooperative binding. The breakdown of the additivity may arise from cooperative interactions within a single binding site or cooperative interactions among different binding sites. As an example of the former case, even small conformational changes in the DNA brought about by a base substitution could affect substitutions at other positions in the same site. In particular, the substitution of a base pair will change the local conformation or flexibility of DNA, depending on which base pair is adjacent to the mutated base pair. Also, long-range conformation changes such as DNA bending may be caused by specific sequences. On the other hand, eukaryotic transcription factors such as c-Myb have multiple binding sites in promoters, and they may interact with themselves, other transcription factors, and coactivators in the transcriptional machinery. Therefore, a certain level of cooperativity is likely to occur in the binding to DNA. In the case of c-Myb, it has been reported that an oligonucleotide sequence containing a duplicated Myb-binding motif showed a higher affinity than the sequence with only a single Myb-binding motif [61]. Furthermore, chromatin structure in the nucleus may introduce more complex cooperativity than in the in vitro situation. The role of cooperativity in DNA–protein interactions requires further investigation. The relationship between cooperativity and specificity in some DNA–protein complexes will be discussed in the next section.

4. STRUCTURE-BASED METHOD

Owing to the progress of X-ray crystallogphic and NMR spectroscopic techniques, structural data on the protein–DNA complex have been rapidly increasing, and more than 600 complexes have been registered in the Protein Data Bank (PDB). These structural data provide us with a rich source of information about the interactions between amino acids and base pairs at the atomic level. However, we have not fully utilized the functional information of these data. Several classes of DNA-binding proteins with distinct DNA-binding motifs have been revealed. They show a variety of interactions between proteins and DNA. Some interactions such as Asn–A and Lys–G are frequently observed in the complex structures. For members of a single structural family or for a group of families that interact in similar ways with DNA, some rules about the base–amino acid interactions have been proposed. However, statistical analyses of structural databases have shown considerable degrees of redundancy in the specific interactions between amino acids and bases, that is, the same amino acids often interact with different bases and vice versa [24,25]. Furthermore, the analysis of structural data also shows that the position and conformation of amino acid side chains are widely distributed in space around base pairs, as shown in Figure 5. Thus, strict code-like rules are not likely to exist for protein–DNA recognition [26], and a rule-based approach [27,28] may not be effective in the prediction of binding sites. The spatial distributions of side chains around base pairs indicate a possibility that the distribution may be converted to energy potential similarly to the contact potential between amino acids in protein structures and can be used for target prediction. In this section we describe the derivation of the potential and its application to target prediction.

4.1. Database Analysis and Derivation of Contact Potential

To derive the statistical potential of interactions between bases and amino acids, we analyzed the amino acid distributions around each base using 52 selected structures of protein–DNA complex [25]. As shown in Figure 5, we defined a coordinate system by taking as the origin the N-9 atom for A and G and the N-1 atom for T and C. We considered the amino acids within a given box, and the box was divided into grids. Because the sample numbers are not

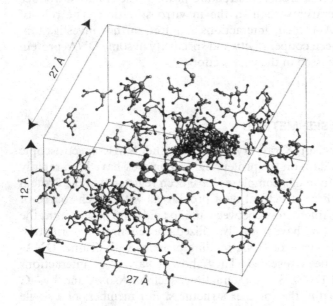

FIGURE 5 Distribution of Asn side chains around A. C_α of the amino acid is shown by a larger sphere. The definition of a coordinate system for amino acids around a base is shown by the three axes. Amino acids within the box are considered for creating potentials.

yet very large, we first considered the information about C_α atoms. Then we transformed the distributions of C_α atoms of amino acids into statistical potentials defined by the equations [29]

$$\Delta E^{ab}(s) = -RT \ln\left[\frac{f^{ab}(s)}{f(s)}\right]$$

$$f^{ab}(s) = \frac{1}{1+m_{ab}w}f(s) + \frac{m_{ab}w}{1+m_{ab}w}g^{ab}(s)$$

where m_{ab} is the number of pairs a and b observed, w is the weight given to each observation, $f(s)$ is the relative frequency of occurrence of any amino acids at grid point s, and $g^{ab}(s)$ is the equivalent relative frequency of occurrence of amino acid a against base b. R and T are the universal gas constant and absolute temperature, respectively. Here we used a box of $|x| = |y| = 13.5$ Å and $|z| = 6$ Å and a grid interval of 3 Å, which was determined by examining various intervals.

Figure 6 shows the typical potential maps of Asn-A and Asp-A projected onto the purine plane. In Figure 6, the dark regions show that the C_α atom of Asn frequently appears at coordinates $(0, -6)$ and $(9, -6)$ in the minor groove as well as $(0, 9)$, $(3, 9)$, and $(6, 9)$ in the major groove. By contrast, the distribution for Asp does not show a strong positional preference around A. Adenine has one acceptor (N7) and one donor (N6) in the major groove and one acceptor (N3) in the minor groove. Asparagine can form a double hydrogen bond with A in the major groove, so that the distribution of C_α of As is highly restricted to the dark region. On the other

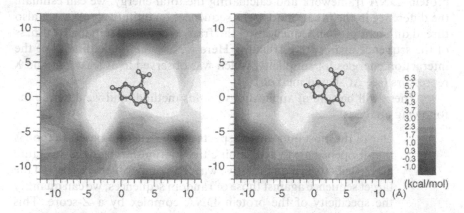

FIGURE 6 The potential maps of Asn-A and Asp-A projected onto the purine plane. Dark regions show preferable C_α positions of the amino acids.

hand, Arg and Lys, which often interact with bases, are widely distributed around the bases because their long side chains can accommodate to arrange their tip atoms to form a hydrogen bond with a base [25]. In this way, different amino acids show different distributions around each base. The potentials derived from the distributions for all combinations of amino acids and base pairs will be useful in predicting target DNA sequences by regulatory proteins.

4.2. Sequence–Structure Threading

Here we will describe how the potentials can be used to predict target DNA sequences for regulatory proteins. Statistical interaction potentials between amino acids derived from the protein structure have been used for screening potential sequences that fit with a given structure (3D–1D matching) or for screening structures that fit with a given sequence (threading method). These methods have been successfully applied to the prediction of the 3D structure of proteins. We surmised that the structures of protein–DNA complexes could be used to predict DNA target sites for regulatory proteins because determining DNA sequences that bind to a particular protein structure should be similar to finding amino acid sequences that fold into particular structures. We can calculate relative energy changes by using the potentials for any sequences against a given protein–DNA framework, where the following coordinates are used: C_α of protein; N9, C4, and C5 of A and G; and N1, C2, and C4 of T and C. An energy value summed over the sequence in the framework represents a kind of fitness of the DNA sequence with the complex structure. By threading different DNA sequences on the protein–DNA framework and calculating the total energy, we can estimate the difference in the fitness among the sequences quantitatively. We can also thread different protein sequences on the framework and estimate the fitness of the sequence against the structure. Here we assume the additivity of the interaction energies for simplicity, which was observed in at least Cro and λ repressors (see ΔG-based method).

There will be several applications of this method, and we describe the following analyses:

1. One of the most interesting aspects of the structure-based method is that it opened a way to examine the relationship between structure and specificity in a quantitative way. By comparing the energy of the target sequence against those of random sequences, we can quantify the specificity of the protein–DNA complex by a Z-score. This enables us to examine the structural effects such as DNA deformation and cooperativity on the specificity.

2. We can scan the genome sequence base by base by using a particular protein–DNA complex structure, calculate the energy at each position, and find potential binding sequences.

3. We can design sequences of both proteins and target sites for the purpose of changing the specificity.

4.3. Relationship Between Structure and Specificity

Protein–DNA binding is usually accompanied by certain conformational changes or deformation of protein and DNA. Thus, conformational flexibility of protein and DNA should be an important factor in determining a good structural match upon complex formation. In the case of c-Myb oncoprotein, the flexibility and stability of its DNA-binding domain have been shown to affect the DNA-binding activity [30]. The flexibility of DNA is sequence-dependent [31,32], and it can affect the binding affinity with protein as well. For some DNA and proteins, the structures of both free and complex forms have been solved. They show that DNA in the complex form is usually deformed from the canonical B form, more than the proteins in the complex form [33,34]. Here we show by several examples that the statistical potential is sensitive enough to detect the difference in structural deformation as the specificity difference.

To quantify the specificity, we evaluated the Z-score by calculating energy against 50,000 random DNA sequences. The Z-score is defined by $(X - m)/\sigma$ in the histogram, where X is the total energy of a DNA sequence in a complex form, m is the mean energy over 50,000 different combinations, and σ is the standard deviation of the energy. For example, a Z-score of -3.0 means that there are potentially one or two DNA sequences that are better fit to the framework among 1000 random sequences. We calculated the histogram for nine different protein–DNA frameworks (Fig. 7). The average Z-score was -3.1.

4.3.1. Specificity Difference in Cognate and Noncognate Binding

It is interesting to compare two structures, cognate and noncognate complex structures, in order to understand what is important for specific binding and what is different between them. For the target DNA sequence of glucocorticoid receptor (GR), the cognate (PDB: 1GLU) [35] and noncognate (PDB: 1LAT) [36] structures were solved. In the noncognate structure, GR has mutated to an estrogen receptor (ER)–like DNA-binding domain by swapping GR's amino acids in the binding site with the corresponding ER amino acids. As a whole the two complex structures are similar, and the proteins seem to interact with the same DNA in the same manner. However, it was

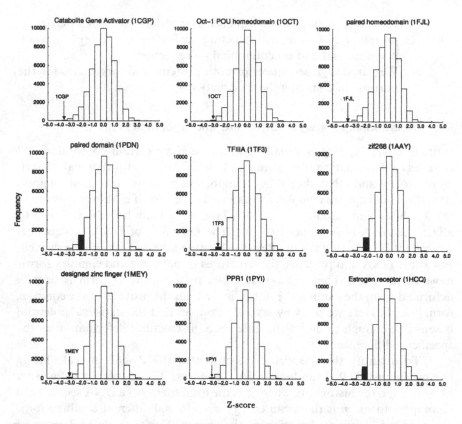

FIGURE 7 Histograms of Z-scores of nine different protein–DNA complexes for 50,000 random DNA sequences. In each histogram, the positions of the cocrystallized DNA sequences with protein are shown as filled bars and arrows. In each case, the database includes 52 complex structures, except where the test complex was itself in the database, in which case it has been excluded. The considered DNA sites for wild type are as follows: 5'-GTCACACTTT-3' for 1CGP; 5'-ATGCAAAT-3' for 1OCT; 5'-TAATCTGATTA-3' for 1FJL; 5'-CGTCA CGGTTGA-3' for 1PDN; 5'-GGATGGGAGACC-3' for 1TF3; 5'-GCGTGGGCGT-3' for 1AAY; 5'-TGAGGCAGAACT-3' for 1MEY; 5'-CGGCAATTGCCG-3' for 1PYI; 5'-AGGTCACAGTGACCT-3' for 1HCQ.

found that the spatial distributions of amino acids around A − 5 and T5 (1GLU residue number) were different when two structures are superimposed. In the cognate binding, C5M of T5 and Cγ2 of Val462 have a favorable van der Waals interaction, whereas the corresponding residue Ala in the noncognate binding does not. The loss of the interaction seems to cause the shift of the helix in the binding interface and changes the distribution of amino acids around the base pair (Fig. 8). These changes are reflected as changes in the potentials and Z-scores. When the noncognate complex was used as a template, the GR recognition site was not detected ($Z = -0.1$). On the other hand, when a cognate complex structure was used, a more favorable Z-score ($Z = -1.9$) was obtained.

Another example of cognate/noncognate binding is NF-κB p50, which is a transcription factor of great importance in cellular signal transduction, particularly in the immune system [37]. One complex structure is an NF-κB p50 homodimer bound to a duplex oligonucleotide with an 11 bp consensus recognition site located in the major histocompatibility complex class I enhancer [38]; the other is bound to a 10 bp idealized motif elated, but not identical, to the natural sites [39]. These complex structures are different in that the former DNA is bent by 15° although they both have a B-DNA like structure. It has been known that NF-κB binds 30 times as strongly to the former as to the latter [40,41]. These structures were used to test whether the subtle differences in specificity could be detected by the analysis of energy potentials. As shown in Figure 9, we obtained the lower Z-score for the former (-3.7 compared with -1.5 for the latter), which is consistent with the experimental result. The difference in Z-score is attributed to the DNA bending, which gives the different spatial distribution of amino acids around DNA.

The cognate and noncognate structures of EcoRV present an interesting example of the cooperative effect of sequence and structure on specificity [62].

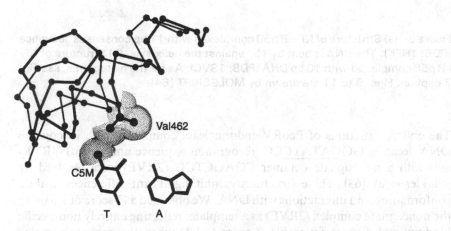

FIGURE 8 The interfaces of the cognate (PDB: 1GLU, heavy line) and noncognate (PDB: 1LAT, light line) glucocorticoid receptor–DNA complex structures. The noncognate structure is superimposed on the cognate structure against A – 5 and T5 (numbering in 1GLU). C_α atoms are shown by black spheres. Val 462 and C5M of T5 in the cognate structure are shown by the dotted spheres.

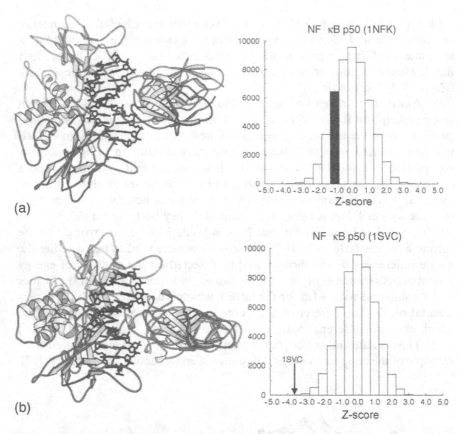

FIGURE 9 (a) Structure of NF-κB p50 complexed with 11 bp consensus sequence (PDB: 1NFK). The DNA is bent by 15° against the helix axis. (b) Structure of NF-κB p50 complexed with 10 bp DNA (PDB: 1SVC). As for the histograms, see Fig. 7 caption. Figs. 9 to 11 are drawn by MOLSCRIPT [64].

The crystal structures of EcoRV endonuclease complexed with its cognate DNA decamer GG<u>GATATC</u>CC (recognition sequence underlined) (4RVE) and with a noncognate octamer CGAGCTCG (2RVE) were resolved by Winkler et al. [63]. These structures exhibit significant differences in their conformation and interactions with DNA. We obtained a Z-score of 1.0 using the noncognate complex (2RVE) as a template, reflecting entirely nonspecific binding, and a more favorable Z-score (-1.1) when the cognate complex (4RVE) was used as a template, reflecting the specific binding within that complex. To examine the respective effects of sequence and structure on the specificity of *Eco*RV-DNA recognition in more detail, we considered virtual states of *Eco*RV in which original DNA sequences from 4RVE (cognate DNA) and 2RVE (noncognate DNA) were swapped for one another [these are designated as 4RVE (noncognate DNA) and 2RVE (cognate DNA), respectively] and calculated the Z-scores. The results indicate that sequence substitution in the structure of 2RVE [i.e., 2RVE (noncognate DNA) → 2RVE (cognate DNA)] caused the Z-score to change by −0.7, while that in the structure of 4RVE [i.e., 4RVE (noncognate DNA) → 4RVE (cognate DNA)] resulted in a much larger change, −2.1. On the other hand, the Z-score change associated with the structural change from 2RVE to 4RVE for the non-cognate sequence [i.e., 2RVE (noncognate DNA) → 4RVE (noncognate DNA)] was negligible (0.0), and that for the cognate sequence [i.e., 2RVE (cognate DNA) → 4RVE (cognate DNA)] was much larger (−1.4). Taken together, these findings indicate that the cognate but not the noncognate sequence is sensitive to the structural change and that the structure of 4RVE has a greater ability to discriminate DNA sequences than that of 2RVE.

Although specificity does not always have a relationship with affinity (thermodynamic preference), these results strongly suggest that the present method can detect subtle differences in specificity, which is attributed to the subtle structural differences.

4.3.2. Specificity Difference in Symmetrical and Asymmetrical Binding

A number of proteins bind to DNA as either homodimers or heterodimers. Homodimers often bind to target DNA sequences asymmetrically, leading to quasi-symmetrical structures in which identical subunits adopt similar but different conformations. Such structural differences are frequently observed when the subunits respectively form cognate and noncognate protein–DNA complexes. These examples are of particular interest because comparison of the subtle structural differences among homologous contacts within the two halves of a protein dimer provides valuable information about how specificity is determined. We examined the Z-scores for dimers and individual protein chains within protein–DNA complexes belonging to several structural families, bZip transcription factor GCN4, λ repressor, nuclear receptors, and transcription factors containing a Zn_2Cys_6 binuclear cluster domain such as HAP1, GAL4, and PPR1. These dimer structures exhibit varying degrees of asymmetry, some showing very dramatic asymmetry and others showing subtle asymmetry. It is difficult to determine how the differences in structure are related to the differences in specificity or which regions are responsible for the difference in specificity. The analysis of these structures in terms of Z-scores has revealed clear specificity differences between the individual chains of each homodimer and enabled us to identify those amino acids responsible for the asymmetry in specificity [62].

4.3.3. Effect of Cooperativity on Binding Specificity

Transcription factors usually bind to their target sites in cooperation with other factors, and the combination of different factors allows a complex mode of gene regulation. Thus, cooperative binding should play an important role in protein–DNA recognition. An interesting example of cooperative binding can be found in transcription factors MATa1 and MATα2 homeodomain proteins in yeast. They form a heterodimer in binding to DNA and repress transcription in a cell-specific manner. The binding experiments have revealed that the α2 and a1 proteins individually have only modest affinity for the target DNA, although the a1/α2 heterodimer binds to DNA with higher specificity and affinity. Using the two complex structures MATα2-DNA [42] and MATa1/α2-DNA [43] in the structural database (see Fig. 10), we tested whether the potentials can detect changes in binding specificity manifested in this cooperativity. Discrimination of target DNA by MATa1/α2 ($Z = -5.2$) was significantly higher than that by MATα2 only ($Z = -1.5$), because the heterodimer covers a longer sequence. When α2 was extracted from the a1/α2 structure, the extracted α2 showed a high selectivity of target DNA ($Z = -4.3$). This clearly demonstrates the cooperativity between MATa1 and α2. When the monomer and heterodimer complexes are compared, the most

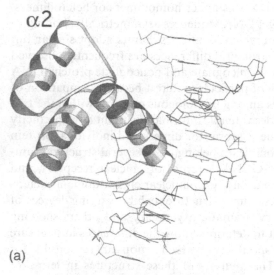

(a)

FIGURE 10   (a) MATα2–DNA complex (PDB: 1APL) and (b) MATa1/α2–DNA complex (PDB: 1YRN) structures. In the heterodimer complex of (b), DNA is bent by 60° against the helix axis.

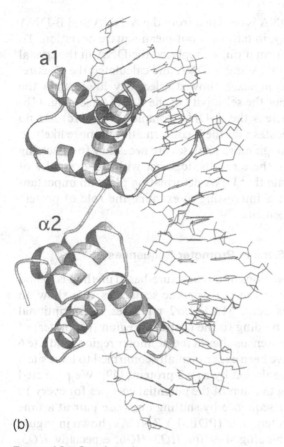

a1

α2

(b)

FIGURE 10 Continued.

obvious difference between them is the DNA bending caused by the hetero-dimer binding, as shown in Figure 10. This DNA bending introduces subtle differences in the spatial distributions of amino acids around each base and thereby affects both energy and specificity.

4.3.4. Effect of DNA Deformation

Deformations of DNA structures are often observed in protein–DNA com-plexes, and they likely enhance the specificity of protein–DNA recognition [44–46]. In the protein–DNA binding interface, DNA often takes an inter-mediate conformation between the canonical A-DNA and B-DNA. As such an example, we considered the structure of Zif268, a zinc finger–DNA complex (Fig. 11a) [47]. The DNA is deviated from the A-DNA and B-DNA by 4.2 Å and 2.9 Å, respectively, in terms of root-mean-square deviation. To examine the effect of DNA deformation, we replaced the DNA in the crystal structure with the modeled A-DNA and B-DNA and calculated the Z-score. Both B-DNA and A-DNA complexes showed selectivity lower than the crystal complex ($Z = -1.9$), but the selectivity of the B-DNA (see Fig. 11b) still remains ($Z = -1.4$), whereas the A-DNA complex (Fig. 11c) has no selectivity ($Z = 0.8$). This indicates that the B conformation is more likely to be recognized by Zif268. Although more analyses are necessary for assessing the role of DNA deformation, these results, together with the examples of MATa1/α2 and NF-κB, indicate that DNA deformation plays an important role in specificity. Likewise, it is interesting to examine the role of protein conformational changes in specificity.

4.4. Prediction of Binding Sites in Promoter Sequences

One of the most practical applications of the structure-based method is to find target sites of regulatory proteins in real genome sequences. We show an example of such applications here. MATa1/α2 regulates transcriptional repression of the *HO* gene by binding to the upstream region (promoter) of the gene [48]. Among the six consensus sites in the promoter region, and site 6 (nucleotides −715 to −761) have been experimentally confirmed to have site 3 (nucleotides −397 to −444) regulation by a1/α2 proteins [49]. We predicted the binding sites by calculating the sum of the potential energies for every 16 base pairs along the promoter sequence by shifting one base pair at a time using a cocrystal structure as a template (PDB: 1 YRN). As shown in Figure 12, our calculation resolved binding sites for *HO2–HO6*, especially *HO3*, *HO4*, and *HO6*, but not the binding site for *HO1*. In fact, deletion analysis revealed that site 1 was not sufficient for regulation by a1-α2 proteins [49]. This demonstrates that the method can be useful to detect candidate binding sites in practice.

4.5. Protein Design

Another application of the method is for the rational design of sequence-specific DNA-binding proteins that will provide reagents for both biological research and gene therapy. In particular, zinc finger proteins appear to provide the most versatile framework for design, because a zinc finger domain usually recognizes three base pairs. Kim and Berg [50] reported the structure of a newly designed protein which is composed of three zinc finger domains and an oligonucleotide corresponding to their consensus DNA sequences. Each of the three zinc finger domains recognized three bases: GAA, (G/T)C(G/A), and

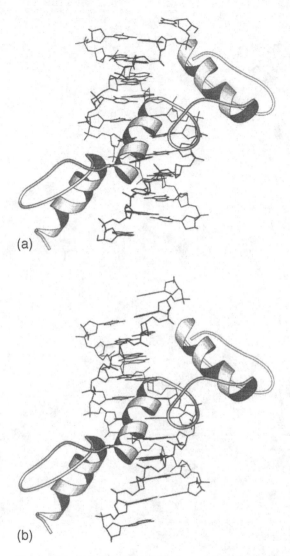

FIGURE 11 (a) Zif268–DNA complex (PDB: 1ZAA). (b) The DNA of 1ZAA was replaced by a standard B-form DNA by least-squares fitting. (c) The DNA of 1ZAA was replaced by a standard A-form DNA by least-squares fitting.

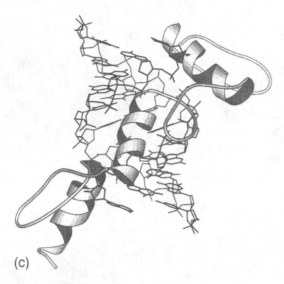

(c)

FIGURE 11 Continued.

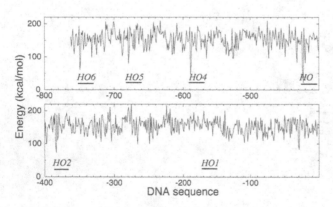

FIGURE 12 Detection of the binding sites by a MATa1/α2 complex in the upstream of *S. cerevisiae* site-specific endonuclease (*HO*) gene (accession No. M14678). The sharp peaks are the predicted binding sites. In the upstream region of the *HO* gene, six consensus patterns (including reverse sequences) exist (shown by underlines): *HO*6, −736 **TGT**ATTCATTCA**CATC** (reverse); *HO*5, −669 **TGT**CTTCAACTGC**ATC**; *HO*4, −576 **TGT**ATTTAGTTA**CATC** (reverse); *HO*3, −411 **TGT**TATTATTTA**CATC**; *HO*2, −371 **TGT**TCACATTAA-**CATC**; *HO*1, −150 TTTAGAACGCTT**CATC** (reverse). Invariant and highly conserved bases are highlighted in boldface.

GGG, [51,52]. The calculated Z-score for the crystal structure with the corresponding DNA sequence GAA, GCA, and GAG was significantly high (-3.2). The method also can give the most preferable DNA sequences for a given protein (Fig. 13). These sequences, (A/G)AA, G(A/C)(G/A), and GGG, agreed well with the experimental results, showing the potential use of the method for the design of DNA-binding proteins to have either increased or altered binding specificity.

4.6. Contribution from Indirect Readout Mechanism

The above method is based on the direct readout mechanism, in which protein recognizes DNA sequence through direct contact between amino acids and base pairs. On the other hand, substitutions of those base pairs not in contact with amino acids often affect binding affinity, indicating that protein may recognize DNA sequence through a particular structure or property of DNA. This indirect readout mechanism may contribute significantly to the specificity of protein–DNA recognition. We can quantify the specificity due to this indirect readout mechanism by estimating the conformational energy of DNA [53,65]. For simplicity, consider six coordinates (three translational and three rotational) to describe the conformation of each DNA base step. We can approximate the conformational energy by harmonic potentials along

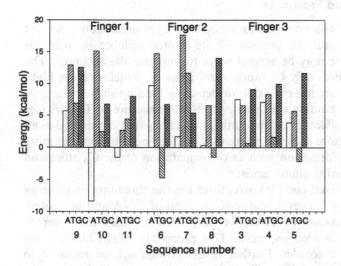

FIGURE 13 Calculated base preference for the three zinc finger positions of 1MEY. (A/G)AA, G(A/C)(G/A), and GGG are preferred by fingers 1, 2, and 3, respectively, in terms of the derived potentials. DNA sequence numbers shown at the bottom of the figure are taken from the crystal structure.

each coordinate. Now we need to know equilibrium conformation and force constants. These parameters can be determined by the empirical analysis of protein–DNA complex structures [54]. It may also be possible to generate a conformational ensemble and estimate these parameters by computer simulations (see Sec. 5). Once the potentials are derived, the conformational energy of DNA can be estimated for a given structure and sequence, and the threading procedure can be used to evaluate the fitness of sequence to the structure of DNA. We can calculate the Z-score for indirect recognition in the same way as for direct recognition. By comparing the two Z-scores we can assess the relative contributions of direct and indirect readout mechanisms. Because both potentials are independent quantities, they can be summed to calculate the total energy and used to find target sites, although a weighting factor needs to be determined because the two potentials were derived from different statistics. We have compared systematically two kinds of Z-scores for many transcription factors and DNA-binding proteins, and found that both the direct and indirect readout mechanisms contribute to the specificity of protein-DNA recognition significantly [65]. We also found that the combination of the two mechanisms increased the accuracy of target prediction [65]. We are applying the combined potentials to the target prediction for transcription factors.

4.7. Limitations and Prospects

The accuracy of the structure-based method for target prediction is not yet satisfactory for practical use because of the limited number of available structural data. There may be several ways to improve the accuracy. The present method requires only C_α atom coordinates of proteins. Thus, high-resolution structures are not essential for detecting binding sites. In fact, we could detect the real binding sites using the NMR structure of TFIIIA (see 1TF3 in Fig. 7). The effects of flexible amino acids are partly considered as an entropic effect. For more accurate estimation, we may need to consider more detailed structural information such as the orientation of the C_β atom and the effects of neighboring amino acids.

We have used fixed complex structures for the threading and energy calculation. In reality, however, the conformation of DNA and/or protein may be sequence-dependent. Thus, we may need to relax the complex structure for each threaded sequence. This procedure might increase the sensitivity of target detection. Further investigations will be required to evaluate the importance of these effects. Despite the current limitations, the structure-based method, which is independent of sequence information, is a promising method for target prediction, because the increase in structural data will be accelerated by structural genomics.

So far we have not considered explicitly the effect of water molecules in protein–DNA interactions. Water molecules often mediate interactions between amino acid and base in the interface of a protein–DNA complex. This can be considered as one of the indirect readout mechanisms. Some of these interactions are involved in the derivation of our statistical potentials. However, because of the small number of such interactions in the structural data, we did not examine the effect of water-mediated interactions separately. When a sufficient number of such interactions become available, we can examine the interaction independently and develop a statistical potential for water-mediated interactions.

5. AB INITIO METHOD

The structure-based method has shown that the distribution of the C_α position of amino acids around base pairs provides important information about specificity in the DNA sequence recognition by proteins. However, the accuracy of this prediction method is limited by the number of available structural data. Thus, it is desirable to complement the method by some other means. Computer simulation is one such possibility. However, computer simulations of very large systems such as DNA–protein complexes are formidable tasks even with the use of the most sophisticated computers. In real DNA–protein interactions, many factors contribute to the recognition process. It will be very difficult and even confusing to include everything at first. A more feasible approach is to start with the simplest system and take additional factors into account step by step. In the present case, the question is how to reproduce the distribution of the C_α position of amino acids around base pairs observed in the experimental DNA–protein complex structures. Thus, it would be reasonable to consider at first the interactions between base pairs and amino acids. In reality, the C_α position is fixed by the main chain of the protein, and the possible range of C_β direction may be restricted. However, such biases, which are context-dependent, are difficult to evaluate a priori. Therefore, at first we will consider intrinsic interactions between base pairs and amino acids.

5.1. Conformational Sampling and Energy Calculation

In order to take account of side-chain flexibility, we have to sample many conformations of side chains for a given C_α position and calculate interaction energy for each conformation. One way to perform conformational sampling is to generate conformations systematically by changing $C_\alpha - C_\beta$ bond orientation and all the dihedral angles. First an amino acid is positioned around the base pair by placing its C_α atom at grid points on the base plane (see Fig. 14).

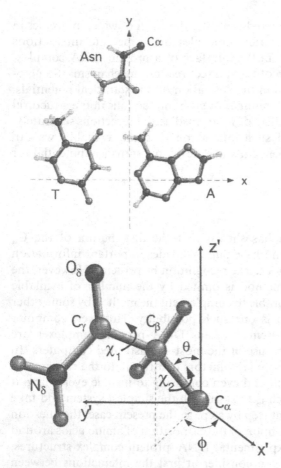

FIGURE 14 Coordinate system of base–amino acid interaction used in the computer simulation. Asn side chain is shown.

Then an initial side-chain orientation is generated by specifying polar angles θ and ϕ formed by the direction cosines of the $C_\alpha–C_\beta$ bond vector. Side-chain rotamers are generated by systematically varying the torsion angles (χ_1 and χ_2 for Asn; amide plane is fixed). For each conformation of the amino acid side chain, bonded and nonbonded interaction energies (van der Waals energies as the 6–12 potential, and electrostatic energies as the Coulomb term) are computed by using the force field parameters of AMBER 4.1 [55]. This process is repeated at each point of the grid. The increments $\Delta\theta$ and $\Delta\phi$ are adjusted to produce a uniform distribution of C_β positions on the spherical surface spanned by the $C_\alpha–C_\beta$ bond vector. Then the partition function can be calculated by the Boltzmann average over the conformational space, and free energy, entropy, and enthalpy can be calculated from the partition function [56]. The number of conformers of amino acids increases rapidly with the size of the side chain. Thus, the systematic sampling method is feasible only for small amino acids. We are also using more efficient conformational sampling algorithms called multicanonical Monte Carlo sampling [57]. This method enables us to sample conformations more efficiently than systematic sampling without sacrificing accuracy. It is also more accurate than the usual canonical Monte Carlo sampling in that the sampling covers a wider energy range. We have tested the accuracy and efficiency of the method by comparing it with other sampling methods for small amino acids [58]. The method has produced the same results as the systematic sampling method and yet the computation time is much shorter. Thus, we are mainly using the multicanonical Monte Carlo sampling for the calculation of free energy maps of base–amino acid interactions.

5.2. Free Energy Landscape of Base–Amino Acid Interactions

By calculating the free energies for different C_α positions and subtracting a reference free energy at a large separation, we can obtain a contour map of interaction free energy, which shows preferable positions of C_α's of amino acids around a base pair. This can be directly compared with the distribution of the C_α positions of amino acids around base pairs derived from the DNA–protein complex database. Figure 15 shows the free energy contour maps for the interactions of Asn with A-T and with G-C base pairs calculated by the above procedure. The preferable position of C_α is localized in a narrow region around A in the case of A-T. In this region, Asn and A form specific double hydrogen bonds, $C=O\cdots HN6$ and $NH\cdots N7$, which are found frequently in the Asn-A pair in the experimental structures of DNA–protein complexes. Also, the distribution of C_α is in agreement with the statistical potential obtained by database analysis (see Fig. 6). On the other hand, Asn tends to be more broadly distributed around G-C. The lowest ΔG values are located in the middle of G-C and extend toward C, which does not have a methyl group. This comparison indicates that the interaction of Asn is more specific toward A-T than toward G-C. This example illustrates how we can quantify the specificity in the base–amino acid interactions and complement the structure-based method.

5.3. Limitations and Prospect

Before applying the ab initio method to general base–amino acid interactions, we need to solve several problems. As noted above, the main chain of proteins will restrain the ranges of C_α positions and C_β directions. Such an effect is

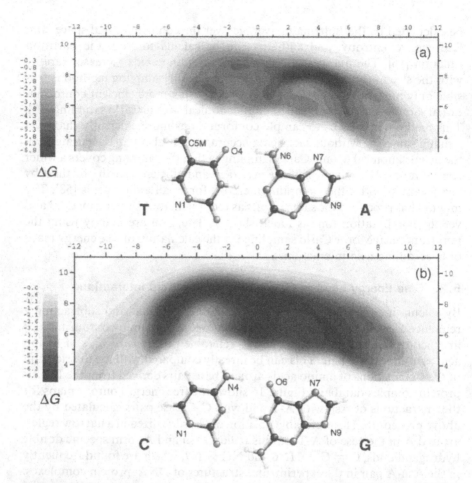

FIGURE 15 Free energy maps of interactions of Asn with A-T (a) G-C (b). The maps were obtained by the systematic sampling method. Darker regions correspond to preferred C_α position of the amino acid.

reflected in the statistical potential derived from a structural database. It may be possible to derive an effective potential for such restraint and combine it with the simulation. Also, the DNA backbone moiety will significantly modify the free energy landscape of base–amino acid interactions because of negatively charged phosphate groups. The contour maps are indeed significantly deformed by the backbone. However, if we are interested only in the difference in the maps for different base–amino acid pairs, such a direct effect of the backbone will be canceled. Adjacent base pairs will affect the map by preventing amino acids from stacking over base pairs and by introducing additional base pair–base pair interactions. Solvent molecules can interfere with the base–amino acid interactions. Upon protein–DNA binding, most water molecules and ions are excluded from the interface. However, some water molecules are trapped inside the interface and mediate the protein–DNA interactions. In such cases the solvent molecules could significantly modify the free energy landscape. Further studies are necessary to assess how important the above effects are on the free energy landscape of base–amino acid interactions, by examining each factor carefully and comparing results with experimental data.

The results obtained so far are consistent with the experimental observations, and the simulation results for all the base–amino acid pairs may be able to complement the statistical potential derived from the structure-based method. Thus, a combination of the structure-based and ab initio methods will become a powerful tool for the target prediction of transcription factors.

Computer simulations may also be used to estimate conformational energy of DNA, which contributes to the indirect readout mechanism. The equilibrium conformations of DNA and force constants of harmonic potentials described in the previous section can be estimated by generating many conformations and fitting the potentials. The effect of sequence changes in DNA on the total energy of the protein–DNA complex can be calculated by molecular dynamics simulations and the free energy perturbation method. However, the calculation of such a large system demands enormous computer time, and currently it is not feasible to consider a large number of sequences. If computer power increases substantially in the future, such large-scale simulation will become a promising method for prediction.

6. APPLICATION TO FUNCTIONAL GENOMICS

So far, complete genomes of many organisms have been sequenced. However, only a fraction of biologically significant information has been extracted from sequence data. Regulation of gene expression is one of the most fundamental processes in life, but its mechanism is not well understood in spite of many known transcription factors. This is because the regulation is achieved by a complicated network among transcription factors, effectors, and their targets. To understand the mechanism of the regulatory network, we first need to identify all the transcription factors and their target sites within a genome. Achieving this goal by experiments alone would be a formidable task, and computer analyses would certainly be required for the target prediction, data archiving, and simulation of the network. Here we describe a strategy for the application of present methods of target prediction to the analysis of gene regulation at the genome level.

Figure 16 illustrates the strategy for target prediction at the genome level. Given a complete genomic sequence of certain organism, one can identify all the putative transcription factors by experimental and computational methods. For some transcription factors, consensus DNA sequences are available from known binding sites. In this case, one can construct a weight matrix and use the sequence-based method to screen potential binding sites and target genes. For more accurate prediction, one can conduct systematic mutational and binding analysis for the transcription factor to derive ΔG data and use the ΔG-based method. If the transcription factor is new or no target sequences are known, one can conduct random oligo screening experiments to derive a consensus sequence and weight matrix and then use the sequence-based method for the prediction.

Another independent route is to ask whether the structure of the transcription factor complexed with DNA is known. If the answer is yes, we can directly apply the structure-based method. On the other hand, if the structure is not known, we can still construct a model structure. If the sequence has homology with another transcription factor whose structure is known, we can use a homology-modeling method to predict the 3D structure. Even if the sequence homology is negative, we can use a so-called threading

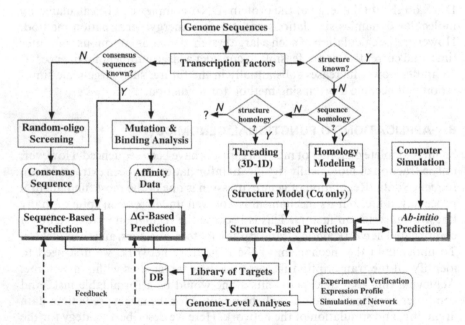

FIGURE 16 A strategy for the application of target prediction to functional genomics.

method for the structure prediction. Then the structure-based method can be used to predict potential target sites of the transcription factor based on its modeled structure. In this case, the model structure requires only the coordinates of C_α atoms, i.e., side-chain modeling is not necessary. The advantage of the structure-based method is that it provides us a possibility to predict potential targets for new transcription factors without carrying out further experiments. The chances for using this method will rapidly increase as more and more structures of transcription factors become available. Furthermore, the emerging structural genomics projects will make this method more promising. Information from the ab initio method will help the structure-based method by increasing the prediction accuracy, as mentioned before. The structure-based method is independent of the sequence-based method in that they rely on different kinds of information. Thus, when combined, they should complement each other. This is in contrast with the situation for different methods based on sequence information, which do not necessarily complement each other [16].

The accuracy of target prediction can be further improved by considering specific information from the genome. For example, one can incorporate the information of binding sites of other transcription factors in the neighborhood of the predicted binding sites. The annotation information of the predicted target gene will be useful, because genes with similar functions are likely to be coregulated. Similarly, neighborhood information of the protein products in the metabolic or signaling pathway will be useful. The evolutionary information of genes among different species will suggest similarities in gene regulation. These pieces of information will significantly enhance the prediction accuracy. Once predictions are made for a set of transcription factors in the genome, we can construct a library of target sites and target genes, together with the library of transcription factors, and feed the information to the database. Then we can carry out genome-level analyses based on the data, such as experimental verification of the predictions, and comparison with expression profile data. Some of these analyses can be fed back to the prediction methods to improve their accuracy. Such cycles of prediction and verification will refine the prediction method. Furthermore, the link information between transcription factors and their target genes together with other network information such as protein–protein interactions would enable simulation of the network, providing insight into the gene regulatory network. Genome-level analyses of this kind would reveal novel mechanisms of gene regulation as systems, which would be impossible to ascertain by the analysis of individual transcription factors. Thus, this kind of systematic analysis of gene regulation by a combination of computational and experimental methods will contribute significantly to functional genomics in the era of post-genome science.

APPENDIX 1: DEFINITIONS OF TERMS

AMBER force field. Molecular mechanics force field widely used for the investigation of interactions and dynamics of biomolecules. http://www.amber.ucsf.edu/amber/amber.html

Bioinformatics. The field of study dealing with management and analysis of data in biological science.

Consensus sequence. A relatively short sequence pattern of DNA, RNA, or protein that is found in various molecules and is associated with the same function. Here we use it for DNA sequence, e.g., binding sites for transcription factors.

Threading. A method for predicting a fold type of protein of unknown structure using statistical potentials derived from the known protein structures. By assigning an amino acid sequence of unknown structure on known protein structures, evaluated is the fitness of the sequence on a given structure. Here, we use the analogy for DNA–protein interaction. By assigning various DNA sequences on a template of DNA–protein structure, the fitness of the DNA is evaluated.

Z-Score. Z-Score is defined as $(X - m)/\sigma$, where X is energy, m is mean energy, and σ is standard deviation. A large value (positive or negative) means high specificity. Threading results are usually evaluated by the Z-score.

Multicanonical sampling. A generalized ensemble algorithm that enables us to explore wider ranges of the phase space than conventional Monte Carlo sampling. This algorithm is also referred to as entropic sampling.

APPENDIX 2: COMPUTER PROGRAMS AND DATABASES

SignalScan (http://bimas.dcrt.nih.gov/molbio/signal/). A program developed to help users find eukaryotic transcription factor elements in a DNA sequence. SignalScan uses both specific sequence elements derived from biochemical characterization and elements from derived consensus sequences to match against a user input DNA sequence.

MATRIX SEARCH (http://bimas.dcrt.nih.gov/molbio/matrixs/). A program developed to facilitate the analysis of DNA sequences for known transcription factor binding sites. It scores input sequences against matrices of transcription factor binding sites using information theory [59].

MatInspector (http://www.gsf.de/biodv/matinspector.html). This tool utilizes a large library (~ 280 entries) of predefined matrix descriptions for protein binding sites to locate matches in DNA sequences of unlimited length [4]. It assigns a quality rating to matches and thus allows quality-based filtering and selection of matches.

ConInspector (http://www.gsf.de/biodv/consinspector.html). This tool predicts potential binding sites in sequences of unlimited length by comparing candidates with predefined consensus descriptions of protein binding sites [60].

Protein Data Bank (PDB) (http://www.rcsb.org/pdb/). This is the database of three-dimensional macromolecular structures primarily determined experimentally by X-ray crystallography and NMR.

TRANSFAC (http://transfac.gbf-braunschweig.de/). A database on transcription factors, their genomic binding sites, and DNA-binding profiles.

REFERENCES

1. Frech K, Quandt K, Werner T. Finding protein-binding sites in DNA sequences: the next generation. Trends Biochem Sci 1997; 22:103–104.
2. Bucher P. Weight matrix descriptions of four eukaryotic RNA polymerase II promoter elements derived from 502 unrelated promoter sequences. J Mol Biol 1990; 212:563–578.
3. Chen QK, Hertz GZ, Stormo GD. MATRIX SEARCH 1.0: a computer program that scans DNA sequences for transcriptional elements using a database of weight matrices. Comput Appl Biosci 1995; 11:563–566.
4. Quandt K, Frech K, Karas H, Wingender E, Werner T. MatInd and MatInspector: new fast and versatile tools for detection of consensus matches in nucleotide sequence data. Nucleic Acids Res 1995; 23:4878–4884.
5. Prestridge DS, Stormo G. SIGNAL SCAN 3.0: new database and program features. Comput Appl Biosci 1993; 9:113–115.
6. Berg OG, von Hippel P. Selection of DNA binding site by regulatory proteins: statistical-mechanical theory and application to operators and promoters. J Mol Biol 1987; 193:723–750.
7. Wingender E, Chen X, Hehl R, Karas H, Liebich I, Matys V, Meinhardt T, Prus M, Reuter I, Schacherer F. TRANSFAC: an integrated system for gene expression regulation. Nucleic Acids Res 2000; 28:316–319.
8. Prestridge DS. SIGNAL SCAN 4.0: additional databases and sequence formats. Comput Appl Biosci 1996; 12:157–160.
9. Frech K, Herrmann G, Werner T. Computer-assisted prediction, classification, and delimitation of protein binding sites in nucleic acids. Nucleic Acids Res 1993; 21:1655–1664.
10. Stormo GD. Consensus patterns in DNA. Methods Enzymol 1990; 183:211–221.
11. Tsunoda T, Takagi T. Estimating transcription factor bindability on DNA. Bioinformatics 1999; 15:622–630.
12. Pickert L, Reuter I, Klawonn F, Wingender E. Transcription regulatory region analysis using signal detection and fuzzy clustering. Bioinformatics 1998; 14:244–251.
13. Karas H, Knuppel R, Schulz W, Sklenar H, Wingender E. Combining structural analysis of DNA with search routines for the detection of transcription regulatory elements. Comput Appl Biosci 1996; 12:441–446.
14. Ponomarenko JV, Ponomarenko MP, Frolov AS, Vorobyev DG, Overton GC,

Kolchanov NA. Conformational and physicochemical DNA features specific for transcription factor binding sites. Bioinformatics 1999; 15:654–668.

15. Kel A, Kel-Margoulis O, Babenko V, Wingender E. Recognition of NFATp/-AP-1 composite elements within genes induced upon the activation of immune cells. J Mol Biol 1999; 288:353–376.

16. Roulet E, Fisch I, Junier T, Bucher P, Mermod N. Evaluation of computer tools for the prediction of transcription factor binding sites on genomic DNA. In Silico Biol 1998; 1:21–28.

17. Takeda Y, Sarai A, Rivera VM. Analysis of the sequence-specific interactions between Cro repressor and operator DNA by systematic base substitution experiments. Proc Natl Acad Sci USA 1989; 86:439–443.

18. Sarai A, Takeda Y. λ Repressor recognizes the approximately 2-fold symmetric half-operator sequences asymmetrically. Proc Natl Acad Sci USA 1989; 86:6513–6517.

19. Tanikawa J, Yasukawa T, Enari M, Ogata K, Nishimura Y, Ishii S, Sarai A. Recognition of specific DNA sequences by the c-Myb proto-oncogene product: role of three repeat units in the DNA-binding domain. Proc Natl Acad Sci USA 1993; 90:9320–9324.

20. Hao D, Ohme-Takagi M, Sarai A. Specific interactions between EREBP and GCC box in plant. J Biol Chem 1998; 273:26857–26861.

21. Beamer LJ, Pabo CO. Refined 1.8 Å crystal structure of the lambda repressor-operator complex. J Mol Biol 1992; 227:177–196.

22. Albright RA, Matthews BW. Crystal structure of lambda-Cro bound to a consensus operator at 3.0 Å resolution. J Mol Biol 1998; 280:137–151.

23. Deng Q, Ishii S, Sarai A. Binding-site analysis of c-Myb: screening of potential binding sites by the mutational matrix derived from systematic binding affinity measurements. Nucleic Acids Res 1996; 24:766–774.

24. Mandel-Gutfreund Y, Schueler O, Margalit H. Comprehensive analysis of hydrogen bonds in regulatory protein-DNA complexes: in search of common principles. J Mol Biol 1995; 253:370–382.

25. Kono H, Sarai A. Structure-based prediction of DNA target sites by regulatory proteins. Proteins 1999; 35:114–131.

26. Matthews BW. Protein-DNA interaction. No code for recognition. Nature 1988; 335:294–295.

27. Suzuki M. A framework for the DNA-protein recognition code of the probe helix in transcription factors: the chemical and stereo-chemical rules. Structure 1994; 2:317–326.

28. Choo Y, Klug A. Selection of binding sites for zinc fingers using rationally randomised DNA reveals coded interactions. Proc Natl Acad Sci USA 1994; 91:11168–11172.

29. Sippl M. Calculation of conformational ensembles for potentials of mean force: an approach to the knowledge-based prediction of local structures in globular proteins. J Mol Biol 1990; 213:859–883.

30. Ogata K, Kanei-Ishii C, Sasaki M, Hatanaka H, Nagadoi A, Enari M, Nakamura H, Nishimura Y, Ishii S, Sarai A. The cavity in the hydrophobic core

of Myb DNA-binding domain is reserved for DNA recognition and transactivation. Nat Struct Biol 1996; 2:178–187.

31. Sarai A, Mazur J, Nussinov R, Jernigan RL. Sequence dependence of DNA conformational flexibility. Biochemistry 1989; 28:7842–7849.

32. Tisne C, Delepierre M, Hartmann B. How NF-κB can be attracted by its cognate DNA. J Mol Biol 1999; 293:139–150.

33. Nadassy K, Wodak SJ, Janin J. Structural features of protein-nucleic acid recognition sites. Biochemistry 1999; 38:1999–2017.

34. Jones S, van Heyningen P, Berman HM, Thornton JM. Protein-DNA interactions: a structural analysis. J Mol Biol 1999; 287:877–896.

35. Koudelka GB, Harrison SC, Ptashne M. Effect of non-contacted bases on the affinity of 434 operator for 434 repressor and cro. Nature 1987; 326:886–888.

36. Gewirth DT, Sigler PB. The basis for half-site specificity explored through a noncognate steroid receptor-DNA complex. Nat Struct Biol 1995; 2:386–394.

37. Luisi BF, Xu WX, Otwinowski Z, Freedman LP, Yamamoto KR, Sigler PB. Crystallographic analysis of the interaction of the glucocorticoid receptor with DNA. Nature 1991; 352:497–505.

38. Kuriyan J, Thanos D. Structure of NF-κB transcription factor: a holistic interaction with DNA. Structure 1995; 3:135–141.

39. Muller CW, Rey FA, Sodeoka M, Verdine GL, Harrison SC. Structure of the NF-κB p50 homodimer bound to DNA. Nature 1995; 373:311–317.

40. Ghosh G, Duyne GV, Ghosh S, Sigler PB. Structure of NF-κB p50 homodimer bound to κB site. Nature 1995; 373:303–310.

41. Chytil M, Verdine GL. The Rel family of eukaryotic transcription factors. Curr Opin Struct Biol 1996; 6:91–100.

42. Wolberger C, Vershon AK, Liu ZB, Johnson A, Pabo CO. Crystal structure of a MATα2 homeodomain-operator complex suggests a general model for homeodomain-DNA interactions. Cell 1991; 67:517–528.

43. Li T, Stark MR, Johnson AD, Wolberger C. Crystal structure of the MATa1/MATα2 homeodomain heterodimer bound to DNA. Science 1995; 270:262–269.

44. Dickerson RE, Chiu TK. Helix bending as a factor in protein/DNA recognition. Biopolymers 1997; 44:361–403.

45. Olson WK, Gorin AA, Lu XJ, Hock LM, Zhurkin VB. DNA sequence-dependent deformability deduced from protein-DNA crystal complexes. Proc Natl Acad Sci USA 1998; 95:11163–11168.

46. Nekludova L, Pabo CO. Distinctive DNA conformation with enlarged major grove is found in Zn-finger-DNA and other protein-DNA complexes. Proc Natl Acad Sci USA 1994; 91:6948–6952.

47. Elrod-Erickson M, Rould MA, Nekludova L, Pabo CO. Zif268 protein-DNA complex refined at 1.6Å: a model system for understanding zinc finger–DNA interactions. Structure 1996; 4:1171–1180.

48. Miller AM, MacKay VL, Nasmyth KA. Identification and comparison of two sequence elements that confer cell-type specific transcription in yeast. Nature 1985; 314:598–603.

49. Russell DW, Jensen R, Zoller MJ, Burke J, Errede B, Smith M, Herskowitz I.

Structure of the *Saccharomyces cerevisiae HO* gene and analysis of its upstream regulatory region. Mol Cell Biol 1986; 6:4281–4294.

50. Kim CA, Berg JM. A 2.2Å resolution crystal structure of a designed zinc finger protein bound to DNA. Nat Struct Biol 1996; 3:940–945.

51. Desjarlais JR, Berg JM. Length-encoded multiplex binding site determination: application to zinc finger proteins. Proc Natl Acad Sci USA 1994; 91:11099–11103.

52. Kim CA, Berg JM. Serine at position 2 in the DNA recognition helix of a Cys_2-His_2 zinc finger peptide is not, in general, responsible for base recognition. J Mol Biol 1995; 252:1–5.

53. Sarai A, Selvaraj S, Gromiha MM, Siebers JG, Prabakaran P, Kono H. Target prediction of transcription factors: refinement of structure-based method. Genome Inform 2001; 12:384–385.

54. Olson WK, Gorin AA, Lu XJ, Hock LM, Zhurkin VB. DNA sequence-dependent deformability deduced from protein-DNA crystal complexes. Proc Natl Acad Sci USA 1998; 95:11163–11168.

55. Cornell WD, Cieplak P, Bayly CI, Gould IR, Merz KM Jr, Ferguson DM, Spellmeyer DC, Fox T, Caldwell JW, Kollman PA. A second generation force field for the simulation of proteins, nucleic acids, and organic molecules. J Am Chem Soc 1995; 117:5179–5197.

56. Pichierri F, Aida M, Gromiha M, Sarai A. Free energy maps of base-amino acid interaction for protein-DNA recognition. J Am Chem Soc 1999; 121:6152–6157.

57. Berg BA, Neuhaus T. Multicanonical ensemble: a new approach to simulate first-order phase transitions. Phys Rev Lett 1992; 68:9–12.

58. Sayano K, Kono H, Gromiha M, Sarai A. Multicanonical Monte Carlo calculation of free-energy map for base-amino acid interaction. J Comput Chem 2000; 21:954–962.

59. Hertz GZ, Hartzell GW, Stormo GD. Identification of consensus patterns in unaligned DNA sequences known to be functionally related. Comput Appl Biosci 1990; 6:81–92.

60. Frech K, Dietze P, Werner T. ConsInspector 3.0: new library and enhanced functionality. Comput Appl Biosci 1997; 13:109–110.

61. Howe KM, Reaks CFL, Watson RJ. Characterization of the sequence-specific interaction of mouse c-myb protein with DNA. EMBO J 1990; 9:161–169.

62. Selvaraj S, Kono H, Sarai A. Specificity of protein-DNA recognition revealed by structure-based potentials: symmetric/asymmetric and cognate/noncognate binding. J Mol Biol 2002; 322:907–915.

63. Winkler FK, Banner DW, Oefner C, Tsernoglou D, Brown RS, Heathman SP, Bryan RK, Martin PD, Petratos K, Wilson KS. The crystal structure of EcoRV endonuclease and of its complexes with cognate and non-cognate DNA fragments. EMBO J 1993; 12:1781–1795.

64. Kraulis PJ. MOLSCRIPT: a program to produce both detailed and schematic plots of protein structures. J Appl Cryst 1991; 24:946–950.

65. Gromiha MM, Siebers JG, Selvaraj S, Kono H, Sarai A. Intermolecular and intramolecular readout mechanisms in protein-DNA recognition. J Mol Biol 2004; 337:285–294.

8

Methods of Computational Genomics: An Overview

Frederique Lisacek

Génome and Informatique, Evry, France and Geneva Bionformatics
(GeneBio), Geneva, Switzerland

1. INTRODUCTION

Current computational issues in genome research have not only arisen from
the recent accumulation of genome data but have also been built up from past
experience of computer-assisted sequence analysis. Within the last three
decades, molecular biology and genetics have increasingly relied on the use
of computers for data analysis. At the same time the nature and scope of
problems to be solved with the help of computers have evolved because
technological advances have gradually modified details of laboratory work.
Thus the nature and quality of generated data have changed, as have the
nature and quality of inference from these data.

Molecular biology of the 1950s apparently relied on methods and
techniques adopted from physics* (the only aspect adopted from chemistry
was the term "molecule"). Since then biology has adopted numerous methods

---

* Chromatography, fluorescence and related methods, electron microscopy, and X-ray
visualization.

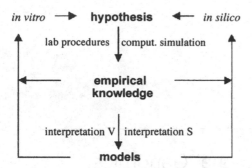

in vitro ⟶ **hypothesis** ⟵ *in silico*

lab procedures | comput. simulation

empirical knowledge

interpretation V | interpretation S

models

FIGURE 1 Overview of experimental biology. An experimental framework for testing hypotheses to build models is the characteristic common to in vitro and so-called in silico biology. The alignment of two or more sequences can be considered a working hypothesis. The identification of a signal in a sequence or a group of sequences is an experimental result. If the signal is meaningful, it will represent a discriminant for computational analysis. The consistency of the resulting model relies on the consistency of both interpretations.

from applied mathematics, computer science, and statistics. An obvious advantage of this acquisition of formal methods was improvement in the clarity of formulating problems as well as in arriving at their solutions. On the other hand, formalisms in biology can (and often do) help to highlight the limitations (or even irrelevance) of ill-conceived problems. An extra benefit (or perhaps a mixed blessing) of formalization of molecular biology is the wide availability of software for computer-assisted sequence analysis, genetics, and evolutionary studies.

There is a definite analogy between laboratory and computational experiments in biology; they both provide an experimental framework for testing hypotheses to build models (see Fig. 1). The similarity, hence the alignment, of two or more sequences can be viewed as a working (null) hypothesis. The identification of a "signal" in a sequence or group of sequences is an experimental result. According to a protocol sometimes called pragmatic inference [1], a signal is believed to be meaningful if its presence in or absence from a sequence is significantly correlated with a well-defined biological function of this sequence. The demonstration of the existence of significant correlation usually relies on laboratory or field experiments. If the signal is meaningful, it can then represent a discriminating criterion for assigning putative biological roles to sequences subjected to computational analysis.*

---

* Ideally, the adequacy of such an assignment should be confirmed by some evidence in vivo. Unfortunately, though, experimentation at the molecular level in vivo does not meet acceptable standards of precision and conclusiveness thus far.

Hypotheses set by the use of a computer program may suggest experiments in vitro or, conversely, in vitro experiments may suggest computational simulations (experiments) and new software tools. At any rate, consistency of the resulting model relies on the consistency of both in vitro and computational interpretations (Fig. 1). More generally, the validity of a model is a function of the experimental means of testing two kinds of hypotheses:

1. Hypotheses involved in actually making (determining) the model
2. Hypotheses resulting from (generated by) the model

It should be mentioned that such methodological issues are generally not included in manuals of bioinformatics. No emphasis is put on the evaluation of the qualitative worth of data. On the other hand, practitioners of computational biology are concerned with assessing the reliability and longevity of hypotheses and models because such evaluations are crucial to the understanding of biology. Unfortunately, these issues are only superficially addressed in sparse paragraphs of existing texts (e.g., Gusfield [2], Baldi and Brunak [3], and Durbin et al. [4]).

The short history of bioinformatics already tells us that some assumptions turned out to be inappropriate. For example, it is now widely accepted that a simple Markov model does not account for a genomic sequence. The underlying assumption of a unique probability distribution constraining the occurrences of nucleotides is clearly inadequate, because it fails to justify observed varying compositional bias in whole chromosomes. More generally, implicit assumptions defining a method can be too loose or too restrictive. However sound a method of analysis is formally, if a hypothesis defining it contradicts an underlying (usually unknown) principle constraining the analyzed sequence, prediction cannot be reliable. For instance, summing entities without verifying their additive properties is a common oversight [5], which fortunately loses momentum in some applications [5a]. Calculations involving energy are indicative. Indeed, if the free energy of an RNA hairpin appears to match the sum of free energies of all base pairs involved, the overall contribution of those amino acids of a protein binding to a ligand is not necessarily equated to the sum of contributions of each individual amino acid [6].

Summing and subtracting (filtering) pieces of information in sequences are common operations. It is tempting, of course, to adopt a classical approach to problem solving for sequence analysis, that is, divide a problem into more or less independent subproblems and solve those subproblems first. But piecing solutions together encounters obstacles. Because independence is a strong assumption that is not easy to prove, summing may turn out not to be either associative $[(a + b) + c \neq a + (b + c)]$ or commutative $(a + b \neq b + a)$ due to a possible underlying hierarchy of subproblems. Moreover, the partition into subproblems may not be adequate because it cannot be matched with a biological reality. Finally, given that a problem P was subdivided into N subproblems, assuming that an optimal solution to P is the sum of all optimal solutions to subproblems $1-N$ may be inadequate. Indeed, the minimal free energy of an RNA molecule is not the sum of all helices with a minimal free energy.

The previous statement shows how a formally sound method such as dynamic programming (DP) yields an elegant but inapplicable model of RNA structure. The original DP algorithm designed by Zuker and Sankoff [7] is simple and is suited to strings of symbols but does not accurately predict RNA structure. There is some justification for looking for stable helices, and parts of the structure are indeed identified by such a program, but only parts. Amendments introduced in Ref. 8 improved prediction by somehow accounting for the nonadditive properties of minimal free energy. Progress was limited, however, by the assumption of relative independence between helices. Moreover, the existing dependencies were not detectable because sequences were processed linearly from the 5′ end to the 3′ end. On paper, a sequence does begin at the 5′ end and finish at the 3′ end, but in action this is not true. The underlying organization of information is not given by the order of appearance in reading the sequence linearly but by the order in which segments are being solicited to perform the overall function of the sequence (i.e., the order of appearance in action). Analysis must respect such an order* to maximize chances of identifying relevant constraints. Local alignment methods provide efficient tools because the most constrained segments are considered first wherever they lie in the sequence.

This example confirms that the formulation of a problem is as determinant as the method used to solve it. Figure 2, as the reverse of Figure 1, exemplifies the purposes of setting and checking hypotheses.

The delineation of a question in bioinformatics over a short period of time can be made difficult. Indeed, the constant transformation of the experimental setup goes along with the transformation of knowledge.† Con-

---

* Such a statement is tautological when the structure is known, but it is common enough that only sequence data are available.

† To confirm this, consider the difference between the two major types of sequences. What is known about proteins today is only more precise than what was known 25 years ago. The repertoire of structures is broader and more detailed. Indeed, the first X-ray crystallographic data were collected roughly 40 years ago, and this technique remains the main source of protein structural data (the use of NMR is also over 20 years old). Computer graphics contributed to better precision. What is known about nucleic acids today is significantly different from what was known 25 years ago. Hypotheses have been refuted (the genetic code is universal; RNA has no catalytic activity; etc.), others have emerged (genes have many parts; nucleotides can be modified to modulate the activity of nucleic acids; etc.), and many cases known as exceptions have accumulated to finally become part of a proper mechanism. Alternative splicing is a classical example, and, more generally, because many sites are cryptic, the variety of solutions chosen by Nature is far from being uncovered.

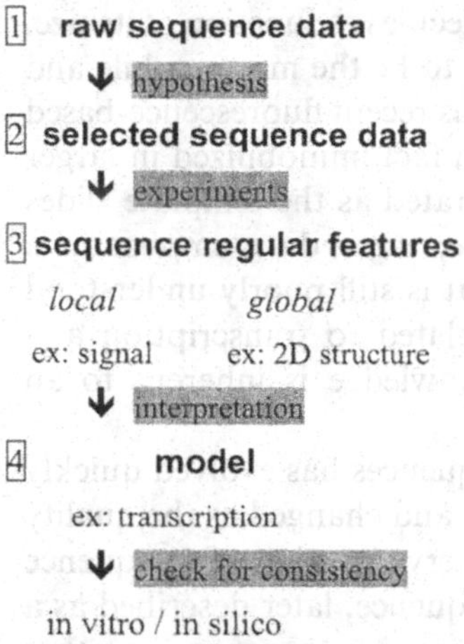

1️⃣ **raw sequence data**

 ⬇ hypothesis

2️⃣ **selected sequence data**

 ⬇ experiments

3️⃣ **sequence regular features**

 local   *global*

 ex: signal   ex: 2D structure

  ⬇ interpretation

4️⃣   **model**

 ex: transcription

  ⬇ check for consistency

 in vitro / in silico

FIGURE 2 Specific situation in silico biology. A set of sequences is selected according to a working hypothesis such as "same function" (1 → 2). Regular features are extracted from the sequences following a certain protocol depending on the chosen computer processing (2 → 3). The program is meant to identify local features such as patterns and/or a global characteristic of sequences such as the overall structure that need to be related to a biological property (3 → 4). When the correlation between the presence or absence of regular features and a biological phenomenon is confirmed, a model can be defined. Consistency can be demonstrated in silico by identifying new instances of sequences verifying the model and in vivo by testing the activity of newly identified sequences.

sequently, questions set to computers have known various fates, depending on how timeproof the related knowledge in biology was. This is illustrated in the brief historical background given in the first section of this chapter. In fact, for many years, rigorous methods have tended to be more useful to formalize problems in biology rather than solve them. A formal point of view and an experimental point of view take a while to reach a common formulation.

 Solving or understanding a problem in biology can be viewed as deriving a model from empirical data. A phenomenon can be formalized even when some data are missing, but the resulting model is more likely to require future revisions. Convincing illustrations of such ongoing transformations of models are reassessments of mechanisms of various cellular processes. For instance, replication and transcription were long thought to result from the binding of a polymerase complex to an origin or a promoter in DNA. In the absence of experimental evidence, the model was justified by a common-sense assumption: The mobility of a molecule is a function of its size. The smaller DNA polymerase was thus believed to be the more mobile and thought to track along the DNA template. Various recent fluorescence-based methods have shown that DNA polymerase is in fact immobilized in larger structures, such that a transcript would be generated as the template slides through the fixed polymerase [9]. Furthermore, these data emphasize a collective behavior of groups of polymerases that is still poorly understood but the existence of which renews questions related to transcription and replication mechanisms. The reappraisal of knowledge is inherent to an empirical science such as biology.

More specifically, knowledge of genetic sequences has evolved quickly since the early 1980s. Models and concepts arose and changed as the quality and quantity of sequence data increased. A conserved region of a sequence alignment was first formalized as a consensus sequence, later described as a weight matrix, then represented as a regular expression, among many other options as discussed in the following.

Sequence databases and analytical methods have mainly been shaped by successive versions of representation, comparison, and classification problems and their corresponding solutions (comparing and clustering sequences or structures and the various parameters used to describe them). Databases were first expanded, and the unmanageable resulting collections were split into more specialized and curated sets. In parallel, analytical tools initially designed to perform rough classifications and produce pairwise alignments were then adapted to specific searches, generalized to yield multiple alignments, and geared to cope with massive data sets.

The availability of complete genomic sequences definitely shed a different light on our knowledge of sequences. New problems emerged, and new tools were designed for the analysis of totally sequenced chromosomes. Previous efforts to develop sequence analysis tools were integrated into the wider picture of genomic annotation. Most of the current automatic parsers contribute to labeling genomic sequences. However, automated analysis of the genomic text still cannot be equated to understanding it. It essentially helps to break the genome into more analyzable parts.

In this context, Sec. 3 focuses on two relevant issues: on the one hand the representation and characterization of sequences, and on the other hand the validation of sequence analysis methods. Assessing these aspects of bioinformatics leads to a banal statement: Biomolecules are active, and however precise sequence characterization is, the knowledge of the conditions within which a gene or a protein is active is necessary. Then we address issues currently germinating: the definition of contextual rules (in particular, specify how sequences interact with their environment) and the use of chronological rules to specify further constraints defining sequence activity.

A "permanent" task of sequence analysis is to determine the most representative data set and the most suitable way to represent data in order to bring information out. The cycling arrows in Figure 3 indicate this. Sequence data sets and descriptions are usually defined and redefined several times to improve analysis. A widespread strategy consists in gathering nucleotide or amino acid sequences likely to share common features (in a set called S in the following). For instance, to characterize transcription promoters in bacteria, S would typically involve examples of the region situated before the Start codon (at least 50 nucleotides long). Given this set, representation evolves toward equating sequences to features such as composition, signals, and specific folding (for RNA or protein).

In principle, analytical tools perform two distinct, though related, functions. First, identify motifs, that is, regions common to most sequences of S,

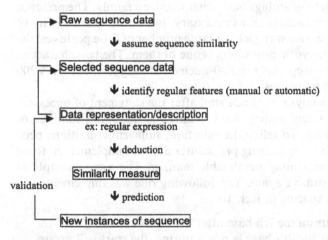

FIGURE 3   Usual tasks of sequence analysis. A "permanent" task of sequence analysis is to determine the most representative data set and the most suitable way to represent data in order to bring information out. This is indicated by the cycling arrows. Sequence data sets and descriptions are usually defined and redefined several times to improve analysis. A widespread strategy consists in gathering nucleotide or amino acid sequences likely to share common features (in a set called S). In principle, analytical tools perform two distinct, though related, functions. First, identify motifs, that is, regions common to most sequences of S, that are unexpectedly regular or irregular with respect to a given feature such as sequence composition. This identification task is either manual or automated. Second, use this acquired knowledge on motifs and/or other assumptions to determine a similarity measure and filter sequences that contain identified motifs in larger sets (databanks or genomes).

unexpectedly regular or irregular with respect to a given feature such as sequence composition. This identification task is either manual or automated as discussed later in the chapter. In prokaryotic promoters, specific features were visible in small data sets. They minimally include the presence of the Pribnow and the Maniatis boxes, respectively located about 10 and 35 nucleotides upstream of ATG. The occurrence of these boxes is further constrained by the number of nucleotides (16 to 18) lying in between them and the helical arrangement of the contact nucleotides (see, e.g., Ref. 10). Second, use this acquired knowledge on motifs and/or other assumptions to determine a similarity measure and filter sequences that contain identified motifs in larger sets (databanks or genomes).

Most families of sequences are characterized by a motif as a result of automatic or manual sequence analysis. A motif is represented either as a consensus (sequence, matrix, regular expression, etc.) or as a set of rules allowing the identification of various consensuses. In the best of cases, the set of rules is good enough to unambiguously characterize a family. The presence of specific motifs in sequences is set as necessary. But in the vast majority of cases, filtering sequences this way yields a large number of false positives; the above example of prokaryotic promoters is one of them. The two characteristic signals were shown to co-occur in a 30-nucleotide-long segment every 200 nucleotides in random sequences.

Many sequence analysis methods stall after the statement of necessary conditions, so quite a few widely used definitions qualifyng families of sequences are incomplete. To refine the selection, sufficient conditions need to be set. A variety of ad hoc filtering procedures are then implemented to do so, most of the time in a nongeneralizable manner. The same example of promoters will be illustrative again. The following rule was implemented to predict the start point spacing in Ref. 10:

> The start will occur on the 7th base after the −10 sequence if the 7th base is a purine; if the 7th base is not a purine, the start will occur on the 8th base if the 8th base is a purine; if neither is a purine, the start defaults to a purine at either position 9 or 6; if a purine is lacking in all these positions, the start defaults to a C at position 8.

What if a new promoter sequence lacks purine in all relevant positions and position 8 corresponds to a start point but is a T? Should the rule be extended to: The start defaults to a pyrimidine at position 8 (though the rule would become tautological)? Such a rule surely does not follow Occam's razor principle, which remains a reasonable guideline when setting rules.

This situation shows that ad hoc filtering procedures are usually data-dependent, and it also reinforces the need to reassess the selection and representation of data as stressed in Figure 3. The validation and optimization of a method are important issues to be addressed further.

To further emphasize the necessity to complement frequency calculations, let us recall that binding is only one of the several events taking place in
a process to which a protein contributes. Moreover, many processes are
performed in parallel in a cell and each involved mechanism implies time-
dependent transformations (binding, cleaving, releasing, etc.). The nature of
biological processes thus appears to be essentially sequential, i.e., most
molecular mechanisms are locally and temporally defined.

For instance, any secretory mechanism is depicted as a succession of
specific steps [11]. Secretory activity is composed of such mechanisms. The
dynamics involved can be described only by chronological rules. Given
another instance of a process such as splicing and answering the simple
question of which snRNP binds first helps to understand the assembly of
proteins in the spliceosome [12]. Data of this type of are currently poorly used
in automated splice site prediction.

Most molecular recognition simulations are based on an attempt to
assign a weight or a probability to each of the amino acids or nucleotides
constituting a binding site or any other type of known site. Sequence analysis
methods generate and/or use rules of pragmatic inference based on concepts
of measure (such as probability), logic, or various relations of order (such as
trees or hierarchies). The rules alone—albeit pragmatically useful—are
generally insufficient to assign biological functions to sequences or sets
thereof. Different dimensions of knowledge may be needed to supplement
pragmatic protocols of sequence analysis and to determine functionally
meaningful correlations between sequences and their alleged biological roles.
Chronology could be a key to setting sufficient conditions to unambiguously
characterize a set of sequences. Moreover, it would also help associate
sequences within a context, which is also governed by chronological rules.

There are cases of explicit sets of necessary and sufficient conditions
used to fine-tune computer analysis of a given family of sequences. For
example, tRNA sequences (provided they are not from organelles) are now
unequivocally identified in genomic DNA. Their rather simple and very
regular secondary structure can be described with various constraints on
symbols. These constraints are necessary. They apply with respect to a definite
hierarchy or chronology, which sets sufficient conditions. As a result, given
any genome, diverse reliable programs assist the automatic annotation of
tRNA genes (see Sec. 3.4 for details).

However, the role and influence of tRNAs in protein synthesis are
ultimately the reason one would consider such molecules. A thorough
understanding of this role entails taking different data into account. First,
to remain at the sequence level, the potential modification of nucleotides is of
significant importance. RNA edition may yield different outcomes from
slowing down translation to shifting reading frames [12a]. For instance, a
single modification of a nucleotide immediately downstream from the anti-

codon bears on the efficiency and/or the sensitivity of translation. Second, various other features such as the relative abundance of tRNA in the cell or the frequency of codons, which are usually organism-dependent, are used to quantify the active role of tRNA in translation. Finally, understanding of a function can be achieved by trying to assess minimal requirements and answering questions such as: What are the effects of deleting an arm of a tRNA? How far can the anticodon be modified to denature translation?

This example demonstrates the need for formalizing the sequence of operations or actions a gene or protein is involved in when active, that is, the context within which a particular molecule interacts with other molecules and contributes to a cellular process. The concept of interaction between macro-molecules has long been considered only from a structural point of view. Early X-ray data of protein–DNA or protein–protein complexes yielded realistic descriptions of binding [13,14]. As such, they provided a solid basis for the design of reliable computer graphics tools. However, the structural con-straints of binding can only partially explain the function of a protein while accounting for the activity but not for prior events determining whether binding will or will not take place. Binding is only one molecular action in a series of many.

More generally, biology is mostly about potential versus realized events. That is why Bayesian models are often more appropriate than any other probabilistic models, and likewise it is what makes the "if . . . then" rules of a knowledge-based system suited to solving some problems in biology, as will be discussed further in Sections 3.3 and 3.4.

Compensatory mechanisms are another type of interaction, which defines a context. Indeed, the compensation of a weak signal by another signal is observed in a number of instances. As such, compensation provides the overall consistency of a process.

Experimental setups more and more acutely account for the fact that genes and proteins cannot be understood without knowledge of the context within which they interact between themselves or with other cellular constit-uents. Consequently, sequence analysis is slowly evolving away from consid-ering collections of gene or protein sequences based on a unique criterion such as having the same function or the same structure. However, because genes contain the information that ultimately encodes for the function of proteins, interest has been focused on genomes, confining interpretation to one direction: from genes to proteins. The difficulty of understanding genomic data confirmed that such a uniquivocal view is too restrictive. Gene expres-sion and protein levels vary with the cellular location as well as the global activity of the cell more often than not in an unrelated way. To account for these variations, concepts of transcriptome (all gene transcripts found in a given tissue at a set developmental stage) and proteome (all proteins found in a given tissue at a set developmental stage) were recently proposed to help reassess the role of genes and proteins in a living environment. So far, there is no obvious way to express a relationship between these two sets. The differential display of controlled variations is the main characteristic of these new approaches.

As a concluding note for this introductory section, the ultimate scope of computational experiments in biology is adequate simulation of a biological entity or a mechanism. Provided hypotheses are well set, the more optimal the algorithm, the more accurate the simulation. The optimality of an algorithm relies on a definite hierarchy of operations following a set chronology. Whether a program simulates RNA folding or successive or parallel steps of a biological process such as splicing, the succession of operations will ideally reflect an identified set of time constraints.

This chapter is devoted to methods of description and interpretation of genomic data. Whenever appropriate the methods are illustrated by examples from either molecular biology or the practice of sequence analysis. The references cited in this chapter constitute a representative selection (often review articles) but are by no means exhaustive.

2. HISTORICAL BACKGROUND

2.1. The 1970s: The Era of Consensus Sequences

The experimental heritage of the 1960s produced more data on protein than nucleic acid sequences. Indeed, Edman and Begg's test [15] was the first means of generating protein sequence data. Dayhoff [16] organized the first compilation of protein sequences more than 10 years before what was to become genetic sequence databanks [GenBank, 1979 (see http://www.ncbi.nlm.nih.gov/Genbank/index.html); EMBL, 1980 (see http://www.ebi.ac.uk/embl/index.html)]. Structural data were collected in the Brookhaven Protein Data Bank [17]. So, until Sanger et al. [18] and Maxam and Gilbert [19] set up the conditions for sequencing DNA, sequence analysis equated with protein comparison.

A computer solution was mainly sought for three types of problems.*

1. Protein Sequence Alignment. Sequence variations of the same protein in different species were studied. Pairwise alignment of protein se-

---

* Phylogeny is intentionally not included in the list. Classification-related problems were set decades before any sequence data became available. Organisms were described in terms of morphological characters.

quences is based on Dayhoff's work on amino acid replaceability [16]. The first computer program aligning protein sequences was based on dynamic programming [20]. Dynamic programming was originally developed to tackle the issue of sequential decision processes [21]. Such a reference to a formally defined computer method was the first step toward importing algorithms not directly related to statistics. The immediate widespread use of this algorithm imposed a standard: The distance between two aligned sequences was to be set as a minimized function of the number of substitution, insertion, and deletion symbols [22,23].

Sequence alignment highlighted conserved regions, which gave rise to the concept of consensus sequences, defined as the longest common subsequence in a set of aligned sequences. A small collection of programs searching consensus sequences was available at the end of the 1970s.

2. Protein Secondary Structure Prediction from Sequence Data. Anfisen et al. [24] established the relevance of the question by showing that a protein folds without further information than that in the sequence. This hypothesis remains the base of most efforts in this area.

The first attempts to predict secondary structure were essentially statistical methods [25,26], based on frequencies of a given amino acid in a given type of secondary structure. α-Helix, β-strand, or coil propensity factors were derived for each amino acid. Directed mutagenesis experiments could confirm propensity values, though in a subjective way (intuitively chosen amino acids were replaced by a few other intuitively selected amino acids in a given protein). Basically, a predicted structure had one chance out of two to be correct.

3. Automated Restriction Mapping. From a formal point of view, the problem is simply combinatorial. Restriction enzymes cleave a DNA sequence at specific sites. Initial data resulting from cleavage are sets of overlapping fragments of various sizes to be mapped to reconstitute the sequence.

The question, seemingly well set, caught the attention of mathematicians and computer scientists in the late 1970s. Formal solutions came from work in pattern matching, programming with constraints, or early applications of artificial intelligence, (see review in Ref. 27). The challenge for mathematicians and computer scientists was to generate a contradiction-free reasoning. However, biologists needed to map a real sequence with real data, and none of the computer methods accounted for experimental errors. Consequently, biologists kept on mapping manually.

2.2. The 1980s: Exponential Growth and Increasing Speed

Changes in the 1980s were both quantitative and qualitative. Not only did sequence databanks grow exponentially, but experimental techniques boosted data production in faster and more systematic ways. Organelle genomes (from 30 to 150 kb in size) were first targeted.

The development of polymerase chain reaction (PCR) methodologies [28] brought complete genomes into the picture. In addition, rapid multiple peptide synthesis [29,30] or cassette mutagenesis [31] provided tools to systematically test the effect of replacing a native amino acid at any given position by any of the 19 other possibilities. Competition experiments in vitro were also transformed by methods such as SELEX (systematic evolution of ligands by exponential enrichment) [32], which tests the affinity of 48 octamers as opposed to a subjective selection of a dozen potential sites.

In this context of massive data increase, the use of computers spread, and two questions corollary to those mentioned in the previous section were set.

2.2.1. How Should Large Sequence Data Sets Be Managed?

The reaction to the increasing difficulty of searching for particular information in exponentially growing sequence databanks was twofold. On the one hand, because databanks are an important resource, efficient search tools were defined (FASTA, by Pearson and Lipman [33]; BLAST, by Altschul et al. [34] with complementary strategies. On the other hand, clean and consistent data sets would be easier to search, so specialized databases began to arise. The first compilations of RNA sequences were released (e.g., that of Sprinzl et al. [35]). The PROSITE database is one of a kind, that inspired a number of subsequent initiatives [36]. It is based on the extraction of protein sequences and the selection of their most significant (sense is correlated to sequence function) parts, and as such constitutes a dictionary or lexicon. Its structure allows fast comparisons, indexing, etc. Dictionaries are eventually destined to create new vocabularies and as such can be used to redescribe sequences. Searching databases and making relevant comparisons requires a high dose of intuition to interpret sequence similarity, as convincingly pointed out by Brenner [37].

2.2.2. How Should Sequence Data Be Represented?

The crucial importance of description in the use of formal methods requires more work in the representation of raw data [38]. Going from a linear type of information reduced in a consensus sequence to a two-dimensional type of information presented in a weight matrix [39] is illustrative. The description of sequence data in general has evolved toward more flexible representations, shifting away from a strong statistical view to a more logical one. In particular, sequence analysis has become a fertile ground for the validation of inductive methods often developed for other purposes (statistical inference, neural networks, pattern recognition methods, symbolic learning methods, etc.).

Analyzing a sequence entails identifying regular features. Regularity can be set as a list of constraints restricting the occurrence of nucleotides or amino acids. To outline regular features in a sequence, most approaches rely on rewriting its content. Exceptions to this rule are examples where an explicit sequence could be isolated as a regular feature just by eye such as the Shine–Dalgarno sequence.

Rewriting involves drawing a correspondence between two alphabets and swapping the alphabets. Compositional bias can be made obvious by swapping {A, T, G, C} and {R, Y}, setting the correspondence as {A,G} → R and {T, C} → Y.

The more appropriate the expression of a sequence, the more apparent regularities will be. In fact, trials of all kinds were made to visualize regular features in genetic sequences. Examples include:

1. The use of geometrical rewriting rules to emphasize symmetry in DNA [40]. Four orthogonal vectors a few millimeters long (as in north, east, south, and west), each corresponding to a nucleotide, were defined to draw a DNA sequence in a two-dimensional space.

2. The use of rewriting rule expressing the frequency of the oligonucleotide YRY(N)iYRY in introns led to the identification of several periodicity classes in nucleic acid sequences [41].

3. With the use of harmonic rules in a sophisticated rewriting script, a melodic score could be derived from gene sequences, thus demonstrating the underlying specificity of genetic information [42].

More common, and constant throughout the years, has been the initial alignment of sequences suspected to share common features to emphasize similar regions. A widespread reductionist approach consists in associating aligned regions with a set of sequences and defining a characteristic motif. Local similarity refines descriptions and alignments [43,44], which gave rise to a package called IDEAS, a precursor of current sequence comparison programs.

Regularities extracted from an alignment can be represented as a frequency matrix [44a] or logical formulas [45] that sum up the information on a functional/binding site or a partial protein structure (HTH, turn, etc.). A family of aligned sequences can also be equated to a profile [46]. The introduction of diversity and flexibility of sequence description led to the performance of various calculations of indices, weight matrices, and scoring functions with numerous ad hoc methods (reviews in Refs. 47–49). Nakai et al., As it is now, hundreds of similarity measures can be used to compare a short unlabeled sequence to a given characterized site. Sorting out the most appropriate is not a trivial problem.

Protein structure prediction can be seen as an attempt to make explicit rewriting rules that allow an amino acid sequence to be transformed into a succession of secondary structures {helix, sheet, turn, coil}. Propensity factors or any other amino-acid-associated index as listed in Nakai et al. [48] have not provided the right terms to express such rules. The introduction of pattern matching approaches [50,51] was in essence directed toward stating explicit rewriting rules. A similar attempt was made with RNA [52]. In both cases, regularities of secondary structures were searched for an correlated to the presence or absence of specific amino acids or nucleotides. However, these attempts have remained limited by the lack of further rules expressing the dynamics of folding. To date, these rules are still very difficult to derive from empirical data.

The variety of examples mentioned here confirms how rewriting procedures can refine knowledge.

2.3. The Era of Genome Projects

The capacity to cope with increasing length and volume of sequences provided a first rough selection criterion for computer software and databases. Such a selection is not necessarily a reflection of the quality of a method. Even a sound algorithm such as that of Smith and Waterman [43] could not keep up with the increase, though optimization criteria were suggested by Gotoh [53] to tackle memory size and running speed problems.

The release of organelle genomes over 100 kb long (such as the mitochondrial genome of *Podospora anserina* [54]), the first yeast chromosome [55], and long *E. coli* contigs [56] provided ideal examples to test the worthiness of programs defined earlier as well as the relevance of the problem they were meant to solve. For instance, the use of codon bias to identify open reading frames or search for *E. coli* promoters revealed its limitations (see discussion in Ref. 57).

Most programs were working with the implicit assumption that the 5′ and 3′ ends of genes or the N- or C- terminal ends of protein sequences were given. Sequence comparison, structure prediction, and most motif and pattern search methods were all more or less based on a one-to-one correspondence between motifs and sequences. As a result, a whole new field of research opened with data generated by the polymerase chain reaction (PCR), where beginning or end meant little.

On the one hand, DNA sequences need to be inferred from "rawer" data. Indeed, new techniques such as random or shotgun sequencing or the generation of expressed sequence tags from cDNA libraries [58], brought about the problem of assembling overlapping DNA fragments to put a gene sequence together. On the other hand, the increased rates of generating false positives and false negatives while searching contigs (compared to searching well-defined sequences) was a new challenge.

Similarly with database size, exponential growth was not always appropriately handled. GNOMIC [59], a dictionary associating an oligonucleotide with a biological context defined as adjacent nucleotides, turned out to be too sensitive to an increasing number of sequences and could not keep up with incoming new information. In contrast, motifs (short consensus sequences), gathered with respect to functional properties such that sequences containing a common listed motif are grouped together in families, do not expand wildly while more sequences become available; the number of families remains manageable. For nucleic acids nonredundant databases containing very large collections of functionally equivalent sequences (FESs; introns, exons, 3′UTRs, 5′UTRs, and contiguous genomic fragments from different sources) have been made available to facilitate efforts in genome annotation [1]. For proteins, early attempts such as PROSITE [36] or PRINTS [60] initiated a trend for defining protein families. In all cases, nonredundant collections of functionally equivalent sequences help to generate standards (such as prior frequency distributions) for statistical analysis of function-associated short sequences.

The analysis of longer sequences also gave rise to various initiatives such as the possible use of spectral analysis methods developed for physics (review in Ref. 61) or the systematic rewriting (data reduction based on the concept of local repetitiveness) of sequences with respect to their informational content-compositional bias, based on the definition complexity [62,63]. Though the application of Shannon communication theory information theory [64] in biology is not new (see, e.g., Refs. 65–67), it is made more interpretable with calculations performed at the scale of entire chromosomes or genomes. It is completed by introducing methods of data compression [68–70].

Within two decades, the three founding questions of bioinformatics listed earlier have followed different tracks.

Sequence alignment tools are of intensive everyday use. Though the gap penalty problem is still subject to various controversies as well as the choice among dozens of scoring matrices, the issue has evolved to a point where regular users know what to expect from the program they are using.

Sequence restriction mapping is outdated in the genome mapping era and has definitely lost relevance as such. A renewal of the topic came recently from a new experimental method reducing the error in assessing fragment length [71].

Until recently, the issue of protein structure prediction evolved independently of genome developments with mitigated success [72]. Threading was thought to have become a resourceful approach in the early 1990s, but the enthusiasm died down [73]. The lack of quick enough experimental feedback has been slowing progress. The most recent developments involving the increase of structural data and the use of genomic data have, however, revitalized the topic (including the comparative assessment of techniques of protein structure prediction, known as CASP [74]).

3. GENOMIC SEQUENCE ANALYSIS

It may be obvious though relevant to specify the variety of automated procedures defined and used in genome research. Computer applications described in the literature can be categorized into three types:

1. Engineering software that reads data from sequencers and translates it into nucleotide sequences. In this case, inputs and outputs are clearly defined, and specifications can be made explicit. For instance, the experimental setup requires programs to cope with noisy spectral data, which is achieved by customizing signal processing methods. This aspect will not be discussed here. Examples of detailed description of methods can be found in Ref. 75.

2. Building software corresponding to algorithmic solutions to a well-determined problem for which the input is a consistent set of sequences and the output is meant to be unique such as an assembled sequence or a genome map. This part will be only briefly introduced in the next section.

3. Analytical software designed to understand the content of genetic sequences with varying assumptions. In this case, neither the input nor the output can be uniquely determined. Indeed, the consistency of a set of sequences input to detect a DNA binding site or to predict a folded state is difficult to prove. On the other hand, the potentially attached to outputs (e.g., is a detected site in an unannotated sequence a real result or an artefact?) gives rise to multiple interpretations. Except for Sec. 3.1, the remainder of the chapter is devoted to describing and discussing the various methods implemented in the context of interpreting genomic data.

3.1. Sequences and Contigs

3.1.1. DNA

The speed and cost of sequencing methods became an issue after the release of a few organelle genomes in the early 1990s. Although the original method was still referenced [76], it was being challenged by attempts made to skip the reading of electrophoretic gels. The idea of piecing short sequences together to reconstitute a larger sequence was more in tune with automation. However, random sequencing was limited to ~ 40 kb long target sequences,* and the reliability of sequence coverage by a collection of independent clones was begging the definition of a consistent model. As clearly explained by Fraser and Fleischmann [77] for bacterial genomes, a model evaluating the probability of a nucleotide not being sequenced given the number of clones alrady processed gave rise to the shotgun approach. It provided the threshold values necessary to set the number of clones for a given sequence length and a given genome length, assuming sequencing was performed from both ends with relatively long inserts. In fact, the shotgun whole-genome sequencing method is an example of evolution of a technology based on a computational model.

Final sequence assembly relies on the alignment of overlapping fragments, which can be made difficult when DNA is very repetitive as it is in eukaryotic genomes. The possible occurrence of gaps between contigs imposes the definition of closure strategies, which involve the use of PCR and/or probing by hybridization of identified gapped regions. Related questions have been discussed for bacteria [77], nematodes [78], and humans [79]. In the latter, the use of quality values of sequences indicating the likelihood of each base being correct is emphasized. Distinguishing sequencing errors from true polymorphism is of crucial importance. Combinatorial and pattern matching methods used in this context are well described by Gusfield [2], who considered sequences exclusively as strings of symbols.

Text search methods are appropriate because DNA sequences are not random. Nonrandomness is demonstrated by the presence of regular features. Periodicity, repeats, and compositional bias were observed (for early work see, e.g., Refs. 65 and 66) and are still the focus of a number of studies (see below for further references). The detection of these features depends on the description of sequences. In the basic representation of strings of symbols, the identification of regular features is based on only two possible parameters: length (of strings or substrings) and frequency (of symbols or groups of symbols). They are combined in different more or less complex ways depending on the type of expected output. Criteria such as length of inserts and length of overlap are used to piece sequences together, generate contigs, and map genomes. Statistical analysis of strings of symbols allows the correlation of regularity with known properties of sequences, which generally yields new

---

* Up to 1995, the size of the genome of phage λ was the upper limit [76].

assumptions or definitions. A basic representation can give rise to a variety of interpretations, depending on the setting of the initial goal.

3.1.2. Expressed Sequence Tags

Expressed sequence tags (ESTs) are cDNA sequences generated by automated single-pass sequencing and as such are rapidly produced at a low cost [58]. Generally, irrespective of their final use, EST sequence data are processed into contigs to reduce redundancy and generate longer sequences. The process of generating contigs involves, first, grouping homologous nucleotide sequences into clusters and, second, assembling clustered sequences into a consensus [80,81].

Assembling contiguous sequences from ESTs may, however, deprive us of useful information. If potential isoforms are clustered together and condensed into a single consensus sequence, information regarding polymorphism is lost. Indeed, when a set of ESTs matches a known gene, the multiple alignment of translated ESTs with the gene often shows variations. The question then becomes that of distinguishing which variations represent isoforms and which simply correspond to sequencing errors [82].

3.1.3. Maps

Physical mapping of a genome involves various techniques, which are well presented in Ref. 83. Only the latest method, optical mapping, will be briefly introduced here. Crystal clear explanations are given in Ref. 71. Large DNA molecules can be digested on an open glass surface and visualized by fluorescence microscopy to generate an ordered set of overlapping maps. Each clone is mapped redundantly, and a maximum likelihood method is implemented to select the most probable one. The probability of computing the correct restriction map is a function of the number of cleavages in the map and the number of DNA molecules used to create the map. Error arises from incomplete digestion or mistaken cleavage sites, unknown orientation, or erroneous sizing. A large statistical sample of maps (thousands) is preferred to the minimal threshold value (hundreds) predicted by a probabilistic model, and a confidence measure is used to rank maps.

A whole genome can be characterized in this way, and the resulting map complements the shotgun approach to tackle the gap problem and overcome the difficulty of mapping repeat-rich regions.

3.2. Sequence Data Description

3.2.1. Raw Sequences

Assuming initially that little is known about genetic sequences, chromosomes or genomes are considered just as sequences of symbols. The study of the two basic parameters frequency and length yields various definitions qualifying sequences.

1. *Frequency.* Compositional bias is usually estimated with respect to random sequences. The design of relevant random sequences for comparison purposes is a whole topic in itself and will not be debated here. Nucleotide or oligonucleotide (di-, tri-, tetra-, etc.) frequencies are examined in a given chromosome or genome. The interpretation of some distributions, among others, gave rise to

   Early Markov models [84,85]
   The definition of isochores in human genes [86]
   The definition of homogeneity or heterogeneity of sequences [87]
   The definition of repetitious (compositionally simple) or non-repetitious (compositionally complex) sequences [1,63]
   Various tables of over- or underrepresented words, simple repeats, or tandem repeats (see, among others, Refs. 88–90)

2. *Length.* Relevant information identified by the study of frequency parameters can be enhanced by working on distance distributions such as

   Length of over- or underrepresented words
   Length of repeats
   Distance between repeats

Sequences are also encoded in binary strings as the simplest representation that allows the assessment of the informational content of DNA based on information theory. Shannon entropy was applied in various conditions (it is valid for all sizes of alphabets) to the study of biological sequences, in particular to tackle the issue of sequence redundancy [1]. More recently, data compression techniques have offered a variation on the theme and better insight into the notion of repeat [69,70].

Assuming that global properties of sequences are known, the same parameters are used to get further information.

Property 1: Sequences are coding.

* Frequency. Nucleotides are grouped as triplets or codons. In chromosomes or genomes, codon and dicodon distributions are used to qualify sequences, for example, coding versus noncoding in *E. coli* [91] or coding versus noncoding in eukaryotes [88] (see also review in Ref. 92).

In coding sequences only, synonym codon use is studied. It is specific to organisms [93] and characteristic of the level of expression of genes in *E. coli* [94].

In noncoding sequences only, nucleotide distributions are examined. An example is noncoding versus intragenic regions in humans [95].

* Length.   intron versus exon length in eukaryotes (review in Ref. 92) and noncoding versus coding region length.

The two parameters can be combined, e.g., correlation of the $G+C$ content and coding sequence length [96]. The identification and annotation of open reading frames in genomes rely on such features.

Property 2: Sequences are folding or binding.

In this case, frequency and length parameters are difficult to decorrelate.

* Frequency and length.   Significant palindromes such as transcription termination sites [97], unlabeled identifiable secondary structures [98], and labeled identifiable secondary structures such as tRNA (see further for references).

3.2.2. Generating New Descriptors

Sequence analysis programs are first and foremost designed and used to generate new knowledge. There are two distinct issues:

1. The discovery of new patterns yielding the definition of new descriptors of sequences
2. The extraction of relevant pieces of knowledge in massive and miscellaneous collections of data, i.e., data mining

Both are addressed in the following sections.

The discovery of new patterns usually results from modeling, and various frameworks in which models are generated are presented below. Most of the time, supplementary information is required to refine the interpretation of identified patterns, and further processing of data is also detailed.

Combining basic frequency and length parameters in mathematical formulas is preliminary, as exemplified above. In this section, it is assumed that raw genomic data have been partially processed so that some knowledge can be associated with a group of sequences to be further analyzed. Assumptions can be related to functional or structural properties.

As discussed earlier, rewriting a sequence provides a new look at data. Moreover, rewriting is usually based on rules expressing a set of constraints, that is, regular features. "Regular" is often equated with "frequent." Many instances of a nucleotide at a particular position in various sequences make it regular. Frequency is not necessarily a reflection of how stringent a regularity is, because a constraint can be strong without being frequent. Such is the case of the letter "q" in most languages derived from Latin. A consensus word containing "q" would be (vowel OR 'c') 'q' ('u' OR '') (vowel OR ''). The low frequency of words containing "q" makes it more difficult to identify constraints. Such a situation is likely to be common enough in biological sequences and begs the question of how representative a set of sequences actually is. In the majority of cases, the answer is rationalized posterior to whatever analytical method is used.

Multiple alignment methods (global, Thompson et al. [99]; local, Morgenstern et al. [100]) provide a common and simple means of revising the definition of similarity. Portions of the alignment corresponding to conserved regions can be used as new descriptors, that is, a motif. Particular care in drawing an equivalence or a lineage of the various definitions of motifs as well as the basis of a flexible motif search is taken in Bucher et al. [101]. Motif descriptors are shown to fall into four categories depending on all possible combinations of two criteria: qualitative/quantitative and variable/fixed length.

Other recent methods do not require sequences to be aligned for the identification of motifs [102,103]. Qualitative patterns of variable length are generated using methods of this type. Two approaches for the discovery of new patterns are detailed: a pattern-driven approach, which is based on enumerating possible patterns and choosing the fittest, and a sequence-driven approach based on a search for common parts in sequences. The two strategies may actually be combined. The definition of pattern boundaries remains an open question, in particular for proteins, when structural data are not available. In that case, recent contributions attempt to provide approximate answers (e.g., Ref. 104).

Generally speaking, systematic rewriting of sequences offers a wider basis to the definition of similarity while suggesting measures and distances between sequences other than the standard editing distance. A short discussion on the matter can be found in Ref. 61.

A common motivation in using formal methods is the desire to identify new descriptors of sequences, which would favor the selection of relevant information from data sets. In that respect, acquiring knowledge from examples is an important step. A learning phase or, more generally, inductive reasoning has become almost unavoidable to identify regular features in a collection of examples. Neural networks and evolutionary and hidden Markov models are the main references. In fact, induction provides a means of reformulating a problem. Once a correlation is inductively brought to the foreground, it is used to define a filtering method, whether in the form of a metric or a scoring function associated with threshold values. In the context of pragmatic inference mentioned earlier, Konopka [1] states three directions along which such filters are suitable: prediction, simulation, and generalization. Successful inference is assessed by the quality of the outcome in any of these cases. More often than not, if the quality is unsatisfactory, new assumptions can be tested and properties of sequences refined.

3.2.3. Learning Signals for Prediction

A signal is a sequence within which the occurrence of nucleotides or amino acids is obviously constrained. In most cases, the automatic identification of a signal is based on inductive reasoning or learning. Examples of a signal are first gathered. The frequency and nature of the nucleotide or amino acid are almost always the chosen initial descriptors. Consider the illustrative case of splice sites.

The frequency of each nucleotide can be recorded in a weight matrix in which each position within the signal is individually considered. However, one nucleotide is likely to be restrained by the presence of others. Weight arrays provide the possibility of accounting for dependencies between adjacent positions [105], which seems appropriate in the case of codons. Dependencies between two nonadjacent positions can be identified by the maximal dependence decomposition method introduced in GENSCAN [106]. Dependencies between two or more nonadjacent positions can be identified by a neural network such as in NetGene [107] or by another learning algorithm such as the system described in LEGAL [108].

Various initial sequence sets are used in all gene-finding systems to adjust parameters. Referred to as learning sets, they are used to derive the information used later for the detection of new signals. This information is a function of the way signals are described. For instance, signals are first described as weight matrices in GENSCAN, or as profiles [109], whereas NetGene or LEGAL require the binary encoding of signals.

Unexpected frequencies of nucleotides at two distinct positions within a signal may indicate a correlation, which is detected by means of χ^2 calculations in GENSCAN. Binary data are processed in NetGene or LEGAL so as to extract the raw information content and combine logical operations with basic counting operations.

Both Baldi and Brunak, [3] and Durbin et al. [4] are quite exhaustive on the formalisms used in learning, and there is no point discussing them here. The topic is here simply to put the matter into perspective and provide a few references.

The Specific Case of Neural Networks. Neural networks were obviously first designed to model functions of the brain* and as such were meant to perform "intelligent" tasks. The early use of neural networks to approach and formalize the definition of biological systems has a long history covering neurology (e.g., Ref. 111), immunology (e.g., Ref. 112), etc.

---

* McCullough and Pitts [110] introduced the assumption of an all-or-nothing model for a neurone and the simple threshold switch of the algebraic sum of inputs.

The first application of a related concept known as a perceptron to the recognition of ribosomal binding signals was implemented by Stormo et al. [113]. A short time later, neural networks were given a fresh start when the back-propagation algorithm was defined [114]. This method was applied almost immediately to other signal recognition factors such as promoter sequences [115] or to protein structure prediction [116].

Neural networks (NNs) are used to express a complex correlation between an input and an output. They are composed of three or more interconnected layers of units called neurones. A weight is associated with a connection between neurones. The stronger the connection, the stronger its associated weight. Examples of inputs bound to a known output are given to calculate weights of connections in an attempt to minimize the error between the expected and obtained output.

Nucleotides or amino acids within a gene or protein sequence are most probably not linearly correlated, justifying the use of models where discontinuous functions can be approximated. Using an NN mainly requires (1) that data be described as subtly as possible so that inputs are informative and (2) some clarity as to what the output should be.

The interesting comparative study of *E. coli* promoter prediction methods of Horton and Kanehisa [117] emphasized the importance of data selection and the corresponding description. The complex architecture of a network used with a careless description of an arbitrary data set could not compete with a simpler design such as a perceptron running with carefully selected sequences and preprocessed descriptors.

Morever, neural networks are mostly used to discriminate between two sets so that the output is a simple yes or no answer to a question. The issue of setting counterexamples is discussed later in this section, but to remain here at a more global level one should just note the lack of work put into defining more sophisticated output layers.

Neural networks are useful tools considering the limitations of human eyesight and the number and size of sequences to be scanned for search purposes. However, they are not likely to generate new knowledge. The black box setup of a neural net prevents the rationalization of an automatic decision made by the program. Whatever site or sequence is supposed to be recognized, the resulting score attributed by a network has no known biological meaning. As such, neural networks do not generate much substance to define explicit rewriting rules (between two different representations, see Sec. 2.2.2).

In a compilation of the application of neural networks to biology, Wu [118] says about some particular NN setup: "The uppermost limit for the accuracy weakly depends on the specific network architecture, and confirms the relevance of the input information as a determining factor." Such a conclusion is tautological: The more we know about a subject, the more efficient a neural net is. So can a neural net really go further than what is known in biology?

3.2.4. Hidden Markov Model and Generalization

Various sequential problems are formalized by using stochastic models, which minimally require a set of states and a set of transitions between states individually associated with a probability value. To represent a given set of related sequences as a probabilistic model is to instantiate the set of states and the set of transitions along with their probabilities. The easiest way to do this with DNA sequence is to consider each nucleotide as a state and calculate the frequency of oligonucleotides of a given length corresponding to the order of the model. Prokaryotic noncoding regions were found to match such a description [91].

The demonstrated heterogeneity of sequences (see Sec. 2.2.2) rules out the existence of a uniform probability distribution of nucleotides along chromosomes—thus the relevance of simple Markov models. Moreover, in terms of sequence representation, selecting the four nucleotides as the set of states confines sequences to a static description.

Hidden markov models (HMMs) introduced new hypotheses as well as a needed change of representation [119]. Interestingly, within this framework, states are no longer static symbols but symbol transformations such as "delete," "insert," and "match" as a hidden mechanism constraining the occurrence of symbols. Such a dynamic description increases the chances of rationalizing some mutation phenomena. Sequences are considered the observable part of such a hidden mechanism, which supposedly corresponds to a succession of states. A mapping between the observation and the hidden mechanism levels is defined; the sequential change of states is governed by transition rules. These rules are first weighed using a training set. A built-in optimization algorithm guarantees the fitting of data to the model. The most common is maximum likelihood (see Durbin et al. [4] for a discussion of the possible alternatives for optimization). Maximum likelihood as the basis of the optimization procedure of an HMM is another formulation of the minimum message length (maximum compression) used in coding theory [120]. A rare attempt to clearly draw equivalence between models is to be noted [101].

Although a large set of parameters has to be managed by an HMM, which appears cumbersome to naïve users, various applications range from the multiple alignment of a set of related sequences to the detection of open reading frames (ORFs) in genomic sequences. The PFAM database [121] of motifs generated by hidden Markov models contributes to widen the scope of PROSITE.

3.2.5. Folding and Simulation

Covariations. As detailed earlier, the identification of dependencies between nucleotides or amino acids is a first step in determining new sequence descriptors. In the particular case of RNA, models of covariance were defined [122–124] to tackle the problem at the nucleotide level. In an attempt to predict RNA structure, sorting relevant from irrelevant dependencies is a tedious task. To date, only one example of manual sorting has led to the specification of structural covariations [125]. Dependencies as such express a compensation mechanism. Two nucleotides covary to maintain the possibility of base pairing.

Compensatory effects are likely to be involved at further levels such as groups of nucleotides. Compensation could be intrinsically recursive. Strong signals compensate for weak ones, but compensation is also observed at the level of structures in terms of distance(s) and angle(s).* The angle formed by the anticodon and the aminoacyl arms of the tRNA molecule is constrained to maintain the distance between the tips of these helices [126].

Amino acid covariations in protein sequences have inspired related questions, though no base-pairing rule holds. However, an amino acid change can be understood as losing or acquiring a property such as charge. The loss of a charged residue can be compensated for by the appearance of another charged residue elsewhere in the sequence [127].

Genetic Algorithms and Simulated Annealing. The understanding of principles of RNA folding motivated a revival of simulation. Genetic algorithms [128] and simulated annealing [129] were introduced to provide a flexible framework and avoid being trapped in local minima as with, for instance, a Monte Carlo method. In a simulation framework, intermediary stages can be looked at and folding pathways can be drawn. The sketch of a chronology is potentially defined. In fact, as discussed by Culberson [130], particular cases where the combination of local searches directs future search seem to be advantageously dealt with by genetic algorithms over other approaches to optimization. However, the setting of fitness functions remains a difficult question, and arguments promoting the superiority of genetic algorithms in general terms are not convincing.

---

* The L-shaped tRNA molecule is maintained by so-called stability bonds, which preserve the right angle.

3.2.6. Interpreting New Descriptors and Generating Models

A common distinction between computer solutions to concrete or real problems (set in biology or any other empirical field) is that they are either data-driven or model-driven. Data-driven solutions correspond to an attempt to rationalize data using whatever means, whereas in a model-driven solution a preset model guides the rationalization of data. The choice of model is obviously crucial, and so far very few seem adequate to express properties of biosequences. The hidden Markov model (HMM) as one of the latest, performs well in a limited way (see Sec. 3.2.4). In each of the frameworks in which a set of sequences can be modeled—primarily grammatical, phylogenetic, and structural—model-driven solutions may never get to be experimentally tested. Experimental feedback cannot be required in all instances, but the option of getting it should always be there.

The ultimate goal of determining new descriptors is to define models that can account for common properties of a family of sequences. Because strong regular features are usually matched with conserved regions of an alignment, sequence analysis tools either require an alignment as an input, as neural nets do, or produce an alignment as an output, as some HMMs do. In a data-driven model an alignment is usually the source, whereas in a model-driven scheme it is the target or an element of rationalization.

The balance between model-driven and data-driven approaches is difficult to set in an empirical science like biology. Our current lack of knowledge often causes us to lean toward data-driven solutions.

Pattern Matching and Grammatical Models. The success of applying string-matching algorithms to analyze genetic sequences lies in the obvious analogy between searching words in long texts and searching patterns in genomic sequences. As such, the results of decades of pure research in computer science were made available for biology. These include a collection of fast and efficient algorithms for search patterns [2]. Patterns recorded in the PROSITE database, for example, are regular expressions. Regular expressions are recognized by finite automata, which have been studied since the 1930s. Moreover, most alignment tools are inspired from related algorithms developed for matching substrings with strings. So there is undoubtedly a strong formal background from which molecular biology has benefited.

A natural interpretation of searching patterns as in words in a text is to consider the language possibly represented in the text. Formal language theory seems to provide a range of models and tools applicable to the genomic text. However, practical issues do not meet formal expressions so easily, and the model-driven approach is debatable in this case, at least in the short term. Attempts to use formal grammars to express regularities in sequences have not enhanced the understanding of a possible DNA syntax. It has been proven in many instances that DNA is not a random text; regularities can not only be found, they can also be expressed in terms of grammatical rules. It is therefore not surprising that attempts to derive some syntactic rules are successful. But the question is, How representative are these rules? Grammars defined with RNA families of sequences are not refined filters [131]. They capture some of the features of RNA but do not provide an unambiguous characterization of RNA molecules.

So far, the predictive power of grammatical methods seems limited by the complexity of context-sensitive grammars. Rules governing the occurrence of symbols in chromosomes sequences are likely to be context-sensitive. The recognition of such a language is an NP-complete problem.

With a lesser goal of simple discovery, most combinatorial pattern discovery methods are data-driven. Although they are restricted to identifying regular expressions, some attempts to identify new patterns have succeeded as described in the review of Brazma et al. [102]. The conclusion of this survey does, however, point out the need for more subtle patterns.

Phylogeny and Evolutionary Models. When sequences are considered merely as strings of symbols and aligned, the editing distance will determine how related sequences are that can be visualized in a tree. No further information except the origin of each sequence is needed to carry out such a task. Mathematics was used very early on for modeling phylogeny, because it mostly requires optimization functions. Building trees can remain very formal when a sequence alignment is considered as an abstraction. That is characteristic of a model-driven approach. In this case, the chance of being contradicted is low, because there is no experimental framework currently available to test the overall worth of a genealogy by mimicking mechanisms of evolution. The quest for the tree of life is of no help for the time being (possibly never, as pointed out by Doolittle [132]), because multiple contradictions keep arising from the incoming complete genome sequences [133].

The strength of evolutionary considerations is given by the knowledge of the function and/or the structure of the genes (or proteins) being aligned. Interpretation is enhanced when species are closely related. Then more practical questions can be solved. For example, in the context of the study of bacterial transcription factors, the evolution of a motif such as a helix–turn–helix DNA binding motif [134] provides guidelines for the identification of other transcription factors containing this motif.

From a very different perspective, in vitro testing of evolutionary mechanisms gives some feedback on regions of a specified set of molecules [135].

Structural Models. If sequence data are easy to generate, structural data are not. Most of all, the task is time-consuming. Consequently, sequences of known structure are aligned with sequences of unknown structure to take advantage of homology and predict structures. The same applies to protein or RNA.

Physical models of thermodynamics or various other areas of physics as well as mathematics have shaped much of our current knowledge of protein and RNA folding. Structural biology is the most model-driven area of biology, and justifiably so, because a large part of the knowledge overlaps with physics. These models combined with empirical data yielded various estimates to make free energy, hydrophobicity, solvent accessibility, and other parameter calculations more precise. They also provided guidelines to state the conformational rules that are commonly used to estimate possible folding.

The relationship between the structure and function of a macromolecule still needs to be thoroughly investigated in order to rationalize the involvement of a particular set of shapes in a particular biochemical activity.

3.2.7. Optimization and Validation

Sequence analysis usually relies on the definition of filtering methods. A filter is a selection operation and as such is associated with two sets of criteria (possibly reduced to only one element). Intrinsic criteria constrain the method itself (determining its quality), whereas extrinsic criteria constrain the way the method is used. Optimizing the selection entails optimizing criteria.

Figure 4 illustrates a succession of filters characterizing sequence analysis.

Selection 1: Choosing a Relevant Data Set. The slow building up of reference sets for sequence analysis created intermediary messy situations. Early compilations of sequences such as *E. coli* promoters [136,137]; transfer RNA [35,138], and structures such as the dictionary of protein secondary structures [139], lasted as standard sets for a number of years but were isolated cases. Most of the time, home-made data sets would be used to determine similarity measures, and a collection of indices, weight or probability matrices, profiles, etc. (see large sample in Ref. 47) was accumulated. Specialized databases now available provide better resources for standardizing data. However, particularly when a method includes a learning step, which is common, choices of a learning set and a test set are neither open nor debated enough. The systematic production of genomic data is only a partial answer to the question of relevance and consistency of data.

The automated version of the Selection 1 operation is usually based on keywords that characterize the function or correspond to various features of

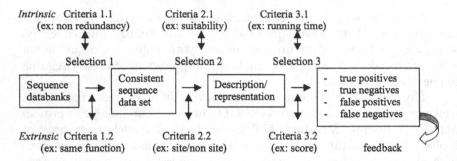

Intrinsic Criteria 1.1   Criteria 2.1   Criteria 3.1
    (ex: non redundancy)   (ex: suitability)   (ex: running time)

       Selection 1   Selection 2   Selection 3

Sequence databanks → Consistent sequence data set → Description/ representation → - true positives - true negatives - false positives - false negatives

Extrinsic Criteria 1.2   Criteria 2.2   Criteria 3.2
    (ex: same function)   (ex: site/non site)   (ex: score)   feedback

FIGURE 4 Successive filters of sequence analysis. Note that the optimization of each step relies on feedback information because these operations are more often than not repeatedly applied, taking into account knowledge acquired at an earlier stage. Selection 1 involves choosing a relevant compilation of sequences. Specialized databases provide good resources for standardizing data. The optimization of a selection of sequences usually involves a criterion of nonredundancy. Other features such as the length of sequences (i.e., getting rid of short fragments) can be added. Selection 2 involves choosing a relevant representation of data. Depending on how much is known about the biological process relative to the data and the problem to be solved, this choice is relatively unlimited. Optimization is therefore still an intuitive process as far as both type of criteria are concerned. Selection 3 involves choosing a relevant measure of similarity. Most analytical methods involve score calculations. Each analyzed sequence is given a score, which will result in deciding whether the sequence is a positive or negative instance. A score is usually optimized to fit the data. Strict computer science criteria assessing the worth of the algorithm can reasonably be included among extrinsic criteria to complete the optimization of the definition of a biological entity.

sequences. Software such as SRS [140] is used for databank interrogation and the extraction of sequence sets.

    The optimization of a selection of sequences based on keywords usually involves a criterion of nonredundancy. Other features such as the length of sequences (i.e., getting rid of short fragments) can be added.

    Selection 2: Choosing a Relevant Representation of Data.   As discussed in previous sections of this chapter, the issue of data representation is quite crucial. Alternatives include, for instance, opting for a deterministic pattern (e.g., a consensus sequence) as opposed to a probabilistic pattern (e.g., a profile).

    To get back to the issue of defining a learning set, it is important to note that to set positive and negative examples is neither uniquivocal nor formally well-founded. It is data-dependent. A positive example is presumably easy to specify. A negative one may cover diverse cases including unknown or undetermined data. There is no consensus definition of a nonsignal. There is no satisfactory definition of a random sequence either. Shuffling nucleotides or generating randomness via a Markov model is usual but barely reliable. Underlying principles characterizing the occurrence of nucleotides being unknown, there is no means of generating an artificial sequence that reproduces the characteristics of a real sequence. Such an issue is debated in Ref. 141.

The Selection 2 operation is poorly automated as a result of such a thin formal background. Optimization is therefore still an intuitive process as far as both type of criteria are concerned.

Selection 3: Choosing a Relevant Measure. Most analysis methods involve calculations of score. Each analyzed sequence is given a score, which will result in deciding whether the sequence is a positive or negative instance. The worth of an algorithm in biology journals has long been and is still often measured only by the rate of success in coming up with the expected results; that is, the worth of the scoring function is tested by whether it

Distinguishes positive from negative results
Distinguishes positive results from those generated with random sequences
Generalizes to other sequences (identifies new instances with possible experimental confirmation)

The unusual contribution of Wootton [142] introduces this too rarely debated problem of evaluating results with a long series of interesting questions* and discusses the relevance of the standard assessment by calculating measures of sensitivity (Sn) and specificity (Sp). A measure of relevance is suggested that is a constructive addition to current evaluations. In this framework, a sequence as an instance of a particular set of features is given a relevance weight depending on the source of knowledge used to establish the relationship between the sequence and the set of features. Indeed, an example of an active site is certainly more convincing if the activity was experimentally demonstrated than if the example was selected only because of similarity to a known active site.

---

* How do methods compare in discriminating a given class of functional features from other, perhaps similar, sequence features? Which methods perform best in recognizing resemblances to known features, albeit very subtle? Which strategies best favor the emergence of unprecedented molecular features and new biological associations? How do methods compare in indicating the existence of significant new associations that are not encoded in prior knowledge and databases? How can optional parameters of methods be varied to achieve different goals? Can the strengths of different algorithms be combined into new strategies that are more powerful than any individual method?

In fact, sensitivity (Sn) and specificity (Sp) should be used mostly as feedback information to improve and optimize a model, that is, to rationalize a posteriori a solution to classification or simple discrimination and refine extrinsic criteria.

The Selection 3 operation usually involves imported methods, which are in principle more or less intrinsically optimized. The worth of an algorithm in computer science is estimated on the basis of complexity and running time calculations (intrinsic criteria). A computer program is optimal if both complexity and running time can be minimized.

Definitions tend to differ according to whether they are set in biology or in computer science; optimization is an illustrative case. Truly, provided a method runs reasonably well, priority should be given to the optimization of extrinsic criteria. Indeed, concern for algorithm efficiency was not originally shared by biologists whose priority was to test ideas and hypotheses. However, memory size became a more pressing issue with growing data sets, even though computer capacities continued to be upgraded at the same time. Running time had been addressed earlier [143] but was really brought into the foreground with programs like FASTA [33].

Pure computer science criteria assessing the worth of the algorithm can reasonably be included among extrinsic criteria to complete the optimization of the definition of a biological entity [144]. The optimization of sequence analysis methods should, in fact, yield the optimized definition of biological entities to be used for unambiguous recognition in any data set. If this goal cannot be achieved, that is, if no such explicit sense can be made of a set of sequences, there are reasons to believe that the set is inappropriately built.

3.3. Advanced Genomic Data Description

3.3.1. Rule-Based and Knowledge-Based Systems

Early artificial intelligence (IA) methods known as expert or rule-based systems were specifically designed to accommodate issues related to problem solving with ill-defined hypotheses and suited to molecular biology (e.g., MOLGEN [145]).

A rule-based system manages likely facts. It is supposed to analyze a new fact by using in-built inference rules and assess how likely the new fact is. Such an assessment is done only if the rules used do not conflict. The knowledge of the system is improved by adding newly assessed facts.

A number of limitations were identified through use:

1. Explicit vs. implicit knowledge: A human expert cannot state all he knows in terms of explicit rules, so the system's knowledge is bound to be lacking.

2. Context dependence: It is not easy to restrict rule firing to specific contextual conditions.
3. Justification: Likelihood is not easy to justify.
4. Likelihood varies with the quantity of knowledge.

Explanation modules were added to cater to the third limitation and slowly evolved to become knowledge bases. The latter provide a more structured environment where context can be drawn and explanations can be built up. For instance, EcoCyc [146] is not a mere description of metabolic pathways of *E. coli*. Further knowledge of chemistry and biochemistry is included in various modules, providing the possibility of rationalizing the putative function of a gene, a lack of carbon, or whatever fact is being considered.

In parallel, to solve conflicting uses of rules and respond to limitation 4, a new architecture, multiagents, began to supersede rule-based systems. Multiagents rely on the division of tasks, which are distributed to a collection of specialized programs called agents [147].

In an expert system, the more likely a fact is, the more likely a rule containing that fact will be used. In a multiagent system, the knowledge of an agent is not more or less likely than that of anoter, but the communication between agents is given a degree of likelihood.

An agent's contribution to reasoning is partial and circumstantial. It is bound to be sustantiated and completed by communication with other agents. A collection of facts is considered as a succession of agreements between agents as opposed to a set whose consistency has to be worked out. Local consistency is preferred in a multiagent scheme, which seems suited to biological sequence analysis.

A multiagent system provides the framework for gradually putting together pieces of information in order to solve a problem. The system is given one or more "scenarios" to consider, that is, guidelines to reach a solution. The system is organized in layers. Information is filtered through a hierarchy of agents. The purpose of an agent is to process information. Lower level agents, called basic agents, start with raw information. In principle, information reaches higher level agents only after being processed by lower level agents.

A modifiable list of properties characterizes an agent. For instance, a basic donor agent is characterized by a position and the sequence surrounding the invariant dinucleotide GT.

To process information means here to sort, revise, and update information. It is achieved with two in-built operations:

1. Selection of the relevant agents
2. Co-operation between agents

In the selection mode, at least one criterion of relevance is defined to filter those agents fulfilling such criterion (possibly criteria). For instance, relevance can be set to be a score of similarity of a signal to consensus above a given threshold. Then it is considered as a new property and added to the list associated with an agent. Consequently, the knowledge of the system is enriched after selection.

In the co-operation mode, agents are merged and so are the associated lists of properties. Information is gathered from distinct sources. An underlying operation of selection is performed, because the merging is done according to a criterion of compatibility. The knowledge of the system is also enriched after co-operation.

The order in which selection and co-operation operations are performed reflects a sketch of resolution (scenario) drawn in accordance with biological models, thus defining a chronology of events. As a result, a higher level agent is the result of a successful step in the resolution process. Missing or erroneous information can be spotted by checking which properties of which list led to selecting the wrong agent or discarding a relevant one.

The chronology can be altered whenever priorities need be modified. Chronological variations of the recognition process can be included among the various recognition steps to test the relevance of a model. Moreover, the poor performance of some agents indicates which property should be modified.

Because the definition of "function" for a gene is becoming more difficult to set, alternative denominations such as "role" or even "agent" are being suggested to step out of the teleonomic frame that is attached to the word "function." The following section emphasizes this trend.

3.3.2. Networks and Graphs

A tendency is definitely emerging that entails focusing on interactions between sequences as opposed to sequences themselves [148,149]. To begin with, the very nature of metabolic data requires considering relationships between genes. Interactions are made explicit with respect to a chosen type of representation, preferably structured as in, for example, biochemical pathways.

Graph theory appears as a well-founded source of hypotheses to be studied, especially as far as transitivity and direction are concerned. Relations in graphs representing biological interactions are still poorly characterized.

Lattices provide a flexible definition of relationships between objects [108,150]. Other logically based networks allow, for instance, a modular expression of functions involved in regulatory mechanisms [151].

Quantifying contextual variations and setting the extent of mechanisms involved in DNA transcription or an immune response are still beyond our reach. Current trials to express data in networks may reveal new hypotheses as optimistically stated by Tavazoie et al. [152].

In fact, boundaries are generally quite difficult to specify in biology. In particular, no obvious practical definition of reversibility is available.

Can a minimal set of interactions be defined? Is it a case of setting thresholds or determining "atomic" elements? At the genomic level, gene knockout experiments reveal not only the apparent contingence of some genes but the often unsuspected consequences of codeletions. Understanding the consequences of the inhibition of some functions helps in estimating the extent of interactions. The same applies at the molecular level. For instance, most eukaryotic tRNAs consist of four helices, but many mitochondrial tRNAs are functional with only three; so what is a minimal tRNA?

3.4. Gene Annotation and Genome Data Interpretation

3.4.1. Gene Identification in Eukaryotic Sequences

The imaginative idea [153] that protein-coding regions in naturally occurring nucleic acid sequences could be identified via statistical analysis has been an inspiration for at least three generations of computational biologists. This initial lack of appreciation stems from the fact that in 1981 very few DNA sequences had been sequenced. In addition a widespread (however erroneous) belief was that all DNA encodes proteins and therefore there appeared to be no practical benefit from knowing that protein-coding regions exhibit interesting statistical properties. The existence of introns was already known, but the DNA sequences available for analysis were mostly those of E. coli and its phages (no introns). In addition, the sequencing techniques of the time allowed sequencing of the complementary DNAs, which were again protein-coding regions. This created a false impression that the sequencing alone would be good enough to find all genes. For these various reasons Shulman and coauthors remain forgotten to this day despite the momentum generated by sequence annotation software tools in the 1990s.

Interestingly the determination of gene structure acquired tremendous popularity, though, clearly computational analyses could not be instrumental in finding details of splicing mechanisms. However, statistical criteria for defining functionally distinct regions in DNA were attractive to their original inventors as potentially shedding some light on actual cellular mechanisms.

The first algorithm (and results of its implementation) for finding approximate locations of introns, exons, and intergenic spacers in long chromosomal fragments (assembled contigs) has been in use since 1988 [154]. The extraction of appropriate data from the available sequence databases (GenBank, EMBL-DNA library, and PIR-Nucleic) was, however, very difficult and time consuming.

Two main categories of currently known methods can be distinguished. "By signal" procedures determine intron–exon structure of protein-coding regions starting with the detection of the exon/intron junctions, based on the recognition of well-known splice sites consensus sequences and regularities. But this leads only to poor detection performance due to large numbers of false signals that are significantly similar to true sites [107,155]. This lack of specificity reflects our incomplete knowledge of the splicing mechanisms. To improve the detection, further analysis of coding regions is generally undertaken [106,156–158]. Coding regions can also be identified as exons with "by (statistical) content" criteria (such as codon use, compositional bias, and periodicity). Such an approach cannot (and does not) rely on a known model of the splicing process.

More generally, gene-finding methods usually do not provide adequate models of splicing. In fact, *splicing is only implicitly simulated.* Parameters are tuned to optimize the output from given inputs, mostof the time through training and a testing phase. These parameters are inaccessible in a neural network or a hidden Markov model but more explicit in a statistical model [106] or a rule-based system [109]. The majority of methods are tailored to match input and output data with currently available sets of sequences and, as such, loosely reflect a biological phenomenon or mechanism; a right or wrong prediction cannot be argued in biological terms. Change started to occur with Burge and Karlin [106], Brendel and Kleffe [159], and Vignal et al. [160].

The undeniably hierarchical nature of the recognition of splice sites and our lack of knowledge of it warrant the choice of a rule-based system. Indeed, if all steps of such recognition were to be precisely defined, conventional methods (e.g., an automaton) would be suitable for simulation. In a rule-based system, reasoning is made difficult because of conflicting sets of rules, and problems cannot be solved. Furthermore, understandable justification of how problems are automatically solved is lacking. To overcome the latter problem, explanatory modules can be added, but the system soon becomes unmanageable. Alternatively, a new architecture where tasks can be divided into subtasks and distributed within the system, called a multiagent system, can be designed. The framework of a multiagent system appears somewhat more suited to tackle the problem. Most of the operations defined in a rule-based system are built into the multiagent system. In particular, filtering operations are predefined, and the instantiation of parameters is left to the user. As in other methods, a preliminary learning phase is required.

The value of using a multiagent system is shown in Vignal et al. [160]. In particular, the system is used to assess the relative importance of each identified component involved in splicing, corresponding to an approach described in Ref. 161 as "understanding and replicating "in silico" the rules by which signals ... are recognised and processed." Results emphasize that sites are differentially rated depending on varying contexts of donor sites. Such variation could be biochemically specified. Furthermore, the poor recognition of some sites seems to indicate a requirement for at least one additional step corresponding to the interaction between an enhancer sequence in the vicinity of the donor site and proteins of the splicing complex.

Recent reviews [162,162a,162b] stress the permanence of the strengths and weaknesses of the methods referred to in the present section.

3.4.2. Data Mining

Exploring Databases and Web Servers. The initiative of specialized databases first introduced in the late 1980s (see Sec. 2.2) was boosted by the spread of computer networks. Within a few years, various types of information (software, sequences, annotations, collections of motifs, structures, sites, images, etc.) became available on servers of the World Wide Web. The analysis and interpretation of genomic data are now synonymous with searching the web. A new challenge is to define intelligent automated strategies of information retrieval from biology servers.

In recent years, themes of servers have evolved in grossly three phases:

General and traditional sequence or structure databases (nucleic acids or proteins) as quoted in Sec. 2.2 (EMBL, GenBank, Swiss-Prot, PIR, etc.).

Specialized databases 1: static objects such as sequence sites, specific sequence families, or collections. Examples: Yeast genome (http://speedy.mips.biochem.mpg.de/mips/yeast), collection of organelle genomes (http://megasun.bch.umontreal.ca/gobase), ribosomal RNA (http://www.psb.ugent.be/RNA/index.html).*

Specialized databases 2 (often more appropriately called knowledge bases): dynamic objects such as sequence functions or relationships between sequences. Examples: Transcription factors in eukaryotes http://transfac.gbf.de/TRANSFAC) or prokaryotes http://www.cifn.unam.mx/Computational_Biology/regulondb), metabolic pathways (http://www.genome.ad.jp/kegg), etc.

Specialized databases 1 were discussed earlier in this chapter. Specialized databases 2 are related to the functional annotation of genomes. Their number is in constant progression. In some instances, the function of a gene can be tested while information is being gathered on metabolic pathways of similar genes and cross-genomic comparisons are being made. Still, a good intuitive knowledge is required to do this. Intuition can be guided in

---

*Documentation (including those molecular biology servers designed to list and categorize existing Web resources) and bibliographical databases are across the board. Example: http://www.ebi.ac.uk or http://www.ncbi.nlm.nih.gov

knowledge bases where various sequence analyses are performed, evaluated, and compared to produce gene annotations [150,163].

Annotation is made difficult when at least 30% of sequences from newly sequenced genomes turn out not to look like any others. The consequences of such an observation are still at an early stage of understanding [164,165]. Recently, attempts to converge sequence annotation from independent institutions gave rise to the Gene Ontology (GO) Consortium, which is striving to develop a common vocabulary applicable to eukaryotes [166]. The objective of GO is to provide controlled vocabularies for the description of the molecular function, biological process, and cellular component of gene products. The controlled vocabularies of terms are structured. GO is not suggested as a way to unify biological databases. Sharing vocabulary is only a step toward unification; it is not sufficient. Many aspects of biology are not included (domain structure, 3D structure, evolution, expression, etc.).

A significant amount of the information gathered in biological servers is hypothetical. The crucial issue of reproducibility of annotation cannot easily be addressed, because there are so few instances of genomes being sequenced and annotated more than once. However, doubts have arise as shown in Ref. 167, and updates are made for some of the most studied organisms [168–170]. Variation in the gene count* is only the most obvious manifestation of subtler changes in gene definition.

Moreover, diverse new experimental advances, such as gene expression analysis (study of the transcriptome, that is, levels of messenger RNA in given conditions) or two-dimensional gel analysis (study of the proteome, that is, all proteins of a given tissue or organism) are likely to quickly bring more information and further modify the landscape of servers. These topics are developed further in this chapter.

Synthesizing Information. Whatever new sequence is released, it is compared to a sequence databank by using a dedicated search program such as BLAST or FASTA. Such comparisons yield a primary annotation of the new sequence as in a set of matching sequences associated with a confidence value. In the absence of such a set, the gene is said to be of unknown function. Gene products ar further documented if they contain previously identified blocks preferably known to be involved in an enzymatic reaction or any known or partially known biochemical activity. Information is accumulated but is not reduced by a synthetic view. A handful of attempts have been made to cross-link data [171–174].

---

*The yeast genome content in genes in both cited publications is downsized from the original range of 6300 to approximately 5500.

Specific competencies are required to generate as well as interpret each type of raw information, whether in genetics, biochemistry, protein chemistry, immunology, etc., so that information piles up in parallel. As a result, a number of different viewpoints, whch are seldom explicitly related to one another, shed different lights on a given type of sequence. A simple example such as the HTH (helix–turn–helix) DNA-binding motif illustrates how diluted the information is. It can be considered as a protein motif defined by structural features [175], or by a regular expression [36], or by a weight matrix [176]. It can be detected with corresponding dedicated search tools or through the use of more general methods [142,177]. It can also be considered as a DNA motif whose sequence is more or less sensitive to mutations [178], etc. If the purpose of characterizing an HTH motif in a protein is the identification of genes potentially regulated by this protein and a better knowledge of regulatory mechanisms, then a combination of all viewpoints is necessary. Some families of regulators have been considered this way [179]. The example of the HTH motif is easily generalized to most sequence-related information.

To list properties of sequences is informative but no more than the list of ingredients to make a cake. It does not give the recipe, that is, quantification and chronology. If the ultimate goal of accumulating information is to discover or reveal the function and related biochemical mechanisms, information has to be weighed and ordered. Such a situation is illustrated in an instructive comparison of the increase in the number of published articles in molecular biology versus the increase in sequence records in general databanks such as GenBank [149]. It shows the overwhelming growth of the latter and stresses how imperative data analysis has become.

There are râre examples of an immediate operational interpretation of similarities between a new sequence and a set of known sequences unless a specific function is suitably targeted. The function need not be explicitly stated but must be circumscribed. For example, a secretion mechanism was revealed by the observation that the cluster of proteins present in pathogenic strains and absent from nonpathogenic strains of some gram-negative bacteria was very similar to the components of the flagellar biosynthesis apparatus. A flagellum-like "syringe" model was defined [180] that explained how proteins are secreted by the bacteria being directly "injected" into the host cytosol. Such an easy-flowing interpretation of sequence similarity is unfortunately exceptional.

The function of a gene or the consistency of a group of similar sequences can be discovered while making educated guesses and searching relevant databases. Intelligent strategies for data mining are yet to be designed. To identify relevant sequence functional features, the subtlety of a handmade selection of sequences from a complete genome is still highly competitive with a blind automated search on servers. A new control mechanism of transcription termination in bacteria was determined from the careful inspection of less than 30 sequences [181], in a way that does not lend itself easily to automation.

3.4.3. Computer-Simulated Functions

Current efforts in sequence annotation are all directed toward automation. The overwhelming volume of sequence data does impose systematic approaches. For the time being, the quality of information found in databases results from either constant updates in a flexible framework (e.g., Ref. 181a) or from preliminary processing that qualify or disqualify sequences for automatic annotation (e.g., Ref. 181b). In fact, reliable automatic annotation is a recognition problem that entails setting necessary and sufficient conditions that uniquely characterize a functional role for a macromolecule, as outlined in the Introduction of this Chapter. These conditions are still too rarely met.

The identification of RNA motifs corresponding to specific molecules is, however, a good example of a satisfactorily solved recognition problem. Reliable and accurate algorithms have been designed in way that is briefly summed up in the following example.

An Example: Tailor-Made Methods for Predicting RNA Motifs. A number of self-contained methods tailor-made for searching for tRNA genes (tRNAscan, [182,183], FAStRNA [144], tRNAscan-SE [184]), *E. coli* transcription terminators [97], self-splicing introns (CITRON [185]), etc. were designed to scan whole genomes. By construction, reported results are usually quite accurate: Less than 3% are false negative and less than 1% false positive. The outline of a common strategy for searching these motifs is given in Ref. 186. A similar approach characterizes the various methods independently defined to identify almost unambiguously different types of RNA molecules in DNA fragments. Such dedicated searches are based on the principle that the more conserved a region, the more easily recognized it is. They depend on the use of weight matrices. Weight matrices make the definition of the RNA motifs more flexible than consensus sequences. Primary and secondary structure features are then used to gradually refine the identification process. A variant definition of patterns as a class of words is also tested in Ref. 144 (Table 1).

Table 1 sums up the search strategy in a more compact way than in Ref. 186.

(A) and (B) are self-explanatory in the light of previous sections of this chapter. (C) Regular distance and length features are used for filtering potential helices. Regular nucleotide and base-pair features are used to assess the quality of a potential helix. A potential molecule is assembled helix by

TABLE 1 Outline of a Common Strategy for Searching RNA Motifs

A. Initial data: RNA sequence alignment
B. Data representation: structure primary and secondary
C. Automated or manual identification of regular features in sequences and secondary structures
    (i) Regular nucleotide occurrence in sequences
    (ii) Regular base pair occurrence in helices
    (iii) Regular size of helices
    (iv) Regular length of interhelical segments
D. Definition of a similarity measure
    (i) Selection rules
    (ii) Evaluation rules
    (iii) Implementation
E. Necessary and sufficient conditions
    (i) Set hierarchy of rules
    (ii) Nonselective local optimization
F. Decision and prediction
    (i) Selective global optimization (expressing compensatory effects)

A similar approach characterizes the various methods independently defined to identify almost unambiguously different types of RNA molecules in DNA fragments. Such dedicated searches are based on the principle that the more conserved a region, the more easily recognized it is. Primary and secondary structure features (A–C) are used to gradually refine the identification process. A potential molecule is assembled helix by helix. Necessary conditions are implemented as selection rules for each potential helix (D(i)). Each selected helix is given a score with evaluation rules (D(ii)). The strategy relies on a succession of search procedures, according to a preset chronology (E) and the use of an overall score to account for compensatory effects (F).

helix. Depending on how stringent constraints are, potential helices are searched locally or on an extended part of a sequence.

(D) Necessary conditions are implemented as selection rules. It is an all-or-nothing filter. Sequences or structures must verify a number of set criteria. Evaluation rules express a statistical bias observed in aligned sequences at one or more positions. Each selected sequence segment is given a score with evaluation rules (bonus or penalty according to whether rules are or are not verified). Likewise, within helices the occurrence of certain base pairs is statistically constrained. Two types of rules can be distinguished:

1. Context-free rules expressing an overall constraint such as the base-pairing rule.
2. Context-sensitive rules expressing a constraint

    a. With a biological interpretation: imposing (selection) or rating positively (evaluation) the presence of (A, U) as the second or third pair in a fixed length helix as a reflection of a known tertiary interaction

b. With no known biological meaning: imposing or rating positively a limit to the number of A's contained in a helix strand as a means to reduce combinatorial explosion (see discussion in Ref. 187).

(E) The strategy relies on a succession of search procedures, according to a preset chronology. The most constrained helix is first searched for and serves as an anchorage point. All other helices are searched for upstream and downstream from this point. For instance, the TΨC arm of a tRNA represents this anchorage point.

(F) One key characteristic in tRNAscan, FAStRNA, or CITRON is the use of an overall score, S, to reduce the number of false positive. S is incremented only if a selected potential helix verifies enough evaluation rules (above a set threshold). Such a setup is a simple way to account for compensatory effects. Indeed, the global S score increases only if the local score of an individual arm is above a given threshold. Let us assume that a number of potential D arms of a tRNA are found in a given DNA region. Each of them is scored locally using evaluation rules. A low-score helix can be selected as a potential D arm in a tRNA provided at least two other helices are stable enough (above threshold).

The recognition of one molecule by another rarely involves a single event. To understand how many events are involved, the chronology of these events—more specifically, how many signals are required and their relative importance—is a step toward more accurate modeling of biological phenomena. How and what can affect chronology is the next step.

Necessary and Sufficient Conditions. Whether a single molecule or a whole mechanism, the concept of sequence seems to dominate in biology. Moreover, the formalism developed to analyze a succession of nucleotides can hold for a succession of operations or actions involved in a molecular process or of events in a cellular process. It seems to point to an intrinsic recursive nature of biological phenomena. Such a hypothesis is illustrated in Figure 5. Assuming that a cellular process is a succession of events, each of these results from molecular activity, and different molecular activities compensate for each other, depending on their relative strength. Assuming that a molecular process is a succession of actions, each of these results from sequence activities, and different sequence activities compensate for each other, depending on their relative strength. Assuming that a gene or a protein is a succession of signals, different signals compensate for each other, depending on the various structural constraints. Furthermore, each signal expresses dependencies between nucleotides or amino acids.

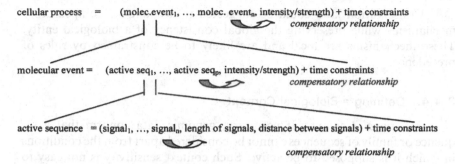

cellular process   =   (molec.event$_1$, ..., molec. event$_n$, intensity/strength) + time constraints

compensatory relationship

molecular event =   (active seq$_1$, ..., active seq$_p$, intensity/strength) + time constraints

compensatory relationship

active sequence   = (signal$_1$, ..., signal$_n$, length of signals, distance between signals) + time constraints

compensatory relationship

FIGURE 5 Sequential nature of biological processes. Even though many processes can take place in parallel in a cell, biological processes appear as sequences within sequences. Compensatory mechanisms and time constraints govern the occurrences of patterns, whether these patterns represent groups of nucleotides or amino acids, groups of sequences, or groups of molecular events.

To make sure that the definition of a function or a mechanism provides a good basis for simulation, ambiguity must be minimized if not completely discarded. A known principle of mathematics is to express a set of necessary and sufficient conditions in order to characterize an object without ambiguity. Examples given above show that sequence analysis methods often stall after the statement of necessary conditions, and quite a few widely used definitions qualifying sequences are incomplete.

The following empirical statement is an attempt to sum up the current discussion:

> To characterize a consistent set of sequences or a sequential process, constraints on symbols are necessary conditions and the order in which these constraints apply provides sufficient conditions.

"Symbol" is considered very broadly; it can range from a single nucleotide or amino acid to a complex event. The set of necessary and sufficient conditions provides a new description of a sequence or a family of sequences.

The uniqueness of such an order is open to discussion. There are examples of alternative processes depending on given conditions. For instance, the two ends of an exon are usually simultaneously recognized by the spliceosome components. However, a two-step mode of recognition of a small vertebrate exon was described by Sterner and Berget [188]. In this process, the miniexon is first recognized as an exon-intron-exon unit, followed by subsequent recognition of the intron.

The variable hierarchy or order in which constraints apply may provide different ways of optimizing the definition of a family of sequences.

Conservation laws could be suggested to account for compensatory mechanisms while preserving the global consistency of a biological entity. These mechanisms are local and are likely to be constrained by rules of precedence.

3.4.4. Defining a Biological Context

Functional Context. If necessary, genomic data confirm that a sequence or family of sequences cannot be considered apart from the conditions in which it is supposed to be active. Such context sensitivity is not easy to study and returns to the notion of similarity to account for the situation where sequences are close in one context but not in another. Attempts to assess the influence of context are under way, as in Ref. 189, where the concept of "neighborhood" is explored. In this framework, relatedness between genes depends on them being involved in the same metabolic pathway or following a similar patterns of codon use or else co-occurring in the same literature reference. Such a variety of neighborhoods may give some clues as to what unannotated genes do.

Cross-genome comparison provides another viewpoint on the functional context of genes. A study of genes involved in a given metabolic pathway shows that diverse organisms have opted for various solutions [164]. This type of information is still at an early stage and curiously does not address the question of defining the concept of "function." This imperative discussion is undertaken, however, when working on a reliable functional classification of genes [190]. In particular, the current classification is based on a set of nonoverlapping classes where any one gene carries a designation for a so-called functional primitive (enzyme, regulator, transport protein, etc.) and a process (electron transport, carbohydrate degradation, macromolecular biosynthesis, etc.). This categorization creates conflicts when many proteins are involved in more than one process or when single subunits of a multiunit enzyme correspond to different processes (this is typically the case of transport proteins that contain a membrane component). The definition of overlapping classes is suggested to preserve the knowledge of non-identified interactions.

Subcellular Localization. Understanding the function of a protein is dependent upon knowing where this protein is meant to be active. This information seems to be deducible from the sequence itself as demonstrated by at least two different approaches.

ANALYSIS OF AMINO ACID SEQUENCES. PSORT is a program designed to predict the subcellular location of proteins from their amino acid sequences [191,191a]. It comprises a collection of methods to identify different features, each characterizing a particular type of protein or part of a protein. As such, it is valuable as an attempt to merge and synthesize diverse pieces of knowledge. Unfortunately, the nature and quality of the various sources are very heterogeneous. PSORT is yet another illustration of a collection of necessary conditions lacking a sufficient counterpart.

Categories used in PSORT are the following:

Signal sequences. PSORT first tests the possible presence of a signal sequence while considering the N-terminal positively charged region (N-region) and the central hydrophobic region (H-region) that characterize a signal sequence. A discriminant score is calculated from three values: the length of the H-region, the peak value of the H-region, and the net charge of the N-region. Because the detection of a signal sequence does not necessarily imply that it is cleaved, the possibility of cleavage is assessed next.

Cleavage sites. A weight-matrix method for signal sequence recognition and the information generated at the previous step are used to detect signal-anchor sequences. Among all candidate positions, a possible cleavage site is the furthermost position of the C-terminal of the signal sequence.

Transmembrane segments. A potential transmembrane segment is identified if its average hydrophobicity (estimated from 17-residue segments) is above a set threshold.

Lipoproteins. The recognition of lipoproteins relies on slightly modified versions of the recognition of signal peptides and transmembrane segments. Species-dependent data are also included.

Mitochondrial, peroxisomal, lysosomal, vacuolar, and chloroplastic proteins. Proteins from organelles show some specificity based partly on amino acid composition. Characteristic motifs are used, though not in full confidence.

ER proteins. Endoplasmic reticulum proteins are still characterized by weak motifs.

Ribosomal proteins. The recognition of ribosomal proteins is reduced to matching a set of motifs.

Posttranslation modified proteins. The recognition of various modification sites can help localize proteins. For example, prenylated proteins are likely to be found in the plasma membrane and the nuclear envelope.

Miscellaneous motifs. The program also relies on PROSITE motifs that are loosely associated with a specified location. Furthermore, searches for particular structural motifs such as coil–coil are implemented.

PSORT is a rare contribution is that it tries to combine and merge various sources of knowledge but suffers from the lack of overlap between these sources.

ANALYSIS OF THE NUCLEOTIDE SEQUENCES. Some proteins, such as ribosomal proteins, can be found both in the cytoplasm and in mitochondria while being encoded in the nuclear genome. Variations of the codon use in these nuclear genes was shown to correlate to the final location of the proteins—cytoplasm or mitochondrion [192].

3.5. The Differential Display Era

3.5.1. Gene Expression

As well detailed in Ref. 193, two technological strategies are used to assess the role of genes. The first is based on the alteration of gene structure by disruption (transposon mutagenesis, referred to as genetic footprinting) or deletion (gene knockout) and is mainly available for microorganisms. These alterations are studied to evaluate their effect on cell viability. Yeast is one of the first organisms for which an extended set of results were generated (see Stanford web site: http://genome-www.stanford.edu/Saccharomyces/).

The second strategy involves the analysis of intact genes to evaluate their behavior in terms of levels of expression given specific conditions. It is implemented via the hybridization of extended collections of cDNA clones using either oligonucleotide chips (each gene is analyzed by 25-mers) or DNA microarrays [each gene is analyzed by a full or partial target DNA sequence (~1 kb)]. Deletion experiments are also carried out using this type of technology. Thousands of genes can be represented on microarrays and chips to be analyzed simultaneously. The same hybridization conditions are guaranteed in mRNA assays as a result of parallel processing. Quantification is measured through the use of fluorescent dyes (green and red).

An overflow of data is being generated by this new technology. Each step of the process involves computer contributions from the design of arrays to the analysis of experimental results. Some problems are hinted at in Ref. 194. The accumulated data are meant to identify patterns and relationships among expression profiles generated in an individual array or a collection of arrays. Solved and unresolved issues of data analysis are presented and debated in Refs. 195 and 196.

3.5.2. Protein Expression

Proteomics is mostly about assessing the consequences of a change of state in a tissue or the whole cell by quantifying proteins and potential modifications.

It is not based on new technology but rather on improved techniques of two-dimensional electrophoresis* and mass spectrometry,[†] which are more formally put together to identify and characterize proteins.[‡]

At least two empirical facts justify the viewpoint of proteomics [201]:

1. The presence of mRNA is a prerequisite to protein synthesis, but all transcripts are not necessarily translated. Consequently, measured levels of mRNA do not account for levels of proteins in a given tissue.

2. It is not possible (currently) to infer posttranslational modifications from a nucleotide sequence.

Protein function is expanded through amino acid modifications. Spatial parameters (localization of proteins in the cell) and/or time-related parameters (among others, the development stage of the organism) influence reactions of glycosylation, methylation, etc. The change in molecular weight of a modified protein is observable by electrophoresis as the corresponding spot on the gel is perceptibly shifted. The shift is increased when the modification affects the charge of the protein.

Proteomics provides a framework within which co- and posttranslational modifications can be studied and related to a particular function. As set, it seems to start from a reasonable basis. Indeed, if phosphorylation or glycosylation sites are predicted solely from amino acid sequences, there is no guarantee other than that supplied by more experimental testing that a predicted site is real. Conversely, rationalizing the potential modification(s) of a protein as observed as a shift on a gel from the variations of peptide masses [202] provides a lead for further interpretation of a function. Contextual information is needed as well [203].

Technical means to separate complex mixtures have been and still are a central concern in physics and chemistry. The physicochemical properties of biological molecules account for the common use of electrophoresis in molecular biology. Alternatively, chromatographic separation can be achieved. Separation, whatever the mode, is coupled with protein digestion (usually with trypsin). Digestion generally follows the electrophoretic approach but precedes the chromatographic approach. Peptide masses are

---

* See discussions of original 2D electrophoresis with carrier ampholytes in Ref. 197 and with immobilized pH gradients in Ref. 198.

[†] Recent improvements in the resolution of mass spectrometers as in Refs. 199 and 200.

[‡] To *identify* a protein is to name it, but to *characterize* a protein is to specify the properties of the named protein. For example, a protein identified as a tyrosine kinase can be characterized as a phosphorylated tyrosine kinase.

directly estimated by a mass spectrometer.* The collection of masses associated with a digest is compared to a protein digest database.† This process is known as peptide mass fingerprinting (PMF). When a sufficient number of matches (threshold-dependent) are found, a protein is identified. Needless to say, this approach makes more sense with completely sequenced organisms for which the set of available proteins is close to exhaustive. A variety of PMF engines [204] can help identify proteins. Tools for characterization from mass data are sparser [202]. In fact, the various forms (splice variant, polymorphic variant, posttranslationally modified, etc.) of a protein are currently distinguished by mining web resources.

Applications in proteomics are often divided into three categories [205].

The Identification of Proteins in a Cellular Extract. The mapping of total cellular proteins mainly involves the study of model organisms, which provides a testing ground for identifying practical problems inherent to proteomics research. A major distinction lies in mapping either all cellular proteins or the subset found in subcellular complexes. As pointed out by Godovac-Zimmermann and Brown [206], the ideal objective of total mapping can be summed up in four points: (a) All protein must be quantitatively extracted from the original biological sample, (b) proteins must be resolved and displayed, (c) each protein must be accurately quantified, and (d) each protein must be identified.

There are still major difficulties in implementing extraction methods. Protein quantification involves a number of issues related to protein expression. Expression reflects a cellular response to a given perturbation, and the distinction between responses may be difficult to establish.

Subcellular complexes such as ribosomes (*E.coli*, yeast, mammalian mitochondrial), organelles, and human or yeast spliceosomes are being studied (see Ref. 206 for review).

The human nucleolus provides another very recent example of a fully characterized organelle (see review in Ref. 207). Some 271 nucleolar proteins were identified on the basis of the human genome data. Approximately 30% of these correspond to uncharacterized genes.

Differential Display for Comparison of Protein Levels. The most common comparison criterion is pathological versus nonpathological. Healthy and diseased tissue samples are processed and compared to identify

---

* For details on mass spectrometry, see, for instance, http://www.micromass.co.uk/basics/.
† For details; see ExPASy Proteomics tools at http:/www.expasy.ch/tools/.

protein expression changes due to expected differential expression. A variety of diseases are considered as discussed in Ref. 208.

   Detection of Protein–Protein Interactions.   A variety of methods provide data on protein–protein interactions. Physical data generated by X-ray crystallography and NMR, antibody-based methods, various assays and screenings, etc., have contributed to the accumulation of information related to interactions that is now stored in databases of interactions (see, e.g., Ref. 209). In the context of automation and high throughput, the main method for detecting the studying protein-protein interactions is the two-hybrid system. Two alternative applications are common: Given two specific proteins, this technique can detect the interaction given a collection of proteins, it can detect which ones are interacting with which. In this case, the method is systematized such that all possible interactions of a given protein can be listed [210].

4. CONCLUSION

The main point of this chapter was to give an overview of essential questions and problems set in the framework of biomolecular sequence analysis, particularly the widely recognsized strengths and weaknesses of data interpretation. Rapid changes in the technology increase the production of data and their consistency, but in the short term efforts are mostly needed in organizing and managing these growing sets. Such an organization requires dependable definitions of biological objects or entities, which can be acquired only from representative and reliable sets of data. The simulation of biological processes is potentially an important part of future experimental biology. Thoughtful insights into quantification methods will be required [211].

   Finally, and at the risk of sounding like a movie producer, scenarios are badly needed. There will be no valuable long-lasting annotation of genomes without a clearer picture of the processes involved at the cellular level. Going beyond identified similarities and correlations between sequences and trying to relate them to each other will help in sketching and testing new processes.

ACKNOWLEDGMENTS

I wish to thank Andrzej Konopka for multiple discussions of topics covered in this chapter as well as in-depth comments and suggestions concerning statistical sequence analysis, sequence annotation, and the history of computational biology. I am also obliged to George L. Miklos for the wise advice he provided on an early version of this manuscript and Antoine Danchin, David Sankoff, and Maxime Crochemore for the very encouraging reports they wrote about an even earlier version originally written in

French. I particularly thank Alain Hénaut for his unconditional support throughout.

GLOSSARY OF TERMS

Bayesian Model. A model that describes a phenomenon with conditional probabilities. It quantifies the likelihood of events. Bayes' theorem expresses the relationship between the probability of an event A occurring knowing event B [denoted $p(A|B)$] and the probability of B occurring knowing A as well as the respective probabilities of event A and event B. The formula states: $p(A|B) = p(B|A) \times p(A)/p(B)$.

Consensus Sequence. The most elementary way of describing a pattern as the most frequent segment containing constrained nucleotides or amino acids.

Curated Set. A collection of nonredundant sequences.

Data-Driven Approach. An attempt to rationalize data using formal means.

Data Mining. The use of computer tools to modify the representation of data and help discover previously unknown relationships among the data.

False Positive. Assuming that a method is defined to select a given type of sequence, a sequence selected by the method that is in fact not of the wanted type.

False Negative. Assuming that a method is defined to select a given type of sequence, a sequence of the wanted type that is not selected by the method.

Family. A set of sequences assumed to share a common property whether structural or functional.

Genetic Algorithm. An optimization procedure. Given a population of entities such as sequences and the definition of a fitness criterion, operations of crossover and random mutations are implemented to simulate an evolutionary process. The process is supposed to converge toward a population of the fittest.

Hidden Markov Model. A probabilistic representation of a succession of events. A biological sequence can be considered as the observable part of a hidden mechanism that supposedly corresponds to a succession of states. A mapping between the observation and the hidden mechanism levels is defined; the sequential change of states is governed by transition rules. These rules are first weighed using a training set. A built: in optimization algorithm guarantees the fitting of data to the model.

In Silico Biology. A new area of experimental biology that uses computer programs to generate new assumptions. Such knowledge discovery can be achieved by mining data (e.g., extracting consistent information from various databases) or by using specialized analytical tools (e.g., a specific motif search).

Knowledge-Based System. An enhanced rule-based system that includes heuristics and strategies that improve the reasoning capacities of the system. Knowledge is often organized in complementary modules.

Model-Driven Approach. An attempt to fit data to a preset model.

Motif. A combination of patterns co-occurring in sequences.

Multiagent System. A system that provides the framework for gradually putting together pieces of information in order to solve a problem. Given one or more guidelines to reach a solution, the purpose of the agent is to process information. Information is filtered through a hierarchy of agents. Lower level agents start with raw information. In principle, information reaches higher level agents only after being assessed by lower level agents. The knowledge of one agent is not more or less likely than that of another, but the communication between agents is given a degree of likelihood. An agent's contribution to reasoning is partial and circumstantial; it is substantiated and completed by communication with other agents. A collection of facts is considered a succession of agreements between agents as opposed to a set whose consistency has to be worked out.

Neural Net. A network used to express a complex correlation between an input and an output. It is composed of three or more interconnected layers of units called neurones. A weight is associated with each connection between neurones; the stronger the connection, the stronger its associated weight. Examples of inputs bound to a known output are given to calculate weights of connections in an attempt to minimize the error between the expected and obtained output.

Pattern. A regular feature in a sequence or structure. It can be a segment within which nucleotides or amino acids consistently occur. It can be a substructure (e.g., a helix in RNA or in proteins) that is consistently present in a structure.

Profile. A frequency matrix that provides a means to describe a sequence alignment. It features the N positions of the alignment in rows and scores based on the frequency of occurrence of the 20 amino acids and of gaps in columns.

Proteome. The set of all proteins found in a given tissue at a set developmental stage of a given organism.

Proteomics. The study of proteomes. It involves either protein separation on two-dimensional electrophoretic gels and the labeling of proteins by measuring their molecular mass or the so-called two-hybrid system, which generates a set of protein-protein interactions.

Regular Expression. A flexible definition of a pattern allowing degenerate positions.

Rewriting. Setting a correspondence between two alphabets and swapping alphabets.

Rule-Based System. A system that performs logical deductions. It contains logical assertions known as facts that are each assigned a degree of likelihood. It also contains built in inference rules that are used to analyze a new fact, that is, assess its consistency with the rest of the facts and its likelihood. Deductive reasoning is performed by an inference engine that chains compatible rules in a nondeterministic way.

Sensitivity. The property of a classification method (usually involving discrimination between positive and negative examples of a given type of sequence) that is quantified by the ratio of the number of true positives to the sum of the number of false positives and the number of true positives.

Specificity. The property of a classification method (usually involving discrimination between positive and negative examples of a given type of sequence) that is quantified by the ratio of the number of true positives to the sum of the number of false positives and the number of false negatives.

Simulated Annealing. An optimization procedure. Given a set of states, transition and probability rules, and a cost function assigned to the states, changes of state take place for as long as the cost of a new state is lower than that of a current state.

Transcriptome. The set of all expressed genes found in a given tissue at a set developmental stage of a given organism.

Weight Matrix. A frequency matrix that provides a means to describe a pattern. It features the four nucleotides or the 20 amino acids in rows and scores based on the frequency of occurrence of the nucleotides or amino acids in columns. The number of columns is set by the length of the pattern described by the matrix.

REFERENCES

1. Konopka AK. Sequences and codes: fundamentals of biomolecular cryptology. In: Smith DW, ed. Biocomputing: Informatics and Genome Projects. Orlando, FL: Academic Press, 1994:119–174.
2. Gusfield D. Algorithms on Strings, Trees and Sequences. London: Cambridge Univ. Press, 1997.
3. Baldi P, Brunak S. Bioinformatics. The Learning Approach. Boston: MIT Press, 1998.
4. Durbin R, Eddy SR, Krogh A, Mitchson G. Biological Sequence Analysis: Probabilistic Models of Proteins and Nucleic Acids. London: Cambridge Univ. Press, 1998.
5. Horovitz A. Non-additivity in protein-protein interactions. J Mol Biol 1987; 196(3):733–735.
5a. Benos PV, Bulyk ML, Stormo GD. Additivity in protein-DNA interactions: how good an approximation is it? Nucleic Acids Res 2002; 30(20):4442–4451.
6. Creighton TE. Proteins: Structures and Molecular Properties. New York: WH Freeman & Co., 1983.
7. Zuker M, Sankoff D. RNA secondary structures and their prediction. Bull Math Biol 1984; 46(4):591–621.
8. Zucker M. On finding all suboptimal foldings of an RNA molecule. Science 1989; 244:48–52.
9. Cook PR. The organisation of replication and transcription. Science 1999; 284:1790–1795.
10. O'Neill MC. *E. coli* promoters. J Biol Chem 1989; 264:5522–5530.
11. Alberts B, Bray D, Lewis J, Raff M, Roberts K, Watson JD. Molecular Biology of the Cell. New York: Garland Publishing, 1994.
12. Rosbash M, Seraphin B. Who's on first? The U1 snRNP-5′ splice site interaction and splicing. Trends Biochem Sci 1991; 16(5):187–190.
12a. Grosjean H, Benne R. Modification and Editing of RNA. Aristotelian Society Monographs Series, Blackwell Publishing, 1998.
13. Steitz TA, Weber IT, Ollis D, Brick P. Crystallographic studies of protein-nucleic acid interaction: catabolite gene activator protein and the large fragment of DNA polymerase I. J Biomol Struct Dyn 1983; 1(4):1023–1037.
14. Amit AG, Mariuzza RA, Philips SE, Poljak RJ. Three-dimensional structure

of an antigen-antibody complex at 2.8 A resolution. Science 1986; 233(4765): 747–753.

15. Edman P, Begg G. A protein sequenator. Eur J Biochem 1967; 1:80–91.
16. Dayhoff MO. Atlas of Protein Sequence and Structure. Washington, DC: Nat Biomed Res Found, 1969–1977.
17. Bernstein, et al. Brookhaven protein data bank. J Mol Biol 1977; 112:535–542.
18. Sanger F, Nicklen S, Coulson AR. DNA sequencing with chain-terminating inhibitors. Proc Natl Acad Sci USA 1977; 74:5463–5467.
19. Maxam AM, Gilbert W. A new method of sequencing DNA. Proc Natl Acad Sci USA 1977; 74:560–564.
20. Needleman SB, Wunsch CD. A general method applicable to the search for similarities in the amino acid sequences of two proteins. J Mol Biol 1970; 48:443–453.
21. Bellman R. Dynamic Programming. Princeton, NJ: Princeton Univ Press, 1957.
22. Levenshtein VI. Binary codes capable of correcting deletions, insertions, and reversals. Cyber Control Theory 1966; 10:707–710.
23. Sellers PH. On the theory and computation of evolutionary distances. SIAM J Appl Math 1974; 26:787–793.
24. Anfisen CB, Haber E, Sela M, White FH. The kinetics of formation of native ribonuclease during oxidation of the reduced polypeptide chain. Proc Natl Acad Sci USA 1961; 47:1309–1314.
25. Chou PY, Fasman GD. Prediction of protein conformation. Biochemistry 1974; 13:222–245.
26. Nagano K. Logical analysis of the mechanism of protein folding. J Mol Biol 1977; 109:235–250.
27. Inglehart J, Nelson PC. On the limitations of automated restriction mapping. Comput Appl Biol Sci 1994; 10(3):249–261.
28. Saiki RK, et al. Primer-directed enzymatic amplification of DNA with a thermostable DNA polymerase. Science 1988; 239:487–491.
29. Geysen HM, Meloen RH, Barteling SJ. Use of peptide synthesis to probe viral antigens for epitopes to a resolution of single amino acid. Proc Natl Acad Sci USA 1984; 81:3998–4002.
30. Houghten RA. General method for the rapid solid-phase synthesis of large numbers of peptides: specificity of antigen-antibody interaction at the level of individual amino acids. Proc Natl Acad Sci USA 1985; 82:5131–5135.
31. Bowie JU, Reidhaar-Olson JF, Lim WA, Sauer RT. Deciphering the message in protein sequences: tolerance to amino acid substitutions. Science 1990; 247(4948):1306–1310.
32. Tuerk C, Gold L. Systematic evoluton of ligands by exponential enrichment: RNA ligands to bacteriophage T4 DNA polymerase. Science 1990; 249:505–510.
33. Pearson WR, Lipman DJ. Improved tools for biological sequence comparison. Proc Natl Acad Sci USA 1988; 85:2444–2448.
34. Altschul SF, Gish W, Miller W, Myers EW, Lipman DJ. A basic local alignment search tool. J Mol Biol 1990; 215:403–410.

35. Sprinzl M, Steegborn C, Huebel F, Steinberg S. Compilation of tRNA sequences and sequences of tRNA genes. Nucleic Acids Res 1996; 24:68–72.

36. Bairoch A. PROSITE: a dictionary of sites and patterns in proteins. Nucleic Acids Res 1991; 19(suppl):2241–2245.

37. Brenner S. The molecular evolution of genes and proteins: a tale of two serines. Nature 1988; 334:528–530.

38. van Bockstaele F. Sequence representation. Biochimie 1985; 67(5):509–516.

39. Staden R. Computer methods to locate signals in nucleic acid sequences. Nucleic Acids Res 1984; 12:505–519.

40. Misraji E, Ninio J. Graphical coding of nucleic acid sequences. Biochimie 1985; 67(5):445–448.

41. Arques DG, Michel CJ. Periodicities in introns. Nucleic Acids Res 1987; 15(18):7581–7592.

42. Ohno S, Ohno M. The all pervasive principle of repetitious recurrence governs not only coding sequence construction but also human endeavor in musical composition. Immunogenetics 1986; 24:71–78.

43. Smith TF, Waterman MS. Identification of common molecular subsequences. J Mol Biol 1981; 147:195–197.

44. Goad WB, Kanehisa MI. Pattern recognition in nucleic acid sequences. I. A general method for finding local homologies and symmetries. Nucleic Acids Res 1982; 10(1):247–263.

44a. Staden R. Methods to define and locate patterns of motifs in sequences. Comput Appl Biosci 1988; 4:53–60.

45. Taylor WR. The classification of amino acid conservation. J Theor Biol 1988; 119:205–218.

46. Gribskov M, McLachlan AD, Eisenberg D. Profile analysis: detection of distantly related protein. Proc Natl Acad Sci USA 1987; 84:4355–4358.

47. Gelfand MS. Prediction of function in DNA sequence analysis. J Comp Biol 1995; 2:87–117.

48. Nakai K, Kidera A, Kanehisa M. Cluster analysis of amino acid indices for prediction of protein structure and function. Protein Eng 1988; 2(2):93–100.

49. Tomii K, Kanehisa M. Analysis of amino acid indices and mutation matrices for sequence comparison and structure prediction of proteins. Protein Eng 1996; 9(1):27–36.

50. Cohen FE, Abarbanel RM, Kuntz ID, Fletterick RJ. Turn prediction in proteins using a pattern matching approach. Biochemistry 1986; 25:266–275.

51. Lathrop RH, Webster TA, Smith TF. ARIADNE: pattern-directed inference and hierarchical abstraction in protein structure recognition. Commun Assoc Comput Machinery 1987; 30:909–992.

52. Gautheret D, Major F, Cedergren R. Pattern searching/alignment with RNA primary and secondary structures: an effective descriptor for tRNA. Comput Appl Biosci 1990; 6(4):325–331.

53. Gotoh S. An improved algorithm for matching biological sequences. J Mol Biol 1982; 162(3):705–708.

54. Cummings, et al. The complete DNA sequence of the mitochondrial genome of *Podospora anserina*. Curr Genet 1990; 17:375–402.

55. Oliver SG, et al. The complete sequence of yeast chromosome. III. Nature 1992; 357:38–46.

56. Blattner FR, Plunkett G III, Bloch CA, Perna NT, Burland V, Riley M, Collado-Vides J, Glasner JD, Rode CK, Mayhew GF, Gregor J, Davis NW, Kirkpatrick HA, Goeden MA, Rose DJ, Mau B, Shao Y. The complete genome sequence of *Escherichia coli* K-12. Science 1997; 277:1453–1474 .

57. Hénaut A, Danchin A. Analysis and prediction from *E. coli* sequences, or *E. coli* in silico in *Escherichia coli* and *Salmonella*. In: Neidhardt F, et al., eds. Cellular and Molecular Biology. Vol. 2. Washington, DC: Am Soc Microbiol, 1996:2047–2066.

58. Adams MD, Kelley JM, Gocayne JD, Dubnick M, Polymeropoulos MH, Xiao H, Merril CR, Wu A, Olde B, Moreno RF, et al. Complementary DNA sequencing: expressed sequence tags and human genome project. Science 1991; 252:1651–1656.

59. Trifonov EN, Brendel V. GNOMIC—Dictionary of Genetic Codes. Rehovot, Philadelpia, PA: Balaban Publishers, 1986.

60. Attwood TK, Beck ME, Bleasby AJ, Parry-Smith DJ. PRINTS—a database of protein motif fingerprints. Nucleic Acids Res 1994; 22(17):3590–3596.

61. Li W. The study of correlation structures of DNA sequences: a critical review. Comput Chem 1997; 21(4):257–271.

62. Salamon P, Konopka AK. A maximum entropy principle for the distribution of local complexity in naturally occurring nucleotide sequences. Comput Chem 1992; 16(2):117–124.

63. Wootton JC, Federhen S. Analysis of compositionally biased regions in sequence databases. Methods Enzymol 1996; 266:554–571.

64. Shannon CE. A mathematical theory of communication. Bell Syst Tech J 1948; 27(379–423):623–656.

65. Gatlin LL. Information Theory and the Living System. New York: Columbia Univ Press, 1972.

66. Yockey HP. An application of information theory to the central dogma and the sequence hypothesis. J Theor Biol 1974; 46:369–406.

67. Konopka AK. Theory of degenerate coding and informational parameters of protein coding genes. Biochimie 1985; 67(5):455–468.

68. Miloslavljevic A, Jurka J. Discovering simple DNA sequences by the algorithmic significance method. Comput Appl Biosci 1993; 9(4):407–411.

69. Rivals E, Dauchet M, Delahaye J-P, Delgrange O. Compression and genetic sequence analysis. Biochimie 1996; 78(5):315–322.

70. Allison L, Edgoose T, Dix TI. Compression of strings with approximate repeats. Proc ISMB'98, Montréal, Canada, 1998:129–132.

71. Aston C, Bud Mishra B, Schwartz DC. Optical mapping and its potential for large-scale sequencing projects. Trends Biotechnol 1999; 17(7):297–302.

72. Rost B, O'Donoghue S. Sisyphus and prediction of protein structure. Comput Appl Biosci 1997; 13(4):345–356.

73. Smith TF, Lo Conte L, Bienkowska J, Gaitatzes C, Rogers RG Jr, Lathrop R.

Current limitations to protein threading approaches. J Comput Biol 1997; 4(3):217–225.

74. Sippl M, Lackner P, Domingues FS, Prlic A, Malik R, Andreeva A, Wiederstein M. Assessment of the CASP4 fold recognition category. Proteins 2001; (suppl 5):55–67.

75. Giddings MC, Severin J, Westphall M, Wu J, Smith LM. A software system for data analysis in automated DNA sequencing. Genome Res 1998; 8(6):644–665.

76. Sanger F, et al. J Mol Biol 1982; 162(4):729–773.

77. Fraser CM, Fleischmann RD. Strategies for whole microbial genome sequencing and analysis. Electrophoresis 1997; 18:1207–1216.

78. Favello A, Hillier L, Wilson RK. Genomic DNA sequencing methods. Methods Cell Biol 1995; 48:551–569.

79. Weber JL, Myers EW. Human whole-genome shotgun sequencing. Genome Res 1997; 7:401–409.

80. Huang X, Madan A. CAP3: a DNA sequence assembly program. Genome Res 1999; 9:868–877.

81. Green P. 1996. http://bozeman.genome.washington.edu/phrap.docs/phrap.html.

82. Lisacek F, Traini M, Sexton D, Harry J, Wilkins M. Strategies for protein isoform identification from EST sequences and its application to peptide mass fingerprinting. Proteomics 2001; 1:186–193.

83. Fonstein M, Haselkorn R. Physical mapping of bacterial genomes. J Bacteriol 1995; 177(12):3361–3369.

84. Almagor H. A Markov chain analysis of DNA sequences. J Theor Biol 1983; 104:633–645.

85. Garden PW. Markov analysis of viral DNA/RNA sequences. J Theor Biol 1980; 82:679–684.

86. Bernardi G. The isochore organisation of the human genome and its evolutionary history—a review. Gene 1993; 135:57–66.

87. Kozhukin CG, Pevzner PA. Genome inhomogeneity is determined mainly by WW and SS dinucleotides. Comput Appl Biosci 1991; 7:39–49.

88. Blaisdell BE. A prevalent persistent nonrandomness that distinguishes coding and noncoding eukaryotic nuclear DNA sequences. J Mol Evol 1983; 19:122–133.

89. Ohno S. On periodicities governing the construction of genes and proteins. Anim Genet 1988; 19(4):305–316.

90. Trifonov EN. The multiple codes of nucleotide sequences. Bull Math Biol 1989; 51(4):417–432.

91. Borodovsky M, McInich J. GenMark: parallel gene recognition for both DNA strands. Comput Chem 1993; 17(2):123–133.

92. Fickett JW. The gene identification problem: an overview for developers. Comput Chem 1996; 20(1):103–118.

93. Nakamura Y, Gojobori T, Ikemura T. Codon usage tabulated from the international DNA sequence databases. Nucleic Acids Res 1998; 26(1):334.

94. Médigue C, Rouxel T, Vigier P, Hénaut A, Danchin A. Evidence for horizontal gene transfer in *Escherichia coli* speciation. J Mol Biol 1991; 222:851–856.

95. Guigo R, Fickett JW. Distinctive sequence features in protein coding genic non-coding, and intergenic human DNA. J Mol Biol 1995; 253(1):51–60.

96. Oliver J-L, Marin A. A relationship between GC content and coding-sequence length. J Mol Evol 1996; 43(3):216–223.

97. d'Aubenton-Carafa Y, Brody E, Thermes C. Prediction of rho-independent E. coli transcription terminators. J Mol Biol 1990; 216:835–858.

98. Hofacker IL, Fekete M, Flamm C, Huynen MA, Rauscher S, Stolorz PE. Automatic detection of conserved RNA structure elements in complete RNA virus genomes. Nucleic Acids Res 1998; 26(16):3825–3826.

99. Thompson JD, Higgins DG, Gibson TJ. ClustalW: improving the sensivity of progressive sequence multiple alignment through sequence weighing, position-specific gap penalties and weight matrix choice. Nucleic Acids Res 1994; 22:4673–4680.

100. Morgenstern B, Frech K, Dress A, Werner T. DIALIGN: finding local similarities by multiple sequence alignment. Comput Appl Biosci 1998; 14(3):290–294.

101. Bucher P, Karplus K, Moeri N, Hoffman K. A flexible motif search technique based on generalised profiles. Comput Chem 1997; 20(1):3–23.

102. Brazma A, et al. Approaches to the automatic discovery of patterns in biosequences. J Comp Biol 1998; 5(2):279–305.

103. Rigoutsos I, Floratos A, Parida L, Gao Y, Platt D. The emergence of pattern discovery techniques in computational biology. Metab Eng 2000; 2(3):159–177.

104. Rigden DJ. Use of covariance analysis for the prediction of structural domain boundaries from multiple protein sequence alignments. Protein Eng 2002; 15(2):65–77.

105. Zhang MQ, Marr TG. A weight array method for splicing signal analysis. Comput Appl Biosci 1993; 9(5)499–509.

106. Burge C, Karlin S. Prediction of complete gene structures in human genomic DNA. J Mol Biol 1997; 268:78–94.

107. Brunak S, Engelbrecht J, Knudsen S. Prediction of human mRNA donor and acceptor sites from the DNA sequence. J Mol Biol 1991; 220:49–65.

108. Mephu N'Guifo E, Sallantin J. Prediction of primate splice junction gene sequences with a co-operative knowledge acquisition system. Proceedings of the First International Conference of Intelligent Systems for Molecular Biology, Washington DC: AAAI Press, 1993:292–300.

109. Guigo R, Knudsen S, Drake N, Smith T. Prediction of gene structure. J Mol Biol 1992; 226:141–157.

110. McCullough WS, Pitts W. Bull Math Biophys 1947; 9:127–147.

111. Changeux J-P, Courrege P, Danchin A. A theory of the epigenesis of neuronal networks by selective stabilisation of synapses. Proc Natl Acad Sci USA 1973; 70(10):2974–2978.

112. Atlan H. Automata network theories in immunology: their utility and their under determination. Bull Mathem Biol 1989; 51:247–253.

113. Stormo GD, Schneider TD, Gold L, Ehrenfeucht A. Use of the 'Perceptron'

algorithm to distinguish translational initiation sites in *E. coli*. Nucleic Acids Res 1982; 10(9):2997–3011.

114. Rumerlhart DE, Hinton GE, Williams RJ. Learning representations by back-propagating errors. Nature 1986; 323:533–536.

115. Lukashin AV, Anshelevich VV, Amirikyan BR, Gragerov AI, Frank-Kamenetskii MD. Neural network models for promoter recognition. J Biomol Struct Dyn 1989; 6(6):1123–1133.

116. Qian N, Sejnowski TJ. Predicting the secondary structure of globular proteins using neural network models. J Mol Biol 1988; 202(4):865–884.

117. Horton PB, Kanehisa M. An assessment of neural network and statistical approaches for prediction of *E. coli* promoter sites. Nucleic Acids Res 1992; 20:4331–4338.

118. Wu CH. Artificial neural networks for molecular sequence analysis. Comput Chem 1997; 21(4):237–256.

119. Krogh A, Brown M, Mian IS, Sjolander K, Haussler D. Hidden Markov models in computational biology. Applications to protein modeling. J Mol Biol 1994; 235(5):1501–1531.

120. Allison L, Yee CN. Minimum message length encoding and the comparison of macromolecules. Bull Math Biol 1990; 52(3):431–453.

121. Sonnhammer E, Eddy S, Durbin R. Pfam: a comprehensive database of protein domain families based on seed alignment. Proteins 1997; 28:405–420.

122. Winker S, et al. Structure detection through automated covariance search. Comput Appl Biosci 1990; 6(4):365–371.

123. Chiu DKY, Kolodziejczak T. Inferring consensus structure from nucleic acid sequences. Comput Appl Biosci 1991; 7(3):347–352.

124. Eddy SR, Durbin R. RNA sequence analysis using covariance models. Nucleic Acids Res 1994; 22(7):2079–2088.

125. Michel F, Westhof E. Modelling of the three-dimensional architecture of group I catalytic introns based on comparative sequence analysis. J Mol Biol 1990; 216:585–610.

126. Steinberg S, Leclerc F, Cedergren R. Structural rules and conformational compensations in the tRNA L-form. J Mol Biol 1997; 266(2):269–282.

127. Fukami-Kobayashi K, Schreiber DR, Benner SA. Detecting compensatory covariation signals in protein evolution using reconstructed ancestral sequences. J Mol Biol 2002; 319:729–743.

128. Gultyaev AP, van Batenburg FHD, Pleij CWA. The computer simulation of RNA folding pathways using a genetic algorithm. J Mol Biol 1995; 250:37–51.

129. Schmitz M, Steger G. Description of RNA folding by "simulated annealing. J Mol Biol 1996; 255(1):254–266.

130. Culberson JC. On the futility of blind search. Tech Rep TR 96-18. Edmonton, Alberta, Canada: Department of Computing Science, University of Alberta, 1996.

131. Searls DB. Linguistics approaches to biological sequences. Comput Appl Biosci 1997; 13:333–344.

132. Doolittle WF. Phylogenetic classification and the universal tree. Science 1999; 284:2124–2129.
133. Pennisi E. Is it time to uproot the tree of life? Science 1999; 284:1305–1307.
134. Rosinski JA, Atchley WR. Molecular evolution of helix-turn-helix proteins. J Mol Evol 1999; 49(3):301–309.
135. Beaudry AA, Joyce GF. Directed evolution of an RNA enzyme. Science 1992; 257:635–641.
136. Hawley DK, McClure WR. Compilation and analysis of *E. coli* promoter DNA sequences. Nucleic Acids Res 1983; 11:2237–2255.
137. Harley CB, Reynolds RP. Analysis of *E. coli* promoter sequences. Nucleic Acids Res 1987; 15:2343–2361.
138. Sprinzl M, Voderwullbecke T, Hartmenn T. Compilation of tRNA sequences. Nucleic Acids Res 1985; 13:r51–r104.
139. Kabsch W, Sander C. Dictionary of protein secondary structure: pattern recognition of hydrogen-bonded and geometrical features. Biopolymers 1983; 22:2577–2637.
140. Etzold T, Argos P. SRS—an indexing and retrieval tool for flat file data libraries. Comput Appl Biosci 1993; 9:49–57.
141. Hénaut A, Rouxel T, Gleizes A, Moszer I, Danchin A. Uneven distribution of GATC motifs in the *E. coli* chromosome, its plasmids and phages. J Mol Biol 1996; 257:574–585.
142. Wootton JC. Evaluating the effectiveness of sequence analysis algorithms using measures of relevant information. Comput Chem 1997; 21(4):191–202.
143. Dumas J-P, Ninio J. Efficient algorithms for folding and computing nucleic acid sequences. Nucleic Acids Res 1982; 10:197–206.
144. el-Mabrouk N, Lisacek F. Very fast identification of RNA motifs in genomic DNA. Application to tRNA search in the yeast genome. J Mol Biol 1996; 264(1):46–55.
145. Friedland P, Iwasaki Y. The concept and implementation of skeletal plans. J Autom Reasoning 1985; 1:161–208.
146. Karp PD, Riley M, Paley SM, Pellegrini-Toole A, Krummenacker M. EcoCyc: encyclopedia of *Escherichia coli* genes and metabolism. Nucleic Acids Res 1999; 27(1):55–58.
147. Huhns MN. editor (1987) Distributed Artificial Intelligence. San Francisco, CA, USA: Morgan Kaufmann.
148. Dachin A. The Delphic boat or what the genomic texts tell us. Bioinformatics 1998; 14(5):383.
149. Boguski M. Biosequence exegesis. Science 1999; 286:453–455.
150. Rechenmann F. Knowledge bases and computational molecular biology. Proceedings of the 2nd International Conference on Building and Sharing Very Large-Scale Knowledge Bases (KBKS), Enschede (Holland), edited by Nicolaas MarsAmsterdam: IOS press, 1995.
151. Yuh C-H, Bolouri H, Davidson EH. Genomic cis-regulation logic: experimental and computational analysis of a sea urchin gene. Science 1998; 279:1896–1902.

152. Tavazoie S, Hughes JD, Campbell MJ, Cho RJ, Church GM. Systematic determination of genetic network architecture. Nat Genet 1999; 22:281–285.

153. Shulman MJ, Steinberg CM, Westmoreland N. The coding function of nucleotide sequences can be discerned by statistical analysis. J Theor Biol 1981; 88(3):409–420.

154. Konopka AK. Towards mapping functional domains in indiscriminantly sequenced nucleic acids: a computational approach. In: Sarma RH, Sarma MH, eds. Human Genome Initiative and DNA Recombination. Vol. 1. Guiderland, NY: Adenine Press, 1990:113–125.

155. Kleffe J, Hermann K, Vahrson W, Wittig B, Brendel V. Logitlinear models for the prediction of splice sites in plant pre-mRNA sequences. Nucleic Acids Res 1996; 24(23):4709–4718.

156. Solovyev V, Salamov A, Lawrence CB. Predicting internal exons by oligonucleotide composition and discriminant analysis of spliceable open reading frames. Nucleic Acids Res 1994; 22:5156–5163.

157. Snyder E, Stormo GD. Identification of protein coding regions in genomic DNA. J Mol Biol 1995; 248:1–18.

158. Hebsgaard SM, Korning PG, Tolstrup N, Engelbrecht J, Rouze P, Brunak S. Splice site prediction in *A. thaliana* pre-mRNA by combining local and global information. Nucleic Acids Res 1996; 24(17):3439–3452.

159. Brendel V, Kleffe J. Prediction of locally optimal splice sites in plant pre-mRNa with applications to gene identification in *Arabidopsis thaliana* genomic DNA. Nucleic Acids Res 1998; 26(20):4748–4757.

160. Vignal L, Lisacek F, Quinqueton J, d'Aubenton-Carafa Y, Thermes C. A multi-agent system simulating human splice site recognition. Comput Chem 1999; 23:219–231.

161. Guigo R. Computational gene identification: an open problem. Comput Chem, 1997; (4):215–222.

162. Mathé C, Sagot M-F, Schiex T, Rouzé P. Current methods of gene prediction, their strengths and weaknesses. Nucleic Acids Res 2002; 30(19):4103–4117.

162a. Guigo R, Agarwal P, Abril JF, Burset M, Fickett JW. An assessment of gene prediction accuracy in large DNA sequences. Genome Res 2000; 10(10):1631–1642.

162b. Parra G, Agarwal P, Abril JF, Wiehe T, Fickett JW, Guigo R. Comparative gene prediction in human and mouse. Genome Res 2003; 13(1):108–117.

163. Gaasterland T. Fully automated genome analysis that reflects user needs and preferences. A detailed introduction to the MAGPIE system architecture. Biochimie 1996; 78:302–310.

164. Koonin EV, Mushegian AR, Galperin MY, Walker DR. Comparison of archaeal and bacterial genomes: computer analysis of protein sequences predicts novel functions and suggests a chimeric origin for the Archaea Mol Microbiol 1997; 25:619–637.

165. Codani J-J, Comet J-P, Aude J-C, Glémet E, Wozniak A, Risler J-L, Hénaut A, Slonimski PP. Automatic analysis of large-scale pairwise alignments of protein sequences. Methods Microbiol 1999; 28:229–244.

166. Ashburner M, Ball CA, Blake JA, Botstein D, Butler H, Cherry JM, Davis AP, Dolinski K, Dwight SS, Eppig JT, Harris MA, Hill DP, Issel-Tarver L, Kasarskis A, Lewis S, Matese JC, Richardson JE, Ringwald M, Rubin GM, Sherlock G. Gene ontology: tool for the unification of biology. The Gene Ontology Consortium. Nat Genet 2000; 25(1):25–29.

167. Tsoka S, Promponas V, Ouzounis CA. Reproducibility in genome sequence annotation: the Plasmodium falciparum chromosome 2 case. FEBS Lett 1999; 451(3):354–355.

168. Blandin G, Durrens P, Tekaia F, Aigle M, Bolotin-Fukuhara M, Bon E, Casaregola S, de Montigny J, Gaillardin C, Lepingle A, Llorente B, Malpertuy A, Neuveglise C, Ozier-Kalogeropoulos O, Perrin A, Potier S, Souciet J, Talla E, Toffano-Nioche C, Wesolowski-Louvel M, Marck C, Dujon B. Genomic exploration of the hemiascomycetous yeasts: 4. The genome of *Saccharomyces cerevisiae* revisited. FEBS Lett 2000; 487(1):31–36.

169. Wood V, Rutherford KM, Ivens A, Rajandream M-A, Barrell B. A re-annotation of the *Saccharomyces cerevisiae* genome. Comp Funct Genomics 2001; 2(3):143–154.

170. Camus JC, Pryor MJ, Medigue C, Cole ST. Re-annotation of the genome sequence of *Mycobacterium tuberculosis* H37Rv. Microbiology 2002; 148(10): 2967–2973.

171. Liang P, Labedan B, Riley M. Physiological genomics of *Escherichia coli* protein families. Physiol Genomics 2002; 9(1):15–26.

172. Das R, Junker J, Greenbaum D, Gerstein MB. Global perspectives on proteins: comparing genomes in terms of folds, pathways and beyond. Pharmacogenomics J 2001; 1(2):115–125.

173. Saqi M-A, Sternberg MJ. A structural census of metabolic networks for *E. coli*. J Mol Biol 2001; 313(5):1195–1206.

174. Akashi H, Gojobori T. Metabolic efficiency and amino acid composition in the proteomes of *Escherichia coli* and *Bacillus subtilis*. Proc Natl Acad Sci USA 2002; 99(6):3695–3700.

175. Wintjens R, Rooman M. Structural classification of HTH DNA-binding domains and protein-DNA interaction modes. J Mol Biol 1996; 262:294–313.

176. Dodd IB, Egan JB. Improved detection of helix-turn-helix DNA-binding motifs in protein sequences. Nucleic Acids Res 1990; 18:5019–5026.

177. Sagot M-F. (1996) Ressemblance lexicale et structurale entre macromolé-cules—formalisation et approches combinatoires. Thèse de Doctorat, Univer-sité de Marne-la-Vallée

178. Niland P, Huhne R, Muller-Hill B. How AraC interacts specifically with target DNAs. J Mol Biol 1996; 264:667–674.

179. Gallegos M-T, Schleif R, Bairoch A, Hofmann K, Ramos JL. AracC/XylS family of transcriptional regulators. Microbiol Mol Biol Rev 1997; 61:393–410.

180. Hueck CJ. Type III protein secretion systems in bacterial pathogens of animals and plants. Microbiol Mol Biol Rev 1998; 62:379–433.

181. Grundy FJ, Henkin TM. The S box regulon: a new global transcription

termination control system for methionine and cysteine biosynthesis genes in Gram-positive bacteria. Mol Microbiol 1998; 30(4):737–749.

181a. Clamp M, Andrews D, Barker D, Bevan P, Cameron G, Chen Y, Clark L, Cox T, Cuff J, Curwen V, Down T, Durbin R, Eyras E, Gilbert J, Hammond M, Hubbard T, Kasprzyk A, Keefe D, Lehvaslaiho H, Iyer V, Melsopp C, Mongin E, Pettett R, Potter S, Rust A, Schmidt E, Searle S, Slater G, Smith J, Spooner W, Stabenau A, Stalker J, Stupka E, Ureta-Vidal A, Vastrik I, Birney E. Ensembl 2002: accommodating comparative genomics. Nucleic Acids Res 2003; 31(1):38–42.

181b. Gattiker A, Michoud K, Rivoire C, Auchincloss AH, Coudert E, Lima T, Kersey P, Pagni M, Sigrist CJ, Lachaize C, Veuthey AL, Gasteiger E, Bairoch A. Automated annotation of microbial proteomes in SWISS-PROT. Comput Biol Chem 2003; 27(1):49–58.

182. Fichant GA, Burks C. Identifying tRNA genes in genomic DNA sequences. J Mol Biol 1991; 220:659–671.

183. Pavesi A, Conterio F, Bolchi A, Dieci G, Ottonello S. Identification of new eucaryotic tRNA genes in genomic DNA databases by a multistep weight matrix analysis of transcriptional control regions. Nucleic Acids Res 1994; 22(7):1247–1256.

184. Lowe TW, Eddy SR. tRNAScan-SE: a program for improved detection of transfer RNA genes in genomic sequence. Nucleic Acids Res 1997; 25(5):955–964.

185. Lisacek F, Diaz Y, Michel F. Automatic identification of group I intron cores in genomic DNA sequences. J Mol Biol 1994; 235:1206–1217.

186. Dandekar T, Hentze MW. Finding the hairpin in the haystack: searching for RNA motifs. Trends Genet 1995; 11(2):45–50.

187. Danchin A, Gascuel O. Protein export in prokaryotes and eukaryotes: indications of a difference in the mechanism of exportation. J Mol Evol 1986; 24:130–142.

188. Sterner D, Berget S. In vivo recognition of a vertebrate mini-exon as an exon-intron-exon unit. Mol Cell Biol 1993; 13:2677–2687.

189. Nitschké P, Guerdoux-Jamet P, Chiapello H, Faroux G, Hénaut C, Hénaut A, Danchin A. Indigo: a World-Wide-Web review of genomes and gene functions. FEMS Microbiol Rev 1998; 22(4):207–228.

190. Riley M. Systems for categorizing functions of gene products. Curr Opin Struct Biol 1998; 8:388–392.

191. Horton PB, Nakai K. A probabilistic classification system for predicting the cellular localisation sites of proteins. Proc ISMB'97, Halkidiki, Greece, 1996: 109–115.

191a. Nakai K. Protein sorting signals and prediction of subcellular localization. Adv Protein Chem 2000; 54:277–344.

192. Chiapello H, Ollivier E, Devauchelle C, Nitschké P, Risler J-L. Codon usage as a tool to predict the cellular location of eukaryotic ribosomal proteins and aminoacyl-tRNA synthetases. Nucleic Acids Res 1999; 27:2848–2851.

193. Kao CM. Functional genomic technologies: creating new paradigms for fundamental and applied biology. Biotechnol Prog 1999; 15:304–311.

194. Hess KR, Zhang W, Baggerly KA, Stivers DN, Coombes KR. Microarrays: handling the deluge of data and extracting reliable information. Trends Biotechnol 2001; 19(11):463–468.
195. Brazma A, Vilo J. Gene expression data analysis. FEBS Lett 2000; 480(1):17–24.
196. Quackenbush J. Computational analysis of microarray data. Nat Rev Genet 2001; 2(6):418–427.
197. O'Farell PH. J Biol Chem 1975; 250:4007–4021.
198. Bjellqvist B, Ek K, Righetti PG, Gianazza E, Gorg A, Westermeier R, Postel W. Isoelectric focusing in immobilized pH gradients: principle, methodology and some applications. J Biochem Biophys Methods 1982; 6(4):317–339.
199. Beavis RC, Chait BT. Rapid, sensitive analysis of protein mixtures by mass spectrometry. Proc Natl Acad Sci USA 1990; 87(17):6873–6877.
200. Brown RS, Lennon JJ. Mass resolution improvement by incorporation of pulsed ion extraction in a matrix-assisted laser desorption/ionization linear time-of-flight mass spectrometer. Anal Chem 1995 Jul 1; 67(13):1998–2003.
201. Williams KL, Hochstrasser DF. Introduction to the proteome. In: Wilkins MR, Williams KL, Appel RD, Hochstrasser DF, eds. Proteome Research: New Frontiers in Functional Genomics. New York: Springer, 1997:1–12.
202. Wilkins MR, Gasteiger E, Gooley AA, Herbert BR, Molloy MP, Binz PA, Ou K, Sanchez J-C, Bairoch A. High-throughput mass spectrometric discovery of protein post-translational modifications. J Mol Biol 1999; 289(3):645–657.
203. Miklos GL, Maleszka R. Protein functions and biological contexts. Proteomics 2001; 1(2):169–178.
204. Fenyo D. Identifying the proteome: software tools. Curr Opin Biotechnol 2000; 11(4):391–395.
205. Pandey A, Mann M. Proteomics to study genes and genomes. Nature 2000; 405:837–846.
206. Godovac-Zimmermann J, Brown LR. Perspectives for mass spectrometry and functional proteomics. Mass Spectrom Rev 2001; 20:1–57.
207. Pederson T. Proteomics of the nucleolus: more proteins, more functions? Trends Biochem Sci 2002; 27(3):111–112.
208. Miklos GL, Maleszka R. Integrating molecular medicine with functional proteomics: realities and expectations. Proteomics 2001; 1(1):30–41.
209. Xenarios I, Fernandez E, Salwinski L, Duan XJ, Thompson MJ, Marcotte EM, Eisenberg D. DIP: The Database of Interacting Proteins: 2001 update. Nucleic Acids Res 2001; 29(1):239–241.
210. Legrain P, Selig L. Genome-wide protein interaction maps using two-hybrid systems. FEBS Lett 2000; 480(1):32–36.
211. Brent R. Genomic biology. Cell 2000; 100(1):169–183.

9

Computational Aspects of Comparative Genomics

Wojciech Makalowski and Izabela Makalowska
Pennsylvania State University, University Park, PA, U.S.A.

1. INTRODUCTION

Genomic science and technology have brought us to the brink of being able to describe the genetic blueprint and molecular evolutionary history of the human species. But we will not be able to fully interpret these data in isolation. This is one of the reasons the Human Genome Project has, from its inception, included the study of so-called model organisms whose biology, experimental advantages, and smaller, simpler genomes have provided not only important biological insights but also steppingstones for technological development.

2. WHAT IS COMPARATIVE GENOMICS?

Biologists have been comparing organisms from the very beginning of the discipline [1]. Development of evolutionary ideas brought explanation and meaning to such comparisons. Two different organisms share some features because they share common origins. With new tools available and new territories of biological research, comparative biology also is exploring new

343

levels of biological organization. It is not a suprise that with the advent of genomic research, a new field—comparative genomics—appeared.

Although, like any young scientific discipline, comparative genomics is not well defined, it is possible to distinguish two different meanings of this approach. Some researchers refer to comparative genomics when they compare different organisms' genetic or physical maps; others define comparative genomics as the large-scale or even whole-genome comparison of the DNA and genes of at least two organisms. Here, we will use the term comparative genomics in the latter sense.

The completion of genomic sequences for multiple prokaryotes and yeast has provided a wealth of information, and the value of comparative analysis of coding sequences from distantly related organisms (e.g., nematode and human) is beyond dispute. Nevertheless, there are limitations to functional inferences based on interspecies comparison of anciently diverged coding sequences. Furthermore, noncoding regions are generally not amenable to comparative analyses across such vast evolutionary distances because sequence divergence is simply too great. Thus it is necessary to study more closely related organisms in order to detect and interpret the conservation of regulatory, noncoding sequences [2,3].

3. EVOLUTIONARY BASIS OF COMPARATIVE GENOMICS

As mentioned in Sec. 2, comparative genomics has its basis in evolutionary biology. We are able to compare and interpret genes because they share some evolutionary history. Some genes may have originated in the past from the same gene, i.e., they share a common ancestor. In fact, most genes share a common ancestor in the recent or more distant past. Such a sequence could be called "molecular Adam." Most probably there was no single molecular Adam existing in the past but rather a group, a whole tribe, of molecular ancestors. Unfortunately, we are not able to reconstruct such a deep molecular evolutionary history. This is simply because at a certain point two sequences mutate (diverge) beyond any similarity recognition, especially if such a divergence is associated with a change in the function of some proteins. From the comparative genomics point of view there are two significant evolutionary events: gene duplication and speciation. Both events lead to proliferation of genes but in different ways. A gene duplication event results in an increased number of genes in a given population or species. However, speciation increases the gene number automatically by increasing the number of species. In the first process, usually only a small number of genes are involved (although exceptions are known); in the latter, whole sets of genes are proliferated.

These two processes are illustrated in Figure 1. Consider a single gene that exists at time t_0; let us call it an "ancestral gene." At time t_1, the first evolutionary event, gene duplication, occurs. As a result we have two related genes that originated in an "ancestral gene," gene A and gene B. At time t_2, another evolutionary event occurs; this time it is speciation. As a result two descendant copies of the ancestral gene exist in two related species: species 1 and species 2. Additionally, there's another duplication of gene A in the lineage leading to species 1. Therefore, we now have five related genes: A1, A3 and B1 in species 1, and A2 and B2 in species 2 (see Fig. 1). All five genes have one thing in common: They originated in an ancestral gene. In evolutionary biology, two features that share a common history (ancestry) are said to be homologous. The same is true at the molecular level: If two genes share a common evolutionary ancestry, they are said to be homologous. It is convenient and desirable to further distinguish paralogous and orthologous genes among the homologous ones. Two genes are orthologous if their last common ancestor existed at the time of speciation. Two genes are paralogous if their last common ancestor existed at the time of gene duplication. Therefore in our example in Figure 1, at present all five genes A1, A2, A3, B1, and B2 are homologous. Furthermore, genes A1 and A2 should be called orthologous because their last common ancestor existed at time t_2 when speciation into species 1 and 2 occurred. Similarly, genes B1 and B2 are orthologous because their last common ancestor existed at speciation time. Gene A1 has four homologs: A2, A3, B1, and B2. Interestingly, gene A2 is orthologous to both A1 and A3 even if A1 and A3 are paralogous in relation to each other. Genes B1 and B2 are paralogous to all A genes because their last common ancestor existed at time t_1, which marks the ancestral gene duplication event. Similarly, we can define relationships between all five hypothetical

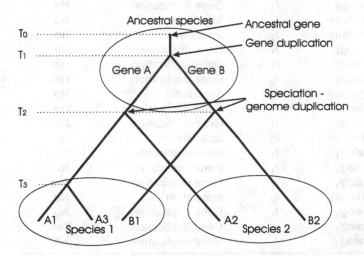

FIGURE 1 Schematic representation of gene evolution. See text for details.

genes at the "present" time (see Table 1). Note that the relation of orthology can be assigned only to genes present in different organisms; paralogy and homology relations apply to genes in either the same or different species.

3.1. Phylogenetic Approach to Data Selection

The evolutionary definition of homologous, especially orthologous, genes suggests that the phylogenetic approach, i.e., plotting phylogenetic trees to infer a relationship between investigated genes, is the only rigorous approach to the selection of such genes. This approach is the only one that allows one to distinguish between paralogous and orthologous genes with high confidence. The procedure usually starts with gene sequence clustering using fast algorithms such as FASTA [4] or BLAST [5] followed by multiple sequence

TABLE 1 Relationships Between Different Homologous Genes Presented in Figure 1

Compared genes	Relationship	Time of last common ancestor	Evolutionary event at the time of last common anc.	Present in the same species?
A,B	Paralogy	t_1	Gene duplication	Yes
A1,A2	Orthology	t_2	Speciationpe	No
A1,A3	Paralogy	t_3	Gene duplicationpe	Yes
A1,B1	Paralogy	t_1	Gene duplication	Yes
A1,B2	Paralogy	t_1	Gene duplication	No
A2,A1	Orthology	t_2	Speciation	No
A2,A3	Orthology	t_2	Speciation	No
A2,B1	Paralogy	t_1	Gene duplication	No
A2,B2	Paralogy	t_1	Gene duplication	Yes
A3,A1	Paralogy	t_3	Gene duplication	Yes
A3,A2	Orthology	t_2	Speciation	No
A3,B1	Paralogy	t_1	Gene duplication	Yes
A3,B2	Paralogy	t_1	Gene duplication	No
B1,B2	Orthology	t_2	Speciation	No
B1,A1	Paralogy	t_1	Gene duplication	Yes
B1,A2	Paralogy	t_1	Gene duplication	No
B1,A3	Paralogy	t_1	Gene duplication	Yes
B2,B1	Orthology	t_2	Speciation	No
B2,A1	Paralogy	t_1	Gene duplication	No
B2,A2	Paralogy	t_1	Gene duplication	Yes
B2,A3	Paralogy	t_1	Gene duplication	No

alignment and phylogenetic tree inference. Next, orthologous genes are selected by looking at their phylogenetic relationships. Obviously, this approach has strong limitations. First of all, it is very time-consuming.

Two steps are extremely slow: phylogenetic tree inference and the selection of orthologous gene groups. Additionally, this method requires data from as many organisms as possible, including outgroup organism data. An outgroup represents an organism that is known to have separated from the analyzed group during an early evolutionary stage. An outgroup organism should be as closely related as possible to the analyzed group of organisms but not actually belong to that group. For example, for mammals, bird and reptile sequences are good outgroups, amphibian and fish sequences not quite as good, and invertebrate genes are phylogenetically too distant to serve as a good outgroup. On top of that, the described method passes on all the theoretical and practical problems of multiple alignments and phylogenetic tree inference. These problems are beyond the scope of the current discussion.

The phylogenetic approach has been successfully applied many times as a means of selecting orthologs, and one of the resources, the HOVERGEN database, is described in detail, in Sec. 4.1.1.

3.2. Heuristic Approach to Data Selection

Because of the problems described above, several heuristic approaches have been developed for both when the whole set of genes for an analyzed genome are known and when only part of the genome is known.

3.2.1. Complete Genome Case

In the case when the complete genome is known, the procedure is based on the simple notion that any group of proteins from distant genomes that are more similar to each other than they are to any other proteins from the same genomes are most likely to belong to an orthologous family. The procedure involves the following steps.

1. All-against-all protein sequence comparison using a fast search program.
2. Selection of the reciprocal best hits between analyzed genomes. For example, if gene A in organism O1 is the most similar to gene B from organism O2, then the reciprocal relation also has to be true; i.e., gene B in organism O2 has to be the most similar to gene A.
3. Combining all the selected pairs of genes into families of orthologous genes.

This approach has been implemented in clusters of orthologous groups of proteins (COGs), which will be described in detail in Sec. 4.1.2 [6].

3.2.2. Partial Genome Case

In a case when the whole proteome is not available, the approach described above cannot be directly applied. To understand the possible problems encountered in dealing with incomplete data, consider again the evolution of a single gene (Fig. 2a). At present, five descendants of an ancestral gene exist (Fig. 2b). Let us assume that information on only two out of the five genes is present in a database (Fig. 2c). Based on this information, only an incomplete gene history can be plotted (Fig. 2d), from which it is impossible to conclude whether the last common ancestor of genes A1 and B2 existed at the speciation or gene duplication time. A gene similarity score is not very helpful here, because different genes evolve at different rates, and similarity is not a good factor for orthology–paralogy discrimination [2]. Therefore, in the case presented in Figure 2 it is very easy to misinterpret data and describe genes A1 and B2 as orthologs. In the case of incomplete genomes, additional information is desired.

Wheelan et al. [7] developed a set of rules to increase the reliability of orthology assignment. In the comparison of human and *Caenorhabditis elegans*, they started with BLAST searches of 1800 human proteins against a nematode database. To avoid assignments based on protein domains only, the "70%" rule was applied, i.e., the initial BLAST alignment had to cover at least 70% of both query and subject sequence. Interestingly, it is not always the top alignment that meets this criterion; sometimes local (domain) alignment gives a slightly higher score in paralogous protein comparison than more global alignment of orthologous proteins. In the next step, selected "best" nematode sequences were used as queries in BLASTp searches against a vertebrate protein data set. This step was based on the assumption that the distance between a human and *C. elegans* protein should be equal (with some small variation) to that of any vertebrate ortholog, e.g., identity between human and nematode orthologs is equal to identity between the same nematode protein and its frog ortholog and is equal to identity between the same nematode protein and its chicken ortholog, and so on. In this step the 10% rule is applied, i.e., only those human–worm sequence pairs for which the BLAST score lies within 10% of the best score for all nematode–vertebrate alignments are considered to be orthologous. This conservative approach leaves some orthologous pairs undiscovered but ensures that no paralogous gene pairs are assigned as orthologs. Using this method, Wheelan et al. compiled 819 human–worm orthologous gene pairs out of 1880 human proteins surveyed.

3.3. Large-Scale Sequence Alignment

There are two aspects of large-scale sequence alignment. First, how to deal with a large number of relatively short and consistent sequences; for example, we have to align a number of proteins or mRNA sequences that are already

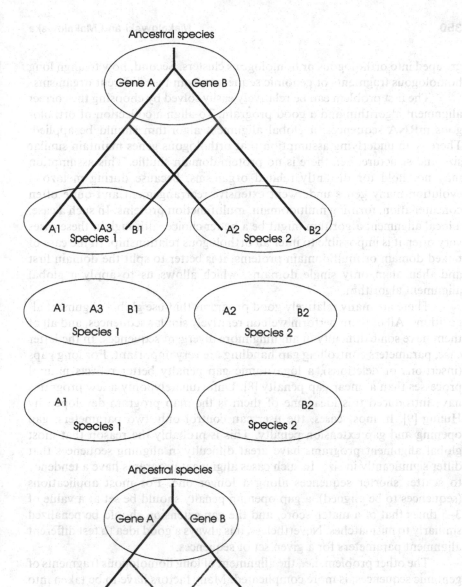

FIGURE 2 The effect of incomplete information on ortholog assignment. See text for details.

grouped into orthologous or homologous clusters. Second, how to align long homologous fragments of genomic sequences from two different organisms.

The first problem can be relatively easily solved by choosing the correct alignment algorithm and a good program. To align a collection of orthologous mRNA sequences, a global alignment algorithm should be applied. There is an underlying assumption that orthologous genes maintain similar size and structure, i.e., there is no protein domain shuffle. This assumption may not hold for distantly related organisms, because during metazoan evolution many genes underwent extensive rearrangement and quite often concatenation, forming multidomain, multifunction proteins. In such a case, a local alignment algorithm might be a better choice, although in these cases very often it is impossible to infer an orthologous relationship. In the case of mixed domain or multidomain proteins, it is better to split the domain first and then align only single domains, which allows us to apply a global alignment algorithm.

There are many relatively good programs that use global alignment algorithms. All of them perform well on relatively similar sequences, and all of them have some difficulty in aligning more divergent sequences. In the latter case, parameters controlling gap handling are very important. For long gaps (insertions or deletions), a logarithmic gap penalty better reflects natural processes than a linear gap penalty [8]. Unfortunately, only a few programs have introduced this idea; one of them is the map program developed by Huang [9]. In most cases, the user can control only two parameters: gap opening and gap extension penalty. This is probably the reason that most global alignment programs have great difficulty in aligning sequences that differ significantly in size. In such cases alignment programs have a tendency to scatter shorter sequences along a longer one. For most applications (sequences to be aligned), a gap opening penalty should be set at a value of 3–5 times that of a match score, and the gap extension should be penalized similarly to mismatches. Nevertheless, it is always a good idea to test different alignment parameters for a given set of sequences.

The other problem, i.e., the alignment of long homologous fragments of genomic sequences, is more complicated. Many factors have to be taken into account. First, the homology has to be recognized. The best markers in this case are probably genes coding for orthologous proteins. Many genomes are occupied by different classes of repetitive elements, which can make up to 50% of a genome [10]. Some of the independent elements can share a significant sequence similarity; they share a remote ancestry, but their insertions are independent events. For example, two mammalian short interspersed elements (SINEs), primate Alu and rodent B1, originated in the same 7SL gene [11] and share extensive sequence similarity. As the most abundant repeats in their respectful lineages, they are very often present in the syntenic locations on primate and rodent chromosomes, but they were inserted there during independent retropositions. Therefore, it is advisable to mask repetitive elements before aligning syntenic sequences. The two most widely used programs for masking repeats are RepatMasker, written by Arian Smit (A. F. A. Smit and P. Green, unpublished) and Jerzy Jurka's CENSOR [12]. Both programs work in a similar way. The user submits a sequence in FASTA format and in return gets several files, among which the most important is a FASTA file with a sequence in which repeats and very often low-complexity regions are replaced by stretches of letters not used in nucleic acid sequence notation, e.g., a series of N's. Both programs also provide additional information such as names of repeats found, their position in the analyzed sequence, alignment of a relevant sequence fragment with a repeat consensus sequence, etc. This additional information is sometimes used in other applications; see, for example, the description of PipMaker in Sec. 4.3.

4. TOOLS FOR COMPARATIVE GENOMICS

4.1. Data Selection Tools

As mentioned above, data selection is a crucial step in gene comparison. With the amount of sequence data growing exponentially, the correct assignment of orthologous or even homologous sequences might be a difficult and time-consuming task. Because sequence identity distributions for paralogous and orthologous genes of any two species overlap significantly on both the protein and DNA levels, simple sequence identity is not a good discriminatory factor for these two types of homologous genes. To ensure pairing of correct genes, different approaches have to be taken. In the following pages we discuss two tools for orthologous sequence selection. One is an example of a formal (phylogenetic) approach to data selection, and the other applies a heuristic approach to the problem.

4.1.1. Homologous Vertebrate Genes Database (HOVERGEN)

HOVERGEN is a database of homologous vertebrate genes, structured under the ACNUC sequence database management system [13]. It allows one to select sets of homologous genes among vertebrate species and to visualize multiple alignments and phylogenetic trees. The database itself contains all vertebrate sequences from GenBank (except ESTs), with some data corrected, clarified, or completed. Homologous coding sequences have been classified into gene families, and protein multiple alignments and phylogenetic trees have been computed for each family. The database is updated every four months. As of April 2000, HOVERGEN contained information on 8626 gene families. HOVERGEN data and software are freely available through anonymous ftp at Universite Claude Bernard, Lyon, France (ftp://pbil.univ-lyon1.fr/pub/hovergen) or at NCBI, Bethesda, MD (ftp://ncbi.nlm.nih.gov/repository/hovergen).

A dedicated graphical interface has been developed to visualize and edit trees. Genes are displayed in color according to their taxonomy. Users have direct access to all information attached to sequences or to multiple alignments simply by clicking on the gene. A screen shot of the main window is presented in Figure 3. This interface was written in C using the Simple User Interface Toolkit (SUIT), which is compatible with many UNIX systems. To run this interface, one needs to use an Xwindows-based computer (either a UNIX computer or a Macintosh or Windows microcomputer with an emulator). Currently, HOVERGEN is available for the following operating systems: Sun workstation using SunOS or Solaris, IBM RS6000, SGI IRIX 5.3, and DEC Alpha.

Advantages of HOVERGEN. The biggest advantage of HOVERGEN is its precomputed data. GenBank records for each sequence are enriched by adding new features such as GC% content or partialness of coding sequences and more consistent labeling of introns, exons, and untranslated regions. Precomputed multiple alignments are stored, so access to them is very fast. It is important to remember that those alignments, as well as phylogenetic trees based on them, are produced in a completely automatic manner and are not manually corrected. Therefore, both alignments and trees should be treated with reservation, and in some cases manual correction may be required.

One of the biggest strengths of HOVERGEN is its graphical interface. Once a gene family has been selected, most of the information is "just a mouse click away." A user can manipulate a phylogenetic tree easily by selecting an outgroup, magnifying a tree, or selecting a subtree (Fig. 3). Sequences for a given systematic group of organisms are color-coded, and the depth of the group can be easily changed by assigning different colors to a single species or to larger groups up to orders. This feature is very useful during the sequence selection process. For example, if one is interested in the selection of human–mouse orthologous gene pairs, a single species may be color-coded. In our example, human sequences are red and mouse sequences are green. If one is

FIGURE 3 A "screen shot" from HOVERGEN. A phylogenetic tree occupies central spaces of the display with a "GenBank information/Protein alignment" window below it. Option controls and information windows are grouped on the right side of the display.

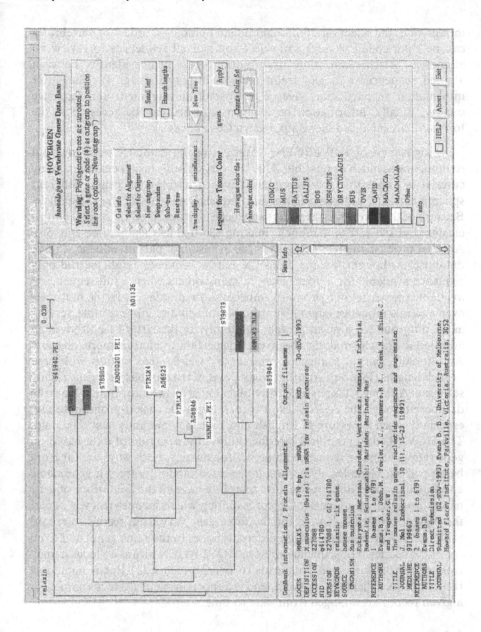

interested in the selection of all mammalian homologs, orders of vertebrates can be color-coded. A user can select some or all sequences to review an alignment. At the same time, information on sequence identity between aligned sequences and gap frequencies is displayed (Fig. 4). The latter information, in particular, gives a glimpse of an alignment and sequence quality. Finally, information on a given sequence can be accessed in the format of an enhanced GenBank record (see above).

Limitations of HOVERGEN. Authors of the database tried to avoid redundancy as much as possible. This is a very difficult task, because in many cases it is impossible to distinguish between paralogous genes, different splicing variants, different alleles, and sequencing errors of the same allele. Because of that, in some cases the same gene is represented by several entries clustered at the end of a branch, making the tree less legible (see Fig. 5a). Another problem in HOVERGEN is the inclusion of partial sequences in the database. Although in some cases the partial sequence might be the only sequence available for a given gene, in many others when a full sequence is available the partial one produces unnecessary redundancy. In the first case, even a partial sequence might give valuable information, for example, serving as an outgroup for a given protein subfamily. In the latter case, it gives an additional branch, making the overall gene family picture more obscure. Additionally, partial sequences produce a "weird" tree section with zero branch lengths (Fig. 5b). Another weak point of the database is a gene family selection. Although it is possible to extract almost all the information from the database through the ACNUC software, it is not the easiest program to use, especially when a mixture of English and French commands has to be used. The HOVERGEN graphical interface allows the selection of data based on gene family name or sequence name. Unfortunately, both ways are far from being perfect. Gene family names are selected randomly from a gene name annotated in a GenBank record and therefore have all the limitations of gene names provided in a GenBank record, among which inconsistency is one of the biggest problems. Names for the same gene vary from record to record (different research groups assign different names to the same gene) and from species to species. The sequence name in HOVERGEN means a "locus name" from a GenBank record. This leads to two problems: (1) Locus name is a rather historical field in GenBank, and is almost obsolete; (2) locus names for a given record are changed from time to time.

FIGURE 4   Sequence alignment in HOVERGEN. The alignment of two groups of paralogous genes of GABA-A receptor (subunit 1 and beta-3) is presented. Note information on sequence divergence and gap frequency in the "Alignments" interface window.

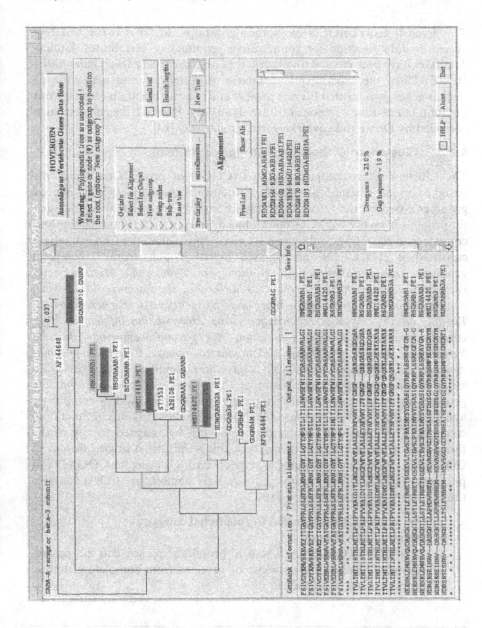

Even if HOVERGEN has some limitations, so far it is the best resource for data selection for comparative genomics of vertebrates. Similar databases have been created for other sets of organisms on a less regular basis. HOVERCEL, the experimental database of vertebrate and nematode *C. elegans* genes, was created for one project and is available upon request from Laurent Duret (duret@biomserv.univ-lyon1.fr). Bacterial genes were also compiled by Duret and coworkers in a similar way [14].

4.1.2. Clusters of Orthologous Groups of Proteins

The Clusters of Orthologous Groups (COGs) of proteins database was designed as an attempt to classify proteins from completely sequenced genomes based on their orthological relationship. The original set included the proteins from five bacterial, one archaeal, and one eukaryotic genome and consisted of 720 COGs; subsequently, as of April 2000, the COG database consisted of 2111 COGs and included 26,919 proteins from 21 complete genomes: *Archaeoglobus fulgidus, Methanococcus jannaschii, Methanobacterium thermoautotrophicum, Pyrococcus horikoshii, Saccharomyces cerevisiae, Aquifex aeolicus, Thermotoga maritima, Synechocystis, Escherichia coli, Bacillus subtilis, Mycobacterium tuberculosis, Haemophilus influenzae, Helicobacter pylori, Mycoplasma genitalium, Mycoplasma pneumoniae, Borrelia burgdorferi, Treponema pallidum, Chlamydia trachomatis, Chlamydia pneumoniae,* and *Rickettsia prowazekii.* COGs have been identified on the basis of an all-against-all sequence comparison of the proteins encoded in complete genomes using the gapped BLAST program after masking low-complexity and predicted coiled-coil regions. The COGs were classified into 17 functional categories [15]. Some of the proteins could not be assigned to any of the functional categories. As a matter of fact, this is the largest single category of COGs. New proteins are assigned to respective COGs using the COGNITOR program. The COG web site (http://www.ncbi.nlm.nih.gov/COG) contains the following principal types of data:

> A list of all COGs organized by functional category
> Individual COG pages
> The COGNITOR page, where a protein sequence can be pasted, searched against the database of proteins from complete genomes, and assigned to a COG

FIGURE 5 Unusual phylogenetic trees in HOVERGEN. (a) Redundant GenBank records cause decrease of tree legibility; (b) inclusion of partial sequences leads to frequent "zero length" branches as in the example of the hexokinase II cluster (AF148513 and HUMHK22 represent full length and partial sequence of the same human protein resulting in "paralogous genes" like tree topology).

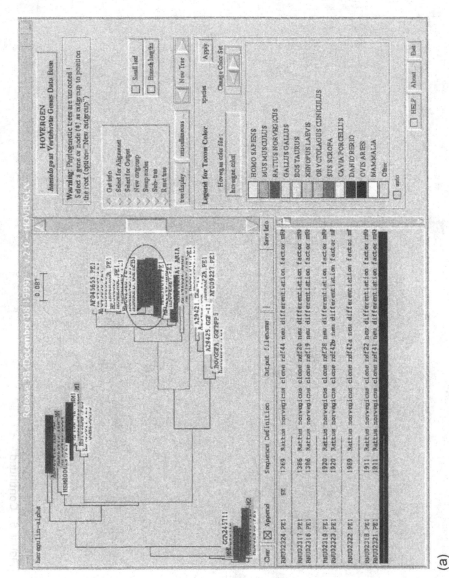

(a)

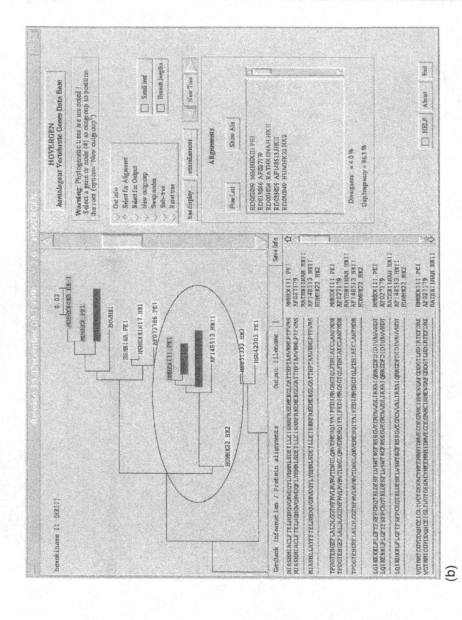

(b)

FIGURE 5 Continued

A phylogenetic pattern search tool
A matrix of co-occurrence of genomes in COGs

COG pages show the respective phylogenetic pattern and are hyperlinked to

Pictorial representations of BLAST search outputs for each member of the COG, including links to the respective GenBank and Entrez-Genomes entries
A multiple alignment of the COG members produced automatically using the ClustalW program
A cluster dendrogram generated by using the BLAST scores as the measure of similarity between proteins

Limitations of COGs. The Clusters of Orthologous Groups of proteins database has some serious limitations. The procedure for its use is limited to complete and rather simple genomes, in which duplicated genes are rather exceptions. It is not clear how it would work for more complicated genomes such as those of *Drosophila* or humans. The *C. elegans* genome was deciphered in 1998 [16], yet this information has not been included in the COG database so far. Another potential problem stems paradoxically from the ease of use of the COGNITOR program. The main goal of the program is to assign a new protein to the relevant COG in the hope that it will be possible to infer the unknown protein function based on the function of known members of the COG. The problem is that function is not always preserved during protein evolution; for example, the lysozyme gene changed its function to that of lactalbumin during mammalian evolution [17,18]. Inferring gene function based only on protein similarity may lead to erroneous prediction (compare Refs. 19 and 20). To deal with this problem some sophisticated methodologies have been proposed, for example, *phylogenomics* [21].

4.2. Alignment Tools

As mentioned earlier, two kinds of large data sets are used in comparative genomics: (1) large sets of relatively short sequences representing genes, proteins, or mRNAs and (2) large segments of syntenic regions of compared genomes (for example, the whole locus of the MHC gene family in the mouse and human genomes [22]. These two types of data require slightly different alignment tools. For the first type of data, almost any good multiple alignment software will work well, but so-called command line software will be more convenient than a "menu"-driven program. This is because it is very tedious to enter the same alignment parameters over and over again when, let's say, we want to compare several hundred or even thousands of homologous genes. In most cases one wants to align all the gene families using the same alignment parameters (mismatch penalties, gap opening and extension penalty, etc.); it is also less error-prone if we define those parameters once and not key them for each sequence set separately. Command line programs are common for the UNIX operating system, which is why this system is so popular among computational biologists.

Several good programs are available on the market right now. Julie Thomson and coworkers critically evaluated some of them using a set of distantly related proteins [23]. It would be useful to have a similar benchmark test for nucleotide sequences because they are more commonly used in comparative genomic studies. Among well-performing programs a few are worth mentioning here. ClustalW is one of the most widely used and is based on the idea of multiple progressive alignment [24]. The program calculates a series of pairwise alignments, comparing each sequence to every other sequence, one at a time. Based on these comparisons a distance matrix is calculated that is used in computing the phylogenetic tree. This tree is used as a guide for the multiple alignment starting with the closest sequence pair and adding more distant sequences to the alignment in the order suggested by the tree. Although the menu-driven version of ClustalW is the most popular, a command line version of the program enables the analysis of a large number of gene families more automatically. Figure 6 presents a simple UNIX shell script to run ClustalW automatically on a number of sequence data sets. Some reliable multiple alignment programs and the web sites at which they can be found are listed in the Appendix.

Alignment of long stretches of syntenic regions is more complicated. In this case regions of significant similarities are very often interlaced with no homologous regions such as different retroposons or by homologous regions mutated eyond recognition. Therefore local alignment programs tend to perform better in this situation. Among these, dot matrix programs are very convenient because they can detect and present very clearly regions of duplication, inversions, and recombination. Because transposable elements and other repeats may further complicate the situation, it is recommended to mask them out before performing an alignment, owing to the high similarity

```
for $ in 'list' do
begin
        clustalw $ parameters
done
```

FIGURE 6 UNIX shell script to execute ClustalW program on number of sequence data sets. Names of files are stored in the list file. Script reads a name of each file and then executes ClustalW program with the current file name as an input file.

of related but not orthologous elements. In most cases, the best solution would be to mask repeats "on the fly" during a search for local similarities and unmask them in the process of alignment extension. Another challenge for this kind of alignment is a visualization of sequence comparison results. For very long sequences, the graphical representation of an alignment is more convenient than the actual alignment of nucleic acid or protein residues.

4.3. Visualization Tools

4.3.1. DOTTER

As mentioned above, a simple yet powerful approach for a two-sequence comparison is a dot matrix algorithm. This allows the user to see the whole sequence alignment at once as a graphic box. An excellent program based on this idea is DOTTER, written by Sonhammer and Durbin [25]. To generate a plot, the program first computes a two-dimentional matrix of scores of all pairwise residue comparisons. To increase the signal-to-noise ratio of a plot, a window is stepped along the diagonals, which assigns a new score calculated by averaging all the points within the window. Each point in the matrix now has a value in the range of 0–255 and corresponds to a gray-scale dot that can be set with the mouse. The Greymap tool provides two thresholds, which can be set by the user. Scores above the maximum are displayed as black, whereas scores below the minimum are white; all scores between these two thresholds are plotted as a gray-scale tones. The Greymap tool allows the dotplot thresholds to be changed dynamically to help find an optimal signal-to-noise ratio and optimal graph resolution. An example of the DOTTER output is presented in Figure 7.

4.3.2. PipMaker

PipMaker computes alignments of similar regions in two DNA sequences [26]. The resulting alignments are summarized with a "percent identity plot," or "pip" for short. PipMaker compares the first and second sequences. Alignments are plotted according to the position in the first sequence file. To generate a pip, PipMaker requires four user-supplied files:

1. First sequence data file. A FASTA file containing the first sequence, with nucleotides given as capital letters:

    One-line header (sequence description)
    ACGTACGTACGTCGTACGTACGTAGTACGTACGTAC-
    TACGTACGTACG

    The maximum length is 2 million nucleotides.

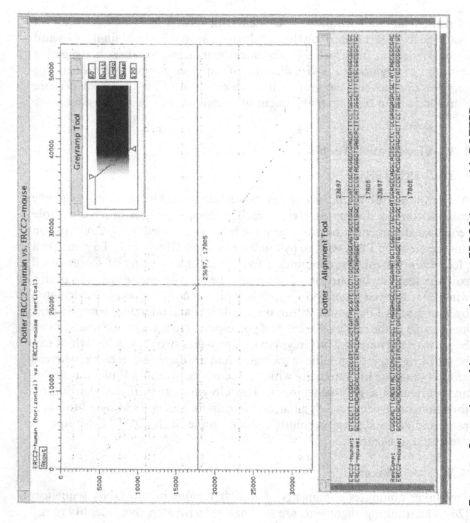

FIGURE 7   Comparison of human and mouse ERCC2 locus with DOTTER.

2. First sequence mask file as produced by RepeatMasker, e.g., a file with "out" extension that contains entries such as

> 4135.60.00.0 HUMAN 1 54 (92195) C La SINE/La (238) 62
> 9 SINE/La (238) 62

3. Exons file for the first sequence. An optional text file providing the positions of transcriptional units in the first sequence. The directionality of a gene (< or >), its start and end positions, and its name should be on one line, followed by separate lines specifying the start positions of each exon as in the following example:

> 100 800 My First Gene
100 200
300 400
600 800
< 1000 2000 My Second Gene
1100 1900
1000 1200
1800 2000

4. Second sequence data file. A FASTA file containing the second sequence. The maximum allowable length of this sequence is 2 Mb.

PipMaker produces three files as output.

1. The percent identity plot. The pip consists of rows that show sequence conservation and features along segments of the first sequence. Each short horizontal line inside the large box corresponds to a section of an alignment bounded by successive gaps. Different types of repetitive elements, annotated genes and exons, and CpG islands associated with the first sequences are visualized with different icons above the alignment box. An example of a pip is presented in Figure 8.
2. A text file with a tabulated form of the alignments. This information is useful for precisely identifying sequence positions corresponding to conserved regions indicated in the pip. An example of such a file is

2870-2926 ↔ 2281-2337 68% (57 nt)
3117-3128 ↔ 2500-2511 83% (12 nt)
3129-3179 ↔ 2513-2563 73% (51 nt)

The first line asserts that positions 2870–2926 in the first sequence (57 base pairs) aligns to positions 2281–2337 of the second sequence, without gaps and at 68% identity.

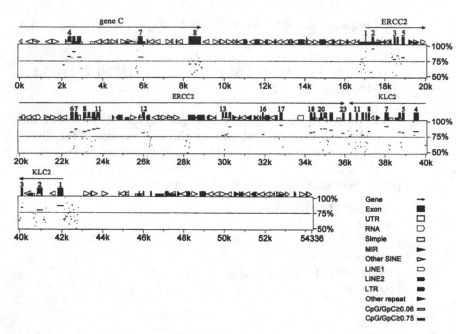

FIGURE 8 The graphical presentation of human and mouse ERCC2 locus comparison with PipMaker.

3. The traditional form of the alignments. For example, the following alignent corresponds to the first line of the compressed form above.

```
2870  GCCCAGGCCTGGGCAGCGAGAGGGCCCTGCTCCCCGCTCAAGGCTCCCAGGACATTC
      ||||||:|  ||::||::|:||||  ||||||||||:||:||||  ::|::::||:|||||||
2281  GCCCAGACATGAACAATGGGAGGCCCCTGCTCTCCACTCACAACCTTTAGAACATTC
```

Some Limitations of PipMaker. PipMaker is a great program for graphical presentation of a long alignment. The main disadvantage is its complete lack of flexibility. The user has to rely on a "black box" approach to an alignment and display problem. It is not clear what program is used for the actual alignment, and the user has no control of any alignment parameters, such as match and mismatch values or gap penalties. Also, the graphical display is not at all flexible. Only regions of similarity between 50% and 100% are presented in a pip graph, but in many cases a user would wish to display more or less stringent hits, as allowedin DOTTER (see Sec. 4.3.1).

5. CONCLUSION

With dozens of bacterial genomes completely sequenced and several eukaryotic genomes, including our own blueprint, deciphered, we are truly in a genomic era in biology. Comparative genomics is a key to understanding all the information hidden in four genomic letters.

Comprehensive understanding of genomes will be possible only when biologists and computer scientists combine forces. The number of computational resources available to biologists interested in genomes is already impressive. Yet many problems are waiting to be solved. One of the biggest challenges will be data presentation. An unprecedented amount of data is flooding biologists, and computer scientists are expected to come up with new, clever ideas to make the data accessible for them. In this chapter we have discussed some computational resources and problems associated with comparative genomics. Although this work is far from being comprehensive and exhaustive, we hope that it will be useful for both parties involved in the comparative analysis of genomes.

APPENDIX

Some of the resources for comparative genomics available on the World Wide Web are given in the following table.

Program	Function	URL
BLAST	Homology search	http://ncbi.nlm.nih.gov/BLAST
FASTA	Homology search	http://fasta.bioch.virginia. edu/fasta/cgi/searchx.cgi
ClustalW	Multiple alignment	http://www2.ebi.ac. uk/clustalw
MULTICLUSTAL	Multiple alignment	http://www.sgi.com/chembio/ resources/clustalw/ parallel_clustal.html
MultAlin	Multiple alignment	http://protein.toulouse.inra. fr/multalin
DIALIGN	Multiple alignment	http://bibiserv.techfak. uni-bielefeld.de/dialign/
MAP	Multiple alignment	http://genome.cs.mtu. edu/map.html
HMMER	Multiple alignment	http://hmmer.wustl.edu
SIM	Pairwise alignment	http://www.expasy. ch/tools/sim.html

Entrez	Data retrieval	http://www.ncbi.nlm.nih.gov/Entrez
Entrez-Genomes	Data retrieval	http://www.ncbi.nlm.nih.gov/Entrez/Genome/main_genomes.html
HOMOLOGENE	Database	http://www.ncbi.nlm.nih.gov/HomoloGene/
UniGene	Database	http://www.ncbi.nlm.nih.gov/
Animal Genome Database	Database	http://ws4.niai.affrc.go.jp/jbase.html
HOVERGEN	Database	http://pbil.univ-lyon1.fr/databases/hovergen.html
COG	Database	http://ncbi.nlm.nih.gov/COG/
MGD	Database	http://www.informatics.jax.org/menus/homology_menu.shtml
DOTTER	Alignment/visualization	http://www.cgr.ki.se/cgr/groups/sonnhammer/Dotter.html
PipMaker	Alignment/visualization	http://globin.cse.psu/pipmaker/
RepeatMasker	Repeat detection and masking	http://ftp.genome.washington.edu/cgi-bin/RepeatMasker
CENSOR	Repeat detection and masking	http://www.girinst.org/Censor_Server.html

REFERENCES

1. Aristotle. *Historia animalium*. London: Heinemann Harvard Univ Press, 1965.
2. Makalowski W, Boguski MS. Evolutionary parameters of the transcribed mammalian genome: an analysis of 2,820 orthologous rodent and human sequences. Proc Natl Acad Sci USA 1998; 95:9407–9412.
3. Makalowski W, Zhang J, Boguski MS. Comparative analysis of 1196 orthologous mouse and human full-length mRNA and protein sequences. Genome Res 1996; 6:846–857.
4. Pearson WR, Lipman DJ. Improved tools for biological sequence comparison. Proc Natl Acad Sci USA 1988; 85:2444–2448.
5. Altschul SF, et al. Gapped BLAST and PSI-BLAST: a new generation of protein database search programs. Nucleic Acids Res 1997; 25:3389–3402.
6. Tatusov RL, Galperin MY, Natale DA, Koonin EV. The COG database: a tool for genome-scale analysis of protein functions and evolution. Nucleic Acids Res 2000; 28:33–36.
7. Wheelan SJ, Boguski MS, Duret L, Makalowski W. Human and nematode orthologs—lessons from the analysis of 1800 human genes and the proteome of *Caenorhabditis elegans*. Gene 1999; 238:163–170.
8. Gu X, Li WH. The size distribution of insertions and deletions in human and

rodent pseudogenes suggests the logarithmic gap penalty for sequence alignment. J Mol Evol 1995; 40:464–473.

9. Huang X. On global sequence alignment. Comput Appl Biosci 1994; 10:227–235.
10. Smit AF. Interspersed repeats and other mementos of transposable elements in mammalian genomes. Curr Opin Genet Dev 1999; 9:657–663.
11. Zietkiewicz E, Labuda D. Mosaic evolution of rodent B1 elements. J Mol Evol 1996; 42:66–72.
12. Jurka J, Klonowski P, Dagman V, Pelton P. CENSOR—a program for identification and elimination of repetitive elements from DNA sequences. Comput Chem 1996; 20:119–121.
13. Duret L, Mouchiroud D, Gouy M. HOVERGEN: a database of homologous vertebrate genes. Nucleic Acids Res 1994; 22:2360–2365.
14. Perriere G, Duret L, Gouy M. HOBACGEN: database system for comparative genomics in bacteria. Genome Res 2000; 10:379–385. *In Process Citation.*
15. Riley M. Functions of the gene products of *Escherichia coli*. Microbiol Rev 1993; 57:862–952.
16. T.C.e.S. Consortium. Genome sequence of the nematode *Caeronabditis elegans*: a platform for investigating biology. Science 1998; 282:2012–2018.
17. Prager EM, Wilson AC. Ancient origin of lactalbumin from lysozyme: analysis of DNA and amino acid sequences. J Mol Evol 1988; 27:326–335.
18. Kumagai I, Takeda S, Miura K. Functional conversion of the homologous proteins alpha-lactalbumin and lysozyme by exon exchange. Proc Natl Acad Sci USA 1992; 89:5887–5891.
19. Everett LA, et al. Pendred syndrome is caused by mutations in a putative sulphate transporter gene (PDS). Nat Genet 1997; 17:411–422.
20. Scott DA, Wang R, Kreman TM, Sheffield VC, Karnishki LP. The Pendred syndrome gene encodes a chloride-iodide transport protein. Nat Genet 1999; 21:440–443.
21. Eisen JA. Phylogenomics: improving functional predictions for uncharacterized genes by evolutionary analysis. Genome Res 1998; 8:163–167.
22. Koop BF, Hood L. Striking sequence similarity over almost 100 kilobases of human and mouse T-cell receptor DNA. Nat Genet 1994; 7:48–53.
23. Thompson JD, Plewniak F, Poch O. A comprehensive comparison of multiple sequence alignment programs. Nucleic Acids Res 1999; 27:2682–2690.
24. Thompson JD, Higgins DG, Gibson TJ. CLUSTAL W: improving the sensitivity of progressive multiple sequence alignment through sequence weighting, position-specific gap penalties and weight matrix choice. Nucleic Acids Res 1994; 22:4673–4680.
25. Sonnhammer EL, Durbin R. A dot-matrix program with dynamic threshold control suited for genomic DNA and protein sequence analysis. Gene 1995; 167:GC1–GC10.
26. Schwartz S, et al. PipMaker—a web server for aligning two genomic DNA sequences. Genome Res 2000; 10:577–586.

10

Computational Methods for Studying the Evolution of the Mitochondrial Genome

Cecilia Saccone
University of Bari, Bari, Italy
Graziano Pesole
University of Milan, Milan, Italy

1. INTRODUCTION

This chapter focuses on the most popular methods used in the study of the evolution of the mitochondrial genome. This theoretically implies the knowledge of the structure and function of this organellar genome, which has several peculiar features. Therefore, its study often requires the adaptation and/or modification of already existing methodologies, and sometimes even the implementation of new tools.

In this light we believe it necessary to report some basic features of the mitochondrial (mt) genome together with the computational methods used in the study of the evolution of this genome. Major attention is paid to the mt genome of metazoa, whose complete DNA sequences are available in the databases.

2. ORIGIN OF MITOCHONDRIA

Mitochondria are cytoplasmic organelles present in all eukaryotic organisms, but in only a small group of phagothrophous or micropinocytotic non-photosynthetic protists, *Giardia*, *Trichomonas*, and *Microsporidia*, called Archaezoa [1] to denote a primitive amitochondriate phase of eukaryotic evolution. The endosymbiotic theory of the origin of mitochondria, already almost a century old, has been strongly supported by the discovery of the structural and functional properties of these organelles, in particular their genetic system. Recently, it has been further supported by the typically bacterial features of the mtDNA of the protozoan *Reclinomonas americana* [2] and by the similarity between the mitochondrial genome and the genome of the α-proteobacterium *Rickettsia prowazekii*, an obligate intracellular para-site evolutionarily rather close to the mitochondrial ancestors [3].

The original endosymbiotic theory, called serial endosymbiotic theory, postulated that a eubacterium had entered into symbiosis with a primitive amitochondriate eukaryotic cell and had established a permanent relation-ship with the primitive nuclear genome. Then, owing to the endosymbiotic life style, the protomitochondrial genome progressively transferred most material into the nucleus. Indeed, several functions were taken over by the host cell or became useless. Hence, the protomitochondrion changed from an occasional endosymbiont to an obligate host, and its genome progressively decreased in size and gene number. Later, it was further proposed that eukaryotes originated through symbiosis between spirochetes and wall-less bacteria [4]. Indeed, the colonization of the primitive cell by eubacteria, which originated the mitochondrion, could be single or multiple. Comparative studies on mtDNA gene organization and content in different organisms and phyloge-netic analyses carried out on mitochondrial genes support the hypothesis that modern mitochondrial genomes all descend from a single common ancestor, and thus their origin is monophyletic (see Ref. 5). However, the genome reduction process has followed different and often diverging evolutionary patterns in the various taxa, which is the foundation of the present diversity in size, gene content, gene organization, and mode of expression of modern mtDNAs [6–11].

Recent studies based on the analysis of the genetic system of some protists, previously scarcely known at the molecular level, provided evidence challenging this serial endosymbiosis theory. The hypothesis has been put forward that mitochondria originated essentially at the same time as the nuclear component of the eukaryotic cell rather than in a subsequent separate event. This is based on the finding that Archaeozoa originally did have mitochondria and lost them during evolution; indeed, genes coding for typically mitochondrial proteins were found in the nucleus of these organisms

[12]. This mitochondrion-driven scenario of the first eukaryote has been further supported by the finding that a genome of mitochondrial descent has also been identified in hydrogenosomes of *Nyctotherus ovalis*, an anaerobic heterotrichous ciliate [13]. Hydrogenosomes are membrane-bound organelles present in some protozoa and chytridiomycete fungi that produce ATP anaerobically and generate molecular hydrogen as a by-product. Indeed, we are now dealing with a new theory, the "hydrogen hypothesis," according to which the eukaryotic nucleus and mitochondria/hydrogenosomes stemmed from a single fusion event between a methanogenic archaebacterium requiring hydrogen and an α-proteobacterium producing hydrogen; then the passage from an anaerobic (hydrogenosomic) to an aerobic (mitochondrial) metabolism took place [4,14].

The recently discovered properties of nuclear genes provide grounds in support of the simultaneous acquisition of the nucleus and the mitochondrion by the primitive eukaryotic cell. It is becoming increasingly clear that the nuclear genome is not a descendant of a primitive archaebacterium but rather a chimera that incorporates contributions from both archaebacterial and eubacterial progenitors. The eubacterial component of the nuclear genome could, however, be much greater than is usually attributed to specific gene transfer from the evolving mitochondrial genome and would include genes that have nothing to do with mitochondrial biogenesis and function. This may be due to the fusion of eubacterial and archaebacterial patterns in the creation of the eukaryotic cell.

3. THE MITOCHONDRIAL GENETIC SYSTEM

3.1. Uniparental inheritance

Among the peculiarities of mitochondrial genetic system is the non-Mendelian uniparental inheritance of mtDNA; in metazoans, only one of the parents, namely the mother, transmits the genome to the offspring.

In female germinal cells, mtDNA has a very special fate. Pedigree analysis of Holstein cows showed that heteroplasmic breeds recover mtDNA homoplasmy in as little as two generations [15,16] owing to the fast segregation of mtDNA molecules. This phenomenon of fast mtDNA segregation is explained by a bottleneck effect in the female germinal line or in the early stages of embryo development.

During maturation of the primary oocyte, mtDNA molecules increase manyfold in number, from 10^3 to 10^8 depending on the species [17,18]. During this process, the selection and then replication of a small subpopulation of mold molecules could cause a rapid change in mtDNA genotype in even a single generation [16]. The alternative hypothesis is based on the remark that the embryo undergoes several cellular divisions before mtDNA replication is started. In mice, at the early blastocyst stage, a small number of cells (10–12), each containing 10^3 copies of mtDNA, forms a small cellular mass, which gives rise to the three germinal layers of the embryo while the other blastocyst cells originate extraembryonic tissues [19]. The uneven distribution of mtDNA in the inner mass cell could contribute to the fast segregation of different mtDNA genotypes.

The transmission of the paternal mitochondrion to the offspring is stopped by a number of molecular and cellular mechanisms that occur over one or several stages of the reproduction process, at prezygotic stage, during fecundation, or at the zygotic stage [20]. Among the prezygotic mechanisms involved in gamete genesis there are uneven cell division and differences in the growth of male and female gametes, producing large female gametes and small male gametes. Hence, female gametes are rich in mitochondria whereas male gametes are poor in them. In the extreme, prawn male gametes completely lack mitochondria due to the uneven cytokinesis during gameto-genesis [21]. In the tunicate ascidia, uniparental inheritance is due to the fact that sperm mitochondria do not enter the oocyte during fecundation [22]. A stochastic zygotic mechanism probably acts in yeast; mtDNA molecules from one of the parents replicate more often than those of the other in a totally random process; hence their rate of transmission to the offspring is higher [20].

In mammals, the whole spermatozoan enters the oocyte during fecun-dation, including the middle portion containing mitochondria. The only known exception is the Chinese hamster (*Cricetulus griseus*). In this species, during fecundation the tail and middle portion of the sperm remain outside the oocyte; hence there are no chances for the paternal mtDNA to be transmitted to the offspring. In all other mammals, the maternal inheritance of mtDNA is due to processes occurring after fertilization [23]. It is well known that in fertilized eggs of the common hamster, paternal mitochondria degenerate during the two-cell stage, when several multivesicular bodies surround and fuse with sperm mitochondria and digest them [24]. In humans, mitochondria from the sperm middle portion have been found in the fecundated oocyte, where they survive to the stage of morula; it is not known whether and when they are finally destroyed (references in Ref. 23). Generally, in mammals maternal inheritance seems to be due to the dilution of the small quantity of mtDNA in sperm into the great quantity of mitochondria present in the oocyte. Thus, the random mtDNA replication in the zygote would nullify the contribution of paternal mtDNA in most individuals [23]. Usually, mammalian spermatozoa contain 50–75 mitochondria in the middle portion, each having a single mtDNA molecule [25], whereas oocytes contain 10^5–10^8 mitochondria [26], each accounting for several copies (up to 10) of mtDNA molecules [18,27].

Paternal mtDNA transmission has been studied in rodents using highly sensitive techniques such as PCR; however, the results remain ambiguous because they essentially cover interspecies crossbreeds rather than intraspecies ones. Gyllensten et al. [28] observed that paternal mtDNA is transmitted in very small quantities to the offspring over several generations in retrobreeding between *Mus musculus* and *Mus spretus*. One paternal mtDNA is inherited on average for each 10^4 maternal mtDNA molecules, which is in agreement with expectations based simply on the dilution of spermatozoan mtDNA into the oocyte. Furthermore, Gyllensten et al. [28] found that, if inherited, paternal mtDNA is distributed more or less evenly in the tissues of the offspring. Completely opposite results were obtained by Shitara et al. [29] from the examination of interspecific retro-crossbreeds among the same species. These studies showed that paternal mtDNA is not distributed in all tissues of the hybrids and neither is it transmitted to the following generations; rather it is present only in the first generation of interspecies crossbreeds.

Polymerase chain reaction analyses carried out on *Mus musculus* intraspecies hybrids on single cells at the early stages of embryogenesis showed that paternal mtDNA is present in these hybrids only to the pronucleus phase and disappears soon afterward when the membrane potential of sperm mitochondria is annihilated [30]. In contrast, in *Mus musculus* and *Mus spretus* interspecies hybrids, paternal mtDNA is found in 70% of the examined hybrids and at different stages of the fecundated oocytes since birth [30]. Based on these results, species-specific mechanisms have been suggested to act on fertilized eggs to recognize and eliminate spermatozoan mitochondria. These mechanisms would involve recognition and interaction of oocyte cytoplasmic factors with the proteins of the spermatozoan middle region. Consequently, only in interspecies crossbreeds, where the above mechanisms can be completely inactivated or have reduced efficiency, could paternal mtDNA transmission occur, because spermatozoan mitochondria would not be recognized as such and thus would not be eliminated [29,30].

Maternal inheritance makes mtDNA crucial in studies on molecular evolution because it allows the evolutionary history of a species to be traced back to the common ancestor following a linear evolutionary pathway. Furthermore, uniparental inheritance excludes the occurrence of recombination events between the genetic pools of the parents, because it occurs in nuclear DNA, which could further bewilder the evolutionary processes. More recently, Awadalla et al. [31] found signs of mixing through recombination between paternal and maternal mitochondrial DNA in humans and chimpanzees. As mtDNA has been widely used as a tool to trace human ancestry and relationships, this finding could have profound implications in the study of the origin of modern humans [32,33].

A correlation between transmission mechanisms and the evolutionary rate of mtDNA has been proposed for mollusks of the genera *Mytilus* and *Anodonta*. They represent a peculiar type of mtDNA transmission called "double uniparental inheritance" (DUI) [34]. Females are homoplasmic and have mtDNA F of maternal origin, which they transmit to both female and male offspring; males are heteroplasmic and have maternally inherited F molecules that they do not transmit to the offspring and paternally inherited M molecules that they transmit to male offspring only. It has been suggested that this transmission mechanism is responsible for the faster evolution of M mtDNA molecules than of F mtDNA [35,36].

3.2. Recombination

Gene recombination is one of the sources of mtDNA variability in plants, fungi, and protists; however, this process does not seem to affect animal mtDNA. Differences in the number of tandem repeat copies in the nematode *Meloidogynes javanica* have been interpreted as the result of possible recombination events [37]. In mammals, the lack of recombination has been demonstrated by the absence of mtDNA crossing-over [38] and by the lack of recombinant mtDNA molecules (that is, hybrids between two different mtDNA donors) in hybrids of somatic cells and of cytoplasm (cytoplasm hybrids or "cybrids" are formed through the fusion of an anucleate cell with a nucleate cell) (references in Ref. 9, but see also below). However, as reported in the previous subsection, findings have been reported that would support some recombination in human and chimpanzee mtDNAs [31].

Generally the absence of recombination in mtDNA can be due to the capability of cells to keep mitochondrial organelles separate from one another and/or to the absence of enzymes for recombination. If each mitochondrion makes up an independent unit in a cell and does not fuse with the others, mtDNA molecules from different organelles will not have the chance to recombine. Conversely, if mitochondria make up a dynamic network, mitochondrial DNA and components from different organelles could mix and make recombination and complementation possible. Although the existence of interconnections between mitochondria in vivo would suggest the occurrence of mitochondrial fusion under well-established physiological conditions, this issue is still much debated.

Hayashi et al. [39] showed by microscopic analysis that two different mitochondrial populations, marked with two different fluorescent DNA-binding dyes and introduced into HeLa cells deprived of mtDNA (rho-0), spread rapidly to all rho-0 HeLa mitochondria. Moreover, coexisting wild-type mtDNA and mutant mtDNA with a large deletion distribute homogeneously throughout mitochondria. This study demonstrated that all mitochondria in a cell behave as a single dynamic cellular unit, which is in agreement with what had been observed in previous studies [40–42].

Yoneda et al. [43] performed several experiments on cytoplasmic hybrids to test the capability of different mtDNAs to mix and complement one another. Indeed, complementation of two mitochondrial mutations can occur only if both mutated mtDNAs are in the same organelle. Hence, two distinct mitochondrial populations, each bearing mtDNA mutated in a different tRNA, are inserted into human rho-0 cells without taking into account the complementation process of the obtained cytoplasmic hybrids; furthermore, no transductional complementation has been observed between CAP^s (chloramphenicol-sensitive) and CAP^r (chloramphenicol-resistant) mitochondria differently from what was previously reported by Oliver and Wallace [42].

Because experiments carried out by Hayashi et al. [39] and Yoneda et al. [43] examined the properties of cytoplasmic hybrids at different cell stages, the conflicting results obtained have been reconciled by hypothesizing a pattern whereby mitochondria fuse in the first stage of hybrid formation and make up compartments that do not communicate during the later stages of cell division and growth [44].

As far as the existence of a mitochondrial enzymatic system for recombination is concerned, studies have demonstrated that mammalian mitochondria possess several DNA repair systems (references in Ref. 45), among which is a repair system for homologous recombination (HR) [46]. This latter could explain the repair of mtDNA from interstring cross-links caused by cisplatin [47]; indeed, this damage is repaired in prokaryotes and in yeast nucleus by homologous recombination [48]. The HR system has been identified in mammalian mitochondria through the analysis of protein extracts obtained from mitochondrial subcellular fractions. These extracts catalyze conservative homologous recombination reactions between plasmid substrates in the presence of ATP or a similar nonhydrolyzable agent, and the process appears to be mediated by a protein homologous to recA bacterial protein [46]. However, it has been suggested that, as in the yeast system, HR activity in mammalian mitochondria is used for DNA repair only and is not associated to gene recombination. Indeed, MHR1 gene of the mitochondrial recombination/repair system in yeast takes part only in gene conversion and not in crossing over [49]; similarly, in mammals, the HR mitochondrial system could be involved in gene conversion only and hence only in DNA repair, which would explain the lack of crossing over in mtDNA [38].

On the whole, presently available data support the absence of gene recombination in animal mtDNA and make this molecule particularly suitable for molecular phylogenetic studies because all mitochondrial genes are inherited as a single linkage group and there are no processes confusing their evolutionary history, which then coincides with the evolutionary history of the species.

3.3. Size and Shape

Because of the different evolutionary pathways that generated the segregation of genetic information in the eukaryotic cell in different cellular compartments such as the nuclei and the mitochondria, the mitochondrial genome shows a great variability in terms of structure, gene content, organization, and mode of expression in the different organisms. Several features, however, are common to the majority of mitochondrial genomes.

The circular double-stranded structure appears to be almost a constant feature of mtDNA, which exhibits an extraordinary variability in length, particularly in the lower eukaryotes and in plants. The size ranges from only about 6000 base pairs (6 kbp) in some protists (e.g., *Plasmodium* and *Theileria*) to 2500 kbp in plants.

In Ciliata, e.g., *Tetrahymena pyryformis* and *Paramecium aurelia*, the genome is linear with a double helix molecule 46 and 40 kbp long, respectively. In the alga *Chlamydomonas reinhardtii*, besides the linear (major species) mitochondrial genome (16 kbp long), a circular (minor species) mtDNA might also be present [50]. In the kinetoplasts of the Tripanosomatidae *Crithidia fasciculata*, *Leishmania tarentolae*, and *Trypanosoma brucei*, the mt genome has a very peculiar structure, being composed of an intricate network containing two types of molecules: many thousands of minicircles (1–3 kbp, depending on the species), accounting for 90–95% of the DNA, and 50–100 maxicircles (20–40 kbp). Minicircles lack long amino acid coding frames, and maxicircles are the equivalent of mtDNA in other organisms [51]. Minicircle-encoded RNAs are necessary to guide the editing process required to generate functional mRNAs from the maxicircle transcripts (see below).

In higher plants, the mt genome ranges from 200 to 2500 kbp, and it has been demonstrated that its size can vary by as much as sevenfold within the same family. This peculiar feature is not related to the number of repeated sequences and detectable translation products. The entire genetic complexity is contained in a circular chromosome (master) that may be resolved into subgenomic circles by recombination through directly repeated sequences. This mechanism is also involved in the rapid and extensive rearrangements that characterize the evolution of plant mtDNA. Moreover, frequent acquisitions of genes from nuclei and chloroplasts take place, resulting in the formation of mosaic genomes [52–56]. Table 1 reports the size and shape of the 47 mt genomes so far completely sequenced in lower eukaryotes and in plants.

TABLE 1 List of Completely Sequenced Mitochondrial Genomes in Lower Eukaryotes and Plants[a,d]

Organism	Accession No.	Size (bp)	Shape
Higher plants			
Arabidopsis thaliana	Y08501	366.924	Circular
Marchantia polymorpha	M68929	186.609	Circular
Algae			
Chlamydomonas eugametos	AF008237	22.897	Circular
Chlamydomonas reinhardtii	U03843	15.758	Linear
Chlorogonium elongatum	Y13644, Y07814, Y13643	22.704	Circular
Chondrus crispus	Z47547	25.836	Circular
Chrysodidymus synuroideus	OGMP-Sequencing Projects	34.119	Circular
Cyanidioschyzon merolae	D89861	32.211	Circular
Nephroselmis olivacea	AF110138	45.223	Circular
Ochromonas danica	OGMP-Sequencing Projects	41.035	Linear
Pedinomonas minor	AF116775	25.137	Circular
Porphyra purpurea	AF114794	36.753	Circular
Prototheca wickerhamii	U02970	55.328	Circular
Rhodomonas salina	OGMP-Sequencing Projects	48.063	Circular
Fungi			
Allomyces macrogynus	U41288	57.473	Circular
Dictyostelium discoideum	AB000109	55.564	Circular
Harpochytrium sp. #105	FMGP Sequencing Project	24.570	Circular
Harpochytrium sp. #94	FMGP Sequencing Project	19.473	Circular
Monoblepharella sp. #15	FMGP Sequencing Project	60.433	Circular
Monosiga brevicollis	FMGP Sequencing Project	76.568	Circular
Mortierella verticillata	FMGP Sequencing Project	58.745	Circular
Neurospora crassa	K03295	3.581	Circular
Phytophthora infestans	U17009	37.957	Circular
Pichia canadensis	D31785	27.694	Circular
Podospora anserina	X55026	100.314	Circular
Rhizophydium sp. #136	FMGP Sequencing Project	68.834	Circular
Rhizopus stolonifer	FMGP Sequencing Project	54.191	Circular
Saccharomyces cerevisiae	M62622	78.520	Circular
Schizophyllum commune	FMGP Sequencing Project	49.705	Circular
Schizosaccharomyces octosporus	FMGP Sequencing Project	44.227	Circular
Schizosaccharomyces pombe	X54421	19.430	Circular
Spizellomyces punctatus	FMGP Sequencing Project	61.300	3 circles[b]
Protozoa			
Acanthamoeba castellanii	U12386	41.591	Circular
Cafeteria roenbergensis	AF193903	43.159	Circular
Chondrus crispus	Z47547	25.836	Circular
Cyanidioschyzon merolae	D89861	32.211	Circular

TABLE 1 Continued

Organism	Accession No.	Size (bp)	Shape
Dictyostelium discoideum	000109	55.564	Circular
Jakoba libera	OGMP-Sequencing Projects	100.252	Circular
Leishmania tarentolae	M10126	20.992	Circular
Malawimonas jakobiformis	OGMP-Sequencing Projects	47.328	Circular
Naegleria gruberi	OGMP-Sequencing Projects	49.838	Circular
Paramecium aurelia	X15917	40.469	Linear
Plasmodium falciparum	M76611	5.967	Linear[c]
Plasmodium yoelii	M29000	5.952	Linear[c]
Reclinomonas americana	AF007261	69.034	Circular
Tetrahymena pyriformis	AF160864	47.296	Linear
Theileria parva	Z23263	5.895	Linear

Source: Unpublished sequences are from the Organelle Genome Megasequencing Project (OGMP, http://megasun.bch.umontreal.ca/ogmpproj.html) and from Fungal MitochondrialGenome Project (FMGP, http://megasun.bch.umontreal.ca/People/lang/FMGP/FMGP.html).
[a] For each genome the relevant accession number, the size, and the shape are reported.
[b] Three "chromosomes" of 58.8, 1.4, and 1.1 kb.
[c] Head-to-tail tandem repeats of 6 kb unit [7].
[d] For an updated list see www.ncbi.nlm.nih.gov/genomes/ORGANELLES/organelles.html

In contrast to such a great variability, the size of the mt genome in metazoans is extremely small compared to that of other eukaryotes, with a roughly constant length (14–19 kbp). Table 2 lists the mt genomes completely sequenced in Metazoa. More recently, several cases in the literature have been described where the length of mtDNA exceeds the normal, rather constant average size found in animal cells. Such a difference in length can range from twice the average size (14–28 kbp) to even more (35 kbp, as in the sea scallop, *Placopecten magellanicus* [57]). In such cases, however, a duplication of some genomic regions seems to have occurred. Length variations are more frequent in invertebrates and in poikilothermic vertebrates, and they appear to generate very rapidly and distribute both within (heteroplasmy) and between individuals. In general, length differences are confined to the control region, but in some cases they are dispersed and/or also include structural genes (for review, see Refs. 11, 58, and 59).

Despite such a variety of structures and sizes, the information content of all mt genomes is not dramatically different in the various organisms, because the information content of the mtDNA is sufficient to code for ribosomal RNA species (two or three), a reduced but complete set of transfer RNAs (with some exceptions), and a small set of proteins (13 in Metazoa). In general (with some exceptions like plants), the differences concern mainly the non-

TABLE 2 Mitochondrial Genomes Completely Sequenced in Metazoa[b]

Taxon	Species	Size[a](bp)	Accession
Chordata			
Mammalia	*Artibeus jamaicensis* (fruit bat)	16651	AF061340
	Balaenoptera musculus (blue whale)	16402	X72204
	Balaenoptera physalus (fin whale)	16398	X61145
	Bos taurus (cow)	16338	V00654
	Canis familiaris (dog)	16728	U96639
	Cavia porcellus (guinea pig)	16801	AJ222767
	Ceratotherium simum (white rhinoceros)	16832	Y07726
	Dasypus novemcinctus (armadillo)	17056	Y11832
	Didelphis virginiana (opossum)	17084	Z29573
	Equus asinus (donkey)	16670	X97337
	Equus caballus (horse)	16660	X79547
	Erinaceus europaeus (hedgehog)	17447	X88898
	Felis catus (cat)	17009	U20753
	Glis glis (dormouse)	16602	AJ001562
	Gorilla gorilla (Western highland gorilla)	16364	D38114
	Halichoerus grypus (grey seal)	16797	X72004
	Hippopotamus amphibius (hippopotamus)	16407	AJ010957
	Homo sapiens (human)	16569	V00662
	Hylobates lar (gibbon)	16472	X99256
	Loxodonta africana (elephant)	16860	AJ224821
	Macropus robustus (wallaroo)	16896	Y10524
	Mus musculus (mouse)	16295	V00711
	Ornithorhyncus anatinus (platypus)	17019	X83427
	Orycteropus afer	16816	Y18475
	Oryctolagus cuniculus (rabbit)	17245	AJ001588
	Ovis aries (sheep)	16616	AF010406
	Pan paniscus (pygmy chimpanzee)	16563	D38116
	Pan troglodytes (common chimpanzee)	16554	D38113
	Papio hamadryas (baboon)	16521	Y18001
	Phoca vitulina (harbor seal)	16826	X63726
	Pongo pygmaeus (Bornean orangutan)	16389	D38115
	Rattus norvegicus (rat)	16300	X14848
	Rhinoceros unicornis (Indian rhinoceros)	16829	X97336
	Sciurus vulgaris (red squirrel)	16507	AJ238588
	Sus scrofa (pig)	16679	AJ002189
Aves	*Aythya americana*	16616	AF090337
	Corvus frugilegus	16931	Y18522
	Falco peregrinus	18068	AF090338
	Gallus gallus (chicken)	16775	X52392
	Rhea americana (greater rhea)	16710	Y16884
	Smithornis sharpei	17344	AF090340
	Struthio camelus (ostrich)	16591	Y12025
	Vidua chalybeata	16895	AF090341
Reptilia	*Alligator mississippiensis* (American alligator)	16646	Y13113
	Chelonia mydas	16497	AB012104

TABLE 2 Continued

Taxon	Species	Size[a] (bp)	Accession
	Chrysemys picta	16866	AF069423
	Dinodon semicarinatus (akamata snake)	17191	AB008539
	Eumeces egregius	17407	AB016606
	Pelomedusa subrufa (African side-neckedturtle)	16787	AF039066
Amphibia	*Xenopus laevis* (African clawed frog)	17553	M10217
Osteichthyes	*Carassius auratus* (goldfish)	16578	AB006953
	Crossostoma lacustre (loach)	16558	M91245
	Cyprinus carpio (common carp)	16575	X61010
	Gadus morhua (Atlantic cod)	16696	X99772
	Gonostoma gracile	16436	AB016274
	Latimeria chalumnae (coelacanth)	16407	U82228
	Oncorhynchus mykiss (rainbow trout)	16642	L29771
	Paralichthys olivaceus (Japanese flounder)	17090	AB028664
	Polypterus ornatipinnis (bichir)	16624	U62532
	Protopterus dolloi (lungfish)	16646	L42813
	Salmo salar (Atlantic salmon)	16665	U12143
	Salvelinus alpinus	16659	AF154851
	Salvelinus fontinalis	16624	AF154850
Chondrichthyes	*Mustelus manazo* (shark)	16707	AB015962
	Raja radiata	16783	AF106038
	Scyliorhinus canicula (smaller spotted catshark)	16697	Y16067
	Squalus acanthias (spiny dogfish)	16738	Y18134
Agnatha	*Lampetra fluviatilis*	16159	Y18683
	Petromyzon marinus (sea lamprey)	16201	U11880
Cephalochordata	*Branchiostoma floridae*	15083	AF098298
	Branchiostoma lanceolatum (amphioxus)	15076	Y16474
Hemichordata	*Balanoglossus carnosus* (acorn worm)	15708	AF051097
Urochordata	*Halocynthia roretzi*	14771	AB024528
Echinodermata	*Arbacia lixula* (black urchin)	15719	X80396
	Asterina pectinifera (starfish)	16260	D16387
	Florometra serratissima (crinoid)	16005	AF049132
	Paracentrotus lividus (common urchin)	15696	J04815
	Stronglylocentrotus purpuratus (sea urchin)	15650	X12631
Arthropoda	*Anopheles gambiae* (African malaria mosquito)	15363	L20934
	Anopheles quadrimaculatus (mosquito)	15455	L04272
	Apis mellifera (common honeybee)	16343	L06178
	Artemia franciscana (brine shrimp)	15822	X69067
	Ceratitis capitata	15980	AJ242872
	Daphnia pulex	15333	AF117817
	Drosophila melanogaster (fruit fly)	19517	U37541
	Drosophila yakuba (fruit fly)	16019	X03240
	Ixodes hexagonus (prostriate hardtick)	14539	AF081828
	Limulus polyphemus	NA	AF002644-53
	Locusta migratoria (migratory locust)	15722	X80245

TABLE 2 Continued

Taxon	Species	Size[a] (bp)	Accession
	Rhipicephalus sanguineus (metastriate tick)	14710	AF081829
Mollusca	*Albinaria coerulea* (land snail)	14130	X83390
	Cepaea nemoralis (banded wood snail)	14100	U23045
	Euhadra herkiotsi	NA	Z71693-701
	Katharina tunicata (black chiton)	15532	U09810
	Mytilus edulis	NA	M83756-62
	Crassostrea gigas (Pacific oyster)	18224	AF177226
	Pupa strigosa	14189	AB028237
Anellida	*Lumbricus terrestris* (common earthworm)	14998	U24570
	Platynereis dumerii (Dumeril's clam worm)	15619	AF178678
Brahiopoda	*Terebratulina retusa* (brachiopod lampshell)	15451	AJ245743
Nematoda	*Ascaris suum* (pig roundworm)	14284	X54253
	Caenorhabditis elegans (free-living worm)	13794	X54252
	Onchocerca volvulus (filarial nematode)	13747	AF015193
Platyhelminthes	*Echinococcus multilocularis*	13738	AB018440
Cnidaria	*Metridium senile* (brown sea anemone)	17443	AF000023
	Sarcophyton glaucum (soft coral)	18453	AF064823, AF063191

[a] NA: Although described in the literature, the complete genome sequence is not yet available in the database.
[b] For an updated list see www.ncbi.nlm.nih.gov/genomes/ORGANELLES/organelles.html

coding and regulatory regions. Indeed, the mt genome is a good example of several different strategies that the eukaryotic cell can use to express the same information content. Among these, the most peculiar is certainly RNA editing (see below), which is particularly active in Protozoa and in plants [60].

Despite the relatively constant gene content, gene order and organization vary strikingly in the various organisms. The mitochondrial genome organization in lower eukaryotes, such as *Saccharomyces cerevisiae*, is very loose (two-thirds contain A-T-rich, noncoding sequences), and several genes, in particular the apocytochrome b, the cytochrome oxidase subunit I, and the 21S rRNA, are discontinuous. The number of introns and G-C- and A-T-rich mini-inserts within the genes are strain-dependent. Some introns of these split genes code for proteins involved in RNA processing or intron transposition [61,62].

In plants, where, as previously reported, the mt genome is much larger than the fungal or animal counterpart, gene organization is highly dispersed. Extensive noncoding sequences separating the coding regions and introns have been detected in several genes. Mitochondrial DNAs show a tri- or multipartite structural organization and, owing to frequent recombination,

TABLE 3 Deviations from the Universal Genetic Code Described for the
Mitochondrial Genomes[a]

		Genetic code	
Code type	Codon	Mitochondrial	Universal code
Ascidian[b]	UGA	Trp	Stop
	AUA	Met	Ile
	AGA	Gly	Arg
	AGG	Gly	Arg
Chlorophycean	UAG	Leu	Stop
Echinoderm[c]	UGA	Trp	Stop
	AAA	Asn	Lys
	AGA	Ser	Arg
	AGG	Ser	Arg
Flatworm (Platyhelminthes)	UAA	Tyr	Stop
	UGA	Trp	Stop
	AAA	Asn	Lys
	AGA	Ser	Arg
	AGG	Ser	Arg
Invertebrate[d]	UGA	Trp	Stop
	AUA	Met	Ile
	AGA	Ser	Arg
	AGG	Ser	Arg
Mold; protozoan; coelenterate mitochondrial and mycoplasma/spiroplasma[e]	UGA	Trp	Stop
Trematode	UGA	Trp	Stop
	AUA	Met	Ile
	AAA	Asn	Lys
	AGA	Ser	Arg
	AGG	Ser	Arg
Vertebrate	UGA	Trp	Stop
	AUA	Met	Ile
	AGA	Stop	Arg
	AGG	Stop	Arg
Yeast[f]	UGA	Trp	Stop
	CUA	Thr	Leu
	CUU	Thr	Leu
	CUG	Thr	Leu
	CUC	Thr	Leu
	CGA	Not used	Arg
	CGC	Not used	Arg
	AUA	Met	Ile

they are continuously rearranged. A consequence of this peculiar evolutionary pattern is that genome organization varies greatly in linear gene orders.
Thus, highly conserved coding sequences are often flanked by completely
different sequences in the mtDNAs of plant species, even closely related
species. Moreover, chloroplast DNA sequences and plasmid-like sequences
are present in the mtDNA [63].

The gene structure and organization of Metazoa differ markedly from
those of yeast and plant mtDNAs. The most distinctive feature of the
metazoan mt genome is its extremely compact gene organization.

In conclusion, the features of the mt genomes can have either a
prokaryotic (e.g., naked DNA, absence of introns in Metazoa) or a eukaryotic
(e.g., presence of introns in lower eukaryotes and plants, presence of 5'-end
polyadenylated mRNAs in Metazoa) nature. This indicates that although
derived from the same ancestral progenitor, the mtDNA has followed
multiple and different evolutionary pathways in the various taxa, thus leading
to the variable situation we observe in extant organisms.

3.4. Genetic Code

As soon as the first mtDNA sequences became available, the comparison
between gene and protein sequences revealed several deviations from the
universal genetic code in mitochondria of different species. Table 3 shows the
deviations from the universal genetic code observed so far in mitochondria. In
mtDNA of most phylogenetic groups, UGA is used as a tryptophan codon
rather than as a termination codon. On the other hand, AGR (R = A or G),
coding for arginine in the universal code, is a stop codon in mtDNA of

---

Notes to Table 3.

[a] A complete list of genetic code deviations and the relevant references can be found at
http://www.ncbi.nlm.nih.gov/Taxonomy/Utils/wprintgc.cgi

[b] Systematic range includes Ascidiacea (sea squirts).

[c] Systematic range includes Asterozoa (star fishes) and Echinozoa (sea urchins).

[d] Systematic range includes Nematoda (*Ascaris, Caenorhabditis*), Mollusca (*Bivalvia,
Polyplacophora*), Arthropoda/Crustacea (*Artemia*), Arthropoda/Insecta (*Drosophila, Apis
mellifera*).

[e] This is the ancestral mitochondrial code, and its systematic range includes Mycoplas-
matales, Fungi (*Emericella nidulans, Neurospora crassa, Podospora anserina, Acre-
monium* sp., *Candida parapsilosis, Trichophyton rubrum, Dekkera/Brettanomyces,
Eeniella* sp., and probably *Ascobolus immersus, Aspergillus amstelodami, Claviceps
purpurea*, and *Cochliobolus heterostrophus*), Protozoa (*Trypanosoma brucei, Leishmania
tarentolae, Paramecium tetraurelia, Tetrahymena pyriformis*, and probably *Plasmodium
gallinaceum*) and Coelenterata (*Ctenophora* and *Cnidaria*).

[f] The systematic range includes *Saccharomyces cerevisiae, Candida glabrata, Hansenula
saturnus, Schizosaccharomyces pombe*, and *Kluyveromyces thermotolerans*.
Source for yeast data: Ref. 198.

vertebrates, and it codes for serine in the mtDNA of echinoderms and for glycine in the ascidian *Ciona intestinalis*. CTN codons (N = A, C, G, U) code for threonine in yeast, and AUA is an additional codon for methionine in most metazoans and in yeast.

Another surprising feature of the mitochondrial genetic system is the use of an oversimplified decoding mechanism allowing translation with a reduced set of tRNA species. In vertebrates this reduction is achieved with a wider wobbling between the third base of the codon and the first base of the anticodon, where a U is able to recognize all bases in the third codon position of four-codon families. This implies that only 22 tRNA species are required to translate all sense codons for the 20 amino acids.

In general, mitochondrial tRNAs are shorter and may present unusual structures with respect to tRNAs involved in cytoplasmic translation [64].

3.5. RNA Editing

The generation of RNA molecules having nucleotide sequences differing from those encoded by genes was first discovered by Benne et al. [65] in trypanosome mitochondria, where the process of uridylate insertion in encoded transcripts generates the functional mRNA. Since then, many other examples of posttranscriptional alterations of the informational content of the mRNA, generally known as "RNA editing" were discovered, mostly located in mitochondria.

RNA editing can be roughly divided into two major mechanisms, insertion/deletion editing (which changes the total number of nucleotides in the RNA) and conversion/substitution editing (which changes nucleotide identity). Insertion/deletion editing has been observed in the mitochondria of kinetoplastid protozoa and of the slime mold *Physarium polycephalum*. This process may create initiation and termination codons, correct frameshifts, and even build entire open reading frames from nonsense sequences. As already said, mitochondrial DNA of trypanosomes and of kinetoplastid protozoa consists of a network of maxicircles and minicircles. Maxicircle transcripts are edited at numerous sites by the insertion (or much less frequently, deletion) of uridylate residues, which in some cases account for more than 50% of the nucleotides of an mRNA.

The information for the correct pattern of insertion/deletions is provided by small guide RNAs (gRNAs), which are antisense with respect to the edited RNA regions and possess 3'U tails of 5–24 residues. These gRNAs are mostly encoded by the minicircle component of the mtDNA.

In the mitochondria of the slime mold *Physarium*, editing sites are spaced at approximately 25 nucleotide intervals, where insertions occur of single cytidines, uridines, and certain dinucleotides containing adenosine and guanosine as well as cytidine and uridine.

Conversion/substitution type editing occurs in plant mitochondria, where in the RNA encoded by the mitochondrial genome many cytidines are converted to uridines. For example, in *Oenothera bertheriana* mitochondrial RNAs, more than 500 editing sites have been observed with C→U conversions (but a few U→C conversions have also been detected). The mechanism of the specific C→U conversions in higher plant mRNA, which probably occurs by oxidative deamination, is largely unknown, but it has been observed to be mostly conservative as it generally improves the degree of conservation at the protein level between plant and non-plant organisms. For example, it has been observed in the *nad3-rps12* mitochondrial genes of some plant species that all three codon positions evolve at comparable dynamics, under a quasi-neutral evolutionary process, and most of the editing events occur in the first and second codon positions, restoring codons for conserved amino acids (see Table 1 in Ref. 66). This implies that RNA editing accounts for most of the selection control in plant mitochondrial RNA.

tRNA editing has been also observed in Gastropoda and in non-eutherian mammals. In gastropods the editing events involve changes from cytidine, thymidine, and guanosine to adenosine residues [67]. In platypus GCU Ser-tRNA, three editing events have been described, C→U, C→A, and A→U [68], as well as a C→U change in marsupial Asp-tRNA [69].

The origin, evolution, and functional role of RNA editing is still rather unclear, although it certainly provides an extra level of regulation of gene expression. A better knowledge of its evolutionary origin could clarify its role and possible selective advantages.

4. EVOLUTION OF THE MITOCHONDRIAL GENOME IN METAZOA

4.1. General Properties

In contrast with the high variability displayed by the mitochondrial genome of lower eukaryotes and plants, evolutionary forces molded metazoa mtDNA into a molecule characterized by compact arrangement, constancy of gene content, and the presence of a main noncoding region. Apart from the replication origin region(s), the genome is saturated with discrete genes lacking intronic sequences and flanking untranslated regions. These genes are often contiguous and sometimes slightly overlapped or separated by only few nucleotides.

Several hypotheses have been put forward to explain why the metazoan mitochondrial genome has attained such a small but constant size. A small genome has several advantages, such as faster replication and a constitutive type of transcription. However, the "race for replication" hypothesis of A. C. Wilson (personal communication), according to which those genomes that replicate the fastest will win, has not found experimental support so far. Moraes and Schon [70] found no difference in the replication rate of a heteroplasmic population of normal and partially deleted human mtDNA genomes in fibroblasts.

In general, metazoan mtDNA consist of a single circular molecule. In the phylum Cnidaria, linear DNA molecules, present as a single 16 kb molecule or two 8 kb molecules, have been reported in species from the classes Hydrozoa (e.g., *Hydra fusca*, *Hydra attenuata*), Scyphozoa (e.g., *Cassiopea* sp.), and Cubozoa (e.g., *Cariodea marsupialis*), whereas Cnidaria species included in the class Anthozoa have circular DNA molecules [71].

The uniformity of gene content is another remarkable feature of the majority of metazoan mtDNAs. The same set of genes is found in all metazoan mtDNA, namely, two ribosomal RNA species, a reduced but complete set of 22 tRNAs, and a set of 13 proteins. In addition, there is at least one region that does not encode any structural gene and that has been shown to include elements for the initiation and control of replication and transcription; this is called the control region and, in vertebrates, the D-loop region.

As rare exceptions to the rule of constant gene content, in Cnidaria only two tRNA genes are found in *Metridium senile* and only one in *Renilla koellikeri* and *Sarcophyton glaucum* [72]. In Nematoda the gene for ATP8 is missing. The mollusk *Mytilus edulis* not only lacks the ATP8 gene but also host an additional tRNA expected to recognize the AUA codons for methionine [73]. Despite the constancy in size and gene content, gene order and organization vary extensively within Metazoa due to gene rearrangements that consist of a different distribution of genes between the two strands (polarity inversions), gene transpositions, and gene losses.

4.2. Genome Features and Organization

Owing to the reduced size of the molecule, the sequencing of the metazoan mt genome has become very popular, providing a lot of information about its variations both between and within phyla. Table 2 lists the metazoan mt genomes that have been completely sequenced.

The length of mtDNA has been reported to be in the range of 13–19 kb. Indeed, when the mt genome is considered without the repeated sequences, its size ranges between 13,738 bp, the smallest size recorded for the platyhelminth *Echinococcus multilocularis*, and 19,517 bp, the largest for *Drosophila melanogaster*. However, genome size can reach up to 20 kb in *Meloidogyne javanica* and 42 kb in *Placopecten magellanicus* due to the presence of repeated sequences. Length variation is more frequent in invertebrates and in poikilothermic vertebrates, and it appears to generate very rapidly and distribute both within (heteroplasmy) and between individuals [11]. Changes in the copy number of tandem repeated sequences have been reported to be entirely consistent with a mechanism of mtDNA recombination, although, as previously pointed out, recombination in mtDNA remains undocumented [37] (see also Sec. 3.2).

It is commonly accepted that gene order varies between lineages and that conservation of gene order is frequently observed among recent evolutionary neighbors. Indeed, the progressive acquisition of new sequences into mtDNA databases has shown an increasing number of mt genomes whose structures deviate from the previous congruent picture of this molecule. The gene organization of some representative animal mtDNA are shown in Figure 1.

In the phylum Cnidaria, generally considered to be one of the most primitive groups of metazoans, there are peculiarities not shared with other mitochondrial systems. In a member of the class Anthozoa subclass Hexacorallia, *Metridium senile*, two mitochondrial protein genes, COI and ND5, contain a group I intron. Furthermore, COI intron encodes a putative homing endonuclease and the ND5 intron contains the ND1 and ND3 genes [74]. Other Anthozoa members of the subclass Octocorallia, *Renilla kolikeri* [72] and *Sarcophyton glaucum* [75,76], do not have introns but an extra gene coding for a protein with similarity to a bacterial mismatch repair protein. Among Anthozoa (Cnidaria), octocorallians have the same gene order, which differs from that of hexacorallians [72,77,78].

In protostomes (Arthropoda, Mollusca, Annelida, and Nematoda), mt gene arrangement is not stable within major groups. Among nematodes, *Ascaris suum* differs from *Caenorhabditis elegans* only in the location of the AT-rich region, but extensive rearrangements have occurred in the mtDNA of *Meloidogyne javanica* [79,80]. Particularly intriguing is the case of Mollusca, where all five available genomes show different gene arrangements. Interestingly, the two species of pulmonate gastropods, *Euhadra kerklotsi* and *Cepaea nemoralis*, although classified in the same superfamily, Helicoidea, have quite different arrangments of tRNA and protein coding genes. On the other hand, the mtDNA of *Albinaria cerulea*, another pulmonate gastropod only distantly related to the two just mentioned, shows in a portion of the genome the same gene order as in *Euhadra* mtDNA and in another portion the gene order of *Cepaea* mtDNA. This demonstrates that a gene rearrangement has occurred in members of the same superfamily and suggests an accelerated rate of gene rearrangement in some pulmonate lineages [81]. The mtDNA gene arrangement of another mollusk, *Katharina tunicata*, is highly unlike that of all other mollusk mtDNAs available so far but notably similar to that of *Drosophila* (Arthropoda). Furthermore, this mollusk contains two additional sequences that can be folded into tRNA-like structures, but it is not clear whether these are functional genes [82]. Finally, the mollusk *Mytilus edulis*, whose difference in mt gene content was mentioned in Sec. 4.1, has a mitochondrial gene arrangement radically different from those reported above [73].

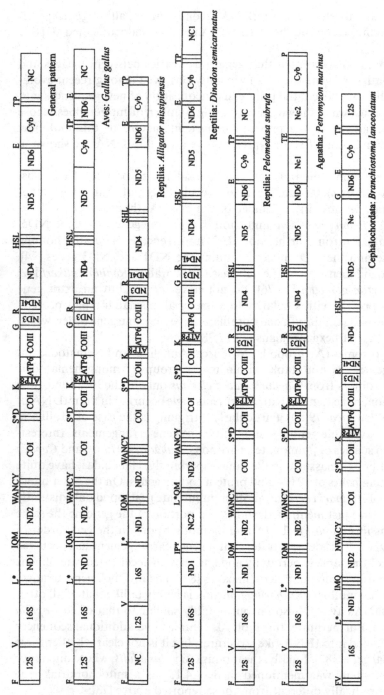

FIGURE 1 Comparative gene organization of metazoan linearized mtDNAs arbitrarily starting at the Phe-tRNA gene. The tRNA genes are indicated according to the transported amino acid: L, Leu(CUN); L*, Leu(UUR); S, Ser(AGY); S*, Ser(UCN); Ψ, a pseudogene. M° and M' denote the two tRNA genes for methionine in *Mytilus edulis*. NC: noncoding region. Introns found in *Metridium senile* are shaded in gray. Gene abbreviations: cytochrome oxidase subunits I, II, and III (COI, COII, COIII); cytochrome b apoenzyme (Cyb); NADH dehydrogenase subunits 1–6, 4L (ND1–6, ND4L); ATP synthase subunits 6, 8 (ATP6, ATP8); small and large rRNA subunits (12S, 16S).

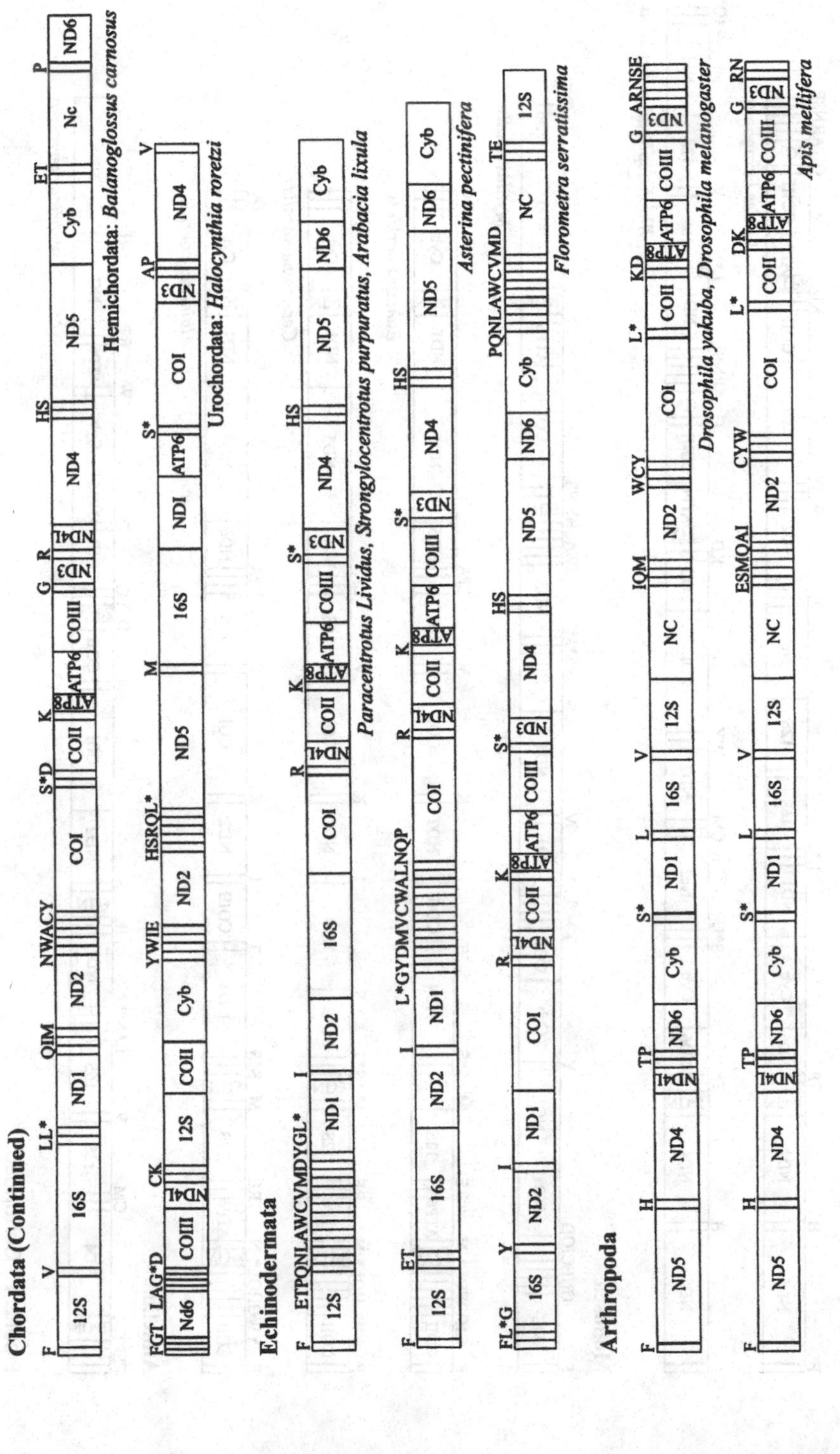

Chordata (Continued)

Hemichordata: *Balanoglossus carnosus*

Urochordata: *Halocynthia roretzi*

Echinodermata

Paracentrotus Lividus, Strongylocentrotus purpuratus, Arabacia lixula

Asterina pectinifera

Florometra serratissima

Arthropoda

Drosophila yakuba, Drosophila melanogaster

Apis mellifera

FIGURE 1 Continued.

Arthropoda (Continued)

Mollusca

Annelida

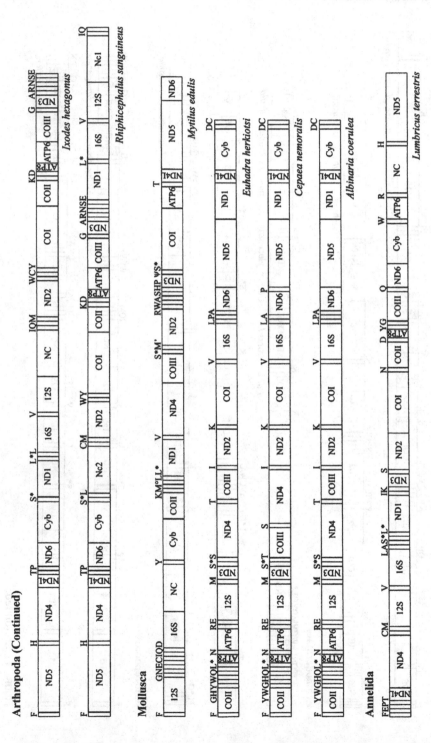

FIGURE 1 Continued.

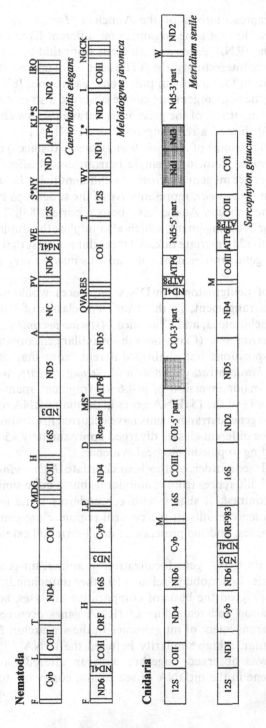

FIGURE 1 Continued.

The only representatives of the Annelida, *Lumbricus terrestris* and *Platynerei dumerii*, have a gene organization different from each other for the location of the tRNA genes and different from that of the other major protostome groups. Interestingly, the ATP8 gene is not immediately upstream of ATP6, a condition found only in pulmonate gastropods [83].

Extensive gene rearrangements can be observed also in Arthropoda. In the Insecta, the comparison of the gene order between *Apis mellifera* [84] and *Drosophila* [85,86] showed a rearrangement of 11 tRNA genes, whereas the organization of the genome of *Artemia franciscana* (Crustacea) is very similar to that of *Drosophila*, showing a single rearrangement affecting only two tRNAs [87]. The entire mt genome from two different ticks, *Ixodes hexagonus* and *Rhipicephalus sanguineus*, representative of the subgroups Prostriate and Metastriate of the subclass Acari, have been reported [88]. The sequences show an extensive rearrangement, which also implies the duplication of the control region in the Metastriate tick. All these data suggest that phylogenetic studies based on gene arrangement comparisons may be very misleading in many cases [89].

A survey of deuterostome mtDNA sequences would suggest a less frequent gene rearrangement. In three different classes of Echinodermata, i.e., Asteroidea, Echinoidea, and Crinoidea, *Asterina pectinifera* (Asteroidea) and *Florometra serratissima* (Crinoidea) show similar gene organizations with few tRNA transpositions but slightly different from that of Echinoidea (*Arbacia lixula*, *Paracentrotus lividus*, and *Strongylocentrotus purpuratus*) on account of a major inversion of a 4.6 kb fragment spanning from the 16S rRNA gene to 13 of the 15 tRNA genes in the main tRNA region [90–92]. Therefore, limited gene rearrangements have occurred in echnoderms, in spite of the fact that the different classes diverged approximately 450–550 million years ago, according to paleontological evidence.

The overall gene order of the hemichordate *Balanoglossus carnosus* shows substantial differences from echinoderms but is quite similar to that of vertebrates. In contrast, it shares with echinoderms some aspects of the genetic code, sequence motifs in the control region, close similarity of the two Leu-tRNA genes, and the presence of an N-terminal extension of ND5 [93].

All studies on tRNA gene localization in animal mt genomes clearly suggest their accelerated mobility relative to other mitochondrial genes [94]. Cantatore et al. [95], on the basis of comparative analyses, suggested that events of duplication and remolding of tRNA genes occurred during the evolutionary rearrangement of mt genomes in the sea urchin *Paracentrotus lividus*. In particular, a high similarity between the tRNA$^{Leu(CUN)}$ and the tRNA$^{Leu(UUR)}$ was observed, together with an altered location of the tRNA$^{Leu(CUN)}$ gene in the mtDNA of sea urchin compared to vertebrates.

In addition, a sequence (72 bp long) containing a trace of the old tRNA$^{Leu(CUN)}$ gene at its original location was observed in the sea urchin, where it coded for an extra amino acid sequence at the ND5 gene amino termini. These data were interpreted by assuming that during evolution a tRNA gene lost its function and became part of a protein-coding gene. This loss was accompanied by the gain of a new tRNA through duplication and divergence from a tRNA gene specific to a different family of codons, namely the tRNA$^{Leu(UUR)}$ gene. On the basis of these assumptions and of other observations indicating that tRNAs are present at the end of duplications and deletions, it has been suggested that tRNAs should be considered as mobile elements involved in gene rearrangement [96,97]. To explain this property we put forward the hypothesis that a gene flanked by two tRNAs can be considered to be very similar to a transposable element, with the two tRNAs corresponding to long terminal repeats (LTRs), each of which has short inverted sequences (amino acid stems) at its end [97]. In addition, it should be considered that as a consequence of the compactness of the metazoan mt genome, tRNA genes were forced to assume multiple roles and regulatory functions. It has been indicated that mitochondrial tRNAs might play a role in the origin of replication [98]. It is known that tRNAs act as recognition signals in vertebrates, where they are scattered along the molecule and make up a sort of punctuation signal for the processing of polycistronic transcripts [99]. On account of the acquisition of such multiple roles by tRNAs, we have speculated that in the course of evolution animal mt tRNAs, which had to cope with the dramatic size reduction of mtDNA, adapted in order to fulfill new tasks. This caused a less stable L-shaped structure and a wider ambiguity in the decoding of the genetic code [100].

It is well known that gene rearrangement events may be used to reconstruct the ancestral gene order and then the phylogenetic interrelationships between organisms (see below). In this case, by assuming that tRNAs move independently of other genes, the possible phylogenetic interrelationships between metazoans can be established because of the transposition events of the ribosomal and messenger genes. The unrooted tree constructed by Cantatore et al. [95] using this criterion with the genomes of Vertebrata, Nematoda, Echinodermata, and Insecta places Echinodermata and Vertebrata more closely in relation, because they are separated by only four events.

4.3. The Noncoding Regions

A peculiar feature of metazoan mt genomes is the presence of a main noncoding region that contains the regulatory elements for the replication and expression of the mtDNA. Regardless of the high degree of conservation of the coding genes, this region shows great variability in length and base

TABLE 4. Nucleotide Composition of the Complete Genome and of the Third Codon Positions of Protein-Coding Genes of Completely Sequenced Metazoan mtDNAs

Organism		Complete genome, nucleotide %				Third codon positions, nucleotide %			
		T	C	A	G	T	C	A	G
Chordata									
Mammalia									
Artibeus jamaicensis	AF061340	32.13	24.78	30.71	12.38	30.47	30.53	41.51	3.55
Balaenoptera musculus	X72204	26.60	27.60	32.80	13.00	19.70	34.30	42.00	4.00
Balaenoptera physalus	X61145	26.70	27.30	32.70	13.30	19.90	33.50	41.50	5.10
Bos taurus	V00654	27.20	25.90	33.40	13.40	20.90	30.80	42.70	5.60
Canis familiaris	U96639	30.47	26.04	29.75	13.74	25.57	28.98	37.93	7.52
Cavia porcellus	AJ222767	28.60	24.80	32.00	14.60	24.40	28.00	40.10	7.50
Ceratotherium simum	Y07726	27.41	28.53	31.06	13.00	18.30	34.03	41.78	5.89
Dasypus novemcinctus	Y11832	27.85	26.70	32.62	12.83	18.62	29.85	46.36	5.16
Didelphis virginiana	Z29573	31.50	21.10	35.30	12.00	31.90	19.10	46.10	2.90
Equus asinus	X97337	27.21	29.33	30.27	13.18	17.65	36.44	39.65	6.26
Equus callus	X79547	25.90	28.50	32.20	13.40	17.70	36.20	39.70	6.40
Erinaceus europaeus	X88898	33.40	20.10	34.00	12.50	37.50	17.50	40.40	4.60
Felis catus	U20753	27.10	26.20	32.60	14.10	21.50	30.70	40.50	7.30
Glis glis	AJ001562	33.39	23.78	30.29	12.54	32.24	23.46	39.93	4.38
Gorilla gorilla	D38114	25.20	30.70	30.90	13.20	18.50	39.40	35.80	6.30
Halichoerus grypus	X72004	25.30	27.50	33.00	14.30	17.80	33.80	40.50	7.90
Hippopotamus amphibius	AJ010957	26.08	29.49	30.58	13.84	15.85	36.27	40.54	7.33
Homo sapiens	D38112	24.70	31.20	30.90	13.10	16.50	41.40	35.70	6.40
Hylobates lar	X99256	23.96	31.76	30.59	13.70	17.11	39.69	33.75	9.45
Loxodonta africana	AJ224821	28.47	25.22	32.78	13.53	40.26	25.95	22.14	11.66
Macropus robustus	Y10524	29.36	27.58	30.31	12.75	22.98	31.92	40.54	4.56

Mus musculus	V00711	28.70	24.40	34.50	12.30	24.10	26.50	45.30	4.10
Ornithorhyncus anatinus	X83427	31.40	23.50	31.40	13.60	31.40	25.00	38.00	5.60
Orycteropus afer	Y18475	28.70	25.41	33.25	12.63	30.74	27.59	42.1	4.77
Oryctolagus cuniculus	AJ001588	30.28	26.92	29.23	13.57	26.07	29.99	37.43	6.51
Ovis aries	AF010406	28.85	26.75	31.66	12.75	22.35	29.86	42.71	5.08
Pan paniscus	D38116	25.30	30.70	31.30	12.70	18.20	39.60	37.10	5.10
Pan troglodytes	D38113	25.20	30.80	31.10	12.90	18.30	39.60	36.30	5.80
Papio hamadryas	Y18001	26.25	31.18	29.71	12.86	17.94	38.72	36.25	7.09
Phoc vitulina	X63726	25.30	27.40	33.00	14.30	17.80	33.80	40.50	7.90
Pongo pygmaeus	D38115	23.80	32.40	30.50	13.20	14.70	43.70	35.40	6.20
Rattus norvegicus	X14848	27.30	26.30	34.00	12.40	21.00	31.90	42.90	4.20
Rhinoceros unicornis	X97336	27.66	28.24	31.40	12.70	18.57	33.35	42.50	5.58
Sus scrofa	AJ002189	27.40	26.85	32.92	12.83	18.25	30.51	46.45	4.79
Aves									
Aythya americana	AF090337	23.70	33.77	26.91	15.61	10.78	43.35	35.74	10.13
Corvus frugilegus	Y18522	26.02	30.75	29.04	14.18	17.79	34.89	40.92	6.39
Falco peregrinus	AF090338	24.81	32.53	29.47	13.19	14.26	39.76	40.63	5.34
Gallus gallus	X52392	23.80	32.40	30.20	13.50	13.80	42.80	37.70	5.70
Rhea americana	Y16884	25.57	33.75	26.05	14.62	15.81	42.97	32.89	8.34
Smithornis sharpei	AF090340	26.21	33.19	27.69	12.89	17.50	41.98	35.49	5.03
Struthio camelus	Y12025	24.89	30.45	30.42	14.24	23.44	30.80	31.12	14.64
Vidua chalybeata	AF090341	23.74	32.10	29.64	14.52	11.32	38.63	43.50	6.56
Reptilia									
Alligator mississippiensis	Y13113	25.73	29.50	31.24	13.53	23.45	29.51	33.35	23.45
Chelonia mydas	AB012104	26.73	28.45	33.25	11.58	17.04	31.97	48.62	2.37
Chrysemys picta	AF069423	28.24	26.88	32.09	12.79	20.28	28.75	46.16	4.81
Dinodon semicarinatus	AB008539	26.76	27.39	33.78	12.07	18.68	29.98	46.31	5.04
Eumeces egregius	AB016606	26.93	29.13	28.11	15.80	18.65	33.45	37.11	10.79
Pelomedusa subrufa	AF039066	28.96	27.43	31.59	12.02	22.69	30.64	42.20	4.47

TABLE 4 Continued

Organism		Complete genome, nucleotide %				Third codon positions, nucleotide %			
		T	C	A	G	T	C	A	G
Amphibia									
Xenopus laevis	M10217	30.00	23.50	33.00	13.50	30.00	22.40	42.70	4.90
Osteichthyes									
Carassius auratus	AB006953	28.13	26.88	29.45	15.54	22.29	27.58	43.14	6.99
Crossostoma lacustre	M91245	25.00	28.60	31.90	16.90	20.10	34.80	35.70	9.50
Cyprinus carpio	X61010	24.90	27.40	31.60	15.80	18.90	31.20	44.10	5.80
Gadus morhua	X99772	32.37	25.69	25.82	16.11	33.53	24.05	34.21	8.22
Gonostoma gracile	AB016274	27.29	33.21	24.06	15.44	22.75	39.81	29.20	8.24
Latimeria chalumnae	U82228	20.85	27.82	35.82	15.51	17.14	28.24	47.70	6.92
Partalicththys olivaceus	AB028664	26.11	29.63	27.43	16.83	22.45	32.06	27.91	17.59
Oncorhynchus mykiss	L29771	26.20	28.90	27.90	17.00	23.80	34.00	33.40	8.80
Polypterus ornatipinnis	U62532	29.90	26.20	30.49	13.41	25.07	27.81	42.69	4.42
Protopterus dolloi	L42813	29.00	26.50	28.80	15.70	28.50	28.50	34.70	8.40
Salmo salar	U12143	26.33	29.01	28.44	16.22	26.91	31.77	30.51	10.81
Salvelinus alpinus	AF154851	28.96	29.11	25.68	16.25	24.94	33.31	33.28	8.47
Salvelinus fontinalis	AF154850	29.23	28.74	25.80	16.23	25.68	32.26	33.46	8.60
Chondrichthyes									
Mustelus manazo	AB015962	33.04	25.04	28.65	13.28	32.17	25.09	38.55	4.20
Raja radiata	AF106038	31.13	27.10	27.85	13.91	27.34	30.33	36.11	6.22
Scyliorhinus canicula	Y16067	33.15	24.30	28.71	13.85	31.74	24.00	38.96	5.30
Squalus acanthias	Y18134	32.18	25.35	28.52	13.95	29.25	26.60	38.75	5.40

Agnatha									
Lampetra fluviatilis	Y18683	29.67	24.40	31.71	14.21	36.88	24.47	29.07	9.57
Petromyzon marinus	U11880	30.40	23.80	32.30	13.50	33.50	21.60	41.10	3.80
Cephalochordata									
Branchiostoma floridae	AF098298	37.23	16.47	24.93	21.37	37.66	11.91	31.91	18.51
Branchiostoma lanceolatum	Y16474	37.10	16.53	24.98	21.39	37.28	12.19	31.90	18.63
Hemichordata									
Balanoglossus carnosus	AF051097	28.03	32.21	22.43	17.33	21.23	39.54	26.63	12.60
Urochordata									
Halocynthia roretzi	AB024528	44.01	8.52	24.26	23.21	49.86	3.09	26.00	21.05
Echinodermata									
Arbacia lixula	X80396	33.00	20.40	29.50	17.00	33.70	18.00	36.30	11.90
Asterina pectinifera	D16387	28.90	24.60	32.40	14.10	27.50	24.30	35.70	12.50
Florometra serratissima	AF049132	45.56	11.97	27.00	15.50	54.26	2.96	33.74	9.04
Paracentrotus lividus	J04815	29.50	22.50	30.80	17.20	26.20	22.20	39.10	12.50
Strongylocentrotus purpuratus	X12631	30.20	22.70	28.70	18.40	27.10	24.60	33.80	14.50
Arthropoda									
Anopheles gambiae	L20934	37.50	12.90	40.00	9.50	47.00	4.30	45.30	3.40
Anopheles quadrimaculatus	L04272	37.10	13.40	40.20	9.30	46.10	5.00	44.30	4.60
Apis mellifera	L06178	41.60	9.60	43.20	5.50	47.60	2.80	47.60	2.00
Artemia franciscana	X69067	33.50	17.90	31.00	17.70	29.90	24.30	35.50	10.30
Ceratitis capitata	AJ242872	44.09	11.94	31.51	12.46	49.23	4.39	43.11	3.27
Daphnia pulex	AF117817	37.04	20.17	23.39	19.40	36.48	20.41	26.78	16.34
Drosophila melanogaster	U37541	40.40	10.30	41.80	7.60	49.00	3.00	45.40	2.60
Drosophila yakuba	X03240	39.10	12.20	39.50	9.20	48.50	3.30	45.30	2.90
Ixodes hexagonus	AF081828	41.24	15.68	29.88	13.19	39.56	15.25	35.90	9.29
Limulus polyphemus	AF00264	30.72	21.85	36.71	10.07	31.25	22.32	37.72	7.94

TABLE 4 Continued

Organism		Complete genome, nucleotide %				Third codon positions, nucleotide %			
		T	C	A	G	T	C	A	G
Locusta migratoria	X80245	30.80	14.60	44.50	10.10	42.30	8.40	44.80	4.50
Rhipicephalus sanguineus	AF081829	43.92	11.75	34.04	10.29	46.33	7.03	43.25	3.39
Mollusca									
Albinaria coerulea	X83390	37.90	13.80	32.70	15.50	42.00	9.60	38.50	9.90
Cepae anemoralis	U23045	33.60	18.90	26.20	21.30	36.30	20.10	26.00	17.60
Crassostrea gigas	AF177226	35.71	14.66	27.64	21.99	39.63	11.56	28.51	30.30
Euhadra herkiotsi	Z71693	39.95	14.13	29.48	16.39	38.75	13.18	30.24	17.83
Katharina tunicata	U09810	39.58	15.35	28.02	16.93	40.57	12.27	36.05	11.01
Mytilus edulis	M83756	34.34	14.00	27.90	23.65	32.81	13.60	29.65	23.81
Pupa strigosa	AB028237	33.70	18.34	27.43	20.53	36.15	16.80	27.34	19.71
Anellida									
Lumbricus terrestris	U24570	31.80	22.50	29.80	15.80	29.90	24.30	35.50	10.30
Platynereis dumerii	AF178678	32.92	20.45	31.22	15.41	34.86	22.16	27.83	15.15
Nematoda									
Ascaris suum	X54253	49.80	7.70	22.20	20.40	62.50	2.50	11.60	23.40
Caenorhditis elegans	X54252	44.80	8.90	31.40	14.90	49.70	4.70	36.60	9.00
Onchocerca volvulus	AF015193	55.45	7.10	16.49	20.96	68.00	1.38	9.17	21.45
Platyhelminthes									
Echinococcus multilocularis	AB018440	48.48	7.60	20.56	23.36	57.51	2.21	16.00	24.29
Cnidaria									
Metridium senile	AF000023	37.56	16.87	25.13	20.44	40.50	14.48	28.67	16.81
Sarcophyton glaucum	AF064823	37.10	16.20	27.80	18.90	41.72	10.70	33.20	14.60

composition. It ranges from only 121 bp in the sea urchin to 4601 bp in *Drosophila* [101].

Differently from Vertebrata, in Invertebrata the structure and evolution of the regulatory region has not yet been fully characterized. In *Ascaris* and *Drosophila* this region is called an AT-rich region for its extremely high A + T content. In *Ascaris* the main AT region is 886 bp long, and a smaller noncoding sequence of 117 bp, which can be folded into a stem-and-loop structure, is also present. In *Drosophila* the AT region is extremely polymorphic; it varies in sequence and length both in different species (1077 bp in *Drosophila yacuba* and 4601 in *Drosophila melanogaster*) and within individuals of the same species or in different mtDNA molecules of a single fly. The putative promoters and the replication origin are contained in two conserved regions, one of which can form a hairpin structure.

In echinoderms the main noncoding region (121–445 bp) is located in the tRNA cluster. It appears as a condensed version of the vertebrate replication origin, and the nascent strand coincides with a very stable stem–loop structure.

In vertebrates the main noncoding region is called the D-loop-containing region, because starting from the strand replication origin (O_H) the heavy strand displaces (D) the parental one, creating a triple-strand structure (see Sec. 5.2). This region, ranging from 879 bp (mouse) to 2134 bp (*Xenopus*), also contains the promoters for both the heavy strands (HSP) and light strands (LSP). The two strands are called heavy (H) and light (L) according to their isopycnic sedimentation in the cesium chloride gradient. The other noncoding region contains the origin of the light strand replication (O_L); it is only 30 bp long and is flanked by five tRNA genes. This region can be folded into a stable stem-and-loop structure that is very conserved in all vertebrates except birds. Indeed, the sequence equivalent to the O_L has not been found at the same position in the bird mt genomes sequenced so far.

In rats and humans the two origins of replication show an intrinsic DNA curvature, correlated with periodic distribution of dinucleotides in the sequence and involved in protein interactions [102,103].

4.4. Base Composition

Table 4 reports the nucleotide composition percentage of the complete genome and of the third codon positions of protein-coding genes of several metazoans. It can be observed that the AT/GC content is highly variable, with the highest GC content observed in *Balanoglossus carnosus* and the lowest in *Apis mellifera*, whose genome is extremely AT-rich. A striking heterogeneity of the base composition can be observed between different lineages as well as within the same lineage. The general compositional pattern is also reflected in

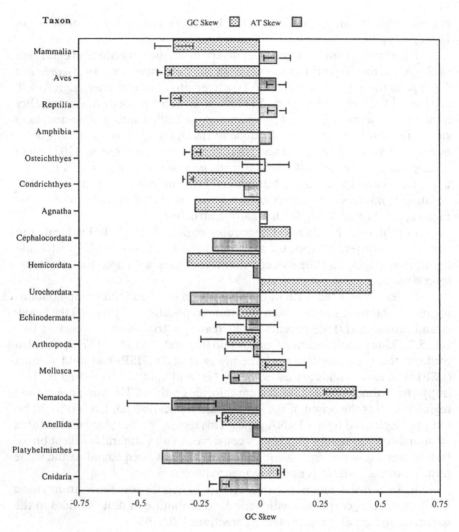

FIGURE 2 Graphical representation of mean and standard deviation of GC and AT skews calculated on the whole mitochondrial genome for all metazoan taxa in Table 2.

the nucleotide composition percentage at the third codon positions. For example, in arthropods the G + C content at the third codon positions ranges between 37% in *Daphnia pulex* and 4.80% in *Apis mellifera*. A conserved compositional pattern is observed in vertebrates where a remarkable asymmetrical distribution of the complementary nucleotides between the two strands can be observed, particularly evident for the GC distribution, with G-ending codons generally avoided [104,105] (see Sec. 5.3). A different pattern is observed in other metazoans whose complete genome base composition displays a relatively high content of G, with the C content surprisingly low in some species.

The compositional features of the metazoan mitochondrial genomes can be represented in terms of the degree of asymmetry between complementary nucleotides (GC and AT skews; see Sec. 5.3). Average AT and GC skews calculated on the base composition of complete genomes for various metazoan lineages are shown in Figure 2.

In Chordata the GC skew is always considerably negative, with the exception of Cephalochordata and Urochordata, whereas the AT skew is very low but generally slightly positive. A taxon-specific pattern can be observed for other organisms both between different taxa and within the same taxon. Urochordata, Nematoda, and Platyhelminthes show a strong compositional asymmetry with a positive GC skew and a negative AT skew. The same pattern, but to a lesser extent, is shown by Mollusca and Cnidaria, where, as in all the remaining taxa, the compositional asymmetry is very low.

5. EVOLUTION OF THE MITOCHONDRIAL GENOME IN CHORDATA

5.1. Genome Features and Organization

The mitochondrial gene organization seems relatively stable only in Chordata, where only a few transpositions are tolerated as reported in Figure 1. As a matter of fact, an identical genome organization has been found among many different eutherian mtDNAs. In particular, it should be noted that the gene arrangement described in Eutheria has been found to be identical to that in Prototheria, Amphibia, Testudines (*Pelomedusa subrufa*), Osteichthyes, and Chondrichthyes. Marsupialia show rearrangements of tRNAs only, whereas the tRNA^{Lys} has been lost and the corresponding sequence has become a pseudogene. Therefore, marsupial mitochondria should import a nuclear coded tRNA^{Lys} to ensure the complete set of tRNAs [106]. In Aves, tRNA^{Glu} and ND6 are translocated immediately adjacent to the control region [107]. Interestingly, a novel arrangement of bird mtDNA was described by Mindell et al. [108], suggesting that the ancestral translocation involving ND6 + tRNA^{Glu} was accompanied by a duplication of the main noncoding region. The presence or absence of the duplicated control region should account for the two alternative arrangements observed in the mtDNA of extant birds investigated so far [6]. It is also remarkable to note that a nontranslated extranucleotide has been found in the ND3 gene of some birds and turtles, thus further supporting diapsid affinity of turtles [109,110]. In *Alligator mississippiensis* (Crocodilia) only three tRNAs moved to different locations [111], and in *Dinodon semicarinatus* (Serpentes) not only is one tRNA transposed but the control region is duplicated and flanked by a pseudogene consisting of the 5′ half-portion of the tRNA^{Pro}, which can be folded into a rather stable secondary structure [112]. In *Petromyzon marinus* (Agnatha), the control region is transposed, and it is interrupted by the presence of two tRNAs [113]. In *Branchiostoma lanceolatum* (Cephalochordata), the control region is translocated and reduced to only about 80 nucleotides, and few tRNAs have been transposed [114].

In comparing the mt gene organization in Vertebrata to that of other animals, the most remarkable feature is the different distribution of tRNA genes. In Vertebrata, tRNA genes are scattered along the molecule and make up a sort of punctuation that functions as signals for RNA processing [99].

5.2.   The D-Loop Region

As noted in Sec. 4.3, vertebrate mtDNA possesses a unique main noncoding region, called the D-loop, that contains the regulatory elements for the replication and expression of the genome. Regardless of its functional importance, this is the most rapidly evolving part of the mtDNA. Detailed comparative analysis of this region has revealed several peculiar well-conserved features in the evolution of vertebrates [115–117]. In particular, the 5′ and 3′ ends contain thermodynamically stable secondary cloverleaf-like structures and short conserved sequence boxes (CSBs) and termination-associated sequences (TASs), which are associated with the start and stop sites of the nascent H strand. How the function of this region is preserved in spite of such primary structure diversity remains to be clarified.

The D-loop region has been extensively studied only in mammals, where it shows great variability in length and base composition. For this reason, the complete alignment of the D-loop region is not possible between mammals even within the same order. Figure 3 shows the structural organization of the D-loop region of 10 mammalian orders. The organization is similar in all the organisms considered, and on the basis of both degree of conservation and base content, the D-loop has been divided into three domains: a highly conserved central domain flanked by two hypervariable regions, the "extend-

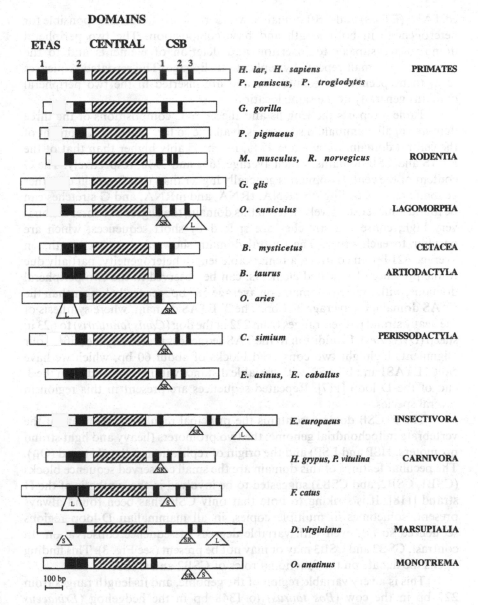

FIGURE 3 General scheme of the organization of the D-loop-containing region in various mammalian orders. ETAS, central, and CSB domains are defined. In the ETAS domain, the shaded boxes represent the conserved sequences. In the CSB domain, the filled boxes correspond to CSB1, CSB2, and CSB3. SR, short repeats; LR, long repeats.

ed TAS" (ETAS) and CSB domains, whose rapid evolution is responsible for heterogeneity in both length and base composition. The two peripheral domains are subject to insertion and deletion of elements and to the generation of small repeats by replication slippage. The repeated sequences seem to be peculiar to each order and are inserted in the two peripheral domains generally at the same locations.

Table 5 reports the lengths and the G + C compositions of the three domains in all mammalian sequences available so far. The G + C content of the central domain, on average 47%, is remarkably higher than that of the ETAS and CSB domains, which average 36% and 39%, respectively. The G content of the central domain is generally higher than that found in the other genomic regions coding for rRNA, tRNA, and mRNA, and G stretches can be found almost exclusively here. In this domain, the degree of conservation is very high; conserved stretches are spaced by short sequences, which are peculiar to each order. The central domain shows a constant length, on average 321 bp. In contrast, a remarkable length heterogeneity, partially due to the presence of repeated elements, can be observed in the two peripheral domains, with the CSB domain, on average 587 bp, generally longer than the ETAS domain, on average 344 bp. The 3' ETAS domain, where synthesis of the heavy strand pauses, ranges from 222 in the dog (*Canis familiaris*) to 623 in sheep (*Ovis aries*). In addition to the TAS sequences previously identified, our alignments highlight two conserved blocks of about 60 bp, which we have called ETAS1 and ETAS2, with variable distance within them and from the 3' end of the D loop [117]. Repeated sequences are present in this region in several species.

The 5' CSB domain contains the principal regulatory elements of the vertebrate mitochondrial genome: the two promoters (heavy and light strand promoters, HSP and LSP) and the origin of replication of the H strand (O_H). The peculiar features of this domain are the small conserved sequence blocks (CSB1, CSB2, and CSB3) suggested to be involved in the synthesis of the H strand [118]. It is striking to note that only CSB1 has been found always present, sometimes in multiple copies, in all mammalian D-loop regions sequenced so far, even with variable degrees of sequence conservation. In contrast, CSB2 and CSB3 may or may not be present (see Fig. 3). This finding opens the debate on the functional roles of CSB2 and CSB3.

This is a very variable region of the genome, and its length ranges from 227 bp in the cow (*Bos taurus*) to 1348 bp in the hedgehog (*Erinaceus europaeus*). The domain length is greatly increased by the presence of repeats, which are frequently found. The region spanning between CSB1 and CSB2 is a hotspot for the insertion of repeated sequences that are found in several species (rabbit, horse, seal, hedgehog, opossum, etc.). Repeated motifs can

TABLE 5　Length and Base Composition of D-Loop Domains in Mammals

Organism	D-loop L (bp)	ETAS domain L (bp)	ETAS domain (C+G)%	Central domain L (bp)	Central domain (C+G)%	CSB domain L (bp)	CSB domain (C+G)%
Artibeus jamaicensis (fruit bat)	1228	289	44	312	53	627	45
Balaenoptera musculus (blue whale)	917	295	32	345	49	277	39
Balaenoptera physalus (fin whale)	936	334	44	325	45	277	39
Bos taurus (cow)	910	345	29	338	48	227	37
Canis familiaris (dog)	1270	222	39	303	45	745	44
Cavia porcellus (guinea pig)	1357	356	36	315	49	686	48
Ceratotherium simum (white rhinoceros)	1381	352	38	317	45	712	40
Dasypus novemcinctus (armadillo)	1604	416	40	323	42	865	29
Didelphis virginiana (opossum)	1615	534	29	311	36	770	15
Equus asinus (donkey)	1207	267	37	319	48	621	51
Equus callus (horse)	1192	268	37	318	48	606	41
Erinaceus europaeus (hedgehog)	1988	323	22	317	45	1348	27
Felis catus (cat)	1560	515	35	325	44	720	45
Glis glis (dormouse)	1157	413	34	315	49	429	28
Gorilla gorilla (Western highland gorilla)	918	240	42	328	52	350	48
Halichoerus grypus (grey seal)	1360	273	44	326	45	761	45
Hippopotamus amphibius (hippopotamus)	960	285	36	338	47	337	39
Homo sapiens (human)	1122	347	45	334	52	442	44
Hylobates lar (gibbon)	1029	362	43	308	54	359	48
Loxodonta africana (elephant)	1449	326	36	317	45	806	44
Macropus robustus (wallaroo)	1428	534	25	311	39	583	37
Mus musculus (mouse)	879	259	29	308	46	312	33
Ornithorhyncus anatinus (platypus)	1559	293	29	319	45	947	26
Orycteropus afer (aardvark)	1364	326	30	316	48	722	36
Oryctolagus cuniculus (rabbit)	1838	365	32	339	48	1134	38
Ovis aries (sheep)	1180	623	32	314	47	243	36
Pan paniscus (pygmy chimpanzee)	1121	370	44	311	52	440	45
Pan troglodytes (common chimpanzee)	1113	370	44	310	52	433	48
Papio hamadryas (baboon)	1076	372	37	313	52	391	48
Phoca vitulina (harbor seal)	1391	274	42	326	45	791	38
Pongo pygmaeus (Bornean orangutan)	917	263	46	333	54	321	46
Rattus norvegicus (rat)	894	231	26	331	47	332	33
Rhinoceros unicornis (Indian rhinoceros)	1376	341	35	317	46	718	41
Sus scrofa (pig)	1246	318	32	316	46	612	38
Average	1251	344	36	321	47	587	39

also be found in the region spanning between CSB3 and tRNA^{Phe} (e.g., rabbit, opossum, and hedgehog).

5.3. Base Composition and Skew

The data in Table 4 show that base composition is rather variable between species, with G being always the less abundant nucleotide. This is particularly evident at the third codon positions of protein-coding genes.

In mammals, a sharp separation, mainly at the level of the third codon positions, is evident between base composition of Primates and that of other mammals with A% > C% in the former and C% > A in the latter. On the whole, the only constant element in gene base composition of the H strand is the low percentage of G, which is avoided at all positions, although with different rates.

All species show a rather similar base composition at the level of the first and second codon positions (data not shown). This is expected because of the strong functional constraints acting on these sites and the high degree of conservation of mitochondrial proteins.

The peculiar amino acid composition of mitochondrial proteins, which are rich in hydrophobic residues because they are membrane proteins, would explain base composition observed at the two first codon positions. Namely, the higher Py(pyrimidine) than Pu(purine) content in the second codon positions (P2) can be easily explained with the fact that, due to their genetic structure, codons with Py in P2 code most hydrophobic acids, whereas those with Pu in P2 code for hydrophilic amino acids. The abundance of Ile, Thr, and Met, amino acids coded by codons having A at first positions (P1), makes A the most represented base in P1, the other bases being present in rather comparable percentages.

As reported earlier, one of the most striking features of the vertebrate mitochondrial genome is the uneven distribution of G and C bases on the two strands (compositional asymmetry). This is so strong as to cause differences in buoyant density in the CsCl gradient between the H strand, rich in G, and the L strand, poor in G. Asymmetry is evident also in other properties of mtDNA, namely gene distribution on the two strands and the replication mechanism. Hence there appears to be a correlation between compositional asymmetry and other forms of mtDNA asymmetry.

The H strand codes most mitochondrial genes (29 out of total 37); among these are 12 of the 13 protein genes (only ND6 gene is coded by the L strand). Owing to the high G content of the H strand, mRNAs coded on this strand are particularly poor in G. This is most evident at the third codon position of protein-coding genes where G-ending codons are avoided.

The degree of compositional asymmetry, expressed in terms of GC and AT skews, can be calculated by using the formulas of Perna and Kocher [119]:

$$GC_{skew} = \frac{G - C}{G + C}, \quad AT_{skew} = \frac{A - T}{A + T}$$

where G, C, A, and T are the occurrences of the four nucleotides. According to these formulas, skew values are in the range -1, $+1$ and compositional asymmetry is greater the closer the skew absolute values approach one (positive or negative sign) and lower the closer the skew values approach zero.

The standard deviations of the skew values can be calculated according to Lobry [120]:

$$SD_{GCskew} = \frac{2}{G + C} \left[\frac{G \cdot C}{G + C} \right]^{1/2} \quad SD_{ATskew} = \frac{2}{A + T} \left[\frac{A \cdot T}{A + T} \right]^{1/2}$$

Figure 2 shows GC and AT skews calculated on the complete genomes of various vertebrate taxa. In all the cases a remarkably negative GC skew is observed, higher in Reptilia, Aves, and Mammalia. To a much lesser extent, a generally positive AT skew is observed. Various models can be suggested to explain the peculiar compositional bias of the vertebrate genome. Metabolic discrimination between nucleotide bases and/or replication erros followed by biased repair could account for this property. Other explanations may be based on the mechanism of mtDNA replication, which, being asymmetrical, leaves one DNA strand as a single, unprotected filament for two-thirds of the replication cycle (see below).

The possibility of damage directly at the level of RNAs cannot be excluded [58]. The need to protect mtDNA transcripts against the attack of free radicals, preferentially affecting G [121], could have originated the dramatic reduction in G content of transcripts and hence an increase in the G percentage on the template strand of transcription (the H strand) and the consequent compositional asymmetry [101].

The strong compositional bias, in particular the trend to avoid G at the third codon position, might also explain several deviations of the vertebrate mt code with respect to the universal one. In vertebrate mitochondria, the most used initiation codon is AUA instead of AUG, which is a G-ending codon. One of the three nonsense codons, UGA, becomes an additional tryptophan codon because of the two contiguous G's in the canonical UGG tryptophan codon, which is very rare in the sense strand of metazoan mtDNA. In other words, the compositional bias was so strong as to influence the genetic code itself.

To gain insight into the cause of mtDNA compositional asymmetry and to identify possible differences in the substitution processes acting on the two DNA strands, the compositional properties of mtDNA from 34 mammalian species have been studied. Figure 4 is a graphical representation of GC and AT skews calculated on the entire genome of 34 mammalian species. In all cases, the GC skew, on average -0.34, was more than twofold (as absolute value) the corresponding AT skew, on average $+0.07$. Primates showed the highest GC skew. To study the compositional bias and asymmetry of mt genomes, we also considered the third positions of the quartet codons (P3Q) because they should better depict genome compositional pressure on account of their more relaxed functional constraints. Indeed, in all cases the compositional bias and asymmetry was much more evident when P3Q positions were considered. In particular, we studied the possible relationships between the compositional features of mammalian genomes and the replication process. It is well known that the replication of vertebrate mtDNA is asymmetrical and unidirectional [122] according to the D-loop mechanism shown in Figure 5. Mitochondrial DNA replication takes place according to an asymmetrical mechanism of asynchronous displacement between the H and L strands, which are synthesized in opposite directions starting from two different replication origins. The replication process starts with the synthesis of the new H strand; the synthesis of the new L strand starts later when the relevant replication origin (O_L) is exposed as a single-helix structure (see Fig. 5), that is, the parental L strand is never left as a single helix during replication, whereas the parental H strand exists as a single helix until the corresponding complementary region of the L strand is newly synthesized. Because replication takes place rather slowly [122], the different regions of the parental H strand remain as a single helix for a rather long time, varying according to their distance from O_L (distance measured in the replication direction of the H strand). During this time, the single helix of the parental H strand is protected only partially by mtSSB proteins; thus it could be more exposed to oxidative or hydrolytic damage and more prone to mutations than the L strand. Therefore, it has been suggested that differences in mutational pressure on the two H and L strands could cause the compositional asymmetry of mtDNA [123]. It is striking to note that DNA polymerase γ, used for mtDNA replication of both strands, is one of the most accurate among eukaryotic polymerases [124–126].

We have found that AT and GC skews, the base composition of P3Q sites, and the degree of gene conservation are correlated to the duration of the single-stranded state of the H-stranded genes during replication (D_{ssH}) calculated according to Reyes et al. [127]. Figure 6 shows the correlation between the base composition on P3Q of the H-strand protein-coding genes and D_{ssH}. We observed a significant increase in A and C frequencies and a corresponding significant decrease in G and T frequency with the D_{ssH}. On the

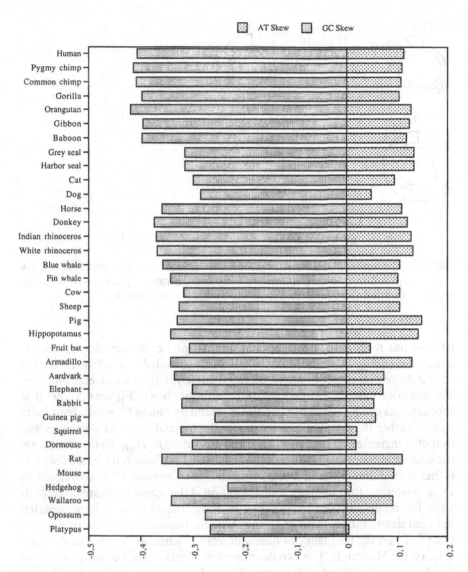

FIGURE 4 Graphical representation of GC and AT skews on the whole mitochondrial genome for 34 mammalian species (see Table 2).

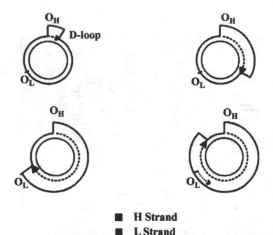

■ H Strand
■ L Strand

FIGURE 5 Schematic representation of the asymmetrical replication of the mammalian mitochondrial genome from H- and L-strand replication origins (O_H and O_L). Dashed lines refer to the newly synthesized strands.

basis of our results we suggested that the hydrolytic deamination of both C and A, depending on the duration of the single-stranded state of the H (heavy) strand during replication, might be responsible for such a correlation [127]. The mutation pattern we hypothesize, in the upper box of Figure 6, shows that spontaneous deamination of C on the H strand produces U, which base-pairs with A rather than G, and consequently the percentage of G decreases and that of A increases on the L (light) strand according to D_{ssH}. In the same way, the deamination of A results in hypoxanthine [128], which base-pairs with C rather than T, resulting in a reduction of T and increase of C correlated to D_{ssH}. Assuming that hydrolytic deamination is the major mechanism responsible for compositional asymmetry in P3Q, the ratio can be calculated between deamination rates of C and A (ratio between the slope of lines A/ G or C/T and the frequency of bases subject to deamination, that is, A% and C%, on the H strand). The two deamination events are not equally probable; the A_H deamination rate is 0.22 and the C_H deamination rate is 1.88. Thus, C deamination is about 8-fold that of A deamination, as reported in studies carried out on human mtDNA only [129].

    If the main factor responsible for the compositional asymmetry and mutation rate of mtDNA is the time for which genes are left as a single helix, most constrained genes should be located closer to O_L so as to be less subject to mutational pressure. The position of a gene in the genome of vertebrates should be the result of a balance created during evolution between gene functional constraints and mutational pressure on that region of the genome.

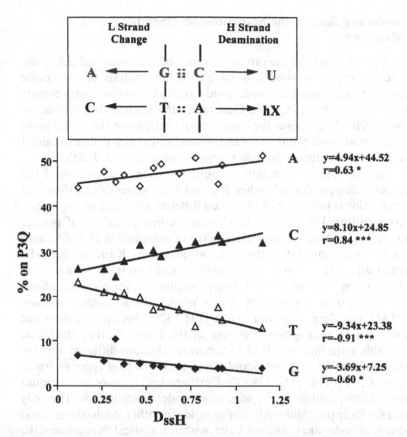

FIGURE 6 Bottom: Correlation between nucleotide percentages calculated on the third position of quartet codons (P_3Q) and the single-stranded state duration for each H-stranded gene (D_{ssH}) as mean values for 34 mammalian species (see Table 2). The linear regression equations are also reported (* $p < 0.05$; *** $p < 0.005$). Top: Suggested deamination processes that may take place in the single-stranded H strand: cytosine (C) into uracil (U) and adenine (A) into hypoxanthine (hX), which would imply changes of G into A and of T into C on the L strand.

By measuring the degree of functional constraint for each gene as the number of variable nucleotide sites, the genes left for a shorter time as a single helix should have fewer variable sites than those left for a longer time as a single helix. Indeed, we found that the degree of gene conservation, with the exception of ATP8 and Cyt b, decreases with the increase of D_{ssH} and that the most conserved genes are located in the regions that remain in single-stranded form for a short time [127].

5.4. Evolutionary Rate of the Mitochondrial Genome in Mammals

In vertebrate genes, owing to the rare occurrence of insertions and deletions, evolution measurement is essentially based on the analysis of nucleotide substitutions. The mathematical model used to calculate evolutionary rates is crucial to obtaining correct results. Generally, any nucleotide substitution model is more reliable the lower the a priori assumptions on the substitution process, because accurate estimates of the evolutionary rates will be obtained only if the actual substitution process meets the assumptions. Furthermore, the model should suitably take into account the base composition of the examined sequences, particularly when it is not homogeneous and there is a strong compositional heterogeneity between different sites and/or sequences, because compositional differences can cause misinterpretation of genetic distances [130,131]. Compositional asymmetry of mammalian mtDNA, with high GC skew values, implies that the probabilities that each nucleotide has be substituted by any of the others are not the same and thus the two transitions and the four transversions are not equiprobable. Because of the above characteristics of mtDNA evolution, a new model, the stationary Markov model (SMM), was devised by our group [131,132]. This method does not impose the a priori assumptions that are at the basis of other stochastic models available in the literature [133,134], because it allows different rates for the two types of transitions (A↔G and C↔T) and the four types of transversions (A↔C, A↔T, C↔G, G↔T). Furthermore, it takes into account the nucleotide composition of the sequences under examination. The only prerequisite for its applicability (which also holds for other stochastic models) is that sequences under analysis must have, within statistical fluctuations, the same base composition at compared sites. The stationarity condition is preliminarily verified on the sequence data by using a simple chi-square test described in Saccone et al. [131]. When the "stationarity condition" is obeyed, the SMM can be reliably used for evolutionary analysis. The stationary Markov model has also been defined as the "general time reversible" (GTR) model and has been included in the new version of the PAUP package. An online version is also available at the web server of the Italian EMBnet node (http://bighost.ba.itb.cnr.it/BIG/Markov/).

The stationary Markov model used in the quantitative analyses of mtDNA makes it possible

1. To determine the four-nucleotide substitution rate matrix, R_{ij}, which describes the propensity of nucleotide j to be substituted by nucleotide i independently from time T
2. To calculate the actual average evolutionary rate of sequence sites (i.e., the rate of silent codon position in mRNA genes)
3. To construct phylogenetic trees

For a given pair of sequences, evolutionary distance can be transformed into absolute rate if divergence time is known, possibly from accurate external sources, between the species to which they belong. Unfortunately, commonly used paleontological dating can be controversial or uncertain, due to both difficulties in the interpretation of dating of fossil remains and to their incompleteness [135]. Therefore they are a relevant source of error in determining absolute evolutionary rates which should be suitably considered.

Using the Markov model, we have calculated the average evolutionary rate of mitochondrial genes and of specific tracts of the D-loop region [136]. To obtain accurate estimates of the absolute nucleotide substitution rate, we considered only pairs of organisms that share a recent ancestor, to avoid the problem of saturation of nucleotide substitutions, and whose times are known with sufficient accuracy from other molecular and nonmolecular sources. Saturation is essentially due to multiple mutational events affecting the same nucleotide position and occurring with an apparent reduction in the number of nucleotide differences observed for longer divergence rates. The overall effect of saturation is thus an underestimate of evolutionary distances, mainly between evolutionarily distant sequences.

Figure 7 reports the absolute nucleotide substitution rates of the mitochondrial functional regions, calculated as mean rate values for 6 closely related mammalian species belonging to the orders Primates, Carnivora, Cetacea, and Perissodactyla. In general, two classes of functional regions can be identified: (1) slow evolving regions including nonsynonymous sites, tRNA and rRNA genes, and D-loop central domain [117]; (2) fast-evolving regions including synonymous sites, CSB, and ETAS D-loop domains, defined according to Sbisá et al. [117]. The non-synonymous sites are the most slowly evolving and the synonymous sites the most rapidly evolving, the latter being about 16-fold faster than the former. tRNA and rRNA genes evolve at a rate slightly higher than the nonsynonymous sites (about twofold), with 16S rRNA about 1.5-fold faster than 12S rRNA. All these regions are subject to strong functional constraints. However, unlike nonsynonymous sites, which are forced to retain a specific primary sequence, tRNA and rRNA genes are mainly constrained to maintain specific secondary and tertiary structures, thus allowing slightly higher variability at the level of their primary structures.

Within the D-loop region, the central domain evolves about twofold faster than the nonsynonymous sites but much slower than the two peripheral D-loop domains. The central domain shows quite a homogeneous evolutionary rate among species, whereas in the CSB and ETAS domains strong rate heterogeneity has been found, as indicated by their high standard deviation (Fig. 7). The species-specific evolution of the D-loop region is due to its peculiar evolutionary pattern, because this region is prone to accept insertions/deletions as well as repeated sequences, particularly in the ETAS and CSB domains [116,117].

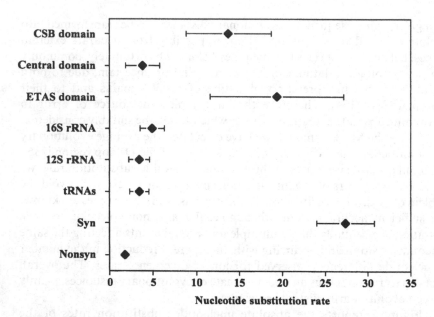

FIGURE 7 Absolute nucleotide substitution rate, in substitutions per site per year $\times 10^{-9}$, of the different mtDNA functional regions as mean for six pairs of closely related mammalian species (human–chimpanzee; pygmy–common chimpanzee; harbor seal–grey seal; horse–donkey; blue whale–fin whale). Mean and standard deviation values are reported. (From Ref. 136.)

In the species considered, similar substitution rates have been found for tRNA, rRNAs, and protein-coding genes at the level of nonsynonymous sites. In contrast, a rather high rate variability has been reported for the protein-coding genes of some mammalian orders [137–139].

The mean synonymous and nonsynonymous absolute rates for each of the 13 protein-coding genes have been calculated for the same above-mentioned species pairs and are reported in Figure 8. As expected, in each gene the synonymous rate is about one order of magnitude higher than the nonsynonymous rate and, taking into account statistical fluctuations, it is approximately uniform between the various genes, with ND3 and ATP8 slightly slower and Cyt b slightly faster. The nonsynonymous rate exhibits wide variations between genes, depending on the functional constraints, the cytochrome oxidase subunits (COI, COII, and COIII) being the slowest and the ATP8 the fastest evolving genes.

To compare the evolutionary dynamics of mtDNA and nDNA, nuclear and mitochondrial encoded tRNA, rRNA, and protein genes in the same

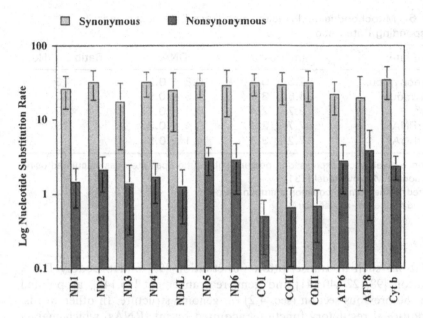

FIGURE 8   Mean absolute nucleotide substitution rate, expressed as a common logarithm, at nonsynonymous and synonymous sites for each of the 13 mitochondrial protein-coding genes as the average for six pairs of closely related mammalian species. Standard deviations are also reported. (From Ref. 136.)

species pairs (human–chimpanzee and rat–mouse) were analyzed. Table 6 compares the evolutionary rate calculated for various mitochondrial and nuclear genes. For noncoding regions, the different structures of the mitochondrial and nuclear genomes makes the comparison quite meaningless. In the protein-coding genes, the synonymous sites evolve about 22-fold faster in the mitochondrial genome than in the nuclear genome, whereas in the nonsynonymous sites a remarkable rate heterogeneity is observed in both genomes, as expected, due to the action of different selective constraints on the genes. The small and large rRNAs evolve about 19-fold and fourfold faster, respectively, in the mitochondrial genome than in the nuclear genome. The highest rate difference has been found for mitochondrial tRNA genes, which evolve about 100-fold faster than their corresponding nuclear genes.

The highest mitochondrial evolutionary rate of synonymous sites and tRNA genes compared to the nuclear counterparts could be due to the relaxed constrains causing a larger wobbling of the codon–anticodon pairing in the mitochondrial system, and the tRNA rate could be related to the additional roles these molecules have acquired during evolution in processes other than

TABLE 6 Mitochondrial and Nuclear Sequence Divergence and Corresponding Rate Ratio

Type of site	mtDNA[a]	nDNA[a]	Ratio (mt/Nuc)
Nonsynonymous[b]	2.6 ± 0.4	0.8 ± 0.2	3
Synonymous[b]	34.6 ± 3.9	1.6 ± 0.9	22
tRNAs[c]	9.7 ± 2.4	0.1 ± 0.1	97
Small rRNA[c]	7.7 ± 2.4	0.4 ± 0.3	19
Large rRNA[c]	17.2 ± 3,8	4.1 ± 0.8	4

[a] Sequence divergence, expressed in percent substitution per site, was calculated using the stationary Markov model [131].
[b] Referred to the human–common chimpanzee pair.
[c] Referred to the rat–mouse pair.

translation, such as transcription punctuation [99] and probably DNA replication [98,122,140,141] and gene rearrangement [95,142], as pointed out in the previous section (sec. 4.2) on genome structure. In other words, the additional regulatory functions acquired by mt tRNAs, which mainly depend on the interaction with nuclear coded products, require a lot of flexibility and adaptation capability. The higher rate of mutation accumulation in mitochondrial tRNAs could also be explained by the "Muller's ratchet" effect, which predicts a higher rate of mutation in asexually propagated genomes [143]. On the other hand, the lower nuclear synonymous substitution rate could be related to very high specific constraints and to the isochore structure of the nuclear genome [144].

5.5. Rate Heterogeneity of the Mitochondrial Genome in Mammals

To verify the existence of a molecular clock in the evolution of mammalian mtDNA we used the relative rate test (RRT) of Muse and Gaut [145] for protein-coding genes and that of Muse and Weir [146] for rRNA genes [147]. The RRT is able to detect substitution rate heterogeneity along different gene lineages. The analyses were carried out on gap-free multialigned supergenes of the 12 H-stranded protein-coding genes and the two ribosomal RNA genes. In the case of protein-coding supergenes (CDS), only the first and second codon positions were taken into account. In the case of ribosomal RNAs, ambiguously aligned sites, mostly adjacent to gaps, were excluded. In total, 7401 sites were analyzed for the CDS and 2136 for the ribosomal supergene. The RRT method compares the evolutionary rate of two species, A and B, called ingroups, using a reference outgroup O. For the ingroup species pair AB diverging from the internal node 1, the relative rate ΔR_{AB} can be calculated as

$$\Delta R_{AB} = \frac{d_{1A}}{d_{1B}} = \frac{d_{OA} - d_{OB} + d_{AB}}{d_{OB} - d_{OA} + d_{AB}}$$

O being the outgroup species and d the genetic distance calculated by the stationary Markov model [131]. The relative rate between species A and a set of species, i.e., B_1, B_2, ..., B_n, can be calculated as the average of the relevant ΔR_{AB_i} values.

The accuracy of rate difference estimates depends on the correct choice of the outgroup, which should be the closest possible to the two ingroup species; on the genetic distance between the two ingroup sequences, which should not be too high so as to prevent the problem of substitution saturation; and on the reliability of the mathematical method used to estimate the genetic distances.

We found significant rate variations not only between orders but even between closely related species of the same order. Figure 9 plots the ΔR values calculated on the P12 sites against those calculated on rRNA sites. A significant correlation was found, suggesting a high level of congruency between ribosomal and nonsynonymous sites. Primates and Proboscidea were found to be the fastest evolving orders, whereas Perissodactyla was the slowest. Indeed, the observed rate variation did not exceed 1.8-fold between the fastest and the slowest orders, thus supporting the suitability of mtDNA for drawing mammalian phylogeny.

At the intraorder level, statistically significant rate differences on P12 CDS sites have been found in several orders. Baboon and orangutan rates were found significantly greater than those of other primates ($\Delta R = 1.30 \pm 0.09$ and 1.30 ± 0.08, respectively). The Indian rhinoceros evolved 1.22 ± 0.08 times faster than other Perissodactyla. Within Artiodactyla, the hippopotamus evolved 1.22 ± 0.08 faster than other ruminants considered. In Rodentia, the squirrel was found to be the slowest evolving species ($\Delta R = 0.82 \pm 0.10$) whereas the dormouse evolved 1.17 ± 0.05 times as fast as the mouse. When considering ribosomal genes, no significant rate difference was observed for the above species at the intraorder level except for the higher rate of the orangutan within Primates ($\Delta R = 1.61 \pm 0.22$).

The existence of a correlation between the evolutionary rate and several physiological and metabolic variables such as body size, generation time, and specific metabolic rate (SMR) was investigated for both P12 CDS and ribosomal sites. For species pairs showing rate differences, the relative values of SMR, generation time, and body size were plotted against the corresponding relative rate (ΔR). No significant correlation was observed

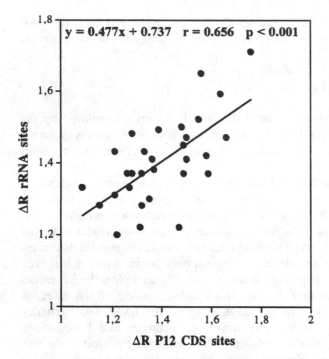

FIGURE 9 Plot of the mean relative rate calculated on P12 protein-coding sites against that calculated on ribosomal sites for the interorder comparisons with statistically significant rate differences reported in Gissi et al. [147]. Regression line and correlation coefficient are also reported.

between these variables and the relative substitution rate, with the exception of a marginal statistical significance observed for the correlation involving generation time [147].

6. PHYLOGENETIC RECONSTRUCTIONS

6.1. Phylogenetic analysis based on the gene order differences of mitochondrial genomes

The phylogenetic relationships between major animal phyla are still not well defined, and several alternative views have been proposed derived from morphological, developmental, ultrastructural, or molecular data [148–150]. The observation that the gene content of the mitochondrial genomes of Metazoa is nearly invariant, with few exceptions, makes the comparison of gene arrangements in different taxonomic groups a promising additional tool for resolving phylogenetic relationships, especially those involving distantly related taxa. Indeed, due to the rather high mutation rate of animal mitochondrial genomes, distant phylogenetic relationships may remain unsettled due to saturation of nucleotide substitutions and compositional bias effects. On the other hand, the great number of potential gene arrangements, even with the relatively small set of 37 genes encoded by the mitochondrial genome of Metazoa, makes it unlikely that different taxa independently evolved identical gene orders. Therefore, the comparison of gene orders in metazoan mitochondrial genomes may significantly contribute to the reconstruction of metazoan phylogeny. Indeed, it has been observed that unique gene arrangements characterize well-established evolutionary lineages such as birds, marsupials, and echinoderms [89].

The availability of over 100 completely sequenced animal mitochondrial genomes (see Table 2) from several animal phyla, including Chordata, Hemichordata, Urochordata, Echinodermata, Arthropoda, Mollusca, Anellida, Nematoda, and Cnidaria, makes it now possible to carry out an extensive comparative analysis of gene arrangements and determine the most likely phylogeny within and between metazoan phyla.

The quantitative comparison of gene order differences is necessary to infer phylogenetic relationships between organisms whose genome structure is known. This requires suitable computational methods to measure the minimum number of edit operations, including gene inversions, transposition, and strand changes, necessary to convert one genome arrangement into another. The use of methods adopting global objective functions able to reconstruct ancestral gene organizations, although in principle more interesting, is not computationally feasible even for moderately sized genomes. However, the "breakpoint distance method" [94] proved to be quite tractable in the case of mitochondrial genomes.

In the case of mitochondrial genomes the most economical explanation for observed differences in gene order between two genomes in terms of the minimum number of rearrangement processes can be reformulated as the problem of calculating an edit distance between two circular permutations of the same set of genes. The elementary edit operations (see Fig. 10) are

1. The inversion or reversal of any number of consecutive genes
2. The transposition of any number of consecutive genes from one position to another with (or without) strand change

A weight can be associated to each kind of move, in order to influence the overall proportion of each kind of operation in the final solution.

To calculate the number of breakpoints between two genomes, A and B, let us consider their gene arrangements,

$$\text{genome A}: a_1 \cdots a_n \qquad \text{genome B}: b_1 \cdots b_n$$

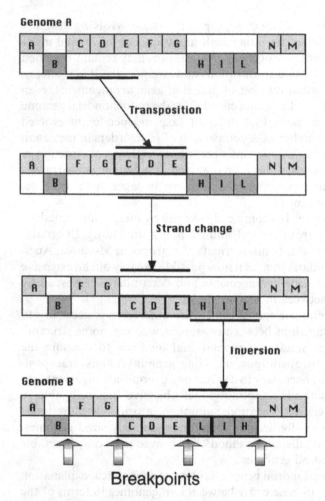

Breakpoints

FIGURE 10 Elementary edit operations including transpositions, strand changes, and inversions that introduce breakpoints in the comparison of the gene order between two genomes. In this example, the breakpoint distance between genome A and genome B is 4.

on the same set of genes $\{g_1, \ldots, g_n\}$, where each gene is signed ($+$ or $-$) depending on its orientation in the genome. If we deal with a circular genome, we say that a_n precedes a_1 or, as illustrated in Figure 10, that gene M precedes gene A. If gene E precedes gene F in genome A but neither does E precede F nor $-$F precede $-$E in genome B, then they determine a breakpoint. Gene strandedness has not been considered here, although if two genes are neighbors a breakpoint is generated when they are located on the same strand in one genome and on different strands in the other. In the example shown in Figure 10, four breakpoints can be observed between genomes A and B, corresponding to two moves, i.e., one inversion and one transposition.

Blanchette et al. [94] used this methodology to infer the phylogenetic tree of some major animal phyla. The phylogenetic tree calculated from pairwise breakpoint distances of mitochondrial genome arrangements of 11 organisms belonging to six different metazoan phyla is quite compatible with the commonly accepted metazoan phylogeny. It significantly supports the deuterostome grouping (Chordata + Echinodermata) but surprisingly places Arthropoda as their sister group. However, it has to be stressed that this methodology also has several drawbacks. The first is that the study of genomic rearrangement inevitably encounters the problem of nonuniqueness because there are often many distinct solutions, all equally optimal. The second problem is that, as for other methods for the reconstruction of phylogeny, rapidly evolving lineages tend to cluster together, producing the so-called long-branch attraction effect. Indeed, it seems evident that in Metazoa some phyla, such as nematodes, snails, and echinoderms, show a higher gene mobility than genomes from other phyla, such as Chordata, which appear more conservative. Finally, there is the problem of saturation, because some lineages may have diverged to randomness (it can be calculated that random genomes with n genes would have on average $n-1/2$ breakpoints between one another). A further striking feature of the evolution of gene order in the mitochondrial genomes of metazoans is the higher mobility of the tRNA genes with respect to rRNA and protein-coding genes [94]. This feature was first described by Saccone et al. [97].

The computer program Derange2 (publicly available by anonymous FTP at ftp.ebi.ac.uk/pub/software/unix) was designed by Mathieu Blanchette to sort a given permutation of genes to identity permutation using inversions, transpositions, and inverted transpositions. It also allows the user to specify a weight associated to each kind of move in order to influence the overall proportion of each kind of move to the final solution.

Also available is the computer program BPAnalysis http://www.cs. washington.edu/homes/blanchem/software.html), which computes minimal breakpoint trees [94] from a set of gene orders. It can be used for phylogenetic reconstructions and for inference of ancestral gene order.

6.2. Phylogenetic Inferences Based on the Evolutionary Analyses of mtDNA Supergenes

The mitochondrial genome, because of its maternal inheritance, lack of recombination, and especially the presence of orthologous genes, is particularly suitable for molecular phylogeny analyses. Conversely, nuclear genes can be paralogous or could evolve under different evolutionary pressures (e.g., insulin in guinea pigs with respect to other mammals [151]), thus leading to controversial results from different gene sets. mtDNA revealed a powerful tool for dealing with the reconstruction of phylogenesis at both interorder and intraorder levels. Either single mitochondrial genes or the entire genome can be used in the analyses. In consideration of the large stochastic fluctuations of the estimates based on a limited number of sites, it is generally advisable to use concatenated genes. For this reason we were the first to propose the use of concatenated protein-coding genes and rRNA genes (commonly denoted as CDS and rRNA supergenes) to carry out evolutionary analyses [131].

The analysis of complete mammalian mt genomes has generated many unexpected and surprising results, concerning mainly the polyphyly of Rodentia and Artiodactyla, the unreliability of the cohort Glires (Lagomorpha and Rodentia), the position of Edentata and Chiroptera, and the grouping of Marsupialia and Monotremata in the same clade.

Evolutionary studies performed on completely sequenced mitochondrial genomes are summarized in Figure 11. Perissodactyla cluster with Carnivora, and Cetacea with Artiodactyla, all of them included in the Ferungulata clade. Lagomorpha and Edentata are sister groups of Ferungulata, while Rodentia diverged before Primates. Insectivora and noneutherian orders (Marsupialia and Monotremata) are placed at a basal position of the mammalian tree.

Rodentia have been unanimously accepted as monophyletic on account of morphological data, but this vision was first challenged by Graur et al. [152], who compared 15 nuclear protein genes from rat, mouse, guinea pig, and other non-rodent mammalian species. The analysis of the complete mt genomes of these three rodent species [153] and the other available mammalian mtDNA genomes strongly supported the hypothesis of rodent polyphyly. Moreover, the addition of the complete sequence of the fat dormouse [154] to the analysis gave further evidence of rodent polyphyly, with dormouse clustering with guinea pig in a different clade from that of rat and mouse (Fig. 11). However, other molecular evidence, not really strongly supported, suggest rodent monophyly, making this issue one of the most debated in mammalian phylogeny (see Ref. 154 and references therein).

A close relationship between Artiodactyla and Cetacea was proposed on the basis of morphological [155], biochemical [156], and molecular grounds [157,158]. Molecular data provided some evidence for the inclusion of Cetacea within the order Artiodactyla [159]. Indeed, the sequencing of pig mitochondrial genome [160] demonstrated the polyphyly of Artiodactyla, as cow clustered with fin whale, taking pig as the outgroup (Fig. 11).

The phylogenetic position of Lagomorpha has not been settled definitely. According to the species sampled or the methodology used for phylogenetic reconstruction, it can be placed in different positions within

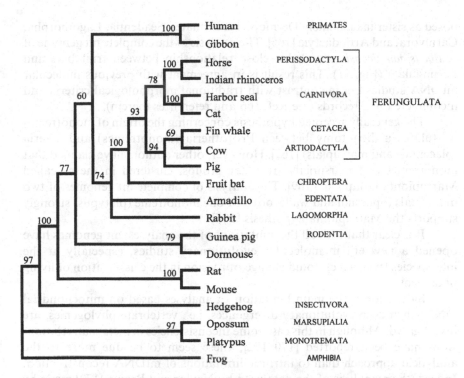

FIGURE 11 Majority-rule consensus tree based on first and second codon positions of the H-strand protein-coding supergene obtained by concatenating all 12 single genes. The tree was obtained with the neighbor-joining method [197], using the distance matrix as calculated with the stationary Markov model [131]. Bootstrap values are based on 100 replications. Frog sequence was used as outgroup. Branch length is not proportional to the genetic distance. For an updated mammalian tree see Ref. 199.

Mammalia [153,154]. Neither mitochondrial [154] nor nuclear genes [161] support Lagomorpha association with the representatives of Rodentia in the cohort Glires (but see [199]).

Paleontological and morphological evidence related the order Edentata (armadillos) to Pholidota [162,163] and placed both of them at a basal position of the eutherian phylogenetic tree. Instead, the analysis of the mt genome suggests a sister group relationship between Edentata and Ferungulata [164]. This is one of the most dramatic inconsistencies reported so far between comprehensive molecular data sets and commonly acknowledged understanding of eutherian phylogeny.

The position of the order Chiroptera has been debated on the basis of both morphological and molecular data, and different orders have been proposed as sister taxa, namely Dermoptera, Primates, Scadentia, Lagomorpha, Carnivora, and Artiodactyla [165]. The analysis of the complete mt genome of *Artibeus jamaicensis* revealed a close relationship between fruit .bats and Ferungulata (Fig. 11). This result is in agreement with previous molecular mtDNA studies but in contrast with traditional morphological criteria and incomplete fossil records (see Ref. 166 and references therein).

The generally accepted hypothesis concerning the origin of monotremes postulates a dichotomy between Prototheria (monotremes) and Theria (placentals and marsupials) [162]. However, other authors have claimed that monotremes and marsupials are sister groups, clustered in the so-called Marsupionta group [167–169]. The analysis of complete mt genomes of two marsupials (opossum and wallaroo) and one monotreme (platypus) strongly supports the Marsupionta hypothesis [106].

It is clear that with all the limitations of the analyses, mt genomes have opened a new era in molecular evolutionary studies, especially at the interspecies level, which could change our vision in the classification of living organisms.

Saturation is the main limitation of analyses based on mitochondrial DNA when deep evolutionary divergences, i.e., vertebrate phylogenies, are investigated. Although in this case some inconsistencies in metazoan relationships have been reported [170–172], these seem to be due more to the analytical approach than to intrinsic limitations of mtDNA reconstruction. Indeed, the reanalysis of the data used by Naylor and Brown [172] made by Arnason (personal communication) recovered the correct tree.

At this point, we should wonder whether mt data are supported by nuclear gene analyses. The studies available in the literature are controversial, and moreover the nuclear genes sequenced so far are still scanty and offer a nonhomogeneous sample. It should also be considered that for nuclear genes there remains the problem of paralogy.

7. EVOLUTION OF THE HUMAN MITOCHONDRIAL GENOME

The study of the variability of human mtDNA, in particular its main control region, has greatly contributed to the problem of the origin of modern humans. The highest intraspecies variability has been observed in the two peripheral domains of the D-loop region, which in humans are denoted as HV1 (ETAS domain) and HV2 (CSB domain) [173]. It is striking to note that the most variable part of the human D-loop, contained in the HV1 region, is a short tract of about 100 bp we call the IS sequence [116], which is present in humans, chimpanzees, and gibbons, but not in gorillas. Figure 12 shows the average rate matrix computed using the stationary Markov model of Saccone et al. [131] for the HV1 regions between African and non-African individuals.

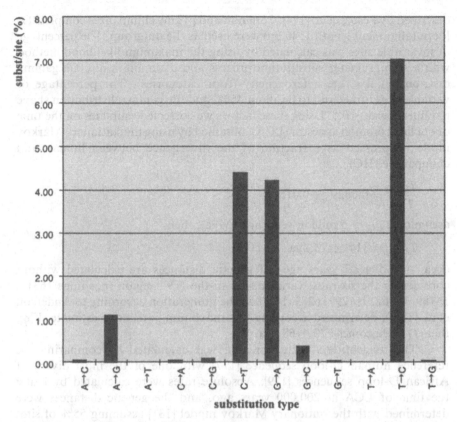

FIGURE 12 Average rate matrix computed by using the stationary Markov model [131] comparing the HV1 region from African and non-African individuals.

It is evident that transitions are much more likely than transversions, but different classes of transitions and transversions show different probabilities. In particular, $C \leftrightarrow T$ transitions are more frequent than $G \leftrightarrow A$ transitions or $G \leftrightarrow C$ transversions with respect to the other transversions.

In 1992, we included in our Markov model a rate heterogeneity-among-sites factor [174]. More simply, when comparing actual and simulated sequences by using chi-square statistics, we calculate the percentage of variable sites and then we correct our measures accordingly. Other models taking into account rate heterogeneity among sites were introduced by Hasegawa and Horai [175] and Tamura and Nei [176].

We report here the evolutionary analysis of 44 complete human D-loop sequences from various continental areas carried out using the method described by Pesole et al. [174]. Common and pygmy chimpanzee complete D-loops (alignment length 1140 bp) were used as the outgroup. The percentage of invariable sites was calculated by using the maximum likelihood method with a six-parameter substitution process and assuming a discrete gamma distribution for rate heterogeneity (four categories). The percentage of variable sites turned out to be about 45%, and the gamma distribution shape parameter was < 0.5. Using these factors we corrected estimates on the time of our last common ancestor (LCA), obtained by using the stationary Markov model, expressed as a fraction of the divergence between human and chimpanzee (HC).

$$\frac{T_{LCA}}{T_{HC}} = 0.0629 \pm 0.0157$$

assuming $T_{HC} = 5$ million years ago (Mya), then

$$T_{LCA} = 314 \pm 78 \text{ kya}$$

(kya = thousand years ago). If genetic distances are calculated without considering the six most variable sites in the HV1 region (positions 16311, 16189, 16362, 16129, 16223, 16278 in the numeration according to Anderson et al. [177]), determined according to a maximum parsimony estimate [178], then T_{LCA} becomes 270 ± 68 kya.

The nucleotide substitution rate was calculated by comparing the reference human D-loop sequence [177] with one of the most divergent African D-loop sequences [179]. Absolute rates were calculated by fixing the time of LCA at 200,000 years ago, and the genetic distances were determined with the stationary Markov model [131] assuming 55% of sites are invariable and a discrete gamma distribution shape parameter of 0.5.

For the whole D-loop,

$$v = 5.7 \pm 1.7 \times 10^{-8} \text{ substitutions/(site.year)}$$

or

$$12.8 \pm 3.8 \times 10^{-8} \text{ substitutions/(variable site.year)}$$

(i.e., 9–16%/My)

For the HV1 D-loop region,

$$v = 8.0 \pm 2.4 \times 10^{-8} \text{ substitutions/(site.year)}$$

or

$$10.7 \pm 3.2 \times 10^{-8} \text{ substitutions/(variable site.year)}$$

(i.e., 8–14%/My)

Our data are not in disagreement with those previously reported by us and other authors and obtained with phylogenetic approaches [180]. But it is well known that the evolutionary rate of the mitochondrial D-loop in humans is a matter of high controversy.

The data available in the literature on this subject can be divided into two classes:

1. Data obtained from phylogenetic studies
2. Data obtained from empirical studies through observation of the number of substitutions in several maternal lineages

The results differ by more than one order of magnitude, the former having a value of about 10^{-7} and the latter about 10^{-6} substitutions per site per year [180], but some factors should be taken into account.

The empirical approach may suffer from the fact that data are still too scanty and thus we cannot exclude heterogeneity in substitution rates among families and/or the possibility that a disease state of the individuals from which the majority of data are derived may affect the mutation rate. In addition, empirical data reveal only the most variable sites, and thus it is clearly not legitimate to extrapolate the results for the whole region.

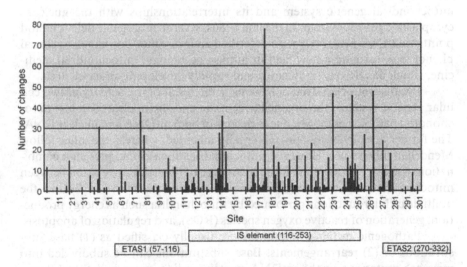

FIGURE 13 Minimum number of changes per site inferred by NEWEST program [178] on 500 HV1 regions of the D-loop (positions 16024–13323 in Anderson et al. [177]). ETAS1, ETAS2, and the primate insertion sequence (IS; see Refs. 116 and 117) are also indicated. A more recent study can be found in Ref. 200 and the related WebVar software at www.pesolelab.it/Tool/

The phylogenetic data set can be obscured by multiple hits and reversions at the same sites and suffer from several a priori assumptions that underlie the method used for the study. For example, some methods consider a fixed ratio between transitions and transversions in the evolution of the mtDNA; others do not take into account the bias in base composition of the mtDNA and/or assume rate homogeneity among sites. Indeed, the real problem is to measure the variability of sites taking into consideration also the type of nucleotide substitution events.

The site variability of the D-loop HV1 region shown in Figure 13 was calculated with an approximate method that, using trees constructed with the neighbor-joining method [197] on Kimura-corrected distances, calculated the minimum number of base changes for each site required by the inferred phylogenetic tree [178]. This method relies on the correctness of the tree and, of course, provides an underestimate of site variability.

8. MITOCHONDRIAL DNA AND HUMAN DISEASES

About 40 years ago, Luft et al. [181] described the first mitochondrial dysfunction. Since then, research in this field has exploded, partially because of progress made in the sequencing of mitochondrial genomes from human and other animal species, mainly mammals, and the understanding of the mitochondrial genetic system and its interrelationships with the nuclear–cytoplasmic genetic system. In the late 1980s, several pathogenic deletions and point mutations in mtDNA were described and were then correlated to clincal phenotypes, opening a new field in human pathology, mitochondrial medicine, which has become a dynamic and rapidly developing research area.

Because mitochondrial biogenesis is the result of the activity of two cellular genetic systems, the organellar genome and the nuclear genome, mitochondrial pathologies may be due to defects of both mtDNA and nuclear DNA. The former are obviously only maternally inherited, whereas the latter follow Mendelian inheritance. However, mitochondrial diseases can present a combination of the two types of inheritance because of the strict correlation between mitochondrial and extramitochondrial metabolic pathways [182] and the multiple cellular functions performed by mitochondria such as energy production, generation of reactive oxygen species (ROS), and regulation of apoptosis.

Pathogenic mtDNA mutations are usually classified as (1) base substitutions or (2) rearrangements. Base substitutions can be subdivided into missense mutations in the mtDNA protein-coding genes and those in mt-coded rRNA and tRNA genes involved in the protein-synthesizing machinery of the organelle. Table 7 reports a list of mutations that have been found to be involved in mitochondrial pathologies. (see www.mitomap.org) The majority include purine transitions or G-involved transversions. This is in line with the

TABLE 7 Mitochondrial DNA Base Substitution Diseases as Reported in MITOMAP and MitBase (updated table available at www.mitomap.org)

Gene	Disease[a]	Position	Change	AA Change
tRNA Phe	MM	618	T→C	
12S	DM	1310	C→T	
12S	DM	1438	A→G	
rRNA 12S	DEAF	1555	A→G	
tRNA Val	AMDF	1606	G→A	
tRNA Val	MELAS	1642	G→A	
RRNA 16S	Rett syndrome	2835	C→T	
rRNA 16S	ADPD	3196	G→A	
tRNA Leu (UUR)	MELAS	3243	A→G	
tRNA Leu (UUR)	DM/DMDF	3243	A→G	
tRNA Leu (UUR)	MM	3243	A→T	
tRNA Leu (UUR)	MM	3250	T→C	
tRNA Leu (UUR)	MM	3251	A→G	
tRNA Leu (UUR)	MELAS	3252	A→G	
tRNA Leu (UUR)	MM	3254	C→G	
tRNA Leu (UUR)	MELAS	3256	C→T	
tRNA Leu (UUR)	MMC	3260	A→G	
tRNA Leu (UUR)	DM	3264	T→C	
tRNA Leu (UUR)	PEM	3271	delT	
tRNA Leu(UUR)	MELAS	3271	T→C	
tRNA Leu (UUR)	DM	3271	T→C	
tRNA Leu (UUR)	MELAS	3291	T→C	
tRNA Leu (UUR)	MM	3302	A→G	
tRNA Leu (UUR)	MMC	3303	C→T	
ND1	MELAS	3308	T→C	M→T
ND1	NIDDM	3316	G→A	A→T
ND1	LHON	3394	T→C	Y→H
ND1	NIDDM	3394	T→C	Y→H
ND1	ADPD	3397	A→G	M→V
ND1	LHON	3460	G→A	A→T
ND1	LHON	4136	A→G	Y→C
ND1	LHON	4160	T→C	L→P
ND1	LHON	4216	T→C	Y→H
tRNA Ile	FICP	4269	A→G	
tRNA Ile	CPEO/MS	4298	G→A	
tRNA Ile	MICM	4300	A→G	
tRNA Ile	FICP	4317	A→G	
tRNA Ile	Mitochondrial encephalo-cardiomyopathy	4320	C→T	
tRNA Gln	ADPD	4336	T→C	
tRNA Met	MM	4409	T→C	

TABLE 7 Continued

Gene	Disease[a]	Position	Change	AA Change
ND2	LHON	4917	A→G	D→N
ND2	LHON	5244	G→A	G→S
ND2	AD	5460	G→A	A→T
ND2	AD	5460	G→T	A→S
tRNA Trp	MM	5521	G→A	
tRNA Trp	MILS	5537	insT	
tRNA Trp	DEMCHO	5549	G→A	
tRNA Asn	CPEO	5692	T→G	
tRNA Asn	CPEO, MM	5703	G→A	
tRNA Cys	Mitochondrial encephalopathy	5814	T→C	
CO1	LHON	7444	G→A	Term→K
tRNA Ser (UCN)	DEAF	7445	A→G	
tRNA Ser (UCN)	PEM/AMDF	7472	insC	
tRNA Ser (UCN)	MM	7497	G→A	
tRNA Ser (UCN)	SNHL	7511	T→C	
tRNA Ser (UCN)	PEM/MERME	7512	T→C	
tRNA Lys	DMDF/MERRF	8296	A→G	
tRNA Lys	MNGIE	8313	G→A	
tRNA Lys	Mitochondrial encephalopathy	8328	G→A	
tRNA Lys	PEO and Myoclonus	8342	G→A	
tRNA Lys	MERRF	8344	A→G	
tRNA Lys	MERRF	8356	T→C	
tRNA Lys	MICM + DEAF/ MERRF	8363	G→A	
ATP6	NARP/Leigh disease	8993	T→C	L→P
ATP6	NARP	8993	T→G	L→R
ATP6	LHON	9101	T→C	I→T
ATP6	FBSN	9176	T→C	L→P
CO3	LHON	9438	G→A	G→S
CO3	LHON	9738	G→T	A→T
CO3	LHON	9804	G→A	A→T
CO3	PEM; MELAS	9957	T→C	F→L
tRNA Gly	MHCM	9997	T→C	
tRNA Gly	CIPO	10006	A→G	
tRNA Gly	PEM	10010	T→C	
tRNA Gly	GER/SIDS	10044	A→G	
ND4L	LHON	10663	T→C	V→A
ND4	MELAS	11084	A→G	T→A
ND4	LHON	11778	G→A	R→H
ND4	DM	12026	A→G	I→V

TABLE 7 Continued

Gene	Disease[a]	Position	Change	AA Change
tRNA Ser (AGY)	CIPO	12246	C→G	
tRNA Ser (AGY)	DMDF	12258	C→A	
tRNA Leu (CUN)	CPEO	12308	A→G	
tRNA Leu (CUN)	CPEO	12311	T→C	
tRNA Leu (CUN)	CPEO	12315	G→A	
tRNA Leu (CUN)	MM	12320	A→G	
ND5	MELAS	13513	G→A	D→N
ND5	LHON	13708	G→A	A→T
ND5	LHON	13730	G→A	G→E
ND6	LDYT	14459	G→A	A→V
ND6	LHON	14484	T→C	M→V
tRNA Glu	MM + DM	14709	T→C	
CYB	PD/MELAS	14787	CTCC→del	I→frameshift
CYB	MM	15059	G→A	G→Term
CYB	Mitochondrial encephalomyo-pathy	15242	G→A	G→Ter
CYB	LHON	15257	G→A	D→N
CYB	Exercise intolerance	15615	G→A	G→D
CYB	MM	15762	G→A	G→E
CYB	LHON	15812	G→A	V→M
tRNA Thr	MM	15915	G→A	
tRNA Thr	LIMM	15923	A→G	
tRNA Thr	LIMM	15924	A→G	
tRNA Thr	MM	15940	delT	
tRNA Pro	MM	15990	C→T	

[a] See Table 8 for disease names.
Source: Refs. 190 and 184.

observation that in mtDNA the propensity of each nucleotide to change is not the same. In particular, changes involving G are very rare, probably avoided.

The phenotypes derived from mtDNA mutations follow complex and peculiar patterns. The degree of penetration of a given pathology differs greatly among the offspring of the same mother. This can be explained by the degree of heteroplasmy in the affected tissue of a particular individual, i.e., the number of mutated molecules compared to normal molecules. It is important to remember that a mammalian cell contains roughly 500–1000 mitochondria, each possessing several (1–10) mtDNA molecules. Thus mitochondrial genetics has many aspects of population genetics in which polyploidy and heteroplasmy, both intramitochondrial and intracellular or intraorgan, play a fundamental role.

TABLE 8   Mitochondrial Diseases

AD: Alzheimer disease
ADPD: Alzheimer disease and Parkinson disease
AMDF: Ataxia, myoclonus, and deafness
CIPO: Chronic intestinal pseudo-obstruction with myopathy and ophthalmoplegia
CPEO: Chronic progressive external ophthalmoplegia
DEAF: Maternally inherited deafness and/or aminoglycoside-induced deafness
DEMCHO: Progressive dementia and chorea
DM: Diabetes mellitus
DMDF: Diabetes mellitus with deafness
Exercise intolerance
Fahr disease: Progressive mitochondrial encephalopathy with cerebral
   calcifications
FBSN: Familial bilateral striatal necrosis
FICP: Fatal infantile cardiomyopathy plus a MELAS-associated cardiomyopathy
GERD: Gastrointestinal reflux disease
Hypertrophic cardiomyopathy with myopathy in adults
Hypertrophic cardiomyopathy with myopathy in children
LDYT: Leber's hereditary optic neuropathy and dystonia
LHON: Leber hereditary optic neuropathy
LIMM: Lethal infantile mitochondrial myopathy
MDM: Myopathy with diabetes mellitus
MELAS-associated cardiomyopathy
MELAS: Mitochondrial encephalomyopathy, lactic acidosis, and stroke-like episodes
MERME: MERRF/MELAS overlap disease
MERRF: Myoclonic epilepsy and ragged red muscle fibers
MHCM: Maternally inherited hypertrophic cardiomyopathy
MICM: Maternally inherited cardiomyopathy
MILS: Maternally inherited Leigh syndrome
Mitochondrial encephalocardiomyopathy
Mitochondrial encephalomyopathy
MM: Mitochondrial myopathy
MMC: Maternal myopathy and cardiomyopathy
MNGIE: Mitochondrial neurogastrointestinal encephalomyopathy
Movement disorders including dystonia and parkinson disease
MS: Multiple sclerosis
NARP: Neurogenic muscle weakness, ataxia, and retinitis pigmentosa
Neurodegeneration, pigmentary retinopathy, and Leigh syndrome
NIDDM: Non-insulin-dependent diabetes mellitus
PD: Parkinson disease
PEM: Progressive encephalopathy
PEO: Progressive external ophthalmoplegia
PME: Progressive myoclonus epilepsy
Rett syndrome
SIDS: Sudden infant death syndrome
SNHL: Sensorineural hearing loss

Another characteristic aspect always connected to the peculiar features of mitochondrial genetics is that the same mutations can be found in different types of diseases as well as in healthy phenotypes, and different mutations can produce similar phenotypes.

On the basis of the above considerations, the great difficulty in the classification of mitochondrial diseases (listed in Table 8) can be easily understood. Most mitochondrial diseases have a delayed onset, which suggests that an age-related factor is required in addition to pathogenic mutations. Since the early 1990s many reports have appeared in the literature demonstrating that in aging the decline in mitochondrial functions and the presence of mt somatic mutations, base substitutions, and rearrangements are common features.

Attardi's group [183] revealed the onset of specific point mutations in the control region of human mtDNA during aging. The cause of somatic mutations is likely to be the damage of mtDNA by ROS production that increases with age. Defects in mtDNA in turn produce a decline in oxidative phosphorylation, which affects the cellular bioenergetic capacity, causing senescence and pathologies.

Mutations in mitochondrial DNA have also been described in cancer cells. The recent discovery that mitochondria regulate apoptosis opens interesting perspectives on the role played by oxidative metabolism in the cell cycle.

In spite of the numerous pathogenic mutations described in the literature and the intensive research carried out by multidisciplinary teams, we still know very little about the mechanisms underlying mitochondrial pathologies, ultimately due to the interrelationship of the two genetic systems of the eukaryotic cells. In this respect, interesting possibilities are offered by studies carried out with cells lacking mitochondria or with animal models, such as the mouse, in which the effects of specific mutations or inactivation of specific genes can be studied.

9. MITOCHONDRIAL SPECIALIZED DATABASES

The great quantity of information related to the mitochondrion at the levels of both DNA and protein for a large number of species and variants is of great interest to the scientific community. Hence it is imperative to have these data available through specialized databases where information is organized in a structured form and accurately annotated by expert researchers. Mitochondrial sequence data can be classified into two categories: data related to mtDNA sequences and data related to nuclear genes involved in mitochondrial bioenergetics and biogenesis. The available mitochondrial data are distributed worldwide in primary databases (as of May 2004, 201,640 mtDNA entries were available in the EMBL data library, release 78, including redundant sequences), and in specialized databases. Several types of data (e.g., human mtDNA variants and fungal mtDNA mutants) are available only in the literature and in specialized databases. Hence a lot of information is available but, as in a puzzle, the dispersed pieces need to be assembled.

This section summarizes the most relevant databases reporting mitochondrial data and lists Web sites where information on the mitochondrion can be accessed. Most databases are simply archives rather than structured collections of data. Archives, due to the organization of the information they contain, allow only data consultation; structured collections, on the other hand, have a high potential for analysis.

Before describing in detail mitochondrial databases, some notes can be useful to the reader to clarify the difference between primary and specialized databases. Primary databases report rough and very general information, i.e., the minimal set of information needed for classifying a nucleotide sequence (the taxonomic data, the bibliographic references, the biological function of the sequence, if known). Moreover, primary databases report nucleic acid sequences related to the most "disparate" classes of organisms and genomic functions. Hence, in order to keep a common structure and have a common guideline for submitting authors, primary databases report only basic information of common interest.

A specialized database is structured with the aim to collect data derived from primary databases or related to them and belonging to a homogeneous set of organisms and/or genes and genomic regions. In specialized databases, in addition to the basic set of information derived from primary databases, value-added information is structured. This information is derived from the literature and in silico analyses performed by the annotators of the database in order to release as exhaustive a set of data as possible.

Below is a synthetic description of the worldwide available mitochondrial databases.

MitBASE. [184] MitBASE has been designed as an integrated and comprehensive database of mitochondrial DNA data that collects, under a single interface, databases for plant, vertebrate, invertebrate, human, protist, and fungal mtDNA and a pilot database on nuclear genes involved in mitochondrial biogenesis in *Saccharomyces cerevisiae*. MitBASE reports all available information from different organisms and from intraspecies variants and mutants. Data have been drawn from the primary databases and from the literature; value-added information has been structured, e.g., editing information on protist mtDNA genomes, pathological information for human mtDNA, variants. Other salient features of MitBASE are the storage of data related to the editing process occurring in plants and protists species, the annotation of fungal mutants, and the definition of a standard gene name classification based on GOBASE gene names with some modifications and on the KEYnet structure [185]. MitBASE gene name classification includes names of fungal intronic open reading frames as assigned by the MitBASE experts on fungi. The different databases have been integrated under ORACLE. The database, funded within the IV EU Framework programme, is no more updated due to lack of funding. However the authors are willing to recover the database with the aim to annotate mitochondrial variation data. The new goal should be to contribute in Biodiversity studies based on mitochondrial DNA variability.

GOBASE. Originating at Montreal, Canada and available on the Web at http://megasun.bch.umontreal.ca/gobase), GOBASE [186] is a taxonomically broad organelle genome database that organizes and integrates diverse data related to organelles. The current version focuses on the mitochondrial and chloroplast subset of data. In a next phase, GOBASE will also include information on representative bacteria that are thought to be specifically related to the bacterial ancestors of mitochondria and chloroplasts. The GOBASE database contains all mitochondrial and chloroplast nucleic acid sequences and deduced protein data obtained from NCBI's Entrez database, taxonomic information extracted from the NCBI's Taxonomy database, standardized gene names and product names assigned by the GOBASE biology experts, and various types of information about gene products. There are links to the Entrez/NCBI database, to the intron and ribosomal RNA secondary structure database [187], to the ENZYME [188] and WIT (Puma) [189] databases, to the OGMP (http://megasun.bch.umontreal.ca/ogmpproj. html) and FMGP mitochondrial databases (http://megasun.bch.umontreal. ca/people/lang/fmgp/fmgp.html), and to the Protist Image Database (PID http://megasun.bch.umontreal.ca/protists/). GOBASE allows sophisticated queries, including completely sequenced genomes, and the use of search criteria concerning the general function of gene products. The majority of data contained in GOBASE have been verified by experts with respect to the consistency of their gene and product nomenclature.

MITOMAP. A Human Mitochondrial Genome Database. Based at Emory University in Georgia, MITOMAP is available on the WEB at http://www.mitomap.org), MITOMAP [190] is a report of all variations detected in human mtDNA samples from subjects affected by mitochondrial pathologies and from subjects whose DNA has been sequenced for genetic population studies. The report is in html format and hence can be browsed using any Internet browser. Moreover, a query system has been implemented that allows a search of the data in a very simplified modality; it does not allow one to navigate through the stored data. The database is regularly updated to contain data available in the most recent literature.

MITOP. A Database for Mitochondria-Related Genes, Proteins, and Diseases. This is available from MIPS, Munchen, Germany, at http:// ihg.gsf.de/mitop2). The MITOP database [191] reports data on nuclear encoded mitochondrial proteins for human, yeast, and *Neurospora crassa.* Each entry in MITOP is a protein. The following information is reported for each protein: function category, EC number, protein class, protein complex, PROSITE motifs, subcellular localization, molecular weight, isoelectric point, disease correlation, pathways and metabolism, and putative orthologs. The site where the database is available also contains the MITOPROT program, which identifies mitochondrial targeting sequences. Direct links are provided to other databases such as the Genome Data Base (GDB), PIR, Mouse Genome Database (MGD), Online Mendelian Inheritance in Man (OMIM), GenBank, and Medline.

MitoNuc. Mitonuc [193] is a database collecting genes and proteins, involved in mitochondrial biogenesis. Each MitoNuc entry defines a nuclear gene coding for a mitochondria-related protein in a given species. Each entry reports a set of defined information: gene description, species name and taxonomic classification, gene name, EC (Enzyme Classification) code in the case of genes coding for enzymes, gene product name, synonymous and functional classification as defined in the KEYnet database [185], metabolic pathways in which the protein product is involved, cellular and submito-chondrial localization of the encoded proteins, and possible presence of tissue-specific isoforms. Cross-references to the EMBL, Swiss-Prot/TrEMBL are also present.

MitoNuc can be retrieved through the SRS server in Bari (http:// bighost.ba.itb.cnr.it/srs) under mitochondrial databases section [193], and specific functional regions of nuclear coded mitochondrial mRNAs can be selected and extracted (5′ and 3′UTRs, signal peptides, etc.).

AMmtDB. AMmtDB multialigned data can be accessed through SRS (http://bighost.ba.itb.cnr.it/srs) under mitochondrial section and can be viewed and managed using multialignment editors such as Genedoc or Seaview. Both these editors can be easily downloaded through the Web. AMmtDB [194] is a database collecting the multialigned sequences of vertebrate and invertebrate mitochondrial genes coding for proteins and tRNAs as well as the multiple alignment of the mammalian mtDNA main regulatory region (D-loop) sequences. For genes coding for tRNAs, multi-alignments based on the primary and secondary structures are both provided. For 27 mammalian D-loop multialignments, the conserved regions of the entire D-loop are reported. The database is organized into three main sections: CDS, tRNA, and D-loop sequences. The genes coding for proteins are multialigned on the translated sequences, and both the nucleotide and amino acid multialignments are provided. The taxonomic classes for the presently available data are mammalian, amphibian, reptilian, birds, osteichthyes and condroichthyes, and agnatha and cephalocordata.

PLMItRNA. PLMItRNA [195] is a database developed to facilitate retrieval of information on the distribution of tRNA molecules and genes in mitochondria of green plants (higher plants and green algae) and Cryptophyta, Rhodophyta and Stramenopiles algae. Retrieval of information or sequences can be accomplished according to several characteristics of the tRNA gene or molecule through the Bari SRS server under "Sequence-related" database section. Currently PLMItRNA contains 610 entries for 577 genes and 33 tRNA sequences identified among 28 higher plants, ten Chlorophyta (green algae), one Cryptophyta, four Rhodophyta (red algae) and four Stramenopiles.

MitoDat. Available on the Web at http://www-lecb.ncifcrf.gov/mito-Dat/, MitoDat is a database collecting nuclear genes specifying the enzymes, structural proteins, and other proteins, many still not identified, involved in mitochondrial biogenesis and function. MitoDat highlights predominantly human mitochondrial proteins, although proteins from other animals in addition to those currently known from yeast and other fungal mitochondria as well as from plant mitochondria are coded. The database consolidates information from the various biological databases such as GenBank, SwissProt, Genome Data Base (GDB), and Online Mendelian Inheritance in Man (OMIM). Because the mitochondrion has a central role in cellular metabolism, it is involved in many human diseases. This database should help users in studying these diseases.

HvrBase. (Available on the Web at http://www.hvrbase.org), HvrBase [196] is a compilation of human and ape mtDNA control region sequences. The collection is also available as a Mac/PC database application with a graphical user interface. The current collection comprises 9388 human sequences from hypervariable region I (HVRI) and 3302 human sequences from hypervariable region ii (HVRII). From apes, 469 HVRI sequences and 13 HVRII sequences are available.

ACKNOWLEDGEMENTS

This work was supported by "Ministero dell'Istruzione, dell'Università e della Ricerca," Italy-Project: "Centro di Eccellenza di Genomica in campo Biomedico e Agrario"(CEGBA).

We would also like to thank Marcella Attimonelli, Carmela Gissi, Alessandra Larizza, Aurelio Reyes, Elisabetta Sbisà, Apollonia Tullo and Gemma Gadaleta who offered their advice and suggestions on several topics covered in this chapter.

REFERENCES

1. Cavalier-Smith T. Kingdom Protozoa and its 18 phyla. Microbiol Rev 1993; 57(4):953–994.
2. Lang BF, Burger G, O'Kelly CJ, Cedergren R, Golding GB, Lemieux C, Sankoff D, Turmel M, Gray MW. An ancestral mitochondrial DNA resembling a eubacterial genome in miniature. Nature 1997; 387(6632):493–497. See comments.
3. Andersson SG, et al. The genome sequence of *Rickettsia prowazekii* and the origin of mitochondria. Nature 1998; 396(6707):133–140. See comments.
4. Lopez-Garcia P, Moreira D. Metabolic symbiosis at the origin of eukaryotes. Trends Biochem Sci 1999; 24(3):88–93.
5. Gray MW, Burger G, Lang BF. Mitochondrial evolution. Science 1999; 283:1476–1481.
6. Boore JL. Animal mitochondrial genomes. Nucleic Acids Res 1999; 27(8):1767–1780.
7. Gray MW, et al. Genome structure and gene content in protist mitochondrial DNAs. Nucleic Acids Res 1998; 26(4):865–878.
8. Paquin B, Laforest MJ, Forget L, Roewer I, Wang Z, Longcore J, Lang BF. The fungal mitochondrial genome project: evolution of fungal mitochondrial genomes and their gene expression. Curr Genet 1997; 31(5):380–395.
9. Shadel GS, Clayton DA. Mitochondrial DNA maintenance in vertebrates. Annu Rev Biochem 1997; 66:409–435.
10. Wolstenholme DR, Fauron CM-R. The molecular biology of plant mitochondria. In: Levings CS, Vasil IK, eds. The molecular Biology of Plant Mitochondria. The Netherlands: Kluwer Dordrecht, 1995:1–59.
11. Wolstenholme DR. Animal mitochondrial DNA: structure and evolution. In Rev Cytol 1992; 141:173–216.
12. Embley TM, Hirt RT. Early branching eukaryotes? Curr Opini Genet Dev 1998; 8:624–629.
13. Akhmanova A, Voncken F, van Allen T, van Hoek A, Boxma B, Vogels G, Veenhuis M, Hackstein JHP. A hydrogenosome with a genome. Nature 1998; 396:527–528.
14. Martin W, Muller M. The hydrogen hypothesis for the first eukaryote. Nature 1998; 392(6671):37–41. See comments.
15. Ashley MV, Laipis PJ, Hauswirth WW. Rapid segregation of heteroplasmic bovine mitochondria. Nucleic Acids Res 1989; 17:7325–7331.
16. Hauswirth WW, Laipis PJ. Mitochondrial DNA polymorphism in a maternal lineage of Holstein cows. Proc Natl Acad Sci USA 1982; 79(15):4686–4690.

17. Wang G, Yan S. Mitochondrial DNA content and mitochondrial gene transcriptional activities in the early development of loach and goldfish. Int J Dev Biol 1992; 36(4):477–482.

18. Piko L, Taylor KD. Amounts of mitochondrial DNA and abundance of some mitochondrial gene transcripts in early mouse embryos. Dev Biol 1987; 123(2):364–374.

19. Handyside AH, Hunter SA. A rapid procedure for visualizing the inner cell mass and trophectoderm nuclei of mouse blastocysts in situ using polynucleotide-specific fluorochromes. J Exp Zool 1984; 231:429–434.

20. Birky CW. Uniparental inheritance of mitochondrial and chloroplast genes: mechanisms and evolution. Proc Natl Acad Sci USA 1995; 92:11331–11338.

21. Moses MJ. J Biophys Biochem Cytol 1961; 10:301–333.

22. Ursprung H, Schabtach E. J Exp Zool 1965; 159:379–384.

23. Ankel-Simons F, Cummins JM. Misconceptions about mitochondria and mammalian fertilization: implications for theories on human evolution. Proc Natl Acad Sci USA 1996; 93:13859–13865.

24. Hiraoka J-I, Hirano Y. Fate of sperm tail components after incorporation into the hamster egg. Gamete Res 1988; 19:369–380.

25. Hecht NB, Liem H, Kleene KC, Distel RJ, Ho SM. Maternal inheritance of the mouse mitochondrial genome is not mediated by a loss or gross alteration of the paternal mitochondrial DNA or by methylation of the oocyte mitochondrial DNA. Dev Biol 1984; 102(2):452–461.

26. Piko L, Matsumoto L. Number of mitochondria and some properties of mitochondrial DNA in the mouse egg. Dev Biol 1976; 49:1–10.

27. Chen X, Prosser R, Simonetti S, Sadlock J, Jagiello G, Schon EA. Rearranged mitochondrial genomes are present in human oocytes. Am J Hum Genet 1995; 57(2):239–247. See comments.

28. Gyllensten U, Wharton D, Josefsson A, Wilson AC. Paternal inheritance of mitochondrial DNA in mice. Nature 1991; 352:255–257.

29. Shitara H, Hayashi JI, Takahama S, Kaneda H, Yonekawa H. Maternal inheritance of mouse mtDNA in interspecific hybrids: segregation of the leaked paternal mtDNA followed by the prevention of subsequent paternal leakage. Genetics 1998; 148(2):851–857.

30. Kaneda H, Hayashi J, Takahama S, Taya C, Lindahl KF, Yonekawa H. Elimination of paternal mitochondrial DNA in intraspecific crosses during early mouse embryogenesis. Proc Natl Acad Sci USA 1995; 92(10):4542–4546.

31. Awadalla P, Eyre-Walker A, Smith JM. Linkage disequilibrium and recombination in hominid mitochondrial DNA. Science 1999; 286(5449):2524–2525. See comments.

32. Cann RL, Stoneking M, Wilson AC. Mitochondrial DNA and human evolution. Nature 1987; 325(6099):31–36.

33. Paabo S. Human evolution. Trends Cell Biol 1999; 9(12):M13–M16.

34. Zouros E, Oberhauser Ball A, Saavedra C, Freeman KR. An unusual type of mitochondrial DNA inheritance in the blue mussel Mytilus. Proc Natl Acad Sci USA 1994; 91(16):7463–7467.

35. Rawson PD, Hilbish TJ. Evolutionary relationships among the male and female mitochondrial DNA lineages in the *Mylilus edulis* species complex. Mol Biol Evol 1995; 12:893–901.

36. Stewart DT, Saavedra C, Stanwood RR, Ball AO, Zouros E. Male and female mitochondrial DNA lineages in the blue mussel (*Mytilus edulis*) species group. Mol Biol Evol 1995; 12:735–747.

37. Lunt DH, Hyman BC. Animal mitochondrial DNA recombination. Nature 1997; 387:247.

38. Clayton DA, Doda JN, Friedberg EC. The absence of a pyrimidine dimer repair mechanism in mammalian mitochondria. Proc Natl Acad Sci USA 1974; 71:2777–2781.

39. Hayashi J, Takemitsu M, Goto Y, Nonaka I. Human mitochondria and mitochondrial genome function as a single dynamic cellular unit. J Cell Biol 1994; 125(1):43–50.

40. Hayashi J, Ohta S, Kikuchi A, Takemitsu M, Goto Y, Nonaka I. Introduction of disease-related mitochondrial DNA deletions into HeLa cells lacking mitochondrial DNA results in mitochondrial dysfunction. Proc Natl Acad Sci USA 1991; 88(23):10614–10618.

41. Gillespie FP, Hong TH, Eisenstadt JM. Transcription and translation of mitochondrial DNA in interspecific somatic cell hybrids. Mol Cell Biol 1986; 6(6):1951–1957.

42. Oliver NA, Wallace DC. Assignment of two mitochondrially synthesized poly-peptides to human mitochondrial DNA and their use in the study of intracellular mitochondrial interaction. Mol Cell Biol 1982; 2(1):30–41.

43. Yoneda M, Miyatake T, Attardi G. Complementation of mutant and wild-type human mitochondrial DNAs coexisting since the mutation event and lack of complementation of DNAs introduced separately into a cell within distinct organelles. Mol Cell Biol 1994; 14(4):2699–2712.

44. Preiss T, Lowerson AS, Weber K, Lightowlers RN. Human mitochondria: distinct organelles or dynamic network? Trends Genet 1995; 11(6):211–212.

45. Bogenhagen DF. Repair of mtDNA in vertebrates. Am J Hum Genet 1999; 64:1276–1281.

46. Thyagarajan B, Padua RA, Campbell C. Mammalian mitochondria possess homologous DNA recombination activity. J Biol Chem 1996; 271(44):27536–27543.

47. Ikeda S, Ozaki K. Action of mitochondrial endonuclease G on DNA damaged by L-ascorbic acid, peplomycin, and cis-diamminedichloroplatinum(II). Biochem Biophys Res Commun 1997; 235(2):291–294.

48. Foury F, Lahaye A. Cloning and sequencing of the PIF gene involved in repair and recombination of yeast mitochondrial DNA. EMBO J 1987; 6(5):1441–1449.

49. Ling F, Makishima F, Morishima N, Shibata T. A nuclear mutation defective in mitochondrial recombination in yeast. EMBO J 1995; 14(16):4090–4101.

50. Ma DP, King YT, Kim Y, Luckett WS Jr, Boyle JA, Chang YF. Amplification and characterization of an inverted repeat from the *Chlamydomonas reinhardtii* mitochondrial genome. Gene 1992; 119(2):253–257.

51. Benne R. Mitochondrial genes in trypanosomes. Trends Genet 1985; 1:117–121.
52. Levings CS, Brown GG. Molecular biology of plant mitochondria. Cell 1989; 56(2):171–179.
53. Gray MW. Origin, evolution of mitochondrial DNA. Annu Rev Cell Biol 1989; 5:25–50.
54. Gray M, Cedergren R, Abel Y, Sankoff D. On the evolutionary origin of the plant mitochondrion and its genome. Proc Natl Acad Sci USA 1989; 86:2267–2271.
55. Newton J. Plant mitochondrial genomes: organization, expression and variation. Annu Rev Plant Physiol Plant Mol Biol 1989; 39:503–532.
56. Palmer JD. Contrasting modes and tempos of genome evolution in land plant organelles. Trends Genet 1990; 6(4):115–120.
57. La Roche J, Snyder M, Cook DI, Fuller K, Zouros E. Molecular characterization of a repeat element causing large-scale size variation in the mitochondrial DNA of the sea scallop Placopecten magellanicus. Mol Biol Evol 1990; 7(1):45–64.
58. Saccone C, Sbisà E. The evolution of the mitochondrial genome. In: Bittar EE, ed. Principles of Medical Biology. Greenwich, CT: JAI Press Inc, 1994:39–72.
59. Saccone C. The evolution of mitochondrial DNA. Curr Opin Genet Dev 1994; 4:875–881.
60. Gray MW, Covello PS. RNA editing in plant mitochondria and chloroplasts. FASEB J 1993; 7(1):64–71.
61. de Zamaroczy M, Bernardi G. Sequence organization of the mitochondrial genome of yeast—a review. Gene 1985; 37(1–3):1–17.
62. Foury F, Roganti T, Lecrenier N, Purnelle B. The complete sequence of the mitochondrial genome of Saccharomyces cerevisiae. FEBS Lett 1998; 440(3):325–331.
63. Gray H, Wong TW. Purification and identification of subunit structure of the human mitochondrial DNA polymerase. J Biol Chem 1992; 267:5835–5841.
64. Yokogawa T, Watanabe Y, Yotsumoto Y, Kumazawa Y, Ueda T, Hirao I, Miura K, Watanabe K. Structure of mitochondrial tRNA. Nucleic Acids Symp Ser 1991; 25:175–176.
65. Benne R, Van den Burg J, Brakenhoff JP, Sloof P, Van Boom JH, Tromp MC. Major transcript of the frameshifted coxII gene from trypanosome mitochondria contains four nucleotides that are not encoded in the DNA. Cell 1986; 46(6):819–826.
66. Pesole G, Ceci LR, Gissi C, Saccone C, Quagliariello C. Evolution of the nad3-rps12 gene cluster in angiosperm mitochondria: comparison of edited and unedited sequences. J Mol Evol 1996; 43(5):447–452.
67. Yokobori S, Paabo S. Transfer RNA editing in land snail mitochondria. Proc Natl Acad Sci USA 1995; 92(22):10432–10435.
68. Yokobori SI, Paabo S. tRNA editing in metazoans. Nature 1995; 377 (6549): 490.
69. Janke A, Paabo S. Editing of a tRNA anticodon in marsupial mitochondria changes its codon recognition. Nucleic Acids Res 1993; 21(7):1523–1525.

70. Moraes CT, Schon EA. Replication of a heteroplasmic population of normal and partially deleted human mitochondrial genomes. In: Palmieri F, et al., eds., Progress in Cell Research. Amsterdam: Elsevier Sci, 1995:209–215.

71. Bridge D, Cunningham CW, Schierwater B, DeSalle R, Buss LW. Class-level relationships in the phylum Cnidaria: evidence from mitochondrial genome structure. Proc Natl Acad Sci USA 1992; 89:8750–8753.

72. Beagley CT, Macfarlane JL, Pont-Kingdon A, Okimoto R, Okada NA, Wolstenholme DR. Mitochondrial genomes of Anthozoa (Cnidaria). In: Palmieri F, et al., eds. Progress in Cell Research. Amsterdam: Elsevier, 1995: 149–153.

73. Hoffmann RJ, Boore JL, Brown WM. A novel mitochondrial genome organization for the blue mussel, *Mytilus edulis*. Genetics 1992; 131:397–412.

74. Beagley CT, Okada NA, Wolstenholme DR. Two mitochondrial group I introns in a metazoan, the sea anemone *Metridim senile*: one intron contains genes for subunits 1 and 3 of NADH dehydrogenase. Proc Natl Acad Sci USA 1996; 93:5619–5623.

75. Pont-Kingdon GA, Okada NA, Macfarlane JL, Beagley CT, Wolstenholme DR, Cavalier-Smith T, Clark-Walker GD. A coral mitochondrial MutS gene. Nature 1995; 375:109–111.

76. Pont-Kingdon GA, Okada NA, Macfarlane JL, Beagley CT, Watkins-Sims CD, Cavalier-Smith T, Clark-Walker GD, Wolstenholme DR. Mitochondrial DNA of the coral *Sarcophyton glaucum* contains a gene for a homologue bacterial MutS: a possible case of gene transfer from the nucleus to the mitochondrion. J Mol Evol 1998; 46:419–431.

77. Beagley CT, Okimoto R, Wolstenholme DR. The mitochondrial genome of the sea anemone *Metridium senile* (Cnidaria): introns, a paucity of tRNA genes, and a near-standard genetic code. Genetics 1998; 148:1091–1108.

78. Beaton MJ, Roger AJ, Cavalier-Smith T. Sequence analysis of the mitochondrial genome of *Sarcophyton glaucum*: conserved gene order among octocorals. J Mol Evol 1998; 47(6):697–708.

79. Okimoto R, Chamberlin HM, Macfarlane JL, Wolstenholme DR. Repeated sequence sets in mitochondrial DNA molecules of root knot nematodes (Meloidogyne): nucleotide sequences, genome location and potential for host-race identification. Nucleic Acids Res 1991; 19(7):1619–1626.

80. Okimoto R, Macfarlane JL, Clary DO, Wolstenholme DR. The mitochondrial genome of two nematodes *Caesnorhabditis elegans* and *Ascaris suum*. Genetics 1992; 130:471–498.

81. Yamazaki N, et al. Evolution of pulmonate gastropod mitochondrial genomes: comparisons of complete gene organization of Euhadra, Cepaea and Albinaria and implications of unusual tRNA secondary structures. Genetics 1997; 145:749–758.

82. Boore JL, Brown WM. Complete DNA sequence of the mitochondrial genome of the black chiton, *Katharina tunicata*. Genetics 1994; 138(2):423–443.

83. Boore JL, Brown WM. Complete sequence of the mitochondrial DNA of the annelid worm *Lumbricus terrestris*. Genetics 1995; 141(1):305–319.

84. Crozier RH, Crozier YC. The mitochondrial genome of the honeybee *Apis mellifera*: complete sequence and genome organization. Genetics 1993; 133(1):97–117.

85. Clary DO, Wolstenholme DR. The mitochondrial DNA molecular of *Drosophila yakuba*: nucleotide sequence, gene organization, and genetic code. J Mol Evol 1985; 22(3):252–271.

86. Lewis DL, Farr CL, Kaguni LS. *Drosophila melanogaster* mitochondrial DNA: completion of the nucleotide sequence and evolutionary comparisons. Insect Mol Biol 1995; 4(4):263–278.

87. Valverde J, Batuecas B, Moratilla C, Marco R, Garesse R. The complete mitochondrial sequence of the crustacean *Artemia franciscana*. J Mol Evol 1994; 39:400–408.

88. Black WC, Roehrdanz RL. Mitochondrial gene order is not conserved in arthropods: prostriate and metastriate tick mitochondrial genomes. Mol Biol Evol 1998; 15(12):1772–1785.

89. Boore JL, Brown WM. Big trees from little genomes: mitochondrial gene order as a phylogenetic tool. Curr Opin Genet Dev 1998; 8:668–674.

90. Cantatore P, Roberti M, Rainaldi G, Gadaleta MN, Saccone C. The complete nucleotide sequence, gene organization, and genetic code of the mitochondrial genome of *Paracentrotus lividus*. J Biol Chem 1989; 264(19):10965–10975.

91. De Giorgi C, Martiradonna A, Lanave C, Saccone C. Complete sequence of the mitochondrial DNA in the sea urchin *Arbacia lixula*: conserved features of the echinoid mitochondrial genome. Mol Phylogenet Evol 1996; 5(2):323–332.

92. Jacobs HT, Elliott DJ, Math VB, Farquharson A. Nucleotide sequence and gene organization of sea urchin mitochondrial DNA. J Mol Biol 1988; 202(2):185–217. Published erratum appears in J Mol Biol 1990; 211(3):663.

93. Castresana J, Feldmaier-Fuchs G, Yokobori S, Satoh N, Paabo S. The mitochondrial genome of the hemichordate *Balanoglossus carnosus* and the evolution of deuterostome mitochondria. Genetics 1998; 150(3):1115–1123.

94. Blanchette M, Kunisawa T, Sankoff D. Gene order breakpoint evidence in animal mitochondrial phylogeny. J Mol Evol 1999; 49(2):193–203. In Process Citation.

95. Cantatore P, Gadaleta MN, Roberti M, Saccone C, Wilson AC. Duplication and remoulding of tRNA genes during evolutionary rearrangement of mitochondrial genomes. Nature 1987; 329:853–855.

96. Moritz C, Dowling T, Brown W. Evolution of animal mitochondrial DNA: relevance for population and systematics. Annu Rev Ecol Syst 1987; 18:269–292.

97. Saccone C, Attimonelli M, De Giorgi C, Lanave C, Sbisà E. The role of tRNA genes in the evolution of animal mitochondrial DNA. In: Quagliariello E, et al., eds. Structure, Function and Biogenesis of Energy Transfer Systems. Amsterdam: Elsevier, 1990; 91–96.

98. Maizels N, Weiner AM. Phylogeny from function: the origin of tRNA is replication, not translation. In: Fitch WM, Ayala FJ, eds. Tempo and Mode in Evolution. Washington, DC: National Academy Press, 1995.

99. Ojala D, Merkel C, Gelfand R, Attardi G. The tRNA genes punctuate the reading of genetic information in human mitochondrial DNA. Cell 1980; 22(2 Pt 2):393–403.

100. De Giorgi C, Martiradonna A, Saccone C. Evolutionary analysis of sea urchin mitochondrial tRNAs: folding of the molecules as suggested by the non-random occurrence of nucleotides. Curr Genet 1996; 30(3):191–199.

101. Saccone C. Structure and evolution of the metazoan mitochondrial genome. In: Papa S, Guerrieri F, Tager JM, eds. Frontiers of Cellular Bioenergetics. New York: Kluwer Academic, 1999:521–551.

102. Gadaleta G, D'Elia D, Capaccio L, Saccone C, Pepe G. Isolation of a 25-kDa protein binding to a curved DNA upstream the origin of the L strand replication in the rat mitochondrial genome. J Biol Chem 1996; 271(23):13537–13541.

103. Pepe G, Gadaleta G, Palazzo G, Saccone C. Sequence-dependent DNA curvature: conformational signal present in the main regulatory region of the rat mitochondrial genome. Nucleic Acids Res 1989; 17(21):8803–8819.

104. Pepe G, et al. Non-random patterns of nucleotide substitutions and codon strategy in the mammalian mitochondrial genes coding for identified and unidentified reading frames. Biochem Int 1983; 4:553–563.

105. Saccone C, Cantatore P, Gadaleta G, Gallerani R, Lanave C, Pepe G, Kroon AM. The nucleotide sequence of the large ribosomal RNA gene and the adjacent tRNA genes from rat mitochondria. Nucleic Acids Res 1981; 9:4139–4148.

106. Janke A, Xu X, Arnason U. The complete mitochondrial genome of the wallaroo (*Macropus robustus*) and the phylogenetic relationship among Monotremata, Marsupialia and Eutheria. Proc Natl Acad Sci USA 1997; 94:1276–1281.

107. Desjardins P, Morais R. Sequence and gene organization of the chicken mitochondrial genome: a novel gene order in higher vertebrates. J Mol Biol 1990; 212:599–634.

108. Mindell D, Sorenson M, Dimcheff D. Multiple independent origins of mitochondrial gene order in birds. Proc Natl Acad Sci USA 1998; 95:10693–10697.

109. Mindell DP, Sorenson MD, Dimcheff DE. An extra nucleotide is not translated in mitochondrial ND3 of some birds and turtles. Mol Biol Evol 1998; 15:1568–1571.

110. Zardoya R, Meyer A. Complete mitochondrial genome suggests diapsid affinities of turtles. Proc Natl Acad Sci USA 1998; 95:14226–14231.

111. Janke A, Arnason U. The complete mitochondrial genome of *Alligator mississippiensis* and the separation between recent Archosauria (birds and crocodiles). Mol Biol Evol 1997; 14:1266–1272.

112. Kumazawa Y, Ota H, Nishida M, Ozawa T. The complete nucleotide sequence of a snake (*Dinodon semicarinatus*) mitochondrial genome with two identical control regions. Genetics 1998; 150(1):313–329.

113. Lee WJ, Kocher TD. Complete sequence of a sea lamprey (*Petromyzon marinus*) mitochondrial genome: early establishment of the vertebrate genome organization. Genetics 1995; 139(2):873–887.

114. Spruyt N, Delarbre C, Gachelin G, Laudet V. Complete sequence of the amphioxus (*Branchiostoma lanceolatum*) mitochondrial genome: relations to vertebrates. Nucleic Acids Res 1998; 26(13):3279–3285.

115. Brown GG, Gadaleta G, Pepe G, Saccone C, Sbisà E. Structural conservation and variation in the D-loop-containing region of vertebrate mitochondrial DNA. J Mol Biol 1986; 192(3):503–511.

116. Saccone C, Pesole G, Sbisà E. The main regulatory region of mammalian mitochondrial DNA: structure-function model and evolutionary pattern. J Mol Evol 1991; 33:83–91.

117. Sbisà E, Tanzariello F, Reyes A, Pesole G, Saccone C. Mammalian mitochondrial D-loop region structural analysis: identification of new conserved sequences and their functional and evolutionary implications. Gene 1997; 205:125–140.

118. Tullo A, Rossmanith W, Imre EM, Sbisà E, Saccone C, Karwan RM. RNase mitochondrial RNA processing cleaves RNA from the rat mitochondrial displacement loop at the origin of heavy-strand DNA replication. Eur J Biochem 1995; 227(3):657–662.

119. Perna NT, Kocher TD. Patterns of nucleotide composition at fourfold degenerate sites of animal mitochondrial genomes. J Mol Evol 1995; 41:353–358.

120. Lobry JR. Asymmetric substitution pattern in the two DNA strands of bacteria. Mol Biol Evol 1996; 13:660–665.

121. Ames BN, Shigenaga MK, Hagen TM. Mitochondrial decay in aging. Biochim Biophys Acta 1995; 1271:165–170.

122. Clayton DA. Replication of animal DNA. Cell 1982; 28:693–705.

123. Brown GG, Simpson MV. Novel features of animal mtDNA evolution as shown by sequences of two rat cytochrome oxidase subunit II genes. Proc Natl Acad Sci USA 1982; 79:3246–3250.

124. Kunkel TA, Alexander PS. The base substitution fidelity of eukaryotic DNA polymerases: mispairing frequencies, site preferences, insertion preferences, and base substitution by dislocation. J Biol Chem 1986; 261:160–166.

125. Kunkel TA, Soni A. Exonucleolytic proofreading enhances the fidelity of DNA synthesis by chick embryo DNA polymerase-gamma. J Biol Chem 1988; 263:4450–4459.

126. Pinz KG, Shibutani S, Bogenhagen DF. Action of mitochondrial DNA polymerase gamma at sites of base loss or oxidative damage. J Biol Chem 1995; 270:9202–9206.

127. Reyes A, Gissi C, Pesole G, Saccone C. Asymmetrical directional mutation pressure in the mitochondrial genome of mammals. Mol Biol Evol 1998; 15(8):957–966.

128. Lindahl T. Instability and decay of primary structure of DNA. Nature 1993; 362:709–715.

129. Tanaka M, Ozawa T. Strand asymmetry in human mitochondrial DNA mutations. Genomics 1994; 22:327–335.

130. Pesole G, Dellisanti G, Preparata G, Saccone C. The importance of base

composition in the correct assessment of genetic distances. J Mol Evol 1995; 41:1124–1227.

131. Saccone C, Lanave C, Pesole G, Preparata G. Influence of base composition on quantitative estimates of gene evolution. Methods Enzymol 1990; 183:570–583.

132. Lanave C, Preparata G, Saccone C, Serio G. A new method for calculating evolutionary substitution rates. J Mol Evol 1984; 20:86–93.

133. Jukes TH, Cantor CR. Evolution of protein molecules. In: Munro MH, ed. Mammalian Protein Metabolism. New York: Academic Press, 1969:21–132.

134. Kimura M. A simple method for estimating evolutionary rates of base substitutions through comparative studies of nucleotide sequences. J Mol Evol 1980; 16(2):111–120.

135. Springer MS. Molecular clocks and the incompleteness of the fossil record. J Mol Evol 1995; 41:531–538.

136. Pesole G, Gissi C, De Chirico A, Saccone C. Nucleotide substitution rate of mammalian mitochondrial genomes. J Mol Evol 1999; 48(4):427–434.

137. Adkins RM, Honeycutt RL, Disotell TR. Evolution of eutherian cytochrome c oxidase subunit II: heterogeneous rates of protein evolution and altered interaction with cytochrome c. Mol Biol Evol 1996; 13:1393–1404.

138. Arnason U, Xu X, Gullberg A, Graur D. The "Phoca standard": an external molecular reference for calibrating recent evolutionary divergences. J Mol Evol 1996; 43:41–45.

139. Honeycutt RL, Nebdal MA, Adkins RM, Janecek LL. Mammalian mitochondrial DNA evolution: a comparison of the cytochrome b and cytochrome c oxidase II genes. J Mol Evol 1995; 40:260–272.

140. Paabo S, Thomas WK, Whitfield KM, Kumazawa Y, Wilson AC. Rearrangements of mitochondrial transfer RNA genes in marsupials. J Mol Evol 1991; 33(5):426–430.

141. Saccone C, Attimonelli M, Sbisà E. Structural elements highly preserved during the evolution of the D-loop containing regions in vertebrate mitochondrial DNA. J Mol Evol 1987; 26:205–211.

142. Macey JR, Larson A, Ananjeva NB, Fang Z, Papenfuss TJ. Two novel gene orders and the role of light-strand replication in rearrangement of the vertebrate mitochondrial genome. Mol Biol Evol 1997; 14(1):91–104.

143. Lynch M. Mutation accumulation in transfer RNAs: molecular evidence for Muller's ratchet in mitochondrial genomes. Mol Biol Evol 1996; 13(1):209–220.

144. Alvarez-Valin F, Jabbari K, Bernardi G. Synonymous and nonsynonymous substitutions in mammalian genes: intragenic correlation. J Mol Evol 1998; 46(1):37–44.

145. Muse SV, Gaut BS. A likelihood approach for comparing synonymous and nonsynonymous nucleotide substitution rates, with application to the chloroplast genome. Mol Biol Evol 1994; 11(5):715–724.

146. Muse SV, Weir BS. Testing for equality of evolutionary rates. Genetics 1992; 132:269–276.

147. Gissi C, Reyes A, Pesole G, Saccone C. Lineage-specific evolutionary rate in mammalian mtDNA. Mol Biol Evol 2000; 17(7):1022–1031.

148. Ayala FJ, Rzhetsky A. Origin of the metazoan phyla: molecular clocks confirm paleontological estimates. Proc Natl Acad Sci USA 1998; 95(2):606–611.

149. Kumar S, Hedges SB. A molecular timescale for vertebrate evolution. Nature 1998; 392(6679):917–920.

150. Wang DY, Kumar S, Hedges SB. Divergence time estimates for the early history of animal phyla and the origin of plants, animals and fungi. Proc Roy Soc Lond B Biol Sci 1999; 266(1415):163–171.

151. Blundell TL, Wood SP. Is the evolution of insulin Darwinian or due to selectively neutral mutation? Nature 1975; 257(5523):197–203.

152. Graur D, Hide WA, Li H-W. Is the guinea-pig a rodent? Nature 1991; 351:649–652.

153. D'Erchia AM, Gissi C, Pesole G, Saccone C, Arnason U. The guinea-pig is not a rodent. Nature 1996; 381:597–599.

154. Reyes A, Pesole G, Saccone C. Complete mitochondrial DNA sequence of the fat dormouse, *Glis glis*: further evidence of rodent paraphyly. Mol Biol Evol 1998; 15(5):499–505.

155. Van Valen L. Detratheridia, a new order of mammals. Bull Am Mus Nat Hist 1966; 132:1–126.

156. Boyden A, Gemeroy D. The relative position of Cetacea among the order of Mammalia as indicated by precipitin tests. Zoologica 1950; 35:145–151.

157. Beintema J, Gaastra W, Lenstra J, Welling G, Fitch W. The molecular evolution of pancreatic ribonuclease. J Mol Evol 1977; 10:49–71.

158. Goldstone A, Smith E. Amino acid sequence of the whale heart cytochrome c. J Biol Chem 1966; 241:4480–4486.

159. Graur D, Higgins D. Molecular evidence for the inclusion of cetaceans within the order Artiodactyla. Mol Biol Evol 1994; 11:357–364.

160. Ursing B, Arnason U. The complete mitochondrial DNA sequence of the pig (*Sus scrofa*). J Mol Evol 1998; 47:302–306.

161. Graur D, Duret L, Gouy M. Phylogenetic position of the order Lagomorpha (rabbits, hares and allies). Nature 1996; 379:333–335.

162. Carroll RL. Vertebrate Paleontology and Evolution. New York: WH Freeman, 1988:698.

163. Novacek NJ. Mammalian phylogeny: shaking the tree. Nature 1992; 379:333–335.

164. Arnason A, Gullberg A, Janke A. Phylogenetic analyses of mitochondrial DNA suggest a sister group relationship between Xenarthra (Edentata) and ferungulates. Mol Biol Evol 1997; 14(7):762–768.

165. Teeling EC, Scally M, Kao DJ, Romagnoli ML, Springer MS, Stanhope MJ. Molecular evidence regarding the origin of echolocation and flight in bats. Nature 2000; 403(6766):188–192. In Process Citation.

166. Pumo DE, Finamore PS, Franek WR, Phillips CJ, Tarzami S, Balzarano D. Complete mitochondrial genome of a neotropical fruit bat, *Artibeus jamaicensis*, and a new hypothesis of the relationships of bats to other eutherian mammals. J Mol Evol 1998; 47(6):709–717.

167. Gregory W. The monotremes and the palimpsest theory. Am Mus Nat Hist Bull 1947; 88:1–52.

168. Kuhne W. The systematic position of monotremes reconsidered (Mammalia). Z Morphol Tiere 1973; 75:59–64.

169. Kuhne W. Marsupium and marsupial bone in Mesozoic mammals and in the Marsupionta. Coll Int CNRS 1975; 218:59–64.

170. Broughton RE, Naylor GJP, Dowling TE. Conflicting phylogenetic patterns caused by molecular mechanisms in mitochondrial DNA sequences. Syst Biol 1998; 47(4):696–701.

171. Naylor GJP, Brown WM. Structural biology and phylogenetic estimation. Nature 1997; 388(6642):527–528.

172. Naylor GJP, Brown WM. Amphioxus mitochondrial DNA, chordate phylogeny, and the limits of inference based on comparisons of sequences. Syst Biol 1998; 47(1):61–76.

173. Vigilant L, Stoneking M, Harpending H, Hawkes K, Wilson AC. African populations and the evolution of human mitochondrial DNA. Science 1991; 253:1503–1507.

174. Pesole G, Sbisà E, Preparata G, Saccone C. The evolution of the mitochondrial D-loop and the origin of modern man. Mol Biol Evol 1992; 9:587–598.

175. Hasegawa M, Horai S. Time of the deepest root for polymorphism in human mitochondrial DNA. J Mol Evol 1991; 32:37–42.

176. Tamura K, Nei M. Estimation of the number of nucleotide substitutions in the control region of mitochondrial DNA in humans and chimpanzees. Mol Biol Evol 1993; 10(3):512–526.

177. Anderson S, et al. Sequence and organization of the human mitochondrial genome. Nature 1981; 290:457–465.

178. Wakeley J. Substitution rate variation among sites in hypervariable region I of human mitochondrial DNA. J Mol Evol 1993; 37(6):613–623.

179. Horai S, Hayasaka K, Kondo R, Tsugane K, Takahata N. Recent African Origin of modern humans revealed by complete sequences of hominoid mitochondrial DNAs. Proc Natl Acad Sci USA 1995; 92:532–536.

180. Parsons TJ, et al. A high observed substitution rate in the human mitochondrial DNA control region. Nat Genet 1997; 15(4):363–368. See comments.

181. Luft R, Ikkos D, Palmieri G, Ernster L, Afzelius B. A case of severe hypermetabolism of nonthyroid origin with a defect in the maintenance of mitochondrial respiratory control: a correlated clinical, biochemical, and morphological study. J Clin Invest 1962; 41:1776–1804.

182. Wallace DC. Mitochondrial diseases in man and mouse. Science 1999; 283(5407):1482–1488.

183. Michikawa Y, Mazzucchelli F, Bresolin N, Scarlato G, Attardi G. Aging-dependent large accumulation of point mutations in the human mtDNA control region for replication. Science 1999; 286(5440):774–779. See comments.

184. Attimonelli M, et al. MitBASE: a comprehensive and integrated mitochondrial DNA database. The present status. Nucleic Acids Res 2000; 28(1):148–152. In Process Citation.

185. Catalano D, Licciulli F, D'Elia D, Attimonelli M. Update of KEYnet: a gene and protein names database for biosequences functional organisation. Nucleic Acids Res 2000; 28(1):372–373. In Process Citation.

186. Korab-Laskowska M, Rioux P, Brossard N, Littlejohn TG, Gray MW, Lang BF, Burger G. The Organelle Genome Database Project (GOBASE). Nucleic Acids Res 1998; 26(1):138–144.

187. Gutell RR, Gray MW, Schnare MN. A compilation of large subunit (23S and 23S-like) ribosomal RNA structures: 1993. Nucleic Acids Res 1993; 21(13): 3055–3074.

188. Bairoch A. The ENZYME database in 2000. Nucleic Acids Res 2000; 28(1):304–305.

189. Overbeek R, Larsen N, Pusch GD, D'Souza M, Selkov E Jr, Kyrpides N, Fonstein M, Maltsev N, Selkov E. WIT: integrated system for high-throughput genome sequence analysis and metabolic reconstruction. Nucleic Acids Res 2000; 28(1):123–125.

190. Kogelnik AM, Lott MT, Brown MD, Navathe SB, Wallace DC. MITOMAP: a human mitochondrial genome database—1998 update. Nucleic Acids Res 1998; 26(1):112–115.

191. Scharfe C, et al. MITOP, the mitochondrial proteome database: 2000 update. Nucleic Acids Res 2000; 28(1):155–158.

192. Pesole G, Gissi C, Catalano D, Grillo G, Licciulli F, Liuni S, Attimonelli M, Saccone C. MitoNuc and MitoAln: two related databases of nuclear genes coding for mitochondrial proteins. Nucleic Acids Res 2000; 28(1):163–165.

193. Etzold T, Ulyanov A, Argos P. SRS: information retrieval system for molecular biology data banks. Methods Enzymol 1996; 266:114–128.

194. Lanave C, Liuni S, Licciulli F, Attimonelli M. Update of AMmtDB: a database of multi-aligned metazoa mitochondrial DNA sequences. Nucleic Acids Res 2000; 28(1):153–154.

195. Volpetti V, Gallerani R, De Benedetto C, Liuni S, Licciulli F, Ceci LR. PLMItRNA, a database for tRNAs and tRNA genes in plant mitochondria: enlargement and updating. Nucleic Acids Res 2000; 28(1):159–162.

196. Handt O, Meyer S, von Haeseler A. Compilation of human mtDNA control region sequences. Nucleic Acids Res 1998; 26(1):126–129.

197. Saitou N, Nei M. The neighbor-joining method: a new method for reconstructing phylogenetic trees. Mol Biol Evol 1987; 4:406–425.

198. Clark-Walker GD, Weiller GF. The structure of the small mitochondrial DNA of Kluyveromyces thermotolerances is likely to reflect the ancestral gene order in fungi. J Mol Evol 1994; 38:593–601.

199. Reyes A, Gissi C, Catzeflis F, Nevo E, Pesole G, Saccone C. Congruent mammalian trees from mitochondrial and nuclear genes using Bayesian methods. Mol Biol Evol 2004; 21:397–403.

200. Pesole G, Saccone C. A novel method for estimating rate variation among sites in a large dataset of homologous DNA sequences. Genetics 2001; 157:859–865.

Appendix 1

Computational Biology: Annotated Glossary of Terms*

Compiled by Andrzej K. Konopka,
Philipp Bucher, M. James C. Crabbe,
Maxim Crochemore, Jaap Heringa,
Frederique Lisacek, Izabela Makalowska,
Wojciech Makalowski, Graziano Pesole,
Cecilia Saccone, Marie-France Sagot,
Akinori Sarai, Peter Schuster, Peter Stadler,
and William R. Taylor

A

A. (1) Symbol for alanine (Ala) in the one-letter naming convention for amino acids. (2) Symbol for nucleotide adenine in DNA or RNA. (3) Symbol of one of four alleles of human blood types (the other three being B, AB, and O).

Ab initio. (1) From the most elementary building blocks (literally "from the beginning") imaginable in a given experimental or model-making situation.

---

* Cited references appear in Appendix References following Appendix 2.

For instance, computational models of organic molecules that take into account only the input data concerning properties of carbon, hydrogen, oxygen, and nitrogen could be considered ab initio. Derivation of properties of matter from elementary particles could also be considered ab initio. (2) Model of a system that does not take into account specialized experimental data concerning the system, just the basic, commonsense knowledge thereof. (3) Calculation based on the simplest model available while experimental description is completely missing.

A-DNA. One of three best-known conformational variants of 3-D structure of DNA (the two others are B-DNA and Z-DNA).

Acceptor site (of an intron). The 3′ end of an intron (often the dinucleotide AG).

Acid. A substance that can release a proton (hydrogen cation H^+) in solution. Solutions of acids in water display pH lower than 7.0. The lower the pH (i.e., the closer to 0), the stronger the acid.

Activation energy. (1) Difference between the energy of substrates of a chemical reaction and the energy of the hypothetical transition state (activated complex) of this reaction. (2) Minimum amount of energy needed to activate atoms or molecules to a condition in which it is equally as likely that they will undergo chemical reaction or transport as that they will return to their original (nonactivated) state.
     The transition state theory of chemical reactions postulates the existence of a high-energy transition state between the low-energy initial conditions and the (also low-energy) product conditions of a one-step reaction (or each single step of a multistep reaction.) Within this theory the activation energy is the amount of energy required to boost the initial materials (substrates) "uphill" to form the transition state (consistent with meaning 1, above). The chemical reaction then proceeds "downhill" of the energy barrier to form the intermediate (in a multistep reaction) or final (in a one-step reaction) products. Catalysts (such as enzymes) lower the activation energy by altering the transition state.

Active site. The region of an enzyme where the substrate binds to form the enzyme–substrate complex.

Adaptation. (1) A heritable feature of an individual's phenotype that improves its chances of survival and reproduction in the existing environment. (2) An activity of a system (such as an organism, a population, or an ecosystem) that leads to better chances of its own survival in a given environment.

Adenine. One of four nucleotides found in nucleic acids. In RNA or DNA sequences adenine is symbolized by the letter A.

Adenosine triphosphate (ATP). A compound containing adenine, ribose, and three phosphate groups. When it is formed, useful energy is stored; when it is broken down (to ADP or AMP), energy is released to drive endergonic reactions. ATP is an energy storage compound.

Algorithm. A systematic list of a finite number of step-by-step instructions for accomplishing some task that has an end result.

Alignment. Juxtaposition of two or more sequences in a way that emphasizes maximum similarity among them according to predefined criteria of similarity.

Alternative splicing. Selection of different sets of donor and acceptor splice sites to produce two or more different mRNA molecules from the same pre-mRNA.

Allele. One of possible alternative forms of the same gene occupying a given locus (position) on a chromosome.

Allele frequency. The relative proportion of a particular allele in a specific population.

Alphabet. An ordered finite set of symbols (letters). Symbols can be elementary (not fractionable into "smaller" symbols) or composite (fractionable into smaller units.) The alphabet that contains only elementary symbols is referred to as an *elementary alphabet*.

Amine. An organic compound that contains an amino group, $-NH_2$ (see also *Amino acid*).

Amino acid (α-amino acid). Organic compound of the general formula:

$$H_2N-\overset{\overset{\displaystyle H}{|}}{\underset{\underset{\displaystyle R}{|}}{C^{\alpha}}}-COOH$$

Protein primary structures (polypeptides) are formed of amino acid residues bound linearly to each other by peptide bonds (result of condensation between carboxyl group of one amino acid with amino group of another). Of the 20 α-amino acids usually present in naturally occurring proteins, 19 have the general structure shown in I. The twentieth protein residue, proline, is an imino acid (amino group NH_2 in formula I is replaced by the imino group $-NH$). Some amino acids are involved in other than protein biosynthesis cellular processes or functions such as regulation of protein turnover and signal transduction.

Except for glycine (H_2N-CH_2-COOH), the α carbon is asymmetrical in all α-amino acids. Therefore, each of the 19 (out of 20; the 20th being glycine) naturally occurring amino acids can exist as either a dextro- (D-) or levo- (L-) optical isomer (enantiomer). However, in naturally occurring proteins one finds only L-amino acids. [This fact is considered extraordinary because from the chemical (in vitro) point of view there should be no reason for such selection of one enantiomer over another. Biology is apparently full of such (physical) symmetry-breaking phenomena].

Amino terminal, N-terminal (of a peptide).   The first amino acid residue in the peptide. The amino group $-NH_2$ is present but not involved in a peptide bond.

Aminoacyl tRNA.   Transfer RNA (tRNA) carrying an amino acid. The amino group of the amino acid is covalently bound to one of the hydroxyl groups (3′ or 2′) of the terminal nucleotide of tRNA.

Analogy.   (1) A resemblance in function or structure that is due to convergence in evolution but not to common ancestry. (Resemblance due to common ancestry is referred to as homology.) (2) A relation between two phenomena that can be described by means of the same model. For example, a mechanical pendulum and an electric condenser are analogous to each other because their behavior can be adequately described by the same mathematical formalism.

Animal.   A member of the kingdom Animalia. In general, a multicellular eukaryote that obtains its food by ingestion.

Annotation.   An explanation or comment appended to a database entry.

Antibody.   One of millions of proteins produced by the immune system that specifically recognizes a foreign substance and initiates its removal from the body.

Anticipatory system.   A system containing a predictive model of itself and/or its environment.

Anticodon.   A trinucleotide in transfer RNA that is able to pair with a complementary triplet of nucleotides (a *codon*) in messenger RNA.

Antigen. Any substance that stimulates the production of antibodies in the immune system of a vertebrate.

Antisense strand. The noncoding strand in a protein-coding region of double-stranded DNA. The sequence of this strand is complementary to the sequence of pre-mRNA resulting from its transcription catalyzed by a DNA-dependent RNA polymerase.

Apoptosis. A series of genetically determined events leading to cell death.

Archaea. One of the three domains of living organisms (the other two being Bacteria and Eukarya). Their separate phylogenetic position has been confirmed by small subunit rRNA sequence analysis and also by investigation of their physiological, biochemical, and genetic features.

ATP. See Adenosine triphosphate.

ATP synthase. An integral membrane protein that couples the transport of proteins with the formation of ATP.

Automaton. (1) A completely or partly self-operating machine or mechanism. (2) A system whose activity can be described by a set of subsequent transitions between states.

Autosome. Any chromosome in a eukaryote other than a sex chromosome.

B

Bacteria. One of the three domains of life (the other two being Archaea and Eukarya).

Bacteriophage. A virus that infects bacteria.

Base. (1) A substance that can accept a proton (hydrogen ion; H^+) in water solution. A water solution of a base has pH greater than 7. (2) In nucleic acids, a nitrogen-containing base (according to meaning 1) that is attached to each sugar in the backbone. Bases found in cellular DNA and RNA are purines (adenine and guanine) and pyrimidines (cytosine and thymine in DNA; uracil in RNA).

Basal factors. Transcription factors that interact directly with DNA in a transcription complex. Most transcription factors in a eukaryotic transcription complex are not basal.

Base pairing. (1) Formation of dimers (pairs) of purine or pyrimidine molecules (bases) via hydrogen bonds. (2) Formation of hydrogen bonds between bases residing in two different single strands of the same double-stranded nucleic acid molecule (DNA or RNA). The energetics of base pairing between strands of double stranded nucleic acid follows the so called Watson–Crick–Chargaff rules of complementarity: Pairing A with T (or U in RNA) leads to more stable complex than other possibilities for A and T while C preferentially pairs with G. (See also *Complementary base pairing*.)

B cell. A type of lymphocyte involved in the immune response of vertebrates. Upon recognizing an antigenic determinant, a B cell develops into a plasma cell, which secretes an antibody. (Other major kinds of cells in immune system are T cells and macrophages.)

B-DNA. A conformational variant of 3-D structure of DNA. This variant is best known as a double-helical, right-handed Watson–Crick structure. (The double-helical structure of B-DNA is today used as an icon that signifies DNA.) Two other well-known 3-D structures of DNA are A-DNA (right-handed) and Z-DNA (left-handed).

Beta-pleated sheet. Type of protein secondary structure.

Binomial. Consisting of two names. For example, the binomial nomenclature of biology gives the name of the genus followed by the name of the species.

Binomial distribution. A discrete probability distribution of obtaining an exact number (say n) of successes out of N Bernoulli trials. Each Bernoulli trial is a success with a fixed probability p and failure with (also fixed) probability $q = 1 - p$.

Bioinformatics. (1) Information technology–based trend in biology-related academic and industrial activities of computer scientists, engineers, mathematicians, chemists, physicists, medical professionals, and biologists. Activities include archiving, searching, displaying, manipulating, and otherwise integrating life science–related data. Examples of bioinformatics tasks include mapping genomes, high-throughput contig annotation, interpretation of gene expression data, and creation of databases. (2) A subfield of computer science that is devoted to biological applications of computer programming, databases, and related activities.

In both meanings, bioinformatics differs from computational biology (the field of biology in which some methods require the use of computers). The difference is primarily in the motivation of practitioners of bioinformatics (desire to demonstrate better software than the competition) as opposed to the motivation of practitioners of computational biology (curiosity and the desire to learn and understand biological systems).

Biological species (concept). A population or group of populations within which there is a significant amount of gene flow under natural conditions but that is genetically isolated from other populations.

Biota. All of the organisms, including animals, plants, fungi, and micro-organisms, found in a given area.

Biotechnology. Application of molecular biology to produce food, medi-cines, and other chemicals usable for controlling the living conditions of humans or other animals.

Body plan. A hypothetical spatiotemporal design that includes an entire animals, its organ systems, and the integrated functioning of its parts.

C

cAMP (cyclic AMP). A compound formed from ATP that mediates the effects of numerous animal hormones. Also needed for the transcription of catabolite-repressible operons in bacteria. cAMP functions in the capacity of a "signal" molecule for communication between cellular slime molds to form organized colonies that superficially resemble multicellular organisms.

Capsid. The protein coat of a virus.

Carboxylic acid. An organic acid whose structure contains the carboxyl group $-COOH$. A given carboxylic acid $R-COOH$ dissociates in water to the carboxylate anion $R-COO^-$ and a proton H^+.

Catalyst. A chemical substance that accelerates a reaction and can be detected as unchanged after this reaction. In the transition state theory of chemical reactions, catalysts lower the activation energy of a reaction via restructuring its transition state to a lower energetic level than the transition state would assume without catalysis.

Cation. An ion with one or more positive charges.

cDNA. See *Complementary DNA*.

Cell cycle. The stages through which a cell passes between one division and the next. Includes all stages of interphase and mitosis.

Cell division. The reproduction of a cell to produce two new cells. In eukary-otes, this process involves nuclear division (mitosis) and cytoplasmic division (cytokinesis).

Cell theory (Wirchov rule). The assumption that all living things consist of one or more cells and that all cells can be made only from other cells.

Central dogma. (1) A postulate that every individual act of protein bio-synthesis is irreversible because translation is irreversible. In other words, a "message" encoded in a DNA protein-encoding region (a gene) can be (sometimes reversibly) transcribed into mRNA and then translated from mRNA to protein, but the protein cannot encode its own protein-coding DNA region (because it cannot be translated back into its own mRNA). (2) A restatement of the known mathematical fact that many-to-one mappings (mathematical functions) are irreversible. The translation code for protein biosynthesis (the genetic code) is a many-to-one mapping whose domain is a set of 64 symbols of trinucleotides (codons) and whose counterdomain is a set of 20 symbols of amino acid residues. One can uniquely encode 20 symbols by using 64 symbols, but one cannot uniquely encode all 64 symbols by using only 20 symbols. In other words, a given protein-coding region in DNA (or in mature mRNA to be precise) can encode a unique polypeptide sequence, but a given polypeptide sequence cannot be uniquely decoded into a DNA sequence. (3). Original (historic) formulation: "Once information has passed into protein, it cannot get out again. In more detail, the transfer of information from nucleic acid to nucleic acid, or from nucleic acid to protein may be possible, but transfer from protein to protein, or from protein to nucleic acid is impossible" (Crick, 1958).

cGMP (cyclic guanosine monophosphate). Compound that serves as an intracellular "signal" in metabolic pathways that involve G proteins.

Channel. A membrane protein that forms an aqueous passageway though which specific solutes may pass by simple diffusion. Some channels are gated; they open and close in response to the binding of specific molecules.

Chaperone protein. A protein that "assists" a newly produced protein in folding to form a functionally appropriate tertiary structure.

Chemical bond. An attractive force stably linking two atoms within a molecule of a chemical compound.

Chemical reaction. (1) A chemical process in which substances are changed into different ones (with different properties) via rupture or rearrangement of the chemical bonds between atoms in a way that does not affect atomic nuclei. (2) Based on IUPAC recommendation: A process that results in the inter-conversion of *chemical species*. Chemical reactions may be *elementary reactions* (one-step reactions) or *stepwise reactions* (multistep reactions).
    Noted that definition 2 includes experimentally observable interconversions of conformers of the same compound. Detectable chemical reactions normally involve sets of molecular entities, but it is often conceptually conve-

nient to also use the term to denote changes involving single molecular entities (i.e., "microscopic chemical events").

Every chemical reaction can be studied in terms of its mechanism (series of subsequent intermediary compounds between the substrates and products.) Reactions occur at a particular rate that depends on several parameters such as the temperature and concentrations of the reactants. Many chemical reactions go to completion—i.e., attain equilibrium—over a period of minutes or hours and can be monitored by classical techniques such as pressure change or electrochemistry. Chemical dynamics explores the detailed behavior of molecules during the most crucial moments of reactions, for example, when bonds are being broken and new bonds are being formed.

Chemical kinetics. Studies of rates of chemical reactions and inference of possible mechanisms based on such studies. (Knowledge of a reaction mechanism comes in part from a study of the rate of a reaction.) The fact that a reaction is thermodynamically favorable does not necessarily mean that it will take place quickly. Many reactions that do proceed are endothermic. Therefore, the enthalpy change is not the ultimate arbiter of the spontaneity of a chemical reaction, and an additional term, the change in free energy, is an important factor. In addition, the rate of the reaction—the change in concentration of a reactant or of a product with respect to time—gives information on how the reaction will proceed. Some of the factors that influence the rate of a reaction are the concentrations of reactants and/or products, temperature, surface area, and pressure. When a catalyst (e.g., an enzyme) is present, then the concentration of the catalyst also influences the rate of reaction.

Chloroplast. An organelle containing the enzymes and pigments that perform photosynthesis. Chloroplasts occur only in eukaryotes.

Chromatin. The nucleic acid–protein complex found in eukaryotic chromosomes.

Chromosome. In most bacteria and viruses, the DNA molecule that contains most or all of the genetic information of the cell or virus. In eukaryotes, a structure composed of DNA and proteins that bears part of the genetic information of the cell.

Citric acid cycle. A set of chemical reactions in cellular respiration. Also known as the *Krebs cycle*.

Class. In taxonomy, the category below the phylum and above the order; a group of related, similar orders.

Clone. Genetically identical cells or organisms produced from a common ancestor by asexual means.

Coacervate.   An aggregate of colloidal particles in suspension that can serve as a physical model of the origin of life and prebiotic evolution.

Code.   (1) A mapping from one symbolic representation (input representation) of a system into another (output representation). If the input and output representations can be expressed in the form of elementary symbols chosen from finite alphabets, then the code can be seen as a relation between input and output alphabets of a device that does encoding (an encoder) or decoding (a decoder). (2) A transformation whereby messages are converted from one representation to another.

Coding region.   A protein-encoding region in a DNA sequence whose non-coding strand (template) undergoes transcription leading to mRNA that in turn can be translated into a polypeptide chain.

Coding strand.   The strand in a protein-encoding region of double-stranded DNA that has the DNA sequence equivalent (Ts in place of Us) of the mRNA sequence resulting from transcription of this region. The term "coding strand" is always relative to a given coding region and should not apply to entire DNA that contains many genes. The strand which is coding in one protein-encoding region can be (and often is) noncoding in another (protein) coding region within the same chromosome.

Codominance.   A condition in which two alleles at a locus produce different phenotypic effects and both effects appear in heterozygotes.

Codon.   A trinucleotide in messenger RNA that, according to the genetic code, corresponds to a specific amino acid.

Coenzyme.   An additional molecule that plays a role in catalysis by an enzyme.

Coevolution.   Concurrent evolution of two or more species that mutually affect each other's evolution.

Collagen.   A fibrous protein found extensively in bone and connective tissue.

Communication.   (1) A process of sharing knowledge or units thereof. (2) A process of sending and receiving messages.

Community.   Any ecologically integrated group of species of microorganisms, plants, and animals inhabiting a given area.

Comparative analysis.   In evolutionary biology, an approach to studying evolution in which hypotheses are tested by measuring the distribution of states among a large number of species.

Comparative genomics. Computer-assisted comparison of DNA sequences from different genomes to reveal functionally significant sequence regions that may play related biological roles.

Complementary base pairing. The A-T (or A-U), T-A (or U-A), C-G, and G-C pairing via hydrogen bonds between bases in double-stranded DNA, in transcription (i.e., between DNA and pre-mRNA), as well as between anti-codon loops of tRNA and codons in mRNA. (See also **Base pairing.**)

Complementary DNA (cDNA). DNA formed by reverse transcriptase (RNA-dependent DNA polymerase) acting with an RNA template; essential intermediate in the reproduction of retroviruses; used as a tool in recombinant DNA technology; lacks introns.

Compound. (1) A substance made up of atoms of more than one element. (2) Made up of many units, as the compound eyes of arthropods (as opposed to the simple eyes of the same group of organisms).

Computing. Executing an algorithm by a symbol-manipulating device placed in appropriate conditions. The nature of "appropriate conditions" usually includes modeling relation between a mechanical device (real or abstract) and the formal system for general description of (all) algorithms. Both the mechanical device and the formal system for algorithms' description can be thought of as idealizations (i.e., models) of the actual symbol-manipulating equipment that does a specific computation. (See also **Modeling relation.**)

Consensus sequences. An artificial sequence (string data structure) that represents the most frequent nucleotide (amino acid) in a given position of a selected region in a properly aligned set of longer sequences. Example: In a selected pentanucleotide region in three aligned (longer) sequences, we have the following pentanucleotides:

P1: ···ACCGA···
P2: ···GTGGA···
P3: ···ATCGT···

It can be seen that in the 5'-most position 1 we have A twice and G once. Therefore the first nucleotide of consensus pentanucleotides is A. In the similar way, the second position is most frequently occupied by T, and so on. The consensus pentanucleotide for the set P1 through P3 is thus ATCGA. It can be noted that the consensus sequence does not need to actually occur in any of the aligned sequences that lead to its determination. (It is an "artificial" sequence.)

Convergent evolution.   The evolution of similar features independently in unrelated taxa from different ancestral structures.

Covalent bond.   A chemical bond formed by sharing electrons between two atoms.

Crossing over.   The reciprocal exchange of corresponding segments between two homologous chromatids.

Cyclic AMP.   See *cAMP*.

Cyclin-dependent kinase (cdk).   A kinase is an enzyme that catalyzes the addition of phosphate groups from ATP to target molecules. Cdk's target proteins are involved in transitions in the cell cycle and are active only when complexed to additional-protein subunits, cyclins.

Cytoplasm.   The contents of the cell, excluding the nucleus.

Cytosine.   A nitrogen-containing base found in DNA and RNA.

D

Data (datum).   (1) Unit of representation (elementary model) of selected aspects of reality provided in a form suitable for symbolic manipulation such as intellectual reflection, discussion, or mechanical transformation (computation). Data can be characterized (described) with the help of at least three properties: (a) content, (b) medium, and (c) structure. (2) Set of unitary representations (of something) organized in a way suitable for symbolic processing. (3) Synonym of "information": a unit of knowledge that can be communicated and processed. (4) A set of units of information or a combination of these units plus some consequences of combining them.

Data structure (information structure).   An organized (and often codified) form in which data are available for symbolic processing with the help of specific algorithms or classes of algorithms.

Degeneracy.   With respect to the genetic code: Many-to-one mapping (encoding) in which an amino acid residue can be coded for by more than one codon of the translation code for protein biosynthesis (the genetic code). Most amino acids can be represented by more than one codon.

Deletion.   (1) Subsequence of one DNA (or polypeptide) sequence that is absent from another, otherwise identical, DNA (or polypeptide) sequence. The opposite of deletion is *insertion*. Two biopolymer sequences are related by an insertion-deletion relation if the same subsequence is present in one and absent from the other. (2) A mutation resulting in deletion. Deletions in DNA can be as small as a single nucleotide, but they can also be so large as to contain several subsequent protein-encoding genes and spacers between them. The larger the deletion (insertion), the easier it is to detect by methods that do not require knowledge of sequences. For instance, missing parts of chromosomes (very large deletions) can be detected by routine microscopic analysis of chromosomes. Deletions of the size of a few protein-encoding genes (medium-size deletions) can be detected by using fluorescent in situ hybridization (FISH) techniques.

Denaturation. Loss of activity of a protein (such as an enzyme) or a nucleic acid as a result of structural changes usually induced by heat or radiation.

Deoxyribonucleic acid. See *DNA*.

Development. Progressive change during the lifetime of an individual organism.

Differentiation. Process whereby originally similar cells follow different developmental pathways.

Diffusion. Random movement of molecules or other particles, resulting in an even distribution of the particles when no barriers are present.

Diploid. Having a chromosome complement consisting of two copies (homologs) of each chromosome.

Directional selection. Selection that favors phenotypes at one extreme of the population distribution of phenotypes.

Disruptive selection. Selection that favors phenotypes at both extremes of the population distribution of phenotypes.

Distance chart. Frequency distribution of distances (number of nucleotides or amino acid residues) separating two sequence motifs in a given long sequence or a collection of functionally equivalent long sequences.

DNA (deoxyribonucleic acid). Major chemical component of hereditary material (chromosomes) of all living organisms. In molecular biology it is assumed that cellular DNA contains encoded instructions about the biosynthesis of all proteins every cell contains or could contain. In eukaryotic organisms DNA is stored in the cell nucleus of every cell but also occurs in cytoplasm and in organelles (mitochondria and chloroplasts).

DNA structures. B-DNA is the well-known right-handed double-helical DNA structure postulated by Watson and Crick in 1953. Other three-dimensional DNA structures (structural isoforms of DNA) that have attracted the attention of researchers since 1953 primarily differ from each other by their handedness (right-handed or left-handed) and, at least in the case of helical isoforms, the number of nucleotides contained in the region between two (360°) turns. Examples of known DNA structures in vitro include

A-DNA: Right-handed antiparallel helix with about 11 nucleotides per helical turn and a helix diameter of about 23 Å.

B-DNA: Right-handed antiparallel double helix with about 10 nucleotides per turn and a helix diameter of about 19 Å.

Z-DNA: Left-handed antiparallel double helix with about 12 nucleotides per helical turn and a helix diameter of about 18 Å.

It is believed (but not proved beyond a doubt) that B-DNA is, in fact, the structure that DNA assumes in vivo whereas its isoforms (such as A-DNA or Z-DNA) exist primarily in vitro.

DNA chip.   A small glass or plastic square onto which thousands of single-stranded DNA sequences are fixed. Cell-derived RNA or DNA can be hybridized to the target sequences. (See *DNA hybridization*.)

DNA hybridization.   A process by which DNAs from two species are mixed and heated so that interspecific double-stranded DNA is formed.

DNA ligase.   Enzyme that unites Okazaki fragments of the lagging strand during DNA replication; also mends breaks in DNA strands. It connects pieces of a DNA strand and is used in recombinant DNA technology.

DNA methylation.   Addition of methyl groups to nucleotides in DNA.

DNA polymerase.   An enzyme that catalyzes the formation of a DNA strand complementary to a nucleic acid (DNA or RNA) template.

Domain.   (1) The largest unit in the current taxonomic nomenclature. Members of the three domains (Bacteria, Archaea, and Eukarya) are believed to have been evolving independently of each other for at least a billion years. (2) Well-defined structural or functional region in nucleic acid or protein (such as the catalytic domain in kinases). (3) A contiguous subsequence of nucleic acid or protein that can clearly be correlated with a functional or structural role in the organism. (4) A region in a chromosome whose supercoiling is independent of other parts of the same chromosome. (5) A region in a chromosome that contains an expressed gene and is hypersensitive to degradation by enzyme DNAse I.

Dominance (allelic).   In genetics, the ability of one allelic form of a gene to determine the phenotype of a heterozygous individual, in which the homologous chromosome carries both itself and a different allele. For example, if A and a are two allelic forms of a gene, A is said to be dominant to a if AA diploids and Aa diploids are phenotypically identical and are distinguishable from aa diploids. The a allele is said to be *recessive*.

Donor site (of an intron). The 5' end of an intron (often the dinucleotide GT).

Duplication, genetic. A mutation resulting from the introduction of an extra copy of a segment of a gene or chromosome into the genome.

Dynamic programming (DP). An optimization technique designed primarily for search problems. It is used extensively in nucleotide and protein sequence analysis (Needleman and Wunsch, 1970) for flexible pairwise string matching. The algorithm makes use of a two-dimensional matrix with the lengths of the two compared strings as its dimensions. The strings are compared by calculating the optimal (best-scoring) alignment over a vast number of possible alignments. To calculate the scores, an exchange table is used whose entries are weights for all pairwise exchanges between the alphabetic symbols occurring in the strings (the exchange table dimensions are 20 × 20 for amino acid comparisons and 4 × 4 for nucleotide sequence comparisons). A gap penalty is assigned for the insertion of gaps in either string. The output of the DP algorithm is the guaranteed optimal alignment, given the residue exchange weights and gap penalties, with the corresponding alignment score.

E

Ecological niche. The set of conditions where the organism lives (the habitat of the organism) together with the functional role of this organism in the ecosystem. Few more specific definitions include: (1) All the places and conditions where an organism of a given species (potentially) could survive. (2) A set of environmental conditions under which specific *functions* of an organism in the community of organisms could assure its survival. (3) A *function* of a species in the community that consists of other species and their environment. (4) A region (volume) in multi-dimensional space of environmental factors that affect the welfare of a species.

Ecology. The scientific study of the interaction of organisms with their environment, including both the physical environment and the other organisms that live in it.

Ecosystem. The organisms of a particular habitat, such as a pond or forest, together with the physical environment in which they live.

Electron.   An elementary particle with a mass of approximately 0.00055 amu and charge of −1.

Electronegativity.   The tendency of an atom to attract electrons when it occurs as part of a compound.

Electrophoresis.   A technique that separates substances from one another on the basis of their electric charges and molecular weights.

Elementary alphabet.   An alphabet whose letters cannot be fractioned into smaller textual units.

Emergent property.   A property of a complex system that is not exhibited by its individual component parts determined from a model of the system.

Enantiomers.   See **Optical isomers**.

3′ End.   The end of a DNA or RNA strand that has a free hydroxyl group at the 3′-carbon of the sugar (deoxyribose or ribose).

5′ End.   The end of a DNA or RNA strand that has a free phosphate group at the 5′-carbon of the sugar (deoxyribose or ribose).

Endergonic reaction.   A reaction for which energy must be supplied. (Contrast with *Exergonic reaction*.)

Endo-.   A prefix used to designate an innermost structure.

Endosymbiosis.   The living together of two species, with one living inside the body (or even the cells) of the other.

Endosymbiotic theory.   A speculative evolutionary scenario in which the eukaryotic cell evolved from a prokaryote that contained other endosymbiotic prokaryotes.

Energetic cost.   The difference between the energy an animal would have expended had it rested and the energy it expended in performing a behavior.

Energy.   The capacity to do work.

Enhancer.   In eukaryotes, a DNA sequence, lying on either side of the gene it regulates, that stimulates a specific promoter.

Entropy.   A state function of a thermodynamic system in equilibrium. Spontaneous irreversible processes in a closed system lead from an equilibrium state with lower entropy to another equilibrium state with higher entropy. See also **Shannon entropy**.

Environment. An organism's surroundings, both living and nonliving; includes temperature, light intensity, and all other species that influence the organism.

Enzyme. A protein that serves as a catalyst for a chemical reaction. (RNA-based catalysts are also called RNA enzymes by some scientists, but in principle the term "enzyme" refers to a protein.)

Epigenesis. A process of interaction between genes and their environment that ultimately results in the distinctive phenotype of an organism.

Epigenetic effect. Also referred to as an *epigenetic phenomenon*, any gene-regulating activity that does not involve changes to the regions in DNA coding for proteins being regulated and that can persist through one or more generations. Two known examples of epigenetic effects are gene silencing and imprinting.

Epigenetic hypothesis. The concept that patterns of gene expression, not genes themselves, define each cell type. (This is the original concept of epigenetics coined by Waddington in the early 1950s to explain differentiation.)

Epigenetic rule. A model of an alleged trend in epigenesis that could channel development in particular directions.

Epigenetics. The study of heritable changes in gene function that occur without a change in the DNA sequence (at least not the one that encoded the studied protein).

Episome. A plasmid that may exist either free or integrated into a chromosome.

Epistasis. An interaction between genes in which the presence of a particular allele of one gene determines whether another gene will be expressed.

Equilibrium. (1) In biochemistry, a state in which forward and reverse reactions are proceeding at counterbalancing rates so there is no observable change in the concentrations of reactants and products. (2) In evolutionary genetics, a condition in which allele and genotype frequencies in a population are constant from generation to generation. (3) In thermodynamics, a state of the system in which all parameters describing the system have constant values that do not change with time.

Estimator. A rule (a random variable) for determining the value of a population parameter based on a random sample from this population. Example: The sample mean is an estimator of the population mean.

Eukaryotes.   Organisms whose cells contain their genetic material inside a nucleus. Include all life forms other than the viruses, Archaeobacteria, and Eubacteria.

Evolution.   Any gradual change. Organic evolution, often referred to as evolution, is any genetic and resulting phenotypic change in organisms from generation to generation.

Exergonic reaction.   A reaction in which free energy is released. (Contrast with *Endergonic reaction*.)

Exon.   A segment of protein-coding region that is incorporated in mRNA after pre-mRNA splicing. The exons of a given protein-coding region (gene) constitute the portion of this region that eventually undergoes translation at the end of a protein biosynthesis event.

Experiment.   A scientific method in which particular factors are manipulated while other factors are held constant so that the potential influences of the manipulated factors can be determined.

F

Family.   In taxonomy, the category below the order and above the genus; a group of related, similar genera.

Feedback control.   Control of a particular step of a multistep process, induced by the presence or absence of a product of one of the later steps.

First law of thermodynamics.   The total energy of a closed thermodynamic system can be neither created nor destroyed and therefore must remain constant. An example of a closed system is the entire universe, which is why the original formulation of the first law was that the energy of the universe cannot be either destroyed or created.

Fission.   Reproduction of a prokaryote by division of a cell into two comparable progeny cells.

Fitness.   The contribution of a genotype or phenotype to the composition of subsequent generations, relative to the contribution of other genotypes or phenotypes.

Frame-shift mutation.   A mutation resulting from the insertion or deletion of a single base pair into or from the DNA sequence of a protein-coding gene. As a result of such mutation, mRNA transcribed from the gene is translated normally until the ribosome reaches the point at which the mutation has occurred. From that point on, codons are read out of proper register, and the amino acid sequence bears no resemblance to the normal sequence.

Free energy. Energy that is available for doing useful work after allowance has been made for the increase or decrease of disorder. Designated by the symbol G (for Gibbs free energy), and defined by $G = H - TS$, where $H =$ heat, $S =$ entropy, and $T =$ absolute (Kelvin) temperature.

G

G protein. A membrane protein involved in signal transduction; characterized by binding guanyl nucleotides. The activation of certain receptors activates the G protein, which in turn activates adenylate cyclase. G protein activation involves binding a GTP molecule in place of a GDP molecule.

Gated channel. A channel (membrane protein) that opens and closes in response to the binding of specific molecules or to changes in membrane potential.

Gel electrophoresis. A method of separating molecules according to their size and electric charge. The medium on which the separation occurs is a semisolid material (gel) suspended in a salty buffer and with electric current passing through it.

Gene. A unit of heredity. The term is used in biology in three alternative variants: (1) The unit of genetic function that carries information for a functionally important unit of biological function. (2) The determinant (Mendelian factor) of an observable characteristic of an organism. (3) The protein-coding region in a cellular DNA (or RNA) sequence. The protein encoded by such a region is referred to as the "gene product."

Variant 3 is favored in molecular biology and its biotechnology-related derivatives.

Gene duplication. A DNA rearrangement that generates a supernumerary copy of a gene in the genome. This would allow each gene to evolve independently to produce distinct functions. Such a set of evolutionarily related genes can be called a "gene family."

Gene family. A set of genes derived from a single parent gene via gene duplication. Individual genes from a family need not be on the same chromosomes.

Gene flow. The exchange of genes between different species (an extreme case referred to as hybridization) or between different populations of the same species caused by migration following breeding.

Gene pool. All of the genes in a population.

Gene structure. The annotated sequence of a coding region in DNA. The annotations indicate in the lexicographic order from the left (5′ end) to the right (3′ end) upstream region, 5′ untranslated region, first exon, first intron, second exon, second intron, and so on up to the last exon followed by the transcription termination sequence (if known) or the 3′ untranslated region or both. The most important issue in determining gene structure is to find correct boundaries (junctions) between introns and exons. Two or more different structures of the same coding region reflect two or more possibilities for gene products resulting from alternative splicing.

Gene therapy. Treatment of a genetic disease by providing patients with cells containing wild-type alleles for the genes that are nonfunctional in their bodies.

Genetic code. Also called the translation code or the universal genetic code. Many-to-one mapping (function) that assigns one of 20 naturally occurring amino acids to each of 61 "sense" trinucleotide code words (codons) during protein biosynthesis. Three additional trinucleotides—TAA, TAG, and TGA—serve as signals for termination of translation during protein biosynthesis but do not encode any amino acid. The initiation of translation is often encoded by the trinucleotide ATG, which is also a codon for the amino acid methionine. (Many protein sequences begin with methionine for this reason.)

Genetic drift. Changes in gene frequencies from generation to generation in a small population as a result of random processes.

Genetics. The study of heredity.

Genome. (1) The complete set of genes of an organism. (2) All the DNA contained in an organism (or a representative cell thereof), which includes both the chromosomes within the nucleus and the DNA in mitochondria. (3) Complete genetic information defining an organism.

Genotype. An exact description of the genetic constitution of an individual, either with respect to a single trait or with respect to a larger set of traits.

Genus. A group of related, similar species.

Germ cell. A reproductive cell or gamete of a multicellular organism.

Glucose. The most common monosaccharide sugar with the formula $C_6H_{12}O_6$.

Glycolysis. The enzymatic breakdown of glucose.

Guanine. A nitrogen-containing base found in DNA, RNA, and GTP.

H

Half-life. The time required for half of a sample of substrates (reactants) in a process to undergo transition into products.

Haploid. Having a chromosome complement consisting of just one copy of each chromosome. This is the normal "ploidy" of gametes or of asexual spores produced by meiosis or of organisms (such as the gametophyte generation of plants) that grow from such spores without fertilization.

Hardy–Weinberg equilibrium. The percentages of diploid combinations expected from a knowledge of the proportions of alleles in the population if no agents of evolution are acting on the population.

α-Helix. A type of protein secondary structure.

Heterochromatin. Chromatin that retains its coiling during interphase; generally not transcribed.

Heterogeneous nuclear RNA (hnRNA). The initial product of transcription of eukaryotic gene, including transcripts of introns. In the older literature hnRNA is often called pre-mRNA.

Heterozygous. Descriptive of a diploid organism having different alleles of a given gene on the pair of homologs carrying that gene. (Contrast with **Homozygous.**)

Hidden Markov model. See **Markov source**.

Histone. Any one of a group of basic proteins forming the core of a nucleosome, the structural unit of a eukaryotic chromosome. (See **Nucleosome.**)

hnRNA. Pre-mRNA. See **Heterogeneous nuclear RNA**.

Homeobox. A 180-base-pair segment of DNA found in a few genes (called *Hox genes* that perhaps regulates the expression of other genes and thus controls large-scale developmental processes.

Homeostasis. The maintenance of a steady state, such as a constant temperature or a stable social structure, by means of physiological or behavioral feedback responses.

Homolog. One of a pair or larger set of chromosomes having the same overall genetic composition and sequence. In diploid organisms, each chromosome inherited from one parent is matched by an identical (except for mutational changes) chromosome—its homolog—from the other parent.

Homology. A similarity between two biopolymer sequences (or structures) that is a result of inheritance from a common ancestor. The sequences (structures) are said to be homologous if they display significant similarity and can be shown to be related to a common ancestral sequence (or structure). Homologous structures or genes have the same evolutionary origin, but their function may differ widely, e.g., the flipper of a seal and the wing of a bat.

Homoplasy. The presence in several species of a trait not present in their most common ancestor. Can result from convergent evolution, reverse evolution, or parallel evolution.

Homozygous. Descriptive of a diploid organism having identical alleles of a given gene on both homologous chromosomes. An organism may be a homozygote with respect to one gene and at the same time a heterozygote with respect to another.

Hox genes. See **Homeobox**.

Hydrocarbon. A compound containing only carbon and hydrogen atoms.

Hydrogen bond. A weak chemical interaction that arises from the attraction between the slight positive charge on a hydrogen atom and a slight negative charge on a nearby atom of another element (such as fluorine, oxygen, phosphorus, sulfur, or nitrogen).

Hydrolyze. To break a chemical bond, as in a peptide linkage, with the insertion of the components of water, $-H$ and $-OH$, at the cleaved ends of a chain. The digestion of proteins is a type of hydrolysis.

Hydrophilic. Having an affinity for water.

Hydrophobic. Not mixing with (repulsing) water.

Hydroxyl group. The $-OH$ group, characteristic of alcohols.

Hypothesis. (1) A tentative assumption (usually about a set of facts) that can be verified as true or false by either experiment or reasoning. (2) A statement assumed to be true for the purpose of argument or further testing. (3) An antecedent H of the conditional statement "if H then P." The P is a prediction or consequence that should be valid if H can be proven to be true.

Hypothesis testing. In statistics, making a decision between rejecting or not rejecting a given null hypothesis on the basis of a set of specific observations.

I

Immunoglobulins. A class of proteins, with a characteristic structure, that are active as receptors and effectors in the immune system.

Immunological memory. The persistence of certain clones of immune system cells made to respond to an antigen. This leads to a more rapid and massive response of the immune system to any subsequent exposure to that antigen.

Imprinting. (1) In genetics, the differential modification of a gene depending on whether it is present in a male or a female. (2) In animal behavior, a rapid form of learning in which an animal comes to make a particular response, which is maintained for life, to some object or other organism.

Inducer. (1) In enzyme systems, a small molecule that when added to a growth medium causes a large increase in the level of some enzyme. (2) In embryology, a substance that causes a group of target cells to differentiate in a particular way.

Information. A vague metaphoric concept that relates what we perceive (by observing, reading, or listening) to a change in our knowledge. There is no single, universally accepted definition of the term.

Informatics. (1) The art of management of data and management-oriented data analysis. (2) Synonym for applied computer science that includes database-related tasks, computer programming, and hardware-related tasks.

Inhibitor. An "anticatalyst" that prevents a catalyst from functioning properly in its catalytic capacity. When the catalyst is an enzyme, an inhibitor is a substance that binds to the surface of the enzyme and interferes with its action on its substrates.

Initiation complex. Combination of a ribosomal light subunit, an mRNA molecule, and the tRNA charged with the first amino acid coded for by the mRNA; formed at the onset of translation.

Initiation factors. Proteins that assist in forming the translation initiation complex at the ribosome.

Inositol triphosphate (IP3). An intracellular second messenger derived from membrane phospholipids.

Insertion. (1) A mutation in which a relatively short segment (sequence) of a biopolymer such as DNA or polypeptide is inserted into another—usually larger—segment of the biopolymer. The inserted sequence itself is sometimes called an insertion. Two sequences that differ only by the inserted sequence are bound by an insertion-deletion relationship. See also **Deletion.** (2) A chromosome abnormality in which a part of one chromosome is inserted into another nonhomologous chromosome.

Interferon. A glycoprotein that is produced by virus-infected animal cells and increases the resistance of neighboring cells to the virus.

Interleukins.   Regulatory proteins produced by macrophages and lympho-cytes, that act upon other lymphocytes and direct their development.

Intermediate filaments.   Fibrous proteins that stabilize cell structure and resist tension.

Interphase.   The period between successive nuclear divisions during which the chromosomes are diffuse and the nuclear envelope is intact. During this period the cell is most active in transcribing and translating genetic information.

Intron.   A portion (intervening sequence) of a protein-coding region in a eukaryotic DNA sequence that is spliced out from pre-mRNA during posttranscriptional modifications. The mature mRNA that participates in translation does not contain introns. Translated portions (exons) of most eukaryotic genes are interrupted by one or more introns. It has long been known that intron-containing and intronless versions of the same gene can be expressed in dramatically different ways. Despite this early knowledge, possi-ble biological roles of introns are still unknown, but it is clear that intervening sequences (and the process of their splicing) can influence many aspects of mRNA metabolism. (See also **Exon**, **Gene structure**, **Transcription**, **Translation**.)

In vitro.   In a test tube ("in glass") after removal (separation) from a living organism.

In vivo.   In a living organism. Many processes that occur in vivo are believed to be reproducible during their in vitro simulations. Such simulations are often termed in vivo as well.

Ion.   A chemical entity (usually an atom or molecule) with electrons either missing (positively charged cation) or present in excess (negatively charged anion).

Ion channel.   A membrane protein that can let ions pass across the mem-brane. The channel can be ion-selective, and it can be voltage-gated or ligand-gated.

Ionic bond.   A chemical bond that arises from the electrostatic attraction between positively and negatively charged ions. Usually a strong bond.

Iso-.   Prefix used to denote similarity or near-identity regarding a set of characteristics.

Isoforms.   (1) Different forms of the same protein encoded by different protein-coding regions in DNA. (2) Different proteins encoded by the same protein-coding region in DNA but translated from different mRNAs resulting from either alternative splicing (after transcription) or selection of different promoters (for transcription initiation). (3) Different structural variants of the same general kind of molecule (such as A-DNA, B-DNA, and Z-DNA).

Isomers.   Molecules consisting of the same numbers and kinds of atoms but differing in the way in which the atoms are combined.

K

k-Gram.   String of k symbols chosen from a given elementary alphabet. In the case of nucleotide sequences, k-grams correspond to oligonucleotides of length k. The most frequently used nucleic acid elementary alphabet is {A, C, G, T (or U)}, where letters stand for adenine, cytosine, guanine, and thymine (or uracil) nucleotides, respectively. Other alphabets used in nucleic acid studies include {K, M}, where K is either guanine or thymine (uracil in RNA) and M is either adenine or cytosine; {R, Y}, where R is a purine (adenine or guanine) and Y is pyrimidine (either cytosine or thymine); and {S, W}, where S is either cytosine or guanine and W is either adenine or thymine (or uracil in RNA).

k-Gram Alphabet.   If the elementary alphabet (say A) contains n symbols and k is fixed, we refer to a set of all n^k k-grams as the k-gram (nonelementary) alphabet over A or as the kth extension A^k of A. If A is an elementary alphabet, we consider a class of all its extensions up to a k-extension (k can be a large integer). We could symbolically represent it as $\{A^k\} = A \ \& \ A^2 \ \& \ A^3 \cdots \ \& \ A^k$, where the symbol & means set-theoretic union. We call $\{A^k\}$ a *generalized k-gram alphabet*.

Karyotype.   The number, forms, and types of chromosomes in a cell.

Kinase.   An enzyme that transfers a phosphate group from ATP to another molecule. Protein kinases transfer phosphate from ATP to specific proteins, playing important roles in cell regulation.

Kinesis.   Orientation behavior in which the organism does not move in a particular direction with reference to a stimulus but instead simply moves at an increasing or decreasing rate until it ends up farther from the object or closer to it. (Contrast with **Taxis.**)

Koch's postulates.   Four rules for establishing that a particular microorganism causes a particular disease.

Krebs cycle.   See **Citric acid cycle.**

L

Lactic acid. The end product of fermentation in vertebrate muscle and some microorganisms.

Lagging strand. In DNA replication, the daughter strand that is synthesized discontinuously.

Law of independent assortment. Nonhomologous chromosomes and genes carried on nonhomologous chromosomes separate randomly during meiosis. Mendel's second law.

Law of segregation. Alleles segregate independently from one another during gamete formation. Mendel's first law.

Leader sequence. A polypeptide sequence that precedes the N-terminal end of a newly synthesized protein. It participates in protein secretion into the destination location in the cell.

Leading strand. In DNA replication, the DNA strand that is synthesized continuously with the help of DNA-dependent DNA polymerase.

Ligand. A chemical entity (such as a molecule, ion, or atom) that is part of a larger molecule or molecular complex. (In biochemical systems, a ligand binds to a receptor site of another molecule.)

Likelihood. A hypothetical estimate of the chance that an event that has already occurred would yield a specific outcome. The concept of likelihood differs from that of probability. (Probability refers to the occurrence of future events with possibly unknown outcomes; likelihood refers to past events with known outcomes.)

Linkage. Association between genetic markers on the same chromosome such that they do not show random assortment and seldom recombine; the closer the markers, the lower the frequency of recombination.

Locus. In genetics, a specific location on a chromosome. May be considered to be synonymous with a gene or an allele.

M

Macroevolution. Evolutionary changes occurring over long time spans and usually involving changes in many traits. (Contrast with **Microevolution.**)

Macromolecule. A giant polymeric molecule. Proteins, polysaccharides, and nucleic acids are macromolecules.

Major histocompatibility complex (MHC).   A complex of linked genes, with multiple alleles, that control a number of immunological phenomena; it is important in graft rejection.

Mammal.   An animal of the class Mammalia, characterized by the production of milk by the female mammary glands and the possession of hair for body covering.

Map unit.   In eukaryotic genetics, one map unit corresponds to a recombinant frequency of 1%.

Mapping.   (1) In genetics, determining the order of genes on a chromosome and the distances between them. (2) In sequence analysis, an older name for sequence annotation. Putative functional or structural domains in long nucleic acid or protein sequences are detected (mapped), and the sequence is labeled (annotated) with names of domains (annotations). (3) In mathematics, a name for a function or transformation that assigns every element of one set (domain) to one and only one element of another set (counterdomain).

Markov process.   A probabilistic model in which the probability of the next state of a system depends solely on the probability of the previous state or a finite sequence of the previous states.

Markov source.   A device that generates symbols with probabilities that could be determined from a Markov process. The terms "Markov source" and "hidden Markov model" are synonymous.

Mass number.   The total number of protons and neutrons in the nucleus of an atom.

Maternal inheritance.   Inheritance in which the phenotype of the offspring depends on factors such as mitochondria or chloroplasts that are inherited from the female parent through the cytoplasm of the female gamete. Also called cytoplasmic inheritance.

Median.   "Middle value" of a sorted list of numbers. The smallest number such that at least half the numbers in the list are no greater than it.

Mean (arithmetic mean).   Given a finite list of N real numbers $x_1, x_2, x_3, ..., x_N$, the mean is a real number M calculated from the formula $M = (1/N) \sum_{i=1}^{N} x_i$. In statistics, the sample mean is an estimator of the expected value of the distribution.

Meiosis.   Division of a diploid nucleus to produce four haploid daughter cells. The process consists of two successive nuclear divisions with only one cycle of chromosome replication.

Mendelian population.   A local population of individuals belonging to the same species and exchanging genes with one another.

Meso-.   A prefix often used to designate a structure located in the middle of another or a stage that appears at some intermediate time. For example, mesoderm, Mesozoic.

Messenger RNA (mRNA).   A transcript of one of the strands of DNA, mRNA carries information (in the form of a sequence of trinucleotide codons) for the synthesis of one or more proteins.

Meta-.   A prefix used in different fields in different ways that often pertain to some concept of "beyond" or "above." (1) In general biology, the prefix meta- can denote a change (as in metabolism) or a shift to a new form or level (as in metamorphosis). (2) In logic and systems science, the prefix meta- can denote description (or other form of "processing" such as formal derivation or computation) from the perspective of a next higher level of organization of thought. For instance, language can be described with the help of a metalanguage, logic can be discussed with the help of metalogic, and so on.

Metabolic pathway.   A sequence (succession) of enzyme-catalyzed reactions in which a product of one reaction is a substrate of the next.

Metabolism.   A set of all chemical reactions that occur in an organism, or a well-defined specific subset of that set (as in "respiratory metabolism").

MHC.   See **Major histocompatibility complex**.

Micro-.   A prefix used to denote something small.

Microbiology.   The scientific study of organisms (usually bacteria and viruses) too small to be visible by the human eye but detectable with the help of optical or electron microscopes.

Microevolution.   The small evolutionary changes typically occurring over short time spans and generally involving a small number of traits and minor genetic changes.

Mimicry.   The superficial resemblance of an organism to an inanimate object in its environment or to another kind of organism.

Missense mutation.   A mutation that changes a codon for one amino acid to a codon for a different amino acid. (Contrast with **Frame-shift mutation**, **Nonsense mutation**, **Synonymous mutation**.)

Mitochondrion.   An organelle that occurs in eukaryotic cells and contains the enzymes of the citric acid cycle, the respiratory chain, and oxidative phosphorylation. A mitochondrion is bounded by a double membrane.

Mitosis. Nuclear division in eukaryotes leading to the formation of two daughter nuclei each with a chromosome complement identical to that of the original nucleus.

Model. With respect to a system or a complex thing, a representation of selected aspects (characteristics) with a simultaneous abstraction-away of all other imaginable or observable points of view and facts.

Modeling relation (MR). A general methodological scheme that binds four relations together: (1) a given natural system of observable objects and phenomena (that we wish to model) to itself, (2) the same natural system to a formal system of inferences, (3) the formal system to itself, and (4) the formal system to the initially mentioned natural system.

Mole. A quantity of a compound whose weight in grams is numerically equal to its molecular weight expressed in atomic mass units. One mole of a substance contains Avogadro's number of molecules (i.e., 6.023×10^{23} molecules).

Molecular biology. Field of life sciences that aims to provide biologically relevant explanations of the structure and function of chemical compounds (i.e., their molecules) found in living cells and tissues. It developed from a tradition that adopted mechanistic methods of thinking from physics and applied them to integrate genetics with cell biology, evolutionary biology, embryology, immunology, and other classical fields of biology (such as systematics).

Molecular clock. An assumption that biopolymers (nucleic acids or proteins) diverge from one another over evolutionary time at an approximately constant rate, which in turn is assumed to be well-correlated with phylogenetic relationships between organisms.

Molecular evolution. An evolutionary process leading to present-day DNA and protein sequences from their ancestral forms.

Molecular weight. Algebraic sum of atomic weights of atoms in a molecule.

Molecule. An electrically neutral chemical entity (particle) made up of two or more atoms joined by covalent or ionic bonds.

Moment (the kth moment). (1) Of a sequence of N numbers (a list), the arithmetic mean of the kth powers of the individual members of the list: $(x_1^k + x_2^k + x_N^k)/N$. (2) Of a random variable X, the expected value $E(X^k)$ of a random variable X^k. The mean of X is the first moment of X.

Mono-. Prefix denoting a single entity.

Monoclonal antibody.   Antibody produced in the laboratory from a clone of hybridoma cells, each of which produces the same specific antibody.

Monomer.   A small molecule, two or more of which can be combined to form oligomers (consisting of a few monomers) or polymers (consisting of many monomers).

Motif (sequence motif).   (1) String or regular expression that significantly often occurs in a collection of similar sequences. Vast majority of "motifs" discussed in molecular biology literature are "signature" sequence patterns (such as Shine–Dalgarno fragments or signature sequences in variable regions of immunoglobulin genes) that distinguish a specific set of protein (or nucleic acid) sequences from all other sequences. The word "motifs" reflects here a tacit assumption of the existence of meaningful functional and structural constraints that are unique and detectable at the sequence level. Some authors also assume the existence of a true homology (as opposed to sequence or structure similarity) relating sequences in a given set of proteins (or just domains) or nucleic acid primary structures. (See also **Motif descriptor** and **Sequence pattern**.) (2) A pattern that can be considered a letter (unit of pattern) in an alphabet that is adequate for studying a specific correlation between sequences and their functional or structural role. (3) An icon or label (such as a "signature" sequence or an essential descriptor in sequence annotation) that abbreviates the essence of results of data integration. Sequence alignment is a special (but not necessarily most important) case of data integration. It pertains to meaning (1).

Motif descriptor.   A data structure used to define a sequence motif. Best known descriptors include consensus sequences, weight matrices, and profiles.

mRNA.   See **Messenger RNA**.

Multicellular.   Consisting of more than one cell, as for example a multicellular organism. (Contrast with **Unicellular**.)

Mutagen.   Radiation or chemical that increases mutation rate.

Mutation.   (1) An inheritable change in one or more genes of an organism that leads to a detectable change in phenotype. (2) Insertion, deletion, or substitution of one or more amino acid residues in a protein. (3) Inheritable modification of a region in DNA sequence. Insertions, deletions, and substitutions are the most frequent kinds of DNA sequence modifications. See also **Silent mutation**, **Sense mutation**, and **Nonsense mutation**.

Mutation pressure.   Change in gene proportion caused by different mutation rates alone.

N

Natural selection. The differential contributions of offspring to the next generation by various genetic types belonging to the same population.

Neutral allele. An allele that does not alter the functioning of the proteins for which it codes.

Neutral theory. A view of molecular evolution that postulates that most mutations are silent mutations, which do not affect the amino acid being coded for. Such mutations are believed to accumulate in a population at rates driven by genetic drift and mutation pressure but not by fitness of the phenotype.

Nonsense (chain-terminating) mutation. Mutation that changes a codon for an amino acid to one of the codons (UAG, UAA, or UGA) that signal termination of translation. The resulting gene product is a shortened polypeptide that begins normally at the amino-terminal end and ends at the position of the altered codon. (Contrast with **Frame-shift mutation, Missense mutation, Synonymous mutation.**)

Nonsynonymous mutation (substitution). A nucleotide substitution in a protein-coding region that leads to a change of an amino acid in the corresponding polypeptide chain.

Nuclear envelope. The surface, consisting of two layers of membrane, that encloses the nucleus of a eukaryotic cell.

Nucleic acid. A chain copolymer of nucleotides. RNA is a copolymer of ribonucleotides; DNA is a copolymer of deoxyribonucleotides.

Nucleoid. Compartment in a prokaryotic cell that stores chromosomes. Unlike the eukaryotic nucleus, it is not surrounded by a membrane.

Nucleolus. Organized subcellular structure found inside the nucleus of eukaryotic cells. The function of the nucleolus is to biosynthesize ribosomal RNA.

Nucleosome. A portion of a eukaryotic chromosome, consisting of part of the DNA molecule wrapped around a group of histone molecules and held together by another type of histone molecule. The chromosome is made up of many nucleosomes.

Nucleotide. The basic chemical unit (monomer) in a nucleic acid. A nucleotide in RNA consists of one of four nitrogen-containing bases linked to ribose, which in turn is linked to phosphate. In DNA, deoxyribose is present instead of ribose.

Nucleus.   (1) In chemistry, a positively charged portion of every atom. (2) In cells, the compartment of eukaryotic cells that contains chromosomes. It is surrounded by a double membrane that physically separates it from the cytoplasm.

Null hypothesis.   In statistics, the hypothesis we wish to falsify (reject) on the basis of the data. The null hypothesis is typically a statement that there is no difference between the result of an experiment and a control.

O

Okazaki fragments.   Separate DNA strands that constitute the lagging strand during DNA replication. With the help of an enzyme (DNA ligase) the Okazaki fragments join together and form one contiguous strand.

Oligomer.   A molecule of polymer or copolymer of intermediate size, made up of few monomers.

Oligopeptide.   See under **Peptide**.

Ontology.   (1) The branch of metaphysics concerned with the nature of being (reality itself) as opposed to the nature of our representations of reality. (The latter is handled by another branch of metaphysics: epistemology.) (2) Name given by a group of computer scientists to an idealized integrated collection of databases of structured vocabularies that can be connected to life-sciences-relevant databases. Each vocabulary is a data structure of a computer-manageable kind such as a directed acyclic graph in which nodes are in a well-defined relation of ancestry (descent). (3) A computer-friendly representation of complex data that can be read and manipulated by computer programs.

Open reading frame (ORF).   A string of amino-acid-encoding trinucleotide codons of the genetic code that begins with a start codon (usually ATG) and ends with a stop codon (TAA, TAG, or TGA) and does not contain any other stop codon (in frame).

Operator.   The region of an operon that acts as the binding site for the repressor.

Operon.   A genetic unit of transcription in bacteria. It usually consists of several structural genes that are transcribed into the same (single) mRNA.

Optical isomers (enantiomers).   Isomers that differ in the configuration of the four different groups attached to the same carbon atom. Two enantiomers are mirror images of one another.

Organelles. Organized and distinguishable structures such as ribosomes, nuclei, mitochrondria, chloroplasts, cilia, and contractile vacuoles that can be found inside cells.

Organic. Pertaining to any aspect of living matter, e.g., to its evolution, structure, or chemistry. The term is also applied to any chemical compound that contains carbon.

Organism. Any living creature.

Origin of replication. A DNA sequence in which helicase unwinds the DNA double helix and DNA polymerase binds to initiate DNA replication.

Orthologs. Genes or sequences that result from a speciation event followed by a sequence divergence. Such genes may not exist side by side in the same organism. The last common ancestor of two orthologous genes existed during the speciation event.

Osmosis. The movement of water through a differentially permeable membrane from one region to another where the water potential is more negative. The letter is often a region in which the concentration of dissolved molecules or ions is higher, although the effect of dissolved substances may be offset by hydrostatic pressure in cells with semirigid walls.

Oxidation. Loss of electrons in a chemical reaction—either outright removal to form an ion or the sharing of electrons with substances having a greater affinity for them, such as oxygen.

Oxidative phosphorylation. ATP formation in the mitochondrion, associated with the flow of electrons through the respiratory chain.

Oxidizing agent. A substance that can accept electrons from another. The oxidizing agent becomes reduced; its partner becomes oxidized.

P

***p*-value.** The probability of erroneously rejecting an acceptable null hypothesis by chance alone. For instance, a p value of <0.05 means that the probability that the evidence supporting rejection of the null hypothesis was due to chance alone is less than 5%.

Paradigm. A general methodological framework along with a set of assumptions, beliefs, and cultural biases within which questions are asked and hypotheses are formed.

Parallel evolution. Evolutionary patterns that exist in more than one lineage, often the result of underlying developmental processes.

Paralogs. Genes or sequences that result from duplication of existing genes followed by sequence divergence. Such genes may descend and diverge while existing side by side in the same organism. If speciation occurs after gene duplication, then two paralogous genes may exist in two different organisms (species). The last common ancestor of two paralogous genes existed during the gene duplication event.

Parsimony. A principle of preferring a minimal change between subsequent steps of a process. For instance, a single-point mutation is a parsimonious change between two generations of evolving DNA sequence.

Pattern. Any logical, geometrical, or (broadly) factual connection between elements of a model that attracts our attention. "Pattern" is usually understood pragmatically by experienced explorers of the same model. (See also **Sequence pattern**.)

Pattern formation. In animal embryonic development, the organization of differentiated tissues into specific structures such as wings.

PCR. See **Polymerase chain reaction**.

Pedigree. The pattern of transmission of a genetic trait in a family.

Penetrance. Of a genotype, the proportion of individuals with that genotype who show the expected phenotype.

Peptide bond. The covalent bond formed in a condensation reaction (removal of a water molecule) between the α-amino group of one amino acid and the α-carboxyl group of another amino acid.

Peptide. A chain of amino acid residues joined together via peptide bonds. Oligopeptide (or just peptide) is a name that is sometimes used for a peptide that contains only a few residues in the chain, whereas "polypeptide" denotes a peptide that is made up of a substantial number of amino acid residues. [By a somewhat different convention, a polypeptide chain is a primary structure of a protein (no matter how many residues it contains), whereas a peptide is understood to be a chemical whose biological role is to be a peptide (not a protein).]

pH. The negative logarithm of the hydrogen ion concentration; a measure of the acidity of a solution. A solution with pH $= 7$ is said to be neutral; pH values higher than 7 characterize basic solutions, and acidic solutions have pH values less than 7.

Phage. See **Bacteriophage**.

Phenotype. The observable properties of an individual that have developed under the combined influences of the genetic constitution of the individual and the effects of environmental factors.

Phenotypic plasticity. The fact that the phenotype of an organism is determined by a complex series of developmental processes that are affected by both its genotype and its environment.

Phosphate group. The functional group $-OPO_3H_2$. The transfer of energy from one compound to another is often accomplished by the transfer of a phosphate group.

Phosphodiester bond. The chemical bond that connects two subsequent nucleotides in one strand of nucleic acid molecule.

Phospholipids. Cellular materials that contain phosphorus and are soluble in organic solvents. An example is lecithin (phosphatidylcholine). Phospholipids are important constituents of cellular membranes.

Phosphorylation. The addition of a phosphate group.

Phylogenetic tree. Graphical representation of the evolutionary relationships of ancestry and descent among a group of genes or species. When species are considered, they are represented as line segments, and points of branching correspond to subsequent speciation events. In the case of genes, points of branching correspond to gene duplication either during speciation or gene duplication during the evolutionary history of a single species.

Phylogeny. The evolutionary history of descent of a group of taxa such as species from their common ancestors, including the order of branching and sometimes absolute ages of divergence. Also the diagram of the "family tree" that shows genetic linkages between ancestors and descendants. [The term also applies to the genealogy of genes derived from a common ancestral gene (homologous genes).]

Phylum. In taxonomy, a high-level category just beneath kingdom and above class; a group of related, similar classes.

Physiology. The scientific study of the functions of living organisms and the individual organs, tissues, and cells of which they are composed.

Plant. A member of the kingdom Plantae. Multicellular, gaining its nutrition by photosynthesis.

Plasmid. A DNA molecule distinct from the chromosome(s); that is, an extrachromosomal element. May replicate independently of the chromosome.

Pleiotropy.   The determination of more than one character by a single gene.

Point mutation.   A mutation that results from a small, localized alteration in the chemical structure of a gene. Such mutations can give rise to wild-type revertants as a result of reverse mutation. In genetic crosses, a point mutation behaves as if it resided at a single point on the genetic map. (Contrast with Deletion.)

Poly-.   A prefix denoting multiple entities.

Polygenes.   Multiple loci whose alleles increase or decrease a continuously variable phenotypic trait.

Polymer.   A large molecule made up of similar or identical subunits called monomers.

Polymerase chain reaction (PCR).   A technique for the rapid production of millions of copies of a particular stretch of DNA.

Polymerization reactions.   Chemical reactions that result in obtaining polymers. There are often three stages of these reactions: initiation, elongation, and termination.

Polycondensation.   Polymerization via the reaction of condensation (release of water molecules).

Polymorphism.   (1) In genetics, the coexistence in the same population of two or more distinct phenotypes that correspond to different alleles. (2) The co-existence of two or more crystal structures for the same chemical entity.

Polypeptide.   See under **Peptide**.

Polyphyletic group.   A group containing taxa that do not all share the most recent common ancestor.

Polyploid.   A cell or an organism in which there are more than two complete sets of chromosomes.

Polysaccharide.   A macromolecule composed of many monosaccharides (simple sugars). Common examples are cellulose and starch.

Polysome.   A complex consisting of a threadlike molecule of messenger RNA and several (or many) ribosomes. The ribosomes move along the mRNA, synthesizing polypeptide chains as they proceed.

Polytene.   An adjective describing giant interphase chromosomes, such as those found in the salivary glands of fly larvae. The characteristic, reproducible pattern of bands and bulges seen on these chromosomes has provided a method for preparing detailed chromosome maps of several organisms.

Population. Any group of organisms coexisting at the same time and in the same place and capable of interbreeding with one another.

Population density. The number of individuals (or modules) of a population per unit area or volume.

Population genetics. The study of genetic variation and its causes within populations.

Population structure. The proportions of individuals in a population belonging to different age classes (age structure). Also, the distribution of the population in space.

Positive control. The situation in which a regulatory macromolecule is needed to turn on the transcription of structural genes. In its absence, transcription will not occur.

Positive cooperativity. A condition that occurs when a molecule can bind several ligands and each one that binds atters the conformation of the molecule so that it can bind the next ligand more easily. The binding of four molecules of O_2 by hemoglobin is an example of positive cooperativity.

Pragmatic. Practical. Dealing with facts and occurrences.

Pragmatics. In linguistics, one of three main aspects of studying languages (two others being syntax and semantics). Pragmatics refers to the use of sentences in the context of other sentences as well as real-world situations.

Pragmatic inference. (1) The art of determining sequence motifs from their instances and the knowledge context to which they pertain. (2) The art of determining an alphabet (vocabulary) of function-associated motifs based on one or more individual patterns and the knowledge of strutures or mechanisms that correlate well with the presence of these patterns.

Primary structure. In biochemistry, the sequence of amino acids in a protein or the nucleotide sequence of a nucleic acid.

Primate. A member of the order Primates, such as a lemur, monkey, ape, or human.

Primer. A short, single-stranded segment of DNA serving as the necessary starting material for the synthesis of a new DNA strand, which is synthesized from the $3'$ end of the primer.

Pro-. A prefix often used in biology to denote a developmental stage that comes first or an evolutionary form that appeared earlier than another. For example, prokaryote, prophase.

Probability.   A measure of the chance (possibility) that a particular event (or set of events) will occur. Values of probability measure are positive real numbers from the closed interval [0,1]. Probability of impossible events equals 0, and probability of sure events equals 1. (The concept of probability is related to but different from the notion of likelihood.)

Probe.   A segment of single-stranded nucleic acid used to identify DNA molecules containing the complementary sequence.

Profile.   A generalized representation of sequence properties derived from a family of functionally or structurally related biopolymer (protein or nucleic acid) fragments. Just like single sequences expressed in appropriate sequence alphabets (nucleotide symbols for DNA and RNA or amino acid residues for primary protein structures), profiles can be used for database searches via dynamic programming or other optimization algorithms. Because profiles allow position-specific scoring systems and gap parameters, profile searches offer a high sensitivity in detecting distant relationships between naturally occurring biopolymer sequences because they allow position-specific scoring as well as simplicity in handling gaps.

Prokaryotes.   Bacteria. Unicellular organisms whose genetic material is not contained within a well-formed cell nucleus.

Promoter.   Part of a protein-coding region in DNA to which DNA-dependent RNA polymerase binds at the beginning of transcription. Binding of RNA polymerase (and transcription factors) to the promoter is considered to be a transcription initiation event.

Proofreading.   The correction of an error in DNA replication just after an incorrectly paired base is added to the growing polynucleotide chain.

Prophage.   A noninfectious unit that is linked with the chromosomes of the host bacteria and multiplies with them but does not cause dissolution of the cell.

Prophase.   The first stage of nuclear division, during which chromosomes condense from diffuse, threadlike material to discrete, compact bodies.

Prosthetic group.   Any nonprotein portion of an enzyme.

Protease.   See **Proteolytic enzyme**.

Protein.   One of the most fundamental building substances of living organisms. A long-chain polymer of amino acids with 20 different common side chains. Occurs with its polymer chain extended in fibrous proteins or coiled into a compact macromolecule in enzymes and other globular proteins.

Protein secondary structure. Local substructures in a protein's three-dimensional structure, each corresponding to a consecutive stretch of amino acids. The two most regular secondary structural elements are the α-helix, with an average number of 3.6 amino acids per turn, and the β-strand, which is the most extended secondary structure seen in proteins. Elements comprising the latter structure typically associate in hydrogen-bonded β-pleated sheets constituted by β-strands in a parallel or antiparallel fashion. Other secondary structure elements that can be classified by the automatic secondary structure assignment algorithm DSSP (Kabsch and Sander, 1983) are the 3/10 helix (tighter than the α-helix), π-helix (wider than the α-helix), β-bulge, hydrogen-bonded turn, bend, and coil.

Proteolytic enzyme (protease). An enzyme (such as trypsin) whose main catalytic function is the digestion of a protein or polypeptide chain.

Proteome. Complete set of all proteins encoded in the nuclear component of the genome of a given organism.

Protobiont. Aggregate of abiotically produced molecules that cannot reproduce but does maintain an internal chemical environment that differs from its surroundings.

Proton. Hydrogen cation H^+ (the nucleus of the hydrogen atom). In particle physics a proton is also known as an elementary particle with an atomic mass of 1 amu and electric charge of +1.

Protozoa. A group of single-celled organisms classified by some biologists as a single phylum; includes the flagellates, amoebas, and ciliates. This volume follows most modern classifications in elevating the protozoans to a distinct kingdom (Protista) and each of their major subgroups to the rank of phylum.

Pseudogene. A DNA segment that displays a significant sequence similarity to a functional gene but whose complete expression cannot be accomplished. A pseudogene usually contains insertions, deletions, or substitutions that alter the sequence of the original (expressible) gene. These sequence modifications can be either a reason for impaired expression or a consequence of not being expressed. The so-called processed pseudogenes are DNA sequences complementary (sometimes with a few mutations) to a mature (processed) mRNA of the functional gene. Their existence is often cited as indirect evidence for potential regulatory roles of noncoding part (3'- and 5'-UTRs, introns, upstream regions, downstream regions) of the corresponding functional genes.

Punctuated equilibrium. An evolutionary pattern in which periods of rapid change are separated by longer periods of little or no change.

Purine.   A type of nitrogenous base. The purines adenine and guanine are found in nucleic acids.

Pyrimidine.   A type of nitrogenous base. The pyrimidines cytosine, thymine, and uracil are found in nucleic acids.

Pyruvate.   A three-carbon acid; the end product of glycolysis and the raw material for the citric acid cycle.

Q

Quantum.   An indivisible unit of energy.

Quaternary structure.   In reference to aggregating proteins, the arrangement of polypeptide subunits.

R

Randomness.   Logical equivalent of a lack of pattern. A situation in which patterns are undetectable or nonexistent.

Random genetic drift.   Evolution (change in gene proportions) by chance processes alone.

Rate constant.   In reference to a particular chemical reaction, a constant that, when multiplied by the concentration(s) of reactant(s), gives the rate of the reaction.

Reactant.   A chemical substance that enters into a chemical reaction with another substance.

Recessive.   See **Dominance**.

Recognition site.   A sequence of nucleotides in DNA to which a restriction enzyme binds and then cuts the DNA. Also called a restriction site.

Recombinant.   An individual, meiotic product or single chromosome in which genetic materials originally present in two individuals end up in the same haploid complement of genes. The reshuffling of genes can be either by independent segregation or by crossing over between homologous chromosomes. For example, a human may pass on genes from both parents in a single haploid gamete.

Recombinant DNA technology.   The application of genetic tools (restriction endonucleases, plasmids, and transformation) to the production of specific proteins by biological "factories" such as bacteria.

Redox reaction. A chemical reaction in which one reactant becomes oxidized and the other becomes reduced.

Reducing agent. A substance that can donate electrons to another substance. The reducing agent becomes oxidized, and its partner, the oxidizing agent, becomes reduced.

Reduction. (1) In chemistry, gain of electrons; the reverse of oxidation. Reductions often lead to the storage of chemical energy, which can be released at any time via an oxidation reaction. (2) In methodology of science, replacement of the modeled complex system by a simpler surrogate system (a model).

Reductionism. (1) Causal: An academic doctrine according to which the only legitimate conclusion about a system can be reached from studying its parts but one should not infer properties of parts from studying the whole system. (Downward causation is "forbidden.") (2) Methodological: An academic trend according to which complex systems should be represented by simple models (surrogate systems), which in turn could be studied with a variety of scientific methods.

Regular expression. (1) In sequence analysis and bioinformatics, a flexible definition of a sequence pattern that allows groups of motifs to reside in the place of a single motif. (2) In information technology (IT), a valid formula of a specialized programming or scripting language dedicated to serve a software tool or database. For example, most information retrieval systems (such as library searchable catalogs) accept queries formulated in a "language" that consists of simple regular expressions. Another example is a scripting language for textual pattern matching in the UNIX operating system. Some programming languages (such as PROLOG or PERL) are entirely designed for textual pattern matching as a way of writing down programs, i.e., regular expressions actually constitute the programming language.

Regular grammar. A formal grammar that is equivalent to a finite-state automaton.

Regulatory gene. A gene that contains the information for making a regulatory macromolecule, often a repressor protein.

Replication fork. A point at which a DNA molecule replicates. The fork is formed by the unwinding of the parent molecule.

Repressible enzyme. An enzyme whose synthesis can be decreased or prevented by the presence of a particular compound. A repressible operon often controls the synthesis of such an enzyme.

Repressor. A protein that can bind to a specific place in a transcription complex and thereby prevent transcription from happening. (For instance, in bacteria the repressor can bind to a specific operator and prevent transcription of the entire operon.) Repressors are usually coded by designated regulatory genes.

Restriction endonuclease. Any of several enzymes, produced by bacteria, that break foreign DNA molecules at very specific sites. Some produce "sticky ends." Extensively used in recombinant DNA technology.

Restriction map. A partial genetic map of a DNA molecule, showing the points at which particular restriction endonuclease recognition sites reside.

Retrovirus. An RNA virus that contains reverse transcriptase. Its RNA serves as a template for cDNA production, and the cDNA is integrated into a chromosome of the mammalian host cell.

Reverse transcriptase. An enzyme (RNA-dependent DNA polymerase) that catalyzes synthesis of DNA (cDNA) using RNA as a template.

Reverse transcription. Synthesis of DNA whose sequence is complementary to a given RNA sequence. The process is catalyzed by reverse transcriptase and sometimes is followed by insertion of reverse-transcribed DNA (complementary DNA, cDNA) into the nuclear DNA genome. [This kind of insertion is a mechanism by which cellular oncogenes (c-oncogenes) can be created from their viral counterparts residing in RNA viruses called retroviruses.]

RFLP (restriction fragment length polymorphism). Coexistence of two or more patterns of restriction fragments (patterns produced by restriction enzymes), as revealed by a probe. The polymorphism reflects a difference in DNA sequence on homologous chromosomes.

Ribonucleic acid. See **RNA**.

Ribosomal RNA (rRNA). RNA molecules that are incorporated into ribosomes via interaction with ribosomal proteins. The evidence exists that rRNAs (at least in bacteria) are actively involved in translation of mRNA and the formation of peptide bonds in growing polypeptide chains.

Ribosome. A small organelle that is the site of protein synthesis.

Ribozyme. An RNA molecule with catalytic activity.

RNA (ribonucleic acid). A nucleic acid that contains the sugar ribose. Various classes of RNA are involved in the transcription and translation of protein-coding regions in DNA. The genetic material of some viruses (such as poliovirus or retroviruses) is made of RNA instead of DNA.

RNA polymerase. An enzyme that catalyzes the formation of RNA from a DNA or RNA template.

RNA splicing. One of the last stages of posttranscriptional RNA processing in eukaryotes, in which transcripts of exons are joined together to form mature mRNA while the transcripts of introns are removed.

rRNA. See **Ribosomal RNA.**

S

S phase. The stage of the cell cycle during which DNA is replicated.

Secondary structure. Local substructure in a biopolymer three-dimensional structure that corresponds to a consecutive sequence of monomer residues (nucleotides in nucleic acids or amino acids in proteins).

Second messenger. A compound, such as cyclic AMP or IP3, that is released within a target cell after a hormone or other "first messenger" has bound to a surface receptor on a cell. The second messenger usually triggers further reactions within the cell.

Segregation, genetic. The separation of alleles or of homologous chromosomes from one another during meiosis so that each of the haploid daughter nuclei produced by meiosis contains one or the other member of the pair found in the diploid mother cell but never both.

Semantics. In linguistics, systematic studies of meaning of linguistic expressions (such as sentences).

Semiconservative replication. The common way in which DNA is synthesized. Each of the two partner strands in a double helix acts as a template for a new partner strand. After replication, each double helix consists of one old and one new strand.

Sense mutation. A mutation that leads to a visible or measurable change in phenotype. At the molecular level a sense mutation in a protein-encoding region leads to the replacement of one amino acid by another in a polypeptide encoded by this gene.

Sequence hypothesis. The first of two basic assumptions of molecular biology (the second is the central dogma). (1) Original formulation: "Specificity of a piece of nucleic acid is expressed solely by the sequence of its bases, and this sequence is a (simple) code for the amino acid sequence of a particular protein" (Crick, 1958). (2) In cellular protein biosynthesis, the amino acid sequence of a polypeptide (primary structure of a protein) is collinear with and determined (solely) by the sequence of a specific protein-encoding region in chromosomal DNA.

Sequence pattern. Given a generalized k-gram alphabet, we select strings of length ranging from 1 to k that we call *motifs* and strings that we call *punctuations*. Usually strings that are not selected to be motifs and their concatenations serve as punctuations. A *sequence pattern* is a string of motifs and punctuations that begins and ends with a motif. A *contiguous pattern* is a sequence pattern that does not contain punctuations, whereas a *noncontiguous pattern* is a sequence pattern that does contain punctuations. *Note*: Some authors define sequence patterns as k-grams only and then talk of motifs as combinations of patterns that are repeated in a sufficiently large number of sequences from a given collection of sequences. In this sense patterns are always k-grams (contiguous), whereas motifs may be both contiguous and noncontiguous.

Sex chromosome. In organisms with a chromosomal mechanism of sex determination, one of the chromosomes involved in sex determination. In humans these are chromosomes X and Y.

Sex linkage. Inheritance controlled by genes located on (or coexpressed with) the sex chromosomes of organisms having a chromosomal mechanism for sex determination.

Shannon entropy. A mathematical function of a random variable that measures deviation of a probability distribution (of this variable) from the discrete uniform distribution. Shannon entropy and its variants have been extensively used in sequence analysis as statistical tools of choice.

Shine–Dalgarno sequence. A "signal" sequence apparently responsible for binding a ribosome to an mRNA molecule just before translation in bacteria (most studies were done on *E. coli*). The most representative (consensus) sequence is a heptanucleotide TAAGGAG (at the DNA level) in which the palindrome AGGA appears to be the most conserved part. The heptanucleotide like that occurs only a few nucleotides (approximatelly 13) upstream of the translation start site. The Shine–Dalgarno sequence is a good example illustrating the definition of motif as a conserved function-associated sequence feature ("signal" sequence).

Signal sequence. The sequence in nucleic acids or proteins that is recognized by other agents (also nucleic acids or proteins) as an indicator of specific function. Examples are Shine–Dalgarno sequences, signal sequences in variable regions of immunoglobulin genes, or a region of a protein that binds necessary substrates to transport this protein through a particular cellular membrane.

Signal transduction pathway. The series of biochemical steps whereby a stimulus to a cell (such as a hormone or neurotransmitter binding to a receptor) is translated into a functionally meaningful response of the cell.

Significance (statistical): (1) The probability that a test or experiment leads by chance alone to an erroneous rejection of the null hypothesis when the null hypothesis is in fact acceptable. (2) The likelihood that a statement is true. (3) The degree of deviation of an observation from its occurrence by chance alone according to a model of chance. The degree of nonconformity to a (presumed correct) model of chance in the foregoing sense.

Silencer. A sequence of eukaryotic DNA that binds proteins that inhibit the transcription of an associated gene.

Silent mutations. Genetic changes that do not lead to a phenotypic change. At the molecular level, these are DNA sequence changes that, because of the redundancy of the genetic code, result in the same amino acids in the resulting protein. See also **Synonymous mutation**.

Simulation. Imitating the behavior of a real system by using properties of its model or a class thereof.

Small nuclear ribonucleoprotein particle (snRNP). A complex of an enzyme and a small nuclear RNA molecule, functioning in RNA splicing.

Somatic. Pertaining to the body or body cells but not to germ cells.

Speciation. Evolution of reproductive isolation within an ancestral species, resulting in two or more descendant species. An evolutionary process leading to the emergence of two or more new species from the ancestral one.

Species. (1) A population or series of populations of closely related and similar organisms. (2) In biology, a set (group) of individual organisms capable of interbreeding freely with each other and at the same time incapable of interbreeding with organisms from outside this set.

Spliceosome. An RNA–protein complex that is involved in splicing out introns from eukaryotic pre-mRNAs.

Splicing. The removal of introns from pre-mRNA (hnRNA) and the simultaneous joining together of exons to form mature mRNA that contains a protein-coding region ready for translation.

Spontaneous reaction. A chemical reaction that will proceed on its own without any outside influence. A spontaneous reaction need not be rapid.

Stability. (1) The capacity for a system to remain within a nominal range of (nonextreme) behavior. (2) Resistance to change, deterioration, or displacement. (3) The ability of a system (or an object) to maintain equilibrium or resume its original state after alteration (such as resuming its original position after displacement).

Stabilizing selection. Selection against the extreme phenotypes in a population, so that the intermediate types are favored. (Contrast with **Disruptive selection.**)

Start codon. The mRNA triplet (AUG) that acts as a signal for the beginning of translation at the ribosome.

Statistical inference. Finding out properties of an unknown statistical distribution from data generated by that distribution.

Standard deviation (SD). Square root of the variance of a distribution. Can be estimated from the sample standard deviation, which is the square root of the sample variance.

Stop codons. Triplets (UAG, UGA, UAA) in mRNA that act as signals for the end of translation at the ribosome.

Structural formula. A representation of the positions of atoms and bonds in a molecule.

Sub-. A prefix often used to designate a structure that lies beneath another or is less than another.

Substrate. One of the chemicals (chemical entities) that enter into a chemical reaction. (Every reaction proceeds from substrates to products.)

Symmetry. (1) The property of the relation between two objects A and B such that if A is in relation with B, B must be in relation with A. (2) Identity of two objects regarding dislocation in space (such as translation or rotation), time, or the generation of a mirror image.

Synonymous mutation (substitution). A nucleotide substitution in a protein-coding gene that does not change the amino acid assigned by the genetic code to the trinucleotide affected by substitution.

Syntax. In linguistics, a set of rules whereby words or other elements of sentence structure are combined to form grammatically correct sentences without regard to their meaning.

System. A group of interacting, interrelated, or interdependent elements forming a complex whole whose properties are not a simple combination of the properties of the elements. Specific additional meanings include the following. (1) A functionally related group of elements, especially (a) the organism regarded as a physiological unit; (b) a group of physiologically or anatomically complementary organs or parts (the immune system, nervous system, and digestive system are representative examples); (c) a group of interacting mechanical or electrical components functioning in a robust manner within a mechanical or electrical device (machine) (an exhaust system or electrical system in an automobile are representative examples here); and (d) a network of objects or structures with indication of connections between them (a Metro or road system in a big city is a representative example here); (2) An organized set of interrelated ideas or principles such as a religion, ideology, or other belief-based general paradigm. A scientific system, legal system, and ethical system are representative examples here. (3) A naturally occurring group of objects or phenomena. The solar system, a (specific) ecosystem, or a pack of wolves are representative examples here. (4) A set of objects or phenomena grouped together for the purpose of classification or analysis. All life forms in the ocean or all elementary particles in the cosmos could be representative examples here. (5) A method, a procedure, or a paradigm that can be systematically reused without changing details each time it is used. A number system (such as binary or decimal), writing system (such as roman script), and cryptographic system are representative examples here.

Systematics. The scientific study of the diversity of organisms via appropriate classification and coding (naming).

T

TATA box. A short consensus sequence (usually octanucleotide) approximately 25 base pairs upstream of the transcription initiation site within the promoter region of protein-coding genes.

Taxis. The movement of an organism in a particular direction with reference to a stimulus. Taxis usually involves the employment of one sense and a movement directly toward or away from the stimulus or else the maintenance of a constant angle to it. Thus positive phototaxis is movement toward a light source, negative geotaxis is movement upward (away from gravity), and chemotaxis is movement toward or away from a chemical.

Taxon. A unit in a taxonomic system.

Taxonomy. The science of classification of organisms.

Telomeres. Repeated DNA sequences at the ends of eukaryotic chromosomes.

Template.   In biochemistry, a molecule or surface upon which another molecule is synthesized in complementary fashion, as in the replication of DNA.

Template strand.   In a protein-coding region of double-stranded DNA during transcription, the strand that is transcribed (antisense strand) to eventually produce the mRNA.

Tertiary structure.   (1) In reference to a protein, the relative locations in three-dimensional space of all the atoms in the molecule. (2) The overall three-dimensional shape of a protein.

Theory.   (1) Systematically organized knowledge applicable in a relatively wide variety of circumstances, especially a system of assumptions, accepted principles, and rules of procedure devised to analyze, predict, or otherwise explain the nature or behavior of a specified set of phenomena. (2) Abstract reasoning; speculation. (3) A belief that guides action or assists comprehension or judgment. (4) An assumption based on limited information or knowledge; a conjecture. (5) A narrative describing a possible scenario (chain) of events that could lead to a given outcome.

Thymine.   A nitrogen-containing base found in DNA.

Tissue.   A group of similar cells organized into a functional unit and usually integrated with other tissues to form a functional part of an organ.

Trait.   An observable form of character. For example, hair color is a character; brown hair and red hair are traits.

Transcription.   The synthesis of RNA, using one (noncoding) strand of DNA as the template.

Transcription factor.   Any protein other than RNA polymerase that participates in a transcription initiation complex and is required for transcription.

Transduction.   Transfer of a portion of genetic material from one cell to another, with a virus, plasmid, or other vector acting as the carrier.

Transfection.   Uptake, incorporation, and expression of foreign DNA by a cell.

Transfer RNA (tRNA).   A category of relatively small RNA molecules (about 75 nucleotides). Each kind of transfer RNA is able to accept a particular activated amino acid from its specific activating enzyme. It is also able to recognize (via an anticodon loop) the mRNA codon for this particular amino acid. Recognition of a codon is usually coordinated with detaching the amino acid and incorporating it into an elongated polypeptide chain at the translation complex (ribosome, mRNA, and aminoacyl-tRNA).

Transformation.   Mechanism for transfer of genetic information in bacteria in which pure DNA extracted from bacteria of one genotype is taken in through the cell surface of bacteria of a different genotype and incorporated into the chromosome of the recipient cell.

Transgenic.   Containing recombinant DNA incorporated into its genetic material.

Translation.   Synthesis of a protein (polypeptide) according to the sequence encoded in mRNA.

Translocation.   In genetics, a rare mutational event that moves a portion of a chromosome to a new location.

Transposable element.   A segment of DNA that can move to, or give rise to copies at, another locus on the same or a different chromosome.

Triplet.   A trinucleotide. (See also **Codon**).

Triplet repeat.   Occurrence of a repeated trinucleotide. It is believed that some genetic diseases can be associated with excessive triplet repeats. For instance, excessive repetition of CGG is associated with the condition called fragile-X syndrome.

tRNA.   See **Transfer RNA**.

U

Unicellular.   Consisting of a single cell; as, for example, a unicellular organism. (Compare **Multicellular**.)

Uniform distribution (discrete uniform distribution) (DUD).   A probability distribution of a random variable in which all events are independent of each other and occur with probabilities p equal to each other. Considered the most conceptually important statistical distribution to the formal foundations of today's probability theory. A good model of DUD is the so-called Bernoulli text, a very long random string of letters from an alphabet of size N in which every letter occurs independently from all other letters and with probability (the same for all letters) equal to $1/N$. Another good model is a sequence of outcomes of a large number of tosses of an unbiased coin. Here both outcomes are independent of each other and occur with probability $1/2$ (same result as with Bernoulli text over two-letter alphabet.)

V

van der Waals interaction. A weak attraction between atoms resulting from the interaction of the electrons of one atom with the nucleus of the other atom. This attraction is about one-fourth as strong as a hydrogen bond (another type of weak interaction).

Variable region. The part of an immunoglobulin molecule or T-cell receptor that includes the antigen-binding site.

Variance of a distribution. A measure of dispersion of a distribution of a random variable X around the expected value $E(X)$. In qualitative terms the value of the variance indicates how representative of the distribution its expected (mean) value is. The variance of a given distribution X can be determined from the formula $\mathrm{var}(X) = E[(X - m)^2]$, where m is the expected value (mean) of X, $E(X)$.

For a finite sample of size N the estimator of variance is the *sample variance,*

$$V = \frac{1}{N} \sum_{i=1}^{N} (x_i - \text{mean})^2$$

where x_i is the ith element of the sample and mean $= (1/N) \sum_{i=1}^{n} x_i$ is the *sample mean.*

Vector. (1) An agent, such as an insect, that carries a pathogen that affects another species. (2) An intermediary object (such as a plasmid or a virus) that carries an inserted piece of DNA into the chromosomal DNA of a cell. (3) In mathematics (algebra), an element $\mathbf{v}$ of a vector space $\mathbf{V}$ over a field $\mathbf{F}$. A one-dimensional array. For a fixed natural number k, any sequence of k real (or complex) numbers can be considered a vector in a k-dimensional vector space. In particular, a sequence of three numbers can be considered a vector in three-dimensional space provided that the appropriate algebraic operations are defined. (4) In physics, a mathematical object characterized by magnitude and direction that can be used to quantitatively represent properties of physical systems such as velocity or force.

Vertebrate. An animal whose nerve cord is enclosed in a backbone of bony segments called vertebrae. The principal groups of vertebrate animals are the fishes, amphibians, reptiles, birds, and mammals.

Virion. The virus particle, the minimum unit capable of infecting a cell.

Viroid. An infectious agent consisting of a single-stranded RNA molecule with no protein coat; produces diseases in plants.

Virus. Ultramicroscopic infectious particle that contains nucleic acid packed inside a capsid (coat) made of protein.

W

Wild-type. Geneticists' term for standard or reference type. Deviants from this standard, even if the deviants are found in the wild, are said to be mutant.

Wirchov rule. See **Cell theory**.

X

X-linked. A character that is coded for by a gene on the X chromosome. (Also called sex-linked.)

Y

Yeast artificial chromosome. A laboratory-made DNA molecule containing sequences of yeast chromosomes (origin of replication, telomeres, centromere, and selectable markers) so that it can be used as a vector in yeast.

Z

Z-DNA. A form of DNA in which the molecule spirals to the left rather than to the right.

z-Value (z-score). The observed value of a Z statistic that is constructed by standardizing some other statistic. The Z statistic is related to the original statistic by measuring the number of standard deviations by which a given data point differs from the expected value:

$$Z = (\text{observed} - \text{expected value of original})/[\text{standard deviation (observed)}]$$

Zygote. The cell created by the union of two gametes, in which the gamete nuclei are combined in one nucleus. Creation of a zygote is the initial stage of development of an individual diploid organism.

Appendix 2

A Dictionary of Programs for Sequence Analysis*

Andrzej K. Konopka

BioLingua Research Inc., Gaithersburg, Maryland, U.S.A.

Jaap Heringa

Centre for Integrative Bioinformatics VU (IBIVU), Faculty of Sciences
and Faculty of Earth and Life Sciences, Free University,
Amsterdam, The Netherlands, and MRC National
Institute for Medical Research, London, England

This listing was compiled in an attempt to enrich description of the methods that are in use in today's sequence analysis. It is written for those brave individuals who do not like to use other people's programs but would certainly like to know what programs have already been written. We hope the compilation will be of use to computational biology software writers as well as to readers who desire to understand details of specific sequence analysis methods in their original ("pure") form. For the purpose of clarity we do not list many software solutions that pertain to artificial learning (such as artificial neural networks or hidden Markov models) that otherwise rely on combinations of "pure" methods.

---

*Cited references are listed in Appendix References following this appendix.

Some of the programs described in this collection do not have names, so they are identified by author name and year of publication. The corresponding research paper is listed in the Appendix References. The fact that some programs have no names is a reflection of the computational biologists' spirit in the early postpioneering period that spanned the 1980s and early 1990s. Writing computer programs was considered the least important part of a research task. With very few exceptions, scientists involved in the field of computer-assisted sequence research actively resented the commerce-oriented attitudes of Microsoft and other commercial software development companies. In fact, scientists of this period wrote a number of first-rate software tools without making much noise about this fact. However, many of these tools are widely used today either in the original stand-alone form or as adaptations within larger commercial or semicommercial packages. This is one reason why we decided to make descriptions of algorithms available to a larger community of readers. We believe these descriptions (primarily included in the original papers cited) will be particularly valuable to the readers who work with software within computational biology and bioinformatics.

The actual origins of computational methods almost always predate by a few years the first publication about them. That is the major reason, we think, why the actual inventors of the method are sometimes not the authors of the initial journal publication concerning it. In compiling this dictionary we attempted to keep the record of pivotal methods as fair to the inventors as possible, but in many instances the only reliable source of the original description of methods is the first journal article.

Argos (1987).   Program by Argos (1987) to generate dot plots for pairwise protein sequence comparison. The method uses classical alignment-based amino acid similarity scores combined with five physicochemical parameters per amino acid in calculating the window scores. Windows of different lengths are tested simultaneously, and ones with the best scores appear in the final dot plot.

Barton and Sternberg (1990).   A flexible profile-searching technique to search with a multiple sequence alignment for sequences that show similarity with the alignment. Significant residue positions are selected on the basis of sequence conservation, functional importance, or the presence of secondary structure. These residues, constituting the pattern, can be separated by gaps that serve to exclude variable regions from the analysis. For each gap, minimal and maximal possible lengths are derived from the initial sequence set. A lookup table similar to a profile is then calculated, which results in scores to compare each element of the pattern with each residue type. The flexible pattern is subsequently compared to every databank sequence using a modified every databank sequence using a modified Needleman and Wunsch (1970) technique. The method is especially recommended for sequence alignments in which crucial elements are separated by long noisy stretches, which are effectively discarded by the method.

BLAST. BLAST (Basic Local Alignment Search Tool) (Altschul et al., 1990) is a widely used rapid program for searching sequence databanks for similarities to a given query sequence. To some extent it mimics the local alignment procedures described by Smith and Waterman (1981) but employs hashing techniques to gain speed. Recent additions to the BLAST suite of programs are PSI-BLAST (Position Specific Iterative BLAST) (Altschul et al., 1997), which uses information from multiple sequences in iterative database searches, and PHI-BLAST (Pattern Hit Initiated BLAST) (Zheng et al., 1998), which combines finding regular expressions (see Sequence pattern in App. 1) with local similarities surrounding the fragments found by the pattern matching.

BLASTN. An adaptation of BLAST designed for comparing DNA sequences with other (or the same) DNA sequences.

BLASTP. An adaptation of BLAST designed to search for related proteins.

BLASTZ. A version of BLAST designed to compare very large nucleotide sequences.

BLITZ. Parallel implementation by Collins and Coulson (1990) of the full Smith and Waterman (1981) local pairwise sequence alignment technique. The computer protocol is devised to perform database searches based on the MPsrch technique (Sturrock and Collins, 1993), which runs on massively parallel computers with SIMD (single instruction, multiple data) processors. Available also is an implementation of the BLITZ server by Compugen.

BLAZE. A database search tool that implements the formalism of Smith and Waterman (1981) applied to a modified version (Gotoh, 1986, 1987) of the Needleman and Wunsch (1970) algorithm. The program was first implemented on the massively parallel computer MassPar by Brutlag et al. (1993).

Boguski et al. (1992). A semimanual program suite that incorporates the space-efficient local alignment routine SIM of Huang et al. (1990) as well as the MSA method for global multiple sequence alignment of Lipman et al. (1989). For each pair of sequences, the highest scoring local alignments containing gaps are determined, from which nongapped regions in each of the sequences are extracted. Whenever meaningful, neighboring blocks of such motifs with the intervening sequence fragments are aligned using the

MSA method, thus allowing gaps. The method of Boguski et al. (1992) also provides a user interface through which parts of the alignment can be manually edited.

Bucher and Hofmann (1996). Statistical local alignment technique for pairwise sequence comparison in which each cell[i, j] in the DP search matrix is interpreted as the total probability that a local alignment would go through it. This is achieved by summing the scores of all local alignments intersecting cell[i, j]. Using this approach, Bucher and Hofmann (1996) reported an increase in pairwise sequence search capabilities.

Chou and Fasman (1974). Early protein secondary structure prediction method for single protein sequences. Predictions are based on differences in residue composition for three states of secondary structure: α-helix, β-strand, and turn. Preferences for each type of amino acid to constitute any of these secondary structures were derived from protein tertiary structures using sliding window approaches. Secondary structures are predicted for each sequence position according to the highest preference values of the structural states. Extensions of secondary structures are made as long as preferences remain above a threshold and certain disruptive residues are not encountered (such as proline, which breaks an α-helix).

Clustal. A widely used global multiple sequence alignment method. The first version of Clustal was first published by Higgins and Sharp (1988) and was specially designed for use on small workstations. Computation was reduced for the pairwise alignments of the sequences by using the Wilbur and Lipman (1983, 1984) algorithm. From the pairwise similarities, a guide tree is constructed using the UPGMA clustering criterion. The sequences are then aligned following the branching order of the tree. For the comparison of groups of sequences, Higgins and Sharp (1988) used consensus sequences to represent aligned subgroups of sequences and also employed the Wilbur–Lipman technique to match these. The Clustal package has been subjected to a number of revision cycles. Higgins et al. (1992) implemented an updated version, ClustalV, in which the memory-efficient dynamic programming routine of Myers and Miller (1988) is used, enabling the alignment of large sets of sequences using little memory. Further, two alignment positions, from different alignments, are compared in ClustalV using the average alignment similarity score of Corpet (1988). The largely extended version ClustalW (Thompson et al., 1994) uses the alternative neighbor-joining (NJ) algorithm (Saitou and Nei, 1987), which is widely used in phylogenetic analysis, to construct a guide tree that can also be used for phylogenetic analysis. Sequence blocks are represented by a profile in which the individual sequences are additionally weighted according to the branch lengths in the NJ tree

(Thompson et al., 1994). An integrated user interface is implemented in ClustalX (Thompson et al., 1997) that is integrated with accessory programs for tree depiction. A WWW server for ClustalW is available at http://www2.ebi.ac.uk/clustalw/.

COILS2. A program to predict coiled-coil structures in globular proteins from sequence information (Lupas et al., 1991; Lupas, 1996). The usual left-handed coiled-coil interaction (superhelical twist) involves a repeated motif of seven helical residues (*abcdefg*), where the *a* and *d* positions are normally occupied by hydrophobic residues constituting the hydrophobic core of the helix/helix interface, whereas the other positions display a high likelihood to comprise polar residues. Furthermore, the heptad *e* and *g* positions are often charged and can form salt bridges. The COILS2 method exploits these facts and compares a query sequence with a database of known parallel two-stranded coiled coils. A similarity score is derived and compared to two score distributions, one for globular proteins (without coiled coils) and one for known coiled-coil structures. The two scores are then converted to a probability for the query sequence to adopt a coiled-coil conformation. Because the program assumes the presence of heptad repeats, probabilities are derived using default window lengths of 14, 21, and 28 amino acids. The program can also use user-defined window lengths for the prediction of extreme coiled-coil lengths. A recently updated scoring matrix, based on data from recent coiled-coil structures and containing amino acid type propensities for various positions in the heptad repeat, shows improved recognition of coiled-coil elements. The COILS2 method accurately recognizes left-handed two-stranded coiled coils but loses sensitivity for coiled-coil structures consisting of more than two strands. The method is not suited to recognize right-handed or buried coiled-coil helices and therefore is not applicable to transmembrane coiled-coil structures. A WWW server for the COILS2 method is available at http://www.ch.embnet.org/software/COILS_form.html.

DCA. Fast implementation of the global multiple sequence alignment method MSA (Lipman et al., 1989) by Stoye et al. (1997). Speed is optimized by using a divide-and-conquer strategy (Stoye et al., 1997). However, the DCA method remains extremely CPU- and memory-intensive and is applicable only to small data sets.

DIAGON. Program by Staden (1982) to generate dot plots for pairwise protein sequence comparison. The program is based on filtering techniques originally published by McLachlan (1971, 1972, 1983). Sequence regions showing significant similarity are identified by using windows of fixed length that are effectively slid over the two sequences to compare all possible stretches of typically five matched residue pairs. The mean value and standard deviation of scores from comparing randomized windows gathered from shuffled sequences are compared with the window scores from the two query sequences (using the Z-score; i.e., the number of standard deviations of the real scores above the random mean). The output values are filtered on the basis of a cutoff value for placing dots in the comparison matrix.

DIALIGN. Local multiple sequence alignment method by Morgenstern (1999). DIALIGN is a segment-based local procedure that constructs a multiple alignment by assembling a collection of high-scoring segments in a sequence-independent progressive manner. The method is thus based on segment-to-segment comparisons rather than the residue-to-residue comparisons used in other programs. The segments are incorporated into a multiple alignment using an iterative procedure. The method aligns only sequence fragments that have sufficient sequence similarity; other regions remain unaligned. A WWW server for the DIALIGN method is available at http://bibiserv.techfak.uni-bielefeld.de/dialign/.

DISTAN. Program originally written as part of the method for frequency analysis of distances between short oligonucleotides (Konopka and Smythers, 1987) in long DNA and RNA sequences and in large collections of such sequences. The algorithm is based on a concept of noncontiguous sequence patterns (contiguous oligonucleotide motifs separated by gaps of variable length).

DISTANP. A program for scoring frequency counts of distances between short oligopeptides expressed in several different alphabets (including the 20 amino acid residue alphabet) in individual polypeptide (protein) sequences as well as in large collections of such sequences. The program was originally an extension of DISTAN adapted for analysis of proteins (Konopka and Chatterjee, 1988). Today it is a stand-alone software tool used for nonroutine sequence analyses.

DSC. A method to predict the protein secondary structure for a set of multiply aligned sequences (King and Sternberg, 1996). The DSC method combines the compositional features of multiple alignments with empirical rules that are important for secondary structure prediction. The information is processed using linear statistics. The empirical rules and concepts used that relate to multiple alignment information are (1) N-terminal and C-terminal sequence fragments normally adopt a coiled structure; (2) alignment positions comprising gaps are indicative for coil regions; (3) periodicity in positions of hydrophobic or conserved residues; and (4) residue ratios in the alignment. These patterns are detected using autocorrelation, feedback of predicted secondary structure information, and some simple filter rules. Prediction occurs in five consecutive steps:

1. The basic prediction of the secondary structure is carried out using the GOR method (see GOR entry), which is used on each of the aligned sequences. The average GOR score for each of the three states is then compiled for each alignment position.

2. For each alignment position a so-called attribute vector is compiled, consisting of 10 attributes: the three averaged GOR scores for H, E, and C from step 1; distance to alignment edge; hydrophobic moment assuming helix; hydrophobic moment assuming strand; number of insertions; number of deletions; conservation moment assuming helix, and conservation moment assuming strand.

3. The positional vectors are doubled in size to 20 attributes by adding the same 10 attributes in a smoothed fashion (using running averages).

4. Seven more attributes are added to the 20 attributes of the preceding step: weights for predicted α-helix and β-strand, based on the 20-attribute vectors of step 3, and the fractions of the five most discriminating residue types, His, Glu, Gln, Asp, and Arg. To convert these 27-attribute vectors to three-state propensities, a linear discrimination function is used. This is effectively a set of weights for the attributes in the positional vector corresponding to each of the secondary structure states, so three sets of 27 attribute weights are used. The optimal weights used in the DSC method were gathered using a training set of known 3-D structures. After applying the weights to the attribute vectors for each alignment position, the secondary structure associated with the highest scoring vector is taken.

5. A set of 11 simple filter rules are used for a final prediction, such as ([E/C]CE[H/E/C][H/C])→C, where [E/C] denotes E or C. These filter rules were derived automatically using machine learning techniques.

The accuracy of the DSC method, as assessed by the authors, is 70.1% (King and Sternberg, 1996). As an additional option, the DSC method can also be used to refine a prediction by the PHD algorithm of Rost and Sander (1993). The average accuracy of this PHD-DSC combinatorial procedure is 72.4% (King and Sternberg, 1996). A WWW server for the DSC method is available at http://bonsai.lif.icnet.uk/bmm/dsc_read_align.html.

DSSP. Widely used protein secondary structure assignment method by Kabsch and Sander (1983). Input for DSSP is a three-dimensional structure (protein coordinate data) from the Protein Data Bank (Bernstein et al., 1977). Assignment is based on hydrogen-bonding patterns and geometrical features of the protein main chain. DSSP groups protein secondary structures into eight classes: α-helix (H), 3/10-helix (G), π-helix (I), β-strand (E), β-bulge (B), bend (S), hydrogen-bonded turn (T) and coil (' ').

Eisenberg et al. (1982). Protocol to measure helix amphipathicity. The measure was named the hydrophobic moment and defined as the vector sum of the individual amino acid hydrophobicities radially directed from the helical axis. In general, the hydrophobic moment provides sufficient sensitivity to discriminate between amphipathic α-helices of globular, surface, and membrane proteins.

FASTA. Program for fast comparison of a given query sequence with a library of sequences created by Pearson and Lipman (1988). For each sequence pair, the highest scoring local alignment is determined. Speed is obtained by delaying the application of the dynamic programming technique to the moment where the most similar segments are already identified by faster and less sensitive techniques. The FASTA routine operates in four steps. The first step searches for identical words of a user-specified length occurring in the query sequence and the target sequence(s) using the algorithm of Wilbur and Lipman (1983, 1984). This technique involves searching for identical words (*k*-tuples) of a certain size within a specified bandwidth along search matrix diagonals. The search is performed by hashing techniques, where a lookup table is constructed for all words in the query sequence, which is then used to compare all words encountered in the target sequence(s). For not-too-distant sequences (>35% residue identity), little sensitivity is lost and speed is greatly increased. Generally, for proteins, a word length of two residues is sufficient (*ktup* = 2). Searching with higher *ktup* values increases the speed but also the risk that similar regions will be missed. For each target sequence, 10 regions with the highest density of ungapped common words are determined. In the second step, these 10 regions are rescored using the Dayhoff PAM250 residue exchange matrix (Dayhoff et al., 1983), and the best scoring region of the 10 is reported under *init1* in the FASTA output. In the third step, regions scoring higher than a threshold value and sufficiently near each other in the sequence are joined, now allowing gaps. The highest score of these new fragments can be found under *initn* in the FASTA output. The fourth and final step performs a full dynamic programming alignment over the final region, widened by 32 residues at either side, because earlier steps tend to cut similar regions short. The final score is written under *opt* in the FASTA output.

Feng and Doolittle (1987). Method for the construction of a phylogenetic tree through progressive global alignment of the sequences. The algorithm uses only strictly pairwise sequence comparisons; it does not use any consensus sequences or averaging of similarities to compare blocks of sequences. Gaps in already aligned sequences are fixed by inserting special gap characters at gap positions. First, a rough branching order is determined using the phylogenetic tree-building method of Fitch and Margoliash (1967). This tree order is basically followed, but the alignment order of nearest neighbors in each obtained subgroup of sequences is reversed and the highest scoring alignment is selected for further comparison. For example, if the initial branch order is ((AB)C)D, then A and B are aligned first. The alignment orders ABC and BAC (alignment taking place successively from left to right) are then checked and the best alignment is taken; for example, BAC. Then, BACD and BADC are examined. Only nearest neighbors are swapped to keep computation manageable; other possible permutations are not considered.

Fickett (1982). One of the first programs to recognize protein-coding regions in naturally occurring nucleotide sequences. The algorithm is based on the calculation of the weighted sum of eight frequency-count-related parameters determined for a given studied sequence.

Frishman and Argos (1992). Profile-based method to delineate conserved sequence blocks and use them to flexibly search sequence databanks. First, a neural network is used to elucidate unknown patterns from a multiple alignment of N sequences. One segment of width W in each position of the alignment is tested, and the net is trained on the alignment of the segment including $N - 1$ sequences, after which the excluded sequence segment is submitted for recognition and the network output recorded. This is repeated with each of the N segments removed. The average net recognition is then used as a measure of conservation for this alignment region. In the second step, the M most conserved protein blocks are used to extensively train M corresponding neural networks, which are then used to scan the protein sequence databank. Variable constraints can be imposed on the distances between the blocks, although the M blocks must be in the same sequential order as in the multiple alignment.

GCG Wisconsin Package. A collection of basic public domain sequence analysis programs that have been integrated into a common graphic interface and input/output format.

GIBBS. Local multiple sequence alignment program by Lawrence et al. (1993). The method is based on the Gibbs statistical method of iterative sampling. The GIBBS algorithm searches for gap-free motifs of a certain preset length W, which are found by a random optimization procedure. Individual sequence segments of given length W are sampled iteratively from a set of N sequences. In the first step the segments are taken from random positions of $N - 1$ sequences, one randomly selected sequence being excluded. A tentative "conserved" region of length L is constructed from these segments, for which observed and statistically expected residue frequencies for each of the L positions are calculated. Then all possible segments from the excluded sequence are tested, one by one, for their consistency with the amino acid probabilities of the generated subalignment. If at least a small fraction of the randomly selected segments are actually related, thus providing a weak information signal, the probability of successful extension of the nascent pattern by related segments from other sequences in the set will be slightly higher than could be expected for a completely random situation. The procedure is repeated iteratively, and the pattern probabilities are recalculated at each step, with the discriminative power of the pattern possibly growing with the inclusion of each new related member.

GOR. Method to predict protein secondary structure for a single protein query sequence. The GOR method quickly became the standard for about a decade after its first appearance (Garnier et al., 1978). The first versions, GOR I and II, predict four states by discriminating between coil and turn secondary structures. GOR III (Gibrat et al., 1987) and the most recent version, GOR IV (Garnier et al., 1996), perform the common three-state prediction. The GOR method relies on the amino acid frequencies observed and uses a 17-residue window (i.e., eight residues N-terminal and eight C-terminal of the central window position) to derive the probability of the window's central residue for each of the three structural states. The amino acid frequencies associated with secondary structure observed in the structural database (PDB) are exploited using an information function based on conditional probabilities for each amino acid to occur in any of the three states H, E, or C. The early versions of the GOR algorithm simply summed the individual propensities associated with each of the 17 residue types in the window and thus did not take the order of the amino acids in the window into account. Considering the amino acid order in full is not feasible, because it would require the sampling of all possible 17-residue fragments directly from the PDB (there are 17^{20} possible fragments). Subsequent versions of the GOR method over the years have explored increasingly detailed approaches to this combinatory problem, along with the growth in the amount of data available in the PDB. The current version, GOR IV (Garnier et al., 1996), supplements the basic propensities with pairwise information over all possible paired positions in the window considered (there are $17 \times 16/2$ possibilities). However, relatively small weights are used for the pairwise propensities compared with the basic propensities. In the GOR IV method, a final filter is implemented to refine the predictions. If a helix shorter than four residues or a strand fragment with less than two amino acids is initially predicted, the method assesses the probabilities of extending the fragment to the minimum associated length or deleting it (i.e., changing it to coil). WWW servers for GOR I, III, and IV can be found at http://pbil.ibcp.fr.

Gotoh (1982). A significant modification of the Needleman–Wunsch algorithm that leads to much better performance of optimization via dynamic programming.

Gotoh (1986, 1987). Algorithm for pairwise sequence alignment. Gotoh (1986, 1987) devised a dynamic programming algorithm that dramatically decreased the storage requirements from order N^2 to order N (assuming that two sequences each N amino acids in length are matched) while keeping speed on the order of N^2.

Gribskov et al. (1987). The first computer algorithm for profile analysis. The technique combines a full representation of a multiple sequence alignment with a sensitive searching algorithm to search for sequences that show similarity with the alignment. The procedure takes as input a multiple alignment of N sequences. First, a profile is constructed from the alignment, i.e., a position-specific scoring matrix (PSSM), that comprises the likelihood of each residue type occurring in each position of the multiple alignment. A Gribskov et al. profile has $L(20 + 1)$ elements, where L is the total length of the alignment, 20 is the number of different amino acid types, and the last matrix row contains gap penalties. As a measure of similarity between different types of residues, a residue exchange matrix is used as in dynamic programming. Each residue position receives as a score the sum of the amino acids at that position, where the contribution for each residue type is weighted with the corresponding residue exchange matrix. For example, if an alignment column contains 3 phenylalanines (F), 2 valines (V), and 5 asparatates (D), the profile value (propensity) for the positional alanine matrix cell would be $3s(A, F) + 2s(A, V) + 5s(A, D)$, where $s(A, F)$ is the residue exchange matrix value for an $A \rightarrow F$ mutation (or vice versa). Gribskov et al. (1987) used a single extra column in the profile to describe the local weight for both the gap opening and the gap extension penalty. For alignment positions not containing gaps, $P_{open} = P_{extend} = 100$, whereas for positions with insertions or deletions these values are lowered depending on the maximum length of any gap crossing a given alignment position. The advantage of such positional gap penalties is that regions with gaps (probably loop regions) will be more likely to attract gaps in a target sequence during profile searching, consistent with structural considerations. Gribskov et al. (1987) use the Smith and Waterman (1981) dynamic programming procedure to align their profile with each individual target sequence. The profile scores are then corrected for sequence length, represented in the form of Z-scores, and ranked to create the final list of databank search hits. Top-scoring sequences with scores above some threshold level are then likely to be related to the multiply aligned sequences used to build the profile. In addition to aligning a single sequence to a profile, it is also possible to align two profiles. In this case two matched profile positions receive a score by summing over the 20 residue types the products of the corresponding propensities from the two profiles.

Hogeweg and Hesper (1984). First integrated progressive algorithm for global multiple sequence alignment. A dendrogram is constructed based on all pairwise similarities of sequences matched by dynamic programming. This dendrogram, also called a guide tree, is then used to progressively align the sequences pairwise, using the Needleman and Wunsch (1970) algorithm, in the order dictated by the tree. Various similarity measures widely used in constructing phylogenetic trees can be applied, such as the Unweighted Pair-Group Mean Average (UPGMA) of Sneath and Sokal (1973), the present-day ancestor method of Blanken et al. (1982), or the neighbor-joining method of Saitou and Nei (1987). During progressive alignment, aligned blocks of sequences are represented by so-called internode sequences, which act as likely ancestral sequences (each internode sequence is associated with the root of the subtree). These are constructed using the subtree covering the sequences of the aligned block. A backtracking algorithm is applied on the subtree associated with the sequence block to infer the most parsimonious amino or nucleic acid at each position; i.e., the acid requiring the fewest mutations at the alignment position considered within the sequence block. The Hogeweg and Hesper (1984) method was also pioneering in that it was the first iterative procedure: From the initial tree based on pairwise alignments, carrying no information yet of related groups of sequences, a multiple alignment is generated from which the associated pairwise sequence similarities are inferred. Using those, a new tree is constructed that is used to create a succeeding alignment, each time based on increased information.

Hopp and Woods (1981). An early sliding-window-based method to predict antigenic sites from protein sequences. Antigenic sites (ASs) are locations on the protein molecule responsible for specific antibody binding. Their detection is an important step in biochemical characterization of a protein. AS prediction techniques are based on the preferred location of antigenic sites on the surface of the protein. The method of Hopp and Woods (1981) calculates a smoothed hydrophilicity plot using a sliding window approach based on hydrophilicity values given by Levitt (1976). Tentative antigenic sites are identified as peaks in the hydrophilicity plot. Hopp and Woods (1981) found that a window length of six residues produced the best results, although many false positive or false negative predictions of antigenic sites occur with their method.

Jameson and Wolf (1988). A synthetic sliding window technique for prediction of antigenic sites from sequence information. The method combines various signals in a single so-called antigenic index, which includes the acid flexibility propensities of Karplus and Schultz (1985), surface probabilities (Janin et al., 1978), and residue hydrophilicities (Hopp and Woods, 1981). These three signals are appropriately weighted to optimize the prediction results. Although this method leads to more reliable predictions than the method of Hopp and Woods (1981), sliding window methods are not applicable to discontinuous antigenic determinants where the residues that constitute the antibody-binding pocket are not close in sequence but are close in proximity in the tertiary structure of the protein.

Jpred. A protocol for running various secondary structure prediction methods for a given multiple alignment and creation of a consensus secondary structure prediction. The Jpred server (Cuff and Barton, 1999) runs state-of-the-art prediction methods such as PHD (Rost and Sander, 1993), PREDATOR (Frishman and Argos, 1995, 1996), DSC (King and Sternberg, 1996), and NNSSP (Salamov and Solovyev, 1995), while the methods ZPRED (Zvelebil et al., 1987) and MULPRED (Barton, unpublished) are also included. The NNSSP method has to be activated explicitly, because it is the slowest of the ensemble and often will not be finished in the computing time slot allocated to the user. The server accepts a multiple alignment and predicts the secondary structure of the sequence on top of the alignment: Alignment positions showing a gap for the top sequence are deleted. A single sequence can also be given to the server. In the latter case, a BLAST search is performed to find homologous sequences, which are subsequently multiply aligned using ClustalX (Thompson et al., 1997) and then processed with the user-provided single sequence on top in the alignment. If a sufficient number of methods predict an identical secondary structure for a given alignment position, that structure is then taken as the consensus prediction for the position. If no sufficient agreement is reached, the PHD prediction is taken. This consensus prediction is somewhat less accurate when the NNSSP method is not included. The Jpred server also accepts a single query sequence, in which case it constructs a set of related sequences by launching a BLAST database search, after which the sequences found are aligned by ClustalX. The resulting multiple alignment is then subjected to the actual Jpred consensus prediction technique. The Jpred server is available at http://barton.ebi.ac.uk/servers/jpred.html.

Karplus and Schultz (1985). A method to predict protein loop flexibility, aimed at delineating antibody–antigen binding sites. The importance of loop flexibility in antibody binding for establishing a so-called induced fit is supported by experimental evidence (Rini et al., 1992). In their method,

Karplus and Schultz (1985) use empirically determined crystallographic temperature factors, which correspond to mean-square atomic displacement.

Konopka (1990). Two highly effective methods to determine the approximate location of putative functional domains (particularly protein-coding regions) in unannotated nucleic acid sequences. The program explores the frequency-count-based measures of sequence heterogeneity: the local compositional complexity (LCC) and periodic asymmetry index (PAI).

Kyte and Doolittle (1982). A technique to identify transmembrane (TM) regions in protein sequences. Although the globular interior of soluble proteins is less apolar than the lipid bilayer, Kyte and Doolittle (1982) used globular protein data to derive their classical hydrophobicity scale. The hydrophobic scale is used to build a smoothed curve, called a hydropathic profile, by averaging over a sliding window of given length. Stretches of hydrophobic amino acids likely to reside in the lipid bilayer then appear as peaks whose lengths should correspond to those expected for transmembrane segments, typically 16–25 residues. The choice of window length should correspond to the expected length of a TM segment. Given that the average membrane thickness is about 30 Å, approximately 20 residues form a helix reaching from one lipid bilayer surface to another. To determine the boundaries of a membrane-spanning segment, a cutoff value for the hydrophobic peaks is required.

LALIGN. A widely used version of the Waterman and Eggert (1987) algorithm for local sequence alignment created by Huang and Miller (1991) that is part of the popular FASTA package (Pearson and Lipman, 1988). A WWW server for LALIGN can be found at http://www.ch.embnet.org/software/LALIGN_form.html.

MACAW. The Multiple Alignment Construction and Analysis Workbench (MACAW) by Schuler et al. (1991) allows the user to lock or shift regions in an alignment while nonlocked subsequences are aligned automatically. It is thus possible to define iteratively conserved regions such that the fraction of poorly defined segments that must be aligned automatically becomes smaller with each iteration. The local similarity analysis method GIBBS of Lawrence et al. (1993) has been incorporated in the MACAW procedure.

MEME. A local multiple sequence alignment method by Bailey and Elkan (1994). The method is able to delineate local motifs occurring in a set of input sequences when it is given the width of the suspected motif. More than one occurrence of the motif can be recognized in individual sequences. The technique makes use of an expectation maximization algorithm that relies on Dirichlet mixtures to estimate the relative frequency of motif occurrences.

Thus recognized related motifs all contain the preset number of amino acids and contain no gaps.

MSA. A method for simultaneous global multiple sequence alignment by Lipman et al. (1989) that performs dynamic programming through a multi-dimensional search matrix. The algorithm MSA (Multiple Sequence Alignment) is based on the approach by Carillo and Lipman (1988), who showed that the optimal alignment path of N sequences is limited to a small region in the N-dimensional search matrix. The upper bounds can be inferred from pairwise comparisons of the sequences. Although this reduces computations, the method is extremely slow, and no more than eight or nine sequences of 200–300 residues in length can be aligned with it in practice. MSA assigns weights to the aligned sequences because similar sequences should not dominate the multiple sequence alignment. Lipman et al. (1989) used the weighting scheme of Altschul et al. (1989) based on phylogenetic trees. A WWW server of the MSA method can be found at http://www.ibc. wustl.edu/ ibc/msa.html.

MultAlin. A global multiple sequence alignment method based on hierarchical clustering by Corpet (1988). MultAlin can refine alignments by performing iteration of the clustering and progressive alignment steps. For matching two prealigned blocks of sequences, the average over the amino acid exchange values associated with all pairwise intercolumn residue comparisons is taken as a score between a pair of matched prealigned block positions i and j,

$$S_{i,j} = \frac{\sum_{m=1}^{M}\sum_{n=1}^{N} D(A_{i,m}, A_{j,n})}{MN}$$

where $A_{i,m}$ is the amino acid type in sequence m of alignment block position i, $A_{j,n}$ is the amino acid type in sequence n of alignment block position j, D is the amino acid exchange weight, and M and N denote the numbers of sequences in the two aligned sequence blocks. A WWW server for the MultAlin method is available at http://www.toulouse.inra.fr/multalin.html.

MULTALIGN. A multiple global sequence alignment method by Barton and Sternberg (1987). MULTALIGN establishes a simple chain order in which the individual sequences are aligned one by one. Initially, all pairwise alignment scores are determined and the two most similar sequences are matched first. During further iterations, the sequence showing the highest alignment score when matched with the prealigned sequence block is added to it. To determine the alignment score, each sequence position i of the kth sequence matched with position j of a prealigned block of $k-1$ sequences receives a score per matched position averaged over the corresponding residue substitution values:

$$S_{i,j} = \frac{\sum_{p=1}^{k-1} D(A_{k,i}, A_{p,j})}{k - 1}$$

where $D(A_{k,i}, A_{p,j})$ is the amino acid exchange weight. The PAM250 substitution matrix (Dayhoff et al., 1983) is used with a constant of 8 added to remove all negative matrix elements. Matched gaps are evaluated by the lowest exchange weight of zero. The resulting multiple alignment can be progressively refined by realigning each sequence with the previous alignment from which that sequence is deleted; i.e., sequence A1 is matched with aligned sequences A2,....,AN, sequence A2 is then realigned with the alignment of A1, A3,....,AN and so forth. This process is repeated until all N sequences are realigned. Barton and Sternberg (1987) recommend two such complete refinement cycles.

MULTAL. A fast global multiple sequence alignment method by Taylor (1988) that constructs a tree during the progressive alignment. MULTAL uses a fast sequential branching method to align the closest pairs of sequences first and then subsequently align the next closest sequences to those already aligned. The order in which the sequences are aligned is largely based on the global amino acid composition of the sequences, which is one of the reasons for the speed of the method. Progressive alignment of the sequences is done by dynamic programming. The MULTAL method can be downloaded from http://mathbio.nimr.mrc.ac.uk.

Myers and Miller (1988). A memory-efficient DP algorithm, in which basically only two rows of the N^2 search matrix linear space algorithm need to be stored (see Needleman and Wunsch, 1970). The algorithm is based on the Gotoh approach and on the "divide-and-conquer" trace-back strategy of back strategy of Hirschberg (1975).

Needleman and Wunsch (1970). Algorithm to align two protein sequences over their full lengths, which is also called global alignment. The algorithm relies on the dynamic programming (DP) technique first introduced by Needleman and Wunsch (1970) to the biological community. Input parameters for the DP algorithm are a set of weights for each possible pairwise amino acid substitution (including self-conservation values), typically given as a so-called amino acid exchange matrix, and a gap penalty value applied each time a gap is inserted in one of the sequences. Based on these input parameters, the DP technique is guaranteed to find the optimal and highest scoring alignment of two given sequences. A DP algorithm operates in two steps. First a search matrix is set, with one sequence displayed horizontally and the other vertically. The matrix is basically traversed from the upper left to the lower right, but a path can start anywhere from the first row or column and end anywhere within the last row or column. Each cell $[i, j]$ in the matrix receives as a score the value composed of the maximum possible value that is the sum of the cell's own substitution value (i.e., the exchange value of the associated matched residue pair of cell $[i, j]$) and the value of the highest scoring cell in row $i - 1$ or column $j - 1$ (with subtraction of the proper gap penalty values). After traversing the matrix, each cell $[i, j]$ therefore contains the maximum score of all possible alignments of the two subsequences up to cell $[i, j]$. Because each step in the DP algorithm is independent of its past, the technique falls in the class of hidden Markov models (HMMs). In the second step of a DP algorithm, usually called the trace-back step, the actual optimal alignment is reconstructed from the matrix cell containing the highest alignment score. The path then follows successively lower scores but each time selects the highest available in the preceding row and column up to the current matrix cell. The Needleman and Wunsch (1970) DP algorithm uses a fixed penalty value for the inclusion of a gap of any length. Most present alignment routines take an intermediate approach by using the formula $P(x) = P_o + P_e x$, where P_o is the penalty placed upon the opening of a gap of length x and P_e is the value for each extension of the gap. Many researchers use a P_o value 10–30 times as large as P_e. The choice of proper gap penalties is also closely connected to the residue exchange values used in the analysis. The Needleman–Wunsch DP algorithm uses a two-dimensional search matrix, so that the algorithmic speed and storage requirements are both of the order NM, when two sequences consisting of N and M amino acids in length are matched.

NNSSP. A protein secondary structure prediction method of Salamov and Solovyev (1995). Its input is a set of multiply aligned sequences. The NNSSP (nearest-neighbor secondary structure prediction) method is an extension of the k-nearest-neighbor approach of Yi and Lander (1993). In the NNSSP method, N- and C-terminal positions of helices and strands and β-turns are explicitly taken as types of additional secondary structures. For each prediction, the database of exemplars (see above) is restricted to sequences similar to the query multiple alignment. This reduces computation time and leads to biologically related nearest neighbors. The NNSSP method combines window sizes of 11, 17, and 23 residues, nearest-neighbor numbers (k) of 50 or 100, and balanced or nonbalanced training. This leads to a total number of $3 \times 2 \times 2 = 12$ different prediction implementations, from which a consensus prediction is established using a simple majority rule. These consensus predictions are subjected to three final filter rules: (1) Helices of less than three residues are deleted (changed to coil), but (EHE) becomes (EEE); (2) strands of length less than three residues are deleted, but (HEEH) becomes (HHHH); (3) helices of four or fewer residues are deleted. The latter rule is applied only after a full cycle of rules (1) and (2). The overall accuracy of the method as reported by the authors is 72.2%. A WWW server for the NNSSP method is available at http://dot.imgen.bcm.tmc.edu:9331/pssprediction/pssp.html.

Parker et al. (1986). A method for protein antigenic site prediction from sequence, using window averaging as in the method of Hopp and Woods (1981). However, the surface profiles (or hydrophilicity plots) are calculated using an alternative set of hydrophilicity values derived from retention times in high-performance liquid chromatography. The Parker et al. (1986) method yields more accurate predictions than that of Hopp and Woods (1981).

Patthy (1987). A method to extract common sequence patterns from a set of protein sequences. As a first step, the sequences are pairwise aligned and the most similar of them grouped. For each group, the alignments are inspected to identify residues conserved in most of the sequences, and an initial pattern is formulated. Then every sequence within the group is optimally aligned with the pattern, resulting in the generation of a multiple alignment. As a next step, the consensus sequences derived for the different groups are amalgamated, each individual sequence realigned with the pattern, an extended multiple alignment generated, and so on. While producing the consensus sequences, the algorithm relies on user-specified thresholds such as the fraction of residues deemed similar or identical according to the Dayhoff PAM250 residue exchange matrix at a given position for it to be included in the consensus.

PHD. A widely used method to predict protein secondary structure when given a set of multiply aligned protein sequences (Rost and Sander, 1993). PHD combines the information from multiple sequence alignments with the optimization strength of the neural network formalism. It makes use of two complete neural networks consecutively: The first network produces a raw three-state prediction for each alignment position by sliding a 13-residue window along the alignment. A second network then refines the first-level predictions by taking the three-state secondary structure propensities produced by the first network and processing the information using a slightly longer 17-residue window. The output of the second network results in three adjusted state probabilities for each alignment position. The PHD method includes a number of such network pairs, which are trained and optimized independently, and feeds the predictions for each alignment position produced by each network pair into a third network to yield a so-called jury decision. The predictions by the jury network are subjected to a final filtering step to simply delete predicted helices of one or two residues, changing those into coils. Initially assessed to predict with 70.8% accuracy, the PHD method has been refined since this assessment and currently attains about 74% correct predictions. If given a single sequence for prediction, the server performs a BLAST search to obtain a set of homologous sequences and aligns those using the MAXHOM alignment program (Sander and Schneider, 1991). The resulting alignment is then fed into the actual PHD neural net algorithm. A Web server is available for the PHD method at http://dodo.cpmc.columbia.edu/predictprotein/.

PILEUP. A global multiple sequence alignment routine from the GCG package (Genetics Computer Group, 1993). The algorithm closely follows ClustalV. It generates a UPGMA-based tree and for the alignment of two sets of matched sequences uses the average alignment similarity score of Corpet (1988).

PRALINE. A method for multiple sequence alignment created by Heringa (1999). It does not use a precalculated search tree like most progressive alignment methods but performs at each alignment step a full profile search and compiles the optimal alignment scores of the most recently aligned sequence block with all other blocks and hitherto unaligned sequences. For the next alignment step PRALINE then selects the highest scoring pair of sequences or blocks of sequences to be aligned. The alignment order and associated tree are thus established during the progressive alignment. The PRALINE method offers a number of strategies based on dynamic programming to optimize the quality of multiple alignment, including profile preprocessing, secondary structure prediction–based alignment, and local alignment–driven global alignment.

The profile preprocessing strategy is aimed at incorporating into each sequence trusted information from other sequences. For each sequence, a multiple alignment is created by stacking other sequences (N-to-1 alignment) that score beyond a user-specified threshold when aligned pairwise with the sequence considered. A low threshold would result in a preprocessed alignment for each sequence comprising all other sequences (where the chance of alignment error is large), while higher thresholds would allow fewer and fewer sequences into the alignment (with fewer alignment errors). A profile is constructed for each of the thus formed preprocessed alignments. PRALINE then performs progressive multiple alignment using the preprocessed profiles, where each sequence is now represented by its preprocessed profile. The preprocessed profile for each of the sequences incorporates knowledge about other sequences (in particular, similar sequences) and comprises position-specific gap penalties. This enables increased matching of distant sequences and likely placement of gaps outside the ungapped core regions in the preprocessed profiles during progressive alignment. The multiple alignment of the preprocessed profiles can also be used to derive consistency scores for each amino acid in the alignment, which for each sequence reflects the consistency among the pairwise alignments used that include that sequence.

The second strategy of exploiting secondary structure prediction to optimize alignments (and vice versa) derives from the fact that state-of-the-art secondary structure prediction methods rely on multiple alignments as input information. This allows an iterative protocol of multiple alignment construction followed by secondary structure prediction using the multiple alignment. A new alignment can then be produced using the predicted secondary structure. The alignment then gives rise to a new secondary structure prediction, and so forth. The secondary structure information is incorporated in the dynamic programming protocol by means of secondary structure–specific exchange matrices (Lüthy et al., 1991) for α-helix, β-strand, and coil. This is done in a conservative manner: Only when the secondary structures in either of the aligned sequences (or sequence blocks) are identical is a secondary structure–specific exchange matrix used to score the matched positions; otherwise the default residue exchange matrix is used.

The local alignment–driven global alignment strategy operates in two steps. First, for each possible positional match between two sequences (or sequence blocks), the score of the optimal local alignment including the match is calculated. Then the optimal global alignment is compiled based on these local alignment scores. This two-step alignment protocol is a variation of the double dynamic programming protocol. The strategy ensures that the global alignment is biased toward matching local motifs and is recommended when local sequence similarity is suspected (for example, in cases of very different sequence lengths). The PRALINE method allows the execution of these strategies in an optionally iterative fashion. It can be obtained at http://mathbio.nimr.mrc.ac.uk.

Pred2ary. A profile- and neural net-based method created by Chandonia and Karplus (1998) for prediction of protein secondary structure. Pred2ary was assessed to have a prediction accuracy of 748% and shows a balanced prediction over the three structural states. It employs a second neural net to filter the raw predictions of the first net, as does the PHD method of Rost and Sander (1993). A recent extended version, which combines in a jury decision the outputs of 120 individually trained networks, is claimed to predict with $75.9 \pm 7.9\%$ accurately. The accuracy is achieved by converting each possible pair of network output weights for helix and strand into an a priori probability for the pair to predict the true structural state. These probabilities are then used for a final prediction corresponding to the highest of the a priori probabilities for each of the three states. The Pred2ary method is accessible through the Web at http://yuri.harvard.edu/~jmc/2ary.html.

PSA. A method of Stultz et al. (1993) that predicts secondary structure for single sequences but employs tertiary structural information using a hidden Markov modeling (HMM) approach. PSA is based on a threading-like approach, in that for a query sequence the goodness of fit is tested with 15 basic tertiary structural models called discrete space models (DSMs). Included as DSMs are, for example, the α-helical globin structure and the flavodoxin-type α/β fold. Each of these models is composed of secondary structure elements chosen from 13 types distinguished by the method: N- and C-cap (for helix); average, buried, and exposed α-helix; buried β-strand; buried and exposed amphipathic β-strand; form β-turn positions; and coil. Using the HMM formalism for modeling the DSMs, the most suitable implementation for each DSM (e.g., the flavodoxin-type DSM can hold five to seven helices) is selected, and the secondary structure of the best-fitting model is then presented to the user as a probability contour plot. Although PSA might not be among the best performers, the approach is interesting, and the method's graphic outputs are useful as a starting point for gaining insight into the probabilities for each of the secondary structures along the sequence. A WWW server for the PSA method is available at http://bmerc-www.bu.edu/psa/.

PSIPRED. A protein secondary structure prediction method based upon the neural network formalism (Jones, 1999). PSIPRED relies on position-specific scoring matrices (PSSMs) as generated by the PSI-BLAST algorithm and feeds those into a two-layered neural network. The PSSMs contain information from local fragments of sequences that are homologous (as assessed by PSI-BLAST) to the query sequence. Only one similar local fragment per homologous sequence is included. Because PSIPRED invokes the PSI-BLAST database search engine to gather information from related sequences, it needs only a single sequence as input. It does not use a third neural net for a jury decision, as the PHD and Pred2ary methods do, but shows a significant accuracy of 76.5% (Jones, 1999). A web server for the method is available at http://insulin.brunel.ac.uk/psipred/.

PREDATOR. A method to predict protein secondary structure using a single query sequence or a set of multiple unaligned sequences (Frishman and Argos, 1995, 1996). The PREDATOR method owes its accuracy mostly to the incorporation of long-range interactions for β-strand prediction. It attains 68% prediction accuracy for single-sequence prediction (Frishman and Argos, 1995). The method uses a k-nearest-neighbor approach and selects each time, for a sliding window of 13 amino acids, 25 exemplars with known secondary structure from a nonredundant database. A total of seven propensities are derived for each position: three general states (P^H, P^E, and P^C) are gleaned from the distribution of the exemplars. An extra α-helix potential (P^{Helix}) is obtained using pairwise hydrogen bonding potentials at a sequence separation of four residues, taken over a seven-residue window. A propensity for β-turn (P^{Turn}) is computed by summing single-residue propensities in classic β-turn positions 1–4 using a four-residue window. Finally, two more propensities for β-strands are determined using tentative long-range β-strand interactions. This is done by assessing the likelihood for each five-residue fragment to form a parallel or antiparallel β-bridge with any other five-residue fragment (separated by more than six amino acids). The propensity calculations are based on summing residue hydrogen bonding propensities obtained from a large collection of β-sheet structures. The maximum corresponding window score was taken as the final parallel and antiparallel β-strand propensity for each residue (P^{Par} and $P^{Antipar}$). For each of the seven independent propensity values, threshold values (T) were calculated and used in five decision rules applied consecutively to get a three-state prediction for each residue.

If the PREDATOR method is applied to multiple sequences, it does not use or construct a multiple alignment but compares the sequences using pairwise local alignments (Smith and Waterman, 1981). The predictions are then carried out for a single base sequence and a number of related sequences. A set of highly scoring local alignments is compiled through matching the base sequence with each of the other sequences. A weight is then calculated for each local fragment based on the local alignment score and the length of its alignment with the base sequence. For each residue in the base sequence, after gathering exemplars using the base sequence and the stacked fragments, the weighted sum over all exemplars is compiled independently for the seven propensities and subjected to the five decision rules to arrive at a three-state prediction. The per-residue accuracy of the method is 74.8% (Frishman and Argos, 1996). The PREDATOR method is accessible via the Web at http://www.embl-heidelberg.de/cgi/predator_serv.pl.

PRRP. A global multiple sequence alignment program by Gotoh (1996). The method optimizes a progressive global alignment by iteratively dividing the sequences into two groups, which are subsequently realigned using a global group-to-group alignment algorithm. Pairwise sequence weights are derived from a tree constructed with the UPGMA cluster criterion and used to calculate the alignment scores when sequence blocks are matched.

PROSITE. A semimanual method for finding characteristic protein patterns developed by Bairoch (1993). The aim of the approach is to make the derived patterns as short as possible but still sensitive enough to recognize the maximum number of related sequences and also sufficiently specific to reject most if not all unrelated sequences (false positives). A large collection of motifs gathered in this way is available in the PROSITE databank (Hofmann et al., 1999). Associated with each motif is an estimate of its discriminative power. The PROSITE databank and related software constitute an invaluable and generally available tool for detecting the function of newly sequenced and uncharacterized proteins. To enhance the discriminatory power of many protein sequence motifs, the PROSITE database also represents many entries using the extended profile formalism of Bucher et al. (1996).

Rost et al. (1995). A method to predict protein transmembrane regions from multiple sequence alignments using a neural network algorithm. Rost et al. used multiple sequence information and trained the protein secondary structure prediction method PHD (Rost and Sander, 1993) on multiple alignments for 69 protein families with known TM helices. The prediction accuracy of the PHD transmembrane prediction method is 95% as assessed by the authors. A WWW server of the method can be found at http://dodo.cpmc.columbia.edu/predictprotein/.

SAGA. A global multiple sequence alignment method by Notredame and Higgins (1996). SAGA uses a genetic algorithm (GA) to gradually optimize a multiple alignment using crossing-over and selection in order to evolve to the best possible alignment. It selects from an evolving alignment population the alignment that optimizes a so-called objective function (OF), that is, a function that reflects the quality of the alignment. SAGA allows any user-defined OF. Provided OFs include the weighted sum of pairs as used in the MSA program (Lipman et al., 1989) and a measure of consistency between the considered multiple alignment and a corresponding library of Clustal pairwise alignments. The latter OF was developed for the COFFEE algorithm (Notredame et al., 1998).

Shepherd (1981). A method to determine the correct reading frame in a potential protein-coding nucleotide sequence. It is based on uneven purine/pyrimidine nucleotide distribution in various positions of trinucleotide codons.

Shulman et al. (1981). One of the first (if not the first) programs to determine the location of protein-coding regions in naturally occurring nucleotide sequence. The main finding from running the Shulman et al. algorithm on a test coding sequence is the three-base quasiperiodicity of short oligonucleotides. This kind of periodicity is much less pronounced in or absent from other functionally important regions of naturally occurring nucleic acids. The specific method proposed by Shulman et al., the diversity index, has been used in many other string analysis tasks under the name "index of coincidence." Many (if not all) other prominent algorithms to locate protein-coding regions explore the three-base quasiperiodicity discovered by Shulman et al.

SIM. A space-efficient version of the Waterman and Eggert (1987) method for local pairwise sequence alignment by Huang et al. (1990). The method calculates a user-defined number of aligned local fragments. The alignments are nonintersecting, i.e., they have no matched amino acid pair in common. The memory requirements of the algorithm are reduced from order N^2 to order N, thereby allowing very long sequences to be searched at the expense of only a small increase in computational time. A Web server for SIM can be found at http://expasy.cbr.nrc.ca/tools/sim-prot.html.

Smith et al. (1990). A technique to derive conserved sequence patterns from sets of homologous sequences. Conserved motifs are identified by listing all common three-residue combinations, with the maximal length of allowed spacers between these three residues set at 24 amino acids. The most frequent occurrences among the group of specific combinations are found and joined into blocks, and a mean score for each column of the block is calculated using the PAM250 residue exchange matrix. Then the best matching subsequences from the rest of the proteins in the group are found, after which the final sequence pattern is specified.

Smith and Waterman (1981). An algorithm to compare two sequences by aligning a best matching local fragment from each sequence only. A problem with global dynamic programming methods (Needleman and Wunsch, 1970) that match complete sequences can arise when highly dissimilar sequences are compared. In such cases global alignment techniques might fail to recognize highly similar internal regions because they are overshadowed by dissimilar regions, and strong gap penalties are normally required to achieve proper global matching. Moreover, many biological sequences are modular and show shuffled domains (Heringa and Taylor, 1997), which can render a global alignment of two complete sequences meaningless. The occurrence of varying numbers of internal sequence repeats (Heringa, 1998) can also severely limit the applicability of global methods. In general, when there is a large difference in the lengths of two sequences to be compared, global alignment routines become unwarranted. To address these problems, Smith and Waterman (1981) developed a so-called local alignment technique in which the most similar regions in two sequences are selected and aligned. For local dynamic programming, the amino acid exchange values used must include negative values. If in the DP search matrix the maximum value (with subtraction of the proper gap penalty) of the highest scoring cell in the row of column preceding the current cell is negative, its contribution is set to zero.

This is done to allow the considered cell to occur as the first cell in any local alignment. For each cell, the following function is evaluated:

$$S[i,j] = s[i,j] + \text{Max} \left\{ \begin{array}{l} S[i-1,j-1] \\ \\ \max_{1<x<i}(S[i-x,j-1] - P(x-1)) \\ \\ \max_{1<y<j}(S[i-1,j-y] - P(y-1)) \\ \\ 0 \end{array} \right\}$$

where Max is the maximum of four terms compared to three for global alignment, which does not feature the zero term. The final highest alignment score value does not have to be in the last row or column as in global alignment routines but can be anywhere in the search matrix. The local alignment algorithm thus relies on dissimilar subsequences producing negative scores that are subsequently discarded by placing zero values in the associated search matrix cells.

SSPRED. A method for predicting protein secondary structure from multiply aligned sequences by Mehta et al. (1995). SSPRED exploits an alternative aspect of the positional information provided by multiple alignments, in that it uses the amino acid pairwise exchanges observed for each multiple alignment position. An advantage of the SSPRED technique is its speed and conceptual clarity. Using the 3D-ALI database (Pascarella and Argos, 1992), which holds multiple alignments of distantly homologous proteins constructed based on structure superpositioning and sequence alignment, amino acid exchange matrices were compiled for helix, strand, and coil. Each matrix simply contains preference values for amino acid exchanges observed at alignment positions with the corresponding secondary structure in the 3D-ALI database. The matrices are used to predict the secondary structure of a query alignment by listing the unique observed residue exchanges for each alignment position and summing the corresponding preference values over each of the three exchange matrices. Each exchange type (e.g., alanine to/from proline) is counted only once for each query alignment position, which provides implicit weighting of the sequences to avoid predominance of redundant sequences. The secondary structure corresponding to the matrix showing the highest sum is then assigned to the alignment position. Following these raw predictions, three simple cleaning rules are applied and completed in three successive cycles: (1) Single-position interruptions are cleaned, e.g., (H[E/C]H) becomes (HHH) and (E[H/C]E) is set to (EEE), where [E/C] indicates E or C. (2) Double-position interruptions are cleaned, e.g., (HH[EE/CC]H) or (H[EE/CC]HH) becomes (HHHHH) and (EE[HH/CC]E) or

(E[HH/CC]EE) is changed to (EEEEE), where [EE/CC] designates EE or CC. (3) Helices of four or fewer residues and strands of two or fewer residues are changed into coils. The accuracy of the method was assessed by the authors at 72% correct prediction, albeit over a relatively small test set of only 38 protein families.

STRIDE. A method, by Frishman and Argos (1995) for secondary structure assignment using protein atomic coordinate data. STRIDE combines many of the features used by protein experts to assign secondary structure, such as hydrogen bonding patterns and stereochemical characteristics. These features are implemented in the program using a knowledge-based approach. STRIDE generally yields assignments in close agreement to those made by crystallographic experts.

Staden (1984). One of the first programs to detect potential protein-coding regions in naturally occurring sequenced nucleotides. The algorithm explores the nonrandomness of frequencies of nucleotide occurrence in different positions of trinucleotide codons.

Sunyaev et al. (1998, 1999). A profile construction method based on a weighting scenario reminiscent of phylogenetic parsimony methods, aimed at increasing the sensitivity of database searches with multiple alignments. To achieve this, amino acid propensities at each alignment position in the alignment profile are weighted according to the probability that identical amino acids occur in more than one sequence at the alignment position. If more alignment positions show identical conservation for a given subset of sequences (not necessarily the same conserved amino acid type over all the alignment positions involved), the occurrence of the amino acids at those positions becomes more expected, which is corrected for by appropriately lowering the weight for the considered position. This approach leads to position-specific sequence weights, which are implemented in the position-specific probabilities for each of the amino acids in a profile. The authors reported increased sensitivity if searches were performed using profiles constructed with this technique.

Taylor (1986). A template-directed method for multiple sequence alignment. The method allows the specification of one or more templates as consensus subsequences that, for example, can be associated with secondary structural elements. Based on these templates the sequence are included in a multiple alignment one by one. The templates are progressively updated to include the variabilities introduced by newly added residues. After a template is created from the initial alignment, it can be extended to include additional related proteins. This process is repeated iteratively until no other protein sequence can be added without giving up essential features.

TMAP. A method of Persson and Argos (1994) to predict protein trans-membrane (TM) segments that relies on information from multiple alignments. TMAP is based on the propensities of amino acids to be positioned in either the central or flanking regions of a transmembrane, calculated using more than 7500 individual TM helices as annotated in the Swiss-Prot sequence databank. Using the residue TM propensities for each segment of a multiple sequence alignment and for each sequence included in the segment, average values of the central and flanking propensities are calculated over sliding windows. The optimal window lengths were found to be 15 and four residues for central and flanking propensities, respectively. If the peak value for a central TM region exceeds a certain threshold, this region is considered a possible candidate to be membrane-spanning. The algorithm then expands this region in either sequence direction until a flanking peak is reached or the central propensity average falls below a certain value. Some further restraints are imposed on the possible length of a tentative TM segment. The additional sensitivity compared to standard sliding window approaches is a result of using multiple alignment information as well as the second propensity for flanking regions. A Web server for the TMAP method is available at http://www.cbb.ki.se/tmap/.

TopPred2. A combinatorial technique for transmembrane (TM) segment prediction based on a standard hydrophobicity analysis supplemented by charge bias analysis (von Heijne, 1992). TopPred2 was devised for prokaryotic protein sequences, because the "positive inside" rule (von Heijne, 1986) (i.e., the overrepresentation of positively charged amino acids within intracellular surface loops positioned in between transmembrane elements) is more pronounced in prokaryotes than eukaryotes (Sipos and von Heijne, 1993). However, the algorithm has been adapted to handle eukaryotic sequences also. A Web server for the method can be found at http://www.biokemi.su.se/~server/toppred2/toppredServer.cgi.

Vingron and Argos (1990). An algorithm to determine all optimal and suboptimal alignments of two sequences. The resulting alignments are depicted in a dot plot. The technique is aimed at recognizing reliably aligned regions, which can be defined as those for which alternative local alignments do not exist.

Vingron and Argos (1991). A method to delineate motifs that are consistently aligned across a given set of sequences. The method is based on all pairwise dot matrices and elucidates consistent and related regions in all matrices through matrix multiplication. A requirement for such a region to be identified is that it be consistently present in all given sequences.

Waterman and Eggert (1987). A local alignment routine for pairwise sequence comparison based on the Smith and Waterman (1981) local alignment algorithm, which allows the calculation of a user-defined number of top-scoring local alignments instead of only the optimal local alignment. The obtained local alignments do not intersect; i.e., they have no matched amino acid pair in common. If during the procedure an alignment is encountered that intersects with any of the top scoring alignments listed thus far, the highest scoring of the conflicting pair is retained in the top list.

Zuker and Steigler (1981). An efficient algorithm for single-stranded RNA folding. Today's version of this popular program has undergone significant modifications, but the optimization principles (variants of dynamic programming) of folding prediction remain the same.

Zuker (1991). An algorithm to determine optimal and suboptimal alignments of two sequences. This algorithm can be used to ascertain the significance of alignments found, because it is entirely possible that an optimal alignment is not the biologically correct alignment. A useful heuristic is that reliably aligned regions are those for which alternative local alignments do not exist.

APPENDIX REFERENCES

Altschul SF, Carrol RJ, Lipman DJ. Weights for data related by a tree. J Mol Biol 1989; 207:647–653.

Altschul SF, Gish W, Miller W, Meyers EW, Lipman DJ. Basic local alignment search tool. J Mol Biol 1990; 215:403–410.

Altschul SF, Madden T, Schäffer L, Zhang AA, Zhang J, Miller Z, Lipman W. Gapped BLAST and PSI-BLAST: a new generation of protein database search programs. Nucleic Acids Res 1997; 25:3389–3402.

Argos P. A sensitive procedure to compare amino acid sequences. J Mol Biol 1987; 193:385–396.

Bailey TL, Elkan C. Fitting a mixture model by expectation maximization to discover motifs in biopolymers. Proceedings of the Second International Conference on Intelligent Systems for Molecular Biology. Menlo Park, CA: AAAI Press, 1994:28–36.

Bairoch A. The PROSITE dictionary of sites and patterns in proteins, its current status. Nucleic Acids Res 1993; 21:3097–3103.

Barton GJ, Sternberg MJE. A strategy for the rapid multiple alignment of protein sequences: confidence levels from tertiary structure comparisons. J Mol Biol 1987; 198:327–337.

Barton GJ, Sternberg MJE. Flexible protein sequence patterns: a sensitive method to detect weak structural similarities. J Mol Biol 1990; 212:389–402.

Bernstein FC, Koetzle TF, Williams GJ, Meyer EF, Brice MD, Rodgers JR, Kennard O, Shimanouchi T, Tasumi M. The protein data bank: a computer-based archival file for macromolecular structures. J Mol Biol 1977; 112:535–542.

Blanken RL, Klotz LC, Hinnebusch AG. Computer comparison of new and existing criteria for constructing evolutionary trees from sequence data. J Mol Evol 1982; 19:9–19.

Boguski MS, Hardison RC, Schwartz S, Miller W. Analysis of conserved domains and sequence motifs on cellular regulatory proteins and locus control regions using new software tools for multiple alignment and visualization. New Biol 1992; 4:247–260.

Bowie JU, Lüthy R, Eisenberg D. A method to identify protein sequences that fold into a known three-dimensional structure. Science 1991; 253:164–170.

Brutlag DL, Dautricourt J-P, Diaz R, Fier J, Moxon B, Stamm R. BLAZE™: an implementation of the Smith-Waterman sequence comparison algorithm on a massively parallel computer. Comput Chem 1993; 17(2):203–207.

Bucher P, Hofmann K. A sequence similarity approach based on a probabilistic interpretation of an alignment scoring system. In: States DJ, Agarwal P, Gaasterland T, Hunter L, Smith RF, eds. Proceedings of the Fourth International Conference on Intelligent Systems for Molecular Biology (ISMB). Menlo Park, CA: AAAI Press, 1996:44–51.

Bucher P, Karplus K, Moeri N, Hofmann K. A flexible motif search technique based on generalized profiles. Comput Chem 1996; 20:3–24.

Carillo H, Lipman DJ. The multiple sequence alignment problem in biology. SIAM J Appl Math 1988; 48:1073–1082.

Chandonia J-M, Karplus M. Protein Struct Funct Genet 1998; 35:293–306.

Chou PY, Fasman GD. Prediction of protein conformation. Biochemistry 1974; 13:211–215.

Collins JF, Coulson AFW. Significance of protein sequence similarities. Methods Enzymol 1990; 183:474–486.

Corpet F. Multiple sequence alignment with hierarchical clustering. Nucleic Acid Res 1988; 16:10881–10890.

Crick FHC. On protein synthesis. Symp Soc Exp Biol 1958; 12:138–163.

Cuff JA, Barton GJ. Evaluation and improvement for multiple sequence methods for protein secondary structure prediction. Proteins Struct Funct Genet 1999; 34:508–519.

Dayhoff MO, Barker WC, Hunt LT. Establishing homologies in protein sequences. Methods Enzymol 1983; 91:524–545.

Eisenberg D, Weiss RM, Terwilliger TC. The helical hydrophobic moment: a measure of the amphiphilicity of a helix. Nature 1982; 299:371–374.

Feng DF, Doolittle RF. Progressive sequence alignment as a prerequisite to correct phylogenetic trees. J Mol Evol 1987; 25:351–360.

Fitch WM, Margoliash E. Construction of phylogenetic trees. Science 1967; 155:279–284.

Fickett JW. Recognition of protein coding regions in DNA sequences. Nucleic Acids Res 1982; 10:5303–5318.

Frishman D, Argos P. Recognition of distantly related protein sequences using conserved motifs and neural networks. J Mol Biol 1992; 228:951–962.

Frishman D, Argos P. Knowledge-based protein secondary structure assignment. Protein Struct Funct Genet 1995; 23:566–579.

Frishman D, Argos P. Incorporation of long-distance interactions in a secondary structure prediction method. Protein Eng 1996; 9:133–142.

Frishman D, Argos P. Seventy-five percent accuracy in protein secondary structure prediction. Proteins 1997; 27:329–335.

Garnier J, Osguthorpe DJ, Robson B. Analysis of the accuracy and implications of simple methods for predicting the secondary structure of globular proteins. J Mol Biol 1978; 120:97–120.

Garnier JG, Gibrat J-F, Robson B. Methods Enzymol 1996; 266:540–553.

Genetics Computer Group. Program manual for the GCG Package, Version 8. Madison, WI, USA: 575 Science Drive, 1993:53711.

Gibrat J-F, Garnier J, Robson B. Further developments of protein secondary structure prediction using information theory. New parameters and consideration of residue pairs. J Mol Biol 1987; 198:425–443.

Gotoh O. An improved algorithm for matching biological sequences. Mol Biol 1982; 162(3):705–708.

Gotoh O. Alignment of three biological sequences with an efficient traceback procedure. J Theor Biol 1986; 121:327–337.

Gotoh O. Pattern matching of biological sequences with limited storage. CABIOS 1987; 3:17–20.

Gotoh O. Significant improvement in accuracy of multiple protein sequence alignments by iterative refinement as assessed by reference to structural alignments. J Mol Biol 1996; 264:823–838.

Gribskov M, McLachlan AD, Eisenberg D. Profile analysis: detection of distantly related proteins. Proc Natl Acad Sci USA 1987; 84:4355–4358.

Heringa J, Taylor WR. Three-dimensional domain duplication, swapping and stealing. Curr Opin Struct Biol 1997; 7:416–421.

Heringa J. Detection of internal repeats: how common are they? Curr Opin Struct Biol 1998; 8:338–345.

Heringa J. Two strategies for sequence comparison: profile-preprocessed and secondary structure-induced multiple alignment. Comput Chem 1999; 23:341–364.

Higgins DG, Sharp PM. CLUSTAL: a package for performing multiple sequence alignment on a microcomputer. Gene 1988; 73:237–244.

Higgins DG, Bleasby AJ, Fuchs R. CLUSTALV: improved software for multiple sequence alignment. Comput Appl Biosci 1992; 8:189–191.

Hirschberg DS. A linear space algorithm for computing longest common subsequences. Commun Assoc Comput Mach 1975; 18:341–343.

Hofmann K, Bucher P, Falquet L, Bairoch A. The PROSITE database, its status in 1999. Nucleic Acids Res 1999; 27:215–219.

Hogeweg P, Hesper B. The alignment of sets of sequences and the construction of phylogenetic trees. An integrated method. J Mol Evol 1984; 20:175–186.

Hopp TP, Woods KR. Prediction of protein antigenic determinants from amino acid sequences. Proc Natl Acad Sci USA 1981; 78:3824–3828.

Huang X, Hardison RC, Miller W. A space-efficient algorithm for local similarities. CABIOS 1990; 6:373–381.

Huang X, Miller W. A time-efficient, linear-space local similarity algorithm. Adv Appl Math 1991; 12:337–357.

Jameson BA, Wolf H. The antigenic index: a novel algorithm for predicting antigenic determinants. Comput Appl Biosci 1988; 4(1):181–186.

Janin J, Wodak S, Levitt M, Maigret M. Conformation of amino acid side-chains in proteins. J Mol Biol 1978; 125:357–386.

Jones DT. Protein secondary structure prediction based on position specific scoring matrices. J Mol Biol 1999; 292:195–202.

Kabsch W, Sander C. Dictionary of protein secondary structure: pattern recognition of hydrogen-bonded and geometrical features. Biopolymers 1983; 22:2577–2637.

Karplus PA, Schultz GE. Prediction of chain flexibility in proteins. Naturwissenschaften 1985; 72:212–213.

King RD, Sternberg MJE. Identification and application of the concepts important for accurate and reliable protein secondary structure prediction. Protein Sci 1996; 5:2298.

Konopka AK. Towards mapping functional domains in indiscriminantly sequenced nucleic acids: a computational approach. In: Sarma H, Sarma MH, eds. Human Genome Initiative and DNA Recombination. Vol. 1. Guiderland, NY: Adenine Press, 1990:113–125.

Konopka AK, Chatterjee D. Distance analysis and sequence properties of functional domains in nucleic acids and proteins. Gene Anal Techn 1988; 5:87–93.

Konopka AK, Smythers GW. DISTAN—a program which detects significant distances between short oligonucleotides. Comput Appl Biosci 1987; 3:193–201.

Kyte J, Doolittle RF. A simple method for displaying the hydropathic character of a protein. J Mol Biol 1982; 157:105–132.

Langosch D, Heringa J. Proteins Struct Funct Genet 1998; 31:150.

Lawrence CE, Altschul SF, Boguski MS, Liu JS, Neuwald AF, Wootton JC. Detecting subtle sequence signals: a Gibbs sampling strategy for multiple alignment. Science 1993; 262:208–214.

Levitt M. A simplified representation of protein conformations for rapid simulation of protein folding. J Mol Biol 1976; 104:59–107.

Lipman DJ, Pearson WR. Rapid and sensitive protein similarity searches. Science 1985; 227:1435–1441.

Lipman DJ, Altschul SF, Kececioglu JD. A tool for multiple sequence alignment. Proc Natl Acad Sci USA 1989; 86:4412–4415.

Lupas A. Prediction and analysis of coiled-coil structures. Methods Enzymol 1996; 266:513–525.

Lupas A, van Dyke M, Stock J. Predicting coiled-coils from protein sequences. Science 1991; 252:1162.

Lüthy R, McLachlan AD, Eisenberg D. Proteins Struct Func Genet 1991; 10:229.

McLachlan AD. Tests for comparing related amino acid sequences: cytochrome c and cytochrome c551. J Mol Biol 1971; 61:409–424.

McLachlan AD. Repeating sequences and gene duplications in proteins. J Mol Biol 1972; 72:417–437.

McLachlan AD. Analysis of gene duplication repeats in the myosin rod. J Mol Biol 1983; 169:15–30.

Mehta PK, Heringa J, Argos P. Protein Sci 1995; 4:2517–2525.

Morgenstern B. DIALIGN 2: improvement of the segment-to-segment approach to multiple sequence alignment [In Process Citation]. Bioinformatics 1999; 15:211–218.

Myers EW, Miller W. Optimal alignment in linear space. CABIOS 1988; 4:11–17.

Needleman SB, Wunsch CD. A general method applicable to the search for similarities in the amino acid sequence of two proteins. J Mol Biol 1970; 48:443–453.

Notredame C, Higgins DG. SAGA: sequence alignment by genetic algorithm. Nucleic Acids Res 1996; 24:1515–1524.

Notredame C, Holm L, Higgins DG. COFFEE: an objective function for multiple sequence alignments. Bioinformatics 1998; 14:407–422.

Parker JMR, Guo D, Hodges RS. New hydrophilicity scale derived from high-performance liquid chromatography peptide retention data: correlation of predicted surface residues with antigenicity and X-ray-derived accessible sites. Biochemistry 1986; 25:5425–5432.

Pascarella S, Argos P. A data bank merging related protein structures and sequences. Protein Eng 1992; 5:121–137.

Patthy L. Detecting homology of distantly related proteins with consensus sequences. J Mol Biol 1987; 198:567–577.

Pearson WR, Lipman DJ. Improved tools for biological sequence comparison. Proc Natl Acad Sci USA 1988; 85:2444–2448.

Persson B, Argos P. Prediction of transmembrane segments in proteins utilizing multiple sequence alignments. J Mol Biol 1994; 237:182–192.

Rini JM, Schulze-Gahmen U, Wilson IA. Structural evidence for induced fit as a mechanism for antibody-antigen recognition. Science 1992; 255:959–965.

Rost B, Sander C. Prediction of protein secondary structure at better than 70% accuracy. J Mol Biol 1993; 232:584–599.

Rost B, Casadio R, Fariselli P, Sander C. Transmembrane helices predicted at 95% accuracy. Protein Sci 1995; 4:521.

Saitou N, Nei M. The neighbor-joining method: a new method for reconstructing phylogenetic trees. Mol Biol Evol 1987; 4:406–425.

Salamov AA, Solovyev VV. Prediction of protein secondary structure by combining nearest-neighbor algorithms and multiple sequence alignments. J Mol Biol 1995; 247:11–15.

Sander C, Schneider R. Database of homology derived protein structures and the structural meaning of sequence alignment. Proteins 1991; 9:56–68.

Schuler GD, Altschul SF, Lipman DJ. A workbench for multiple alignment construction and analysis. Proteins 1991; 9:180–190.

Shepherd JCW. Method to determine the reading frame of a protein from the purine/ pyrimidine genome sequence and its possible evolutionary justification. Proc Natl Acad Sci USA 1981; 78:1596–1600.

Shulman MJ, Steinberg CM, Westmoreland N. The coding function of nucleotide sequences can be discerned by statistical analysis. J Theor Biol 1981; 88:409–420.

Sipos L, von Heijne G. Predicting the topology of eukaryotic proteins. Eur J Biochem 1993; 213:1333–1340.

Smith TF, Waterman MS. Identification of common molecular subsequences. J Mol Biol 1981; 147:195–197.

Sneath PH, Sokal RR. Numerical Taxonomy. San Francisco: Freeman, 1973.

Staden R. An interactive graphics program for comparing and aligning nucleic acid and amino acid sequences. Nucleic Acids Res 1982; 10:2951–2961.

Staden R. Measurement of the effects that coding for a protein has on a DNA sequence and their use for finding genes. Nucleic Acids Res 1984; 12:551–567.

Stoye J, Moulton V, Dress AWM. DCA: an efficient implementation of the divide-and-conquer approach to simultaneous multiple sequence alignment. Comput Appl Biosci 1997; 13:625–626.

Sturrock SS, Collins JF. MPsrch version 1.3. Biocomputing Research Unit. UK: University of Edinburgh, 1993.

Stultz CM, White JV, Smith TF. Structural analysis based on state-space modeling. Protein Sci 1993; 2:305–314.

Sunyaev SR, Rodchenkov IV, Eisenhaber F, Kuznetsov EN. Analysis of the position dependent amino acid probabilities and its application to the search for remote homologues. Proceedings of the 2nd Annual International Conference on Computers in Molecular Biology (RECOMB98). New York: ACM Press, 1998:258–264.

Sunyaev SR, Eisenhaber F, Rodchenkov IV, Eisenhaber B, Tumanyan VG, Kuznetsov EN. PSIC: profile extraction from sequence alignments with position-specific counts of independent observations. Protein Eng 1999; 12:387–394.

Taylor WR. Identification of protein sequence homology by consensus template alignment. J Mol Biol 1986; 188:233–258.

Taylor WR. A flexible method to align large numbers of biological sequences. J Mol Evol 1988; 28:161–169.

Thompson JD, Higgins DG, Gibson TJ. CLUSTAL W: improving the sensitivity of progressive multiple sequence alignment through sequence weighting, positions-specific gap penalties and weight matrix choice. Nucleic Acids Res 1994; 22:4673–4680.

Thompson JD, Gibson TJ, Plewniak F, Jeanmougin F, Higgins DG. The ClustalX windows interface: flexible strategies for multiple sequence alignment aided by quality analysis tools. Nucleic Acids Res 1997; 25:4876–4882.

Vingron M, Argos P. Determination of reliable regions in protein sequence alignments. Protein Eng 1990; 3:565–569.

Vingron M, Argos P. Motif recognition and alignment for many sequences by comparison of dot-matrices. J Mol Biol 1991; 218:33–43.

von Heijne G. The distribution of positively charged residues in bacterial inner membrane proteins correlates with the trans-membrane topology. EMBO J 1986; 5:3021–3027.

von Heijne G. Membrane protein structure prediction. Hydrophobicity analysis and the positive-inside rule. J Mol Biol 1992; 225:487–494.

Waterman MS, Eggert M. A new algorithm for best subsequences alignment with applications to the tRNA-rRNA comparisons. J Mol Biol 1987; 197:723–728.

Wilbur WJ, Lipman DJ. Rapid similarity searches of nucleic acid and protein data banks. Proc Natl Acad Sci USA 1983; 80:726–730.

Wilbur WJ, Lipman DJ. The context dependent comparison of biological sequences. SIAM J Appl Math 1984; 44:557–567.

Yi T-M, Lander ES. Protein secondary structure prediction using nearest-neighbor methods. J Mol Biol 1993; 232:1117–1129.

Zheng Z, Schäffer AA, Miller W, Madden TL, Lipman DJ, Koonin EV, Altschul SF. Protein sequence similarity searches using patterns as seeds. Nucleic Acids Res 1998; 26:3986–3990.

Zuker M. Suboptimal sequence alignment in molecular biology. Alignment with error analysis. J Mol Biol 1991; 221:403–420.

Zuker M, Steigler P. Optimal computer folding of large RNA sequences using thermodynamics and auxiliary information. Nucleic Acid Res 1981; 9:133–148.

Zvelebil MJ, Barton GJ, Taylor WR, Sternberg MJE. Prediction of protein secondary structure and active sites using the alignment of homologous sequences. J Mol Biol 1987; 195:957.

Index

A, 451
Ab initio, 451
  amino acid interactions, 269
  conformational sampling, 267–269
  DNA-protein interactions:
    target prediction, 242, 267–271
  energy calculation, 267–269
  limitations, 269–271
  prospect, 269–271
Absolute nucleotide substitution rates:
  mitochondrial DNA (mtDNA)
      functional regions, 413–414
  mitochondrial protein-coding genes,
    415
A-DNA, 452
Acceptor site, 452
Acid, 452
Activation energy, 452
Active site, 452
Adaptation, 452
Adenine, 453
Adenine triphosphate (ATP), 453
Algorithm, 453
Alignment:
  definition, 99, 453

[Alignment]
  locking or shifting regions, 516. *See
    also* Multiple Alignment
    Construction and Analysis
    Workbench (MACAW)
Alignment quality, 161–164
Alignment tools, 359–361
Alphabet, 453
  acquisition, 8–9, 14–17
    frequency count-based methods,
      8–17
    nucleic acid sequences, 14–17
Alpha-helical transmembrane
    segments: prediction, 165–168
Alpha-helices, 148
Alternative splicing, 453
Allele, 453
Allele frequency, 453
Alphabet, 453
AMBER force field: definition, 274
Amine, 453
Amino acid, 223, 453–454
  distributions, 252
  interactions:
    ab initio method, 269

Printed in the United States
by Baker & Taylor Publisher Services